Texts in Applied Mathematics 18

Texts in Applied Mathematics

John H. Hubbard   Beverly H. West

Differential Equations: A Dynamical Systems Approach

Higher-Dimensional Systems

With 225 Illustrations

Springer-Verlag
New York   Berlin   Heidelberg   London   Paris
Tokyo   Hong Kong   Barcelona   Budapest

John H. Hubbard
Beverly H. West
Department of Mathematics
Cornell University
Ithaca, NY 14853
USA

Series Editors

F. John (*deceased*)
Courant Institute of
    Mathematical Sciences
New York University
New York, NY 10012
USA

J.E. Marsden
Department of
    Mathematics
University of California
Berkeley, CA 94720
USA

L. Sirovich
Division of Applied
    Mathematics
Brown University
Providence, RI 02912
USA

M. Golubitsky
Department of Mathematics
University of Houston
Houston, TX 77204-3476
USA

W. Jäger
Department of Applied Mathematics
Universität Heidelberg
Im Neuenheimer Feld 294
69120 Heidelberg, Germany

Mathematics Subject Classifications: 34xx, 35xx

Library of Congress Cataloging-in-Publication Data
Hubbard, John.
    Differential equations: a dynamical systems approach / John
Hubbard, Beverly West.
        p.   cm. — (Texts in applied mathematics : 5, 18)
        Contents: pt. 1. Ordinary differential equations — pt. 2. Higher-
dimensional systems.
        ISBN 0-387-97286-2 (set). ISBN 0-387-94377-3 (pt. 2)
        1. Differential equations.   2. Differential equations, Partial.
    I. West, Beverly Henderson, 1939–   . II. Title. III. Series.
    QA371.H77   1990
515′.35 — dc20   90-9649

Printed on acid-free paper.

Production managed by Hal Henglein; manufacturing supervised by Jacqui Ashri.
Photocomposed copy prepared using $\mathcal{AMS}$-TEX and LATEX .
Printed and bound by R.R. Donnelley and Sons, Harrisonburg, VA.
Printed in the United States of America.

9 8 7 6 5 4 3 2 1

ISBN 0-387-94377-3 Springer-Verlag New York Berlin Heidelberg

Series Preface

Mathematics is playing an ever more important role in the physical and biological sciences, provoking a blurring of boundaries between scientific disciplines and a resurgence of interest in the modern as well as the classical techniques of applied mathematics. This renewal of interest, both in research and teaching, had led to the establishment of the series: *Texts in Applied Mathematics* (*TAM*).

The development of new courses is a natural consequence of a high level of excitement on the research frontier as newer techniques, such as numerical and symbolic computer systems, dynamical systems, and chaos, mix with and reinforce the traditional methods of applied mathematics. Thus, the purpose of this textbook series is to meet the current and future needs of these advances and encourage the teaching of new courses.

TAM will publish textbooks suitable for use in advanced undergraduate and beginning graduate courses, and will complement the *Applied Mathematical Sciences* (*AMS*) series, which will focus on advanced textbooks and research level monographs.

Preface

As in Part I, this book concentrates on understanding the behavior of differential equations, rather than on solving the equations. Part I focused on differential equations in one dimension; this volume attempts to understand differential equations in n dimensions.

The existence and uniqueness theory carries over with almost no changes. But the behavior of solutions is not nearly so easy to understand; solutions can be thought of as parametrized curves in $\mathbb{R}^n$, which can knot and link in the most complicated ways and usually do.

We begin in Chapter 6 with a number of examples, often of great historical interest, like the two-body problem, which exhibit some of the questions that will occupy the remainder of the book.

Chapter 7 focuses on linear differential equations with constant coefficients, with the usual paraphernalia of eigenvectors and eigenvalues. This material is the main staple of elementary courses on differential equations, and we also study the material at considerable length. But we put more emphasis on how the signs of the eigenvalues determine the stability of solutions and on the relation between existence of bases of eigenvectors and decoupling.

Associated with this chapter are the linear algebra Appendices L1–L8. An attempt to summarize the relevant material grew and grew, until it reached an absurd length and became practically a textbook on the subject. Most of the material is quite standard, but the treatment of the QR method cannot be found in any textbooks that we know of.

Chapters 8 and 8* represent a serious attempt to understand nonlinear autonomous systems in the plane. Logically they should be one chapter (and were one chapter at one time), but pedagogically this did not work very well. There is a body of material that can be understood without entering into too many technical considerations, and this has been collected in Chapter 8. There we stress linearization and the fact that by qualitative analysis with computer graphics, nonlinear differential equations become essentially as easy to study as linear ones.

Chapter 8* represents a considerably deeper understanding of the same material, eventually leading to the Poincaré-Bendixson Theorem and a complete proof of the Pontryagin–Peixoto results, which are the central topics for planar vector fields. Whether it is really wise to include such hard results in a book that hopes to be elementary is something the authors still wonder about.

Chapter 9 is about bifurcations of differential equations in the plane. When modeling a real system, the engineer always imagines he or she has

knobs with which to tune and adjust the system and its models. These
knobs represent the parameters of the system, and as one twiddles them,
sometimes the behavior of the system changes gradually and sometimes
abruptly. These abrupt changes occur at the *bifurcating values* of the parameters; they are quite well understood for differential equations in the
plane. The theoretical underpinnings are built in Chapter 8*, but you can
read Chapter 9 at a more superficial but still highly instructive level with
only Chapter 8 behind you. We have found that students respond quite
well indeed to the material.

Originally, this book was supposed to have four more chapters, cover-
ing electrical circuits, classical mechanics, linear differential equations with
non-constant coefficients, and iteration in higher dimensions. Because the
current volume has become quite lengthy, that material has been relegated
to a new Part III.

How to navigate Chapters 8, 8*, and 9

A course can reasonably end with Chapter 8, but Chapter 9 in bifurcation is
a topic now easy to approach even in an introductory differential equations
course.

Ithaca, New York   John H. Hubbard
                                                                    Beverly H. West

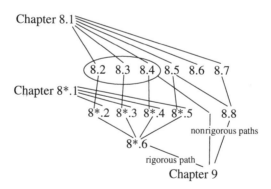

Acknowledgments

In the 1960s, Hubbard participated in seminars run by René Thom and Steve Smale. Their influence pervades this entire book, just as it pervades the entire western school of differential equations.

In a more direct way, the insight of Adrien Douady has shaped our approach to mathematics, and more particularly to differential equations. We are not alone; in France Michèle Artigue and Véronique Gautheron were also influenced by Douady, and their excellent and innovative book *Systèmes Différentiels: Étude Graphique* (1983) first explored many of the same problems we have addressed with a similar spirit.

Over the past ten years, we have been encouraged by many colleagues, students, and editors. Particularly valuable have been enthusiasm, suggestions, and new examples from those who have taught more than once from Part I and from earlier versions of Parts II and III—especially Bodil Branner, Anne Noonburg, Peter Papadopol, and Bob Terrell. On a more technical level, A. Douady provided the proof of a key result in Chapter 8*, and John Guckenheimer provided another. We gratefully acknowledge many additional constructive suggestions from colleagues Allan Back, Fred Barber, Paul Blanchard, Clifford Earle, Bjørn Felsager, G. Roger Livesay, and James E. West.

This book owes its existence to the supporting software *MacMath*. Programmers, some of whom worked on related programs, have included Mike Abato, Yelena Baranova, Fassil Bekele, Doug Campbell, Marc Christensen, Trey Jones, Derek Noonburg, Marc Parmet, Peter Sisson, and Thomas Yan. However, our final skilled programmer/mathematician Ben Hinkle, in his *Extensions of MacMath* to *Planar Systems* and *Planar Iterations*, has allowed us to begin to more thoroughly explore the phenomena, especially of global bifurcations, of which we write.

Several graduate students, particularly Ben Wittner, Jaiqi Luo, Dierk Schleicher, and Ralph Oberste-Vorth, have carefully provided solutions to problems and helped to clarify many sections. Undergraduates Daniel Brown and Tony Yan, respectively, did the original exploration on a number of examples, and provided some of the excellent pictures in Chapter 9 before it was easy to do so. We are grateful to the many other students and readers who have caught errors before publication, including Alexander Hubbard, Tom Palmer, and John Wang.

The reader cannot conceive of how hard it was to create the exercises, most of which are completely new. Many students and colleagues have contributed dozens of them. These exercises, as well as a better understanding of the material, should help the reader to use computer programs, and the geometry they represent, in an effective manner. Although the exercises are extensive, they in no sense exhaust the possibilities, either of the theory or the software. There are simply too many directions in which to go. We will be delighted to receive new exercises and examples that illustrate the material, perhaps for inclusion with appropriate acknowledgment in future versions of the book.

Anne Noonburg has cheerfully worked all existing exercises and provided formal solutions for a large selection. Nevertheless, the authors assume complete responsibility for any errors that occur there or elsewhere in the text; we are grateful if overlooked errors are brought to our attention for correction in future printings, and we shall compose a list of errata for Parts I and II.

The seemingly endless task of preparing the illustrations has been shared by the authors, Kristi Morrison, Heather Park, Dawn Porter, and super-assistant Katrina Thomas, who has also made at least half of the zillions of copies and kept us organized. We are grateful to Graeme Bailey for artful TeXing of the most vexing displays, to Linda Hill for a memorable final send-off, to those staff at the Cornell Mathematics Department who helped with many prepublication versions, and to the staff at Springer-Verlag for helping us to work around innumerable difficulties in finishing all the parts.

John H. Hubbard
Beverly H. West

Contents of Part II

Systems of Ordinary Differential Equations
The Higher-Dimensional Theory $\mathbf{x}' = \mathbf{f}(t, \mathbf{x})$

Contents of Part I

Systems of Ordinary Differential Equations: The One-Dimensional Theory $x' = f(t, x)$

Contents of Part III

Higher-Dimensional Equations continued, $x' = f(t, x)$

Contents of Part IV

Partial Differential Equations
As Linear Differential Equations in Infinitely Many Dimensions; Extension of Eigenvector Treatment
$$\text{e.g., } x'' = c^2(\partial^2 x/\partial s^2) = \lambda x$$

6

Systems of Differential Equations

In many (in fact, most) cases, differential equations of the form $x' = f(t, x)$ are inadequate for the description of a physical system; more variables are needed to specify its state at any time t. Usually a state of the physical system will be specified by the values of several functions, $x_i(t)$, for $i = 1, 2, \ldots, n$. If we know "forces" giving for each of them the derivative $x_i'(t)$ with respect to time, in terms of the values of all the variables (at that particular time) and perhaps also of time, then the evolution of the physical system can be described by a *system of n first order differential equations,*

$$x_i' = f(t, x_1, x_2, \ldots, x_n), \quad \text{for } i = 1, 2, \ldots, n, \tag{1}$$

or as a single *first order differential equation in $\mathbb{R}^n$*, where the vectors are usually written in print as boldface, or by hand with arrows; i.e.,

$$\mathbf{x}'(t) = \mathbf{f}(t, \mathbf{x}) \quad \text{or} \quad \vec{x}' = \vec{f}(t, \vec{x}). \tag{2}$$

In form (2), $\mathbf{x}(t)$ or $\vec{x}(t)$ is a vector $\begin{bmatrix} x_1(t) \\ x_2(t) \\ \vdots \\ x_n(t) \end{bmatrix}$ in $\mathbb{R}^n$, with each component

$x_i(t)$ a real-valued function of t. If f depends explicitly on t, the equation is called *time-dependent* or *nonautonomous*; if f has no explicit dependence on t, the equation is called *autonomous*. Most of our differential equations will be first order, so unless a different order is explicitly stated, this is what we mean.

In previous courses, you may have heard a great deal about higher order differential equations; in fact, we will cover those here in the subject of systems of differential equations because of the following fact:

> *Any higher order differential equation can be expressed as a system of first order equations.*
>
> *That is, an nth order differential equation in $\mathbb{R}$ can be considered as a first order differential equation in $\mathbb{R}^n$.*

Example 6.0.1. The second order differential equation $x'' = -x$ can also be expressed as a system in $\mathbb{R}^2$:

$$x' = y$$
$$y' = -x.$$

as you can confirm by differentiating the first equation and substituting the result in the second.   ▲

We shall follow this example further in succeeding sections, and we shall show in Section 6.4 how *any* higher order differential equation can actually be expressed as a system of first order equations.

Except for *linear* equations *with constant coefficients*, which we shall study at length in Chapter 7,

> *there are very* few *systems of differential equations that can be solved explicitly.*

We shall study in this chapter three systems of differential equations that can nevertheless be analyzed very profitably:

in 6.3, the sharks and sardines equation;

in 6.5, the equation of motion of a particle with one degree of freedom;

in 6.7, the central force problem.

Still, each of these examples is a particular equation, and the methods used do *not* generalize to any substantial class.

The qualitative behavior of solutions of differential equations in $\mathbb{R}^n$, for $n > 1$, is enormously more complicated than the behavior that we examined in Part I , as we shall see in Section 6.1.

In higher dimensions, solutions have ever so much more space in which to get tangled up, and they definitely take advantage of this opportunity. For example, in $\mathbb{R}^2$ a curve can separate the plane into two parts; not so in $\mathbb{R}^3$! Even a closed curve in $\mathbb{R}^3$ encloses nothing—other curves can sneak through and around it.

In Part I of this text we examined a single differential equation in $\mathbb{R}^1$, with graphs of solutions in $\mathbb{R}^2$, the tx-plane. The next simplest system is in $\mathbb{R}^2$, where we seek functions x and y both dependent on t; this will require graphs in $\mathbb{R}^3$, or txy-space . We shall explore those in Section 6.1, and most of our examples will be of systems in $\mathbb{R}^2$ throughout this chapter.

But when we consider systems in $\mathbb{R}^3$, there is virtually no theory anymore, and our understanding of differential equations of dimension n greater than 3 is practically nil (except, we repeat, for the special case of linear equations with constant coefficients, Chapter 7).

So it may come as a pleasant surprise that the numerical methods of Chapter 3 and the theory of Chapter 4 (including the Fundamental Inequality, existence, and uniqueness theorems) remain virtually unchanged when generalized to systems of differential equations, regardless of the dimension n: the statements, the proofs, and all the formulas require nothing more than an arrow over all the vector quantities to become correct. We shall revisit all of these in Section 6.2.

Throughout this chapter the general theory is in $\mathbb{R}^n$, although most of the specific examples are in $\mathbb{R}^2$.

6.1 Graphical Representation of Systems

What kind of drawings can we get for systems of differential equations? For an equation $x' = f(t, x)$ in $\mathbb{R}^1$, the direction field is in $\mathbb{R}^2$; in general, for a differential equation in $\mathbb{R}^n$, the direction field will be in $\mathbb{R}^{n+1}$, which is considerably more difficult to draw and to visualize solutions within, even for $n = 2$. Let us begin with the simplest case beyond a differential equation in $\mathbb{R}^1$.

DIFFERENTIAL EQUATIONS IN $\mathbb{R}^2$; REPRESENTATION IN $\mathbb{R}^3$

Suppose

$$dx/dt = f(t, x, y) \qquad (3)$$
$$dy/dt = g(t, x, y),$$

with solutions

$$x = u(t) \qquad (4)$$
$$y = v(t).$$

A system (3) of differential equations in $\mathbb{R}^2$ gives a direction field in $\mathbb{R}^3$.

Example 6.1.1. Consider the nonlinear, nonautonomous system

$$x' = y$$
$$y' = x^2 - t,$$

A drawing of some solutions in txy-space looks like Figure 6.1.1:

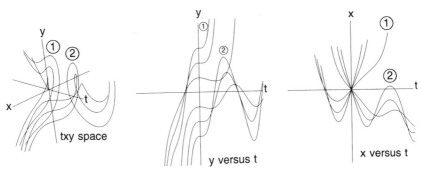

FIGURE 6.1.1. Some solutions to $x' = y$, $y' = x^2 - t$.

As you can see on the left of Figure 6.1.1, the solutions in txy-space appear to form a tangle of curves in $\mathbb{R}^3$, especially, for example, when projected onto the plane of a paper or computer screen. It is a real problem to interpret which parts of the picture are in the foreground and which are in the background. ▲

An individual coordinate function in the tx- or ty-plane can be graphed (simultaneously, with the *MacMath* software) at the right of Figure 6.1.1; these too are tangled. Since the coordinate functions intersect all over the place, these right-hand graphs are quite different (and less helpful) than those produced by *MacMath* for equations in $\mathbb{R}^1$. With sufficient effort you may be able to sort out which curves correspond (e.g., two of them are marked 1 and 2 respectively on all three graphs). But visualizing how in general $x(t)$ and $y(t)$ synthesize to produce the spatial motion of solutions in $\mathbb{R}^3$ is extremely difficult.

To begin to sort out some of these difficulties, to the extent that it is possible, we shall return to the important but overly simple Example 6.0.1 of the system in $\mathbb{R}^2$ resulting from $x'' = -x$: here the result is familiar enough to aid in visualization, and we can discuss the various possible representations of the solutions. Then we shall return to the problem of Example 6.1.1 and give some indication of why it is particularly difficult to represent solutions in a meaningful way.

Example 6.1.2. Consider the system

$$\begin{aligned} x' &= y \\ y' &= -x, \end{aligned} \tag{5}$$

which produces Figure 6.1.2 from a few solutions. This picture is more organized than Figure 6.1.1, but still somewhat jumbled:

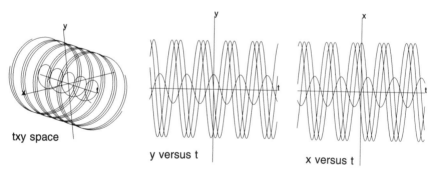

txy space   y versus t   x versus t

FIGURE 6.1.2. Some solutions to $x' = y$, $y' = -x$.

For the purpose of untangling the jumbled first impression of Figure 6.1.2, let us jump ahead to the fact (which you might guess from the second order equation $x'' = -x$, or from the coordinate function graphs on the right of Figure 6.1.2, or else can confirm by substitution) that some of the solutions to the system (5) are

$$\begin{aligned} x &= u(t) = C \sin t \\ y &= v(t) = C \cos t. \end{aligned} \tag{6}$$

For each choice of the constant C, the equations (6) represent in txy-space a *circular spiral*, moving forward in time, as shown in Figures 6.1.3 and 6.1.4 for several different values of C; each solution starts at $t = 0$, with $x_0 = 0$, $y_0 = C$. (To show the coordinate functions most clearly, the time axis in Figure 6.1.3 goes only to 10 in each direction; to show the spirals most clearly, the time axis in Figure 6.1.4 (and 6.1.2) runs to 100 in each direction.)

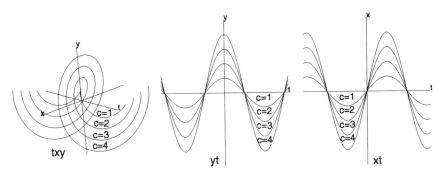

FIGURE 6.1.3. Selected solutions to $x' = y$, $y' = -x$, with $x_0 = 0$, $y_0 = C$.

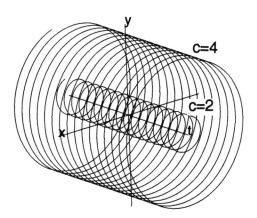

FIGURE 6.1.4. Selected solutions to $x' = y$, $y' = -x$, with $x_0 = 0$, $y_0 = C$, but with time axis running to 100 rather than 10. ▲

DIFFERENTIAL EQUATIONS IN $\mathbb{R}^2$; PHASE PLANE REPRESENTATION

The two-dimensional solution graphs for $x(t)$ and $y(t)$ are far more confusing than the two-dimensional drawings for one-dimensional differential equations we have used in Part I of this text, because these solutions can cross. Even in the comparatively well-organized pictures of Example 6.1.2, solutions are crossing in the tx- and ty-planes, and those pictures are extremely limited, to solutions with initial conditions $t = 0$, $x_0 = 0$, $y_0 = C$.

But there is another two-dimensional picture possible, in the xy-plane, called the *phase plane* (or sometimes the *state space* or *dynamical plane*) for a system in $\mathbb{R}^2$. In Examples 6.1.2 and 6.1.3 we shall see that the phase plane drawing consists of curves that do *not* cross, an important attribute for analyzing the behavior of the system.

Example 6.1.3. Consider again $x' = y$, $y' = -x$. Figure 6.1.5 shows the phase plane portrait (drawn by *MacMath*) of the solutions corresponding to those shown in Figures 6.1.3 and 6.1.4.

FIGURE 6.1.5. Trajectories of selected solutions to $x' = y$, $y' = -x$, with $x_0 = 0$, $y_0 = C$.   ▲

A solution in txy-space (as in the left of Figures 6.1.2, 6.1.3, and 6.1.4) is *projected* onto the xy-phase plane. It is what you would see if you stood high on the t-axis looking down at the xy-plane. This projected curve does *not* correspond to the actual motion of a solution to the system, but rather it is a "track" or *trajectory* of the solution in the phase plane. If you compare Figures 6.1.3 and 6.1.5, you should see that the trajectories in the phase plane are indeed the "tracks" left by the solutions for x and y as functions of t, but

> *the trajectories alone give no information about how a point moves along a trajectory as a function of time.*

Nevertheless, the trajectories in the phase plane provide a very useful way to analyze the system, as we shall demonstrate in the examples of Sections 6.3, 6.5, and 6.7.

As you may have noticed, something especially simple about Examples 6.1.2 and 6.1.3 is the fact that this system of differential equations is *autonomous*, with no explicit dependence on t in the functions for the derivatives. In fact,

> *it is only an autonomous system that will give a meaningful phase plane portrait,*

because then the solutions at different values of t_0 are just time translates of one another, so their projections pile up on the same trajectories.

Look back at Figure 6.1.5 and imagine an ant walking along a solution trajectory. He is always at some point $(x(t), y(t))$, and at $t = 0$ he is at $(x(0), y(0))$. Suppose a second ant starts half an hour later at the same point $(x(0), y(0))$; if the system is autonomous, she will follow the same solution curve, and her later position will always be at $(x(t-30), y(t-30))$. If, on the other hand, the system were nonautonomous, an ant starting at $(x(0), y(0))$ half an hour later than the first ant would feel completely different and would be "blown" along a different trajectory.

For a *non*autonomous system, phase plane trajectories cross over each other and project into an indecipherable mess.

Example 6.1.4. Return to the nonautonomous or time-dependent system $x' = y$, $y' = x^2 - t$ of Example 6.1.1. Some solutions were shown in Figure 6.1.1; the corresponding phase portrait is as follows:

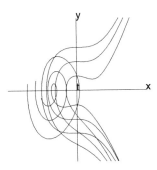

FIGURE 6.1.6. Phase plane projection for the nonautonomous system $x' = y$, $y' = x^2 - t$. ▲

We shall see in Section 6.2 that for autonomous systems, drawings in the xy-phase plane will not only be less of a jumble than Figure 6.1.6, but *will not cross*. As a result, the study of *autonomous* systems in $\mathbb{R}^2$ reduces to the geometry of nonintersecting curves in the plane, and we can study such systems by *phase plane analysis*.

In the remainder of this chapter, we shall strictly limit our study of differential equations in $\mathbb{R}^2$ to autonomous systems, those with phase plane drawings that can be analyzed.

SKETCHING THE PHASE PLANE

An important tool in analyzing an autonomous differential equation in $\mathbb{R}^2$ (especially if you do not have your computer at your fingertips) is to be able to make a quick hand sketch of the phase plane trajectories.

Recall that in Chapter 1 with a differential equation in $\mathbb{R}^1$, the key to hand sketching the solutions is to use *isoclines*. Likewise, for a differential equation in $\mathbb{R}^2$, the first key to hand sketching phase plane trajectories is to use isoclines, particularly those of *horizontal slope* (where $y' = 0$) and those of *vertical slope* (where $x' = 0$). These isoclines of horizontal and vertical slopes are sometimes called *nullclines*. Points where these isoclines cross are called *singular points* or *zeroes*, and there "anything" can happen, as we shall discuss in Chapter 8; in particular, it is possible for any number of trajectories to meet at a singular point.

The second key to hand sketching phase plane trajectories is to *use the signs of x' and y' to tell where trajectories are moving* left or right, up or down.

The following set of steps will lead you through the process:

(i) Write the equation as

$$\frac{dy}{dx} = \frac{dy/dt}{dx/dt},$$

and sketch isoclines of

   (a) *horizontal slope* (where $y' = dy/dt = 0$), marking the isocline with little horizontal slope marks. Mark the regions on either side of this isocline with vertical arrows *up* or *down* where $y' > 0$ and $y' < 0$, respectively.

   (b) *vertical slope* (where $x' = dx/dt = 0$), marking the isocline with little vertical slope marks. Mark the regions on either side of this isocline with horizontal arrows *right* or *left* where $x' > 0$ and $x' < 0$, respectively.

(ii) In each region determined by these isoclines, put together the horizontal and vertical arrows. Then, sketch the *resultant* direction field using these components.

(iii) Trace some sample trajectories through the direction field, following all arrows and slope marks. Remember that at a *singular point* where both $x' = 0$ and $y' = 0$, trajectories may meet or behave in other "bizarre" ways.

Example 6.1.5.

$$x' = y - x - 2 \atop y' = x^2 - y \quad \Rightarrow \quad \frac{dy}{dx} = \frac{x^2 - y}{y - x - 2}.$$

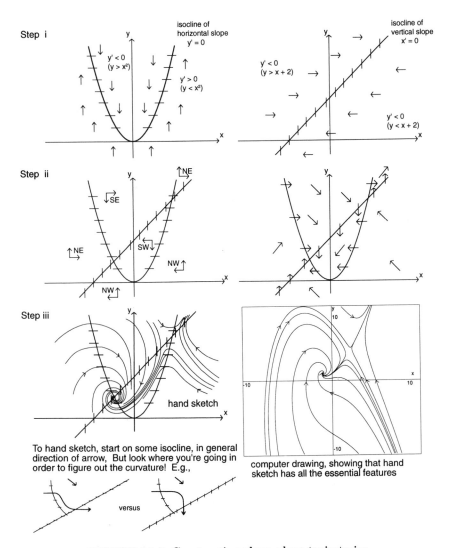

FIGURE 6.1.7. Constructing phase plane trajectories.

DIFFERENTIAL EQUATIONS IN $\mathbb{R}^3$; PHASE SPACE REPRESENTATION

For differential equations in $\mathbb{R}^n$ where $n > 2$, it is not so easy to represent results graphically. For a differential equation in $\mathbb{R}^3$, however, the *MacMath* software can show the three-dimensional *phase space*, with trajectories in an *xyz*-coordinate system. These pictures are more difficult to interpret than those we have discussed in *txy*-space, because the behavior with respect to the independent variable t is hidden. Usually the best aid to understanding such a drawing in three dimensions is being able to watch the trajectories actively being drawn on the screen and/or to rotate the resulting three-dimensional graph.

A popular use of the program for a system in x, y, and z is the famous Lorenz strange attractor, which will be further studied in some detail in Section 8.7.

Example 6.1.6. The following system of equations was used by Edward Lorenz at M.I.T. in the early 1960s for an extremely simplified model of the weather:

$$x' = 10(y - x)$$
$$y' = 28x - y - xz$$
$$z' = -2.66z + xy.$$

The trajectories are immediately attracted to a three-dimensional surface not too hard to imagine from the computer drawings. Figure 6.1.8 shows a single trajectory.

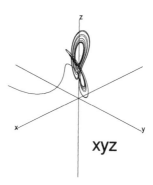

FIGURE 6.1.8. $x' = 10(y - x)$, $y' = 28x - y - xz$, $z' = -2.66z + xy$.

For Figure 6.1.8, each axis runs to 120 in both directions; this picture was begun with stepsize 0.01 at initial condition $z = 10$ and small x, y. It is really fun to watch this evolve on the computer screen, regardless of the orientation of the axes; it is also a good example for experimenting with rotation of axes.

Further understanding can be gained by also studying the various possible two-dimensional graphs: xy, xz, yz, xt, yt, zt. Figure 6.1.9 shows a *MacMath* printout, from *DiffEq, 3D Views* for another single trajectory specified by the Lorenz equations.

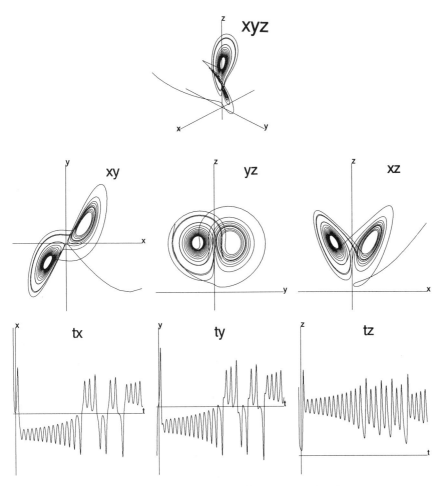

-25.000 < x < 35.000   dx/dt = 10*(y-x)
-30.000 < y < 35.000   dy/dt = 28*x-y-x*z
-10.000 < z < 70.000   dz/dt = -(8/3)*z+x*y

FIGURE 6.1.9. All the possible views for $x' = 10(y - x)$, $y' = 28x - y - xz$, $z' = -2.66z + xy$. Note: we see the 3D view from *below* the xy-plane. ▲

6.2   Theorems for Systems of Differential Equations

We review in this section the major results of Chapters 3 and 4 as they extend to differential equations in $\mathbb{R}^n$. For reference, we have listed in Appendix T the important definitions and theorems from Part I of this text.

The vectors used in describing systems of differential equations are *not restricted* to $\mathbb{R}^2$; the results hold for any $n \geq 1$. Note carefully which quantities become vectors (in boldface) and which remain as scalars (in plain italic type).

NUMERICAL APPROXIMATION

The formulas for numerical solutions to n-dimensional differential equations are as follows, using a stepsize h, so that $t_{i+1} = t_i + h$:

$$\textit{Euler} \qquad \mathbf{x}_{i+1} = \mathbf{x}_i + h\,\mathbf{f}(t_i, \mathbf{x}_i):$$

$$\textit{midpoint Euler} \quad \mathbf{x}_{i+1} = \mathbf{x}_i + h\,\mathbf{m}, \text{ where}$$

$$\mathbf{m} = \mathbf{f}(t_i + \tfrac{h}{2}, \mathbf{x}_i + \tfrac{h}{2}\,\mathbf{f}(t_i, \mathbf{x}_i)):$$

$$\textit{Runge–Kutta} \qquad \mathbf{x}_{i+1} = \mathbf{x}_i + h\mathbf{m}, \text{ where}$$

$$\mathbf{m} = \tfrac{1}{6}(\mathbf{m}_1 + 2\mathbf{m}_2 + 2\mathbf{m}_3 + \mathbf{m}_4)$$

$$\mathbf{m}_1 = \mathbf{f}(t_i, \mathbf{x}_i)$$

$$\mathbf{m}_2 = \mathbf{f}(t_i + \tfrac{h}{2}, \mathbf{x}_i + \tfrac{h}{2}\,\mathbf{m}_1)$$

$$\mathbf{m}_3 = \mathbf{f}(t_i + \tfrac{h}{2}, \mathbf{x}_i + \tfrac{h}{2}\,\mathbf{m}_2)$$

$$\mathbf{m}_4 = \mathbf{f}(t_i + h, \mathbf{x}_i + h\mathbf{m}_3).$$

Notice that the "slopes" $\mathbf{m}$ have all become vector quantities, as well as the dependent variables $\mathbf{x}$ and the functions $\mathbf{f}$. The independent variable t and its stepsize h both remain as scalars.

Example 6.2.1. An Euler's method approximation for

$$x' = y$$
$$y' = x^2 - t,$$

with initial condition $t_0 = 0$, $x_0 = 0$, $y_0 = 2$ and stepsize $h = 0.1$ would begin as shown in Table 6.2.1.   ▲

Table 6.2.1. Setting up a calculation for Euler's method.

t_i	x_i	y_i	$x_i' $ $= y_i$	$y_i' = $ $= x_i^2 - t_i$	x_{i+1} $= x_i + hx_i'$	y_{i+1} $= y_i + hy_i'$
0	0	2	2	0+0 = 0	0+0.2 = 0.2	2+0 = 0
0.1	0.2	2	2	0.04-0.1 = -0.06	0.2+0.2 = 0.4	2-0.006 = 1.994
0.2	0.4	1.994	1.994	0.16-0.2 = -0.04	0.4+0.1994 = 0.5994	1.994-0.004 = 1.990
0.3	0.5994	1.990	→	continue in this manner →		

The order k of the error, Ch^k, of the approximation scheme remains precisely the same in $\mathbb{R}^n$ as in $\mathbb{R}$. Exercise 6.2#5 asks you to show, by partial derivatives and the mean value theorem, that the integration error for the Euler method is in fact bounded by Ch.

The question of finite accuracy error, however, gets complicated for a differential equation in $\mathbb{R}^n$ if $n > 1$. In $\mathbb{R}^1$ you will recall from Volume I, Section 3.4 that systematically rounding down or up simplifies the analysis of error (though it makes the magnitude of the error worse). In $\mathbb{R}^n$ this does not work—rounding errors do not necessarily accumulate, since solution curves may turn and twist— so there is no ordering giving a direction in which to be pushed. Rounding error is probably very equation dependent in $\mathbb{R}^n$, if you round up or down; we suspect that rounding round is okay with the same sort of random walk analysis given in Section 3.4.

The preceding two paragraphs have dealt with what happens as the step-size h decreases, over a *fixed time interval*. An entirely different question is what happens in the long run, as $t \to \infty$, for a *fixed stepsize h*, and the answer is rather complicated, even aside from the introduction of "jag-gies" and other spurious stuff discussed in Section 5.4. In Exercise 6.2#7a, you can compare for $x'' = -x$, the simple system of Examples 6.0.1, 6.1.2, and 6.1.3, the graphical results of the Euler, the midpoint Euler, and the Runge–Kutta methods of numerical approximation; each behaves dramat-ically differently in the long term, using a fixed small stepsize.

THE BASIC THEORETICAL RESULTS

Theorems 6.2.2 and 6.2.3 come straight out of the Fundamental Inequality (Theorem 4.4.1), the Uniqueness Theorem (4.5.1), and the Existence Theorem (4.5.6) for one-dimensional differential equations. (See Appendix T.) The proofs are left as Exercises 6.2#2 and 6.2#3, following exactly the one-dimensional proofs, except in the following ways:

> the quantities x and derivatives of x have now become vector quantities, so x becomes $\mathbf{x}$;

> the length of a vector $\|\mathbf{x}\| \equiv \sqrt{x_1^2 + x_2^2 + \cdots + x_n^2}$, so $|x_1 - x_2|$ becomes $\|\mathbf{x}_1 - \mathbf{x}_2\|$;

> the region U of definition is now a parallelepiped in $\mathbb{R} \times \mathbb{R}^n$, with $t \in [t_a, t_b]$ and $x_i \in [a_i, b_i]$.

Otherwise the only "trick" in carrying over the proofs is to keep straight which quantities will become vectors and which remain scalars. For example, $\gamma = \|\mathbf{u}_1 - \mathbf{u}_2\|$ remains a scalar, so the center of the proof of the Fundamental Inequality remains a one-dimensional problem.

Theorem 6.2.2 (Fundamental Inequality). *If* $\mathbf{x}' = \mathbf{f}(t, \mathbf{x})$ *is defined on a set U in $\mathbb{R} \times \mathbb{R}^n$ with Lipschitz condition*

$$\|\mathbf{f}(t, \mathbf{x}_1) - \mathbf{f}(t, \mathbf{x}_2)\| < K\|\mathbf{x}_1 - \mathbf{x}_2\|$$

for all $(t, \mathbf{x}_1)$ and $(t, \mathbf{x}_2)$ on U, and if, for $\varepsilon_i, \delta \in \mathbb{R}$, $\mathbf{u}_1(t)$ and $\mathbf{u}_2(t)$ are continuous, piecewise differentiable functions on U into $\mathbb{R}^n$ with

$$\|\mathbf{u}_i'(t) - \mathbf{f}(t, \mathbf{u}_i(t))\| \le \varepsilon_i$$

and

$$\|\mathbf{u}_1(t_0) - \mathbf{u}_2(t_0)\| \le \delta,$$

then

$$\|\mathbf{u}_1(t) - \mathbf{u}_2(t)\| \le \delta\, e^{K|t-t_0|} + \frac{\varepsilon_1 + \varepsilon_2}{K}\left(e^{K|t-t_0|} - 1\right).$$

Theorem 6.2.3 (Existence and Uniqueness). *If* $\mathbf{x}' = \mathbf{f}(t, \mathbf{x})$ *is defined on a set U in $\mathbb{R} \times \mathbb{R}^n$ with Lipschitz condition*

$$\|\mathbf{f}(t, \mathbf{x}_1) - \mathbf{f}(t, \mathbf{x}_2)\| < K\|\mathbf{x}_1 - \mathbf{x}_2\|$$

for all $(t, \mathbf{x}_1)$ and $(t, \mathbf{x}_2)$ on U, then there exists a unique solution $\mathbf{x} = \mathbf{u}(t)$ for a given set of initial conditions $\mathbf{x}(t_0)$.

The geometric meaning of Theorem 6.2.3, that unique solutions do not cross, is given by the following:

Corollary 6.2.4. *For a differential equation in $\mathbb{R}^n$ satisfying a Lipschitz condition in* **x**, *solutions* $\mathbf{x} = \mathbf{u}(t)$ *to an initial value problem*

(a) do not cross in $\mathbb{R}^{n+1}$ where they are graphed with respect to time,

and

(b) do not cross in phase space, $\mathbb{R}^n$, if the system is autonomous.

Note: Corollary 6.2.4 makes *no* promises about the individual coordinate functions, $x = u(t)$, $y = v(t), \ldots$, which can indeed cross as you have seen in Figures 6.1.1–6.1.4 . But Corollary 6.2.4 does assure us that for an *initial value* problem in $\mathbb{R}^2$ (satisfying a Lipschitz condition)

(a) we are *not* seeing actual three-dimensional crossings of solutions in the projections of $\mathbb{R}^3$,

and

(b) if the system is *autonomous*, trajectories of solutions will *not* cross in the *phase plane*.

Now that we are dealing with *systems* of differential equations, we must say more about uniqueness. If you have a "general" solution to such a system (that is, a solution with arbitrary constants that does not miss any solutions to the differential equation), this "general" solution, for an equation in $\mathbb{R}^n$, will in general depend on n constants.

Specifying just any old n numbers associated to a solution will *not* necessarily specify a unique solution. Sometimes these numbers will work, and sometimes they will not.

A crucial element to Theorem 6.2.3 and its Corollary 6.2.4 is the requirement that the *initial conditions*, the n values of $x_i(t_0)$, be specified. We can give a glimpse of the potential difficulties associated with trying to specify *boundary values* instead by returning to the equation of Examples 6.1.2 and 6.1.3:

Example 6.2.5. For the simple system $x' = y$, $y' = -x$, recall from Example 6.0.1 that this system in $\mathbb{R}^2$ actually represents the second order differential equation $x'' = -x$, with $x_1 = x$ and $x_2 = y = x'$. Figure 6.2.1 shows just some of the solutions, as one of the coordinate functions of the system, and you can see that these solutions cross each other all over the place.

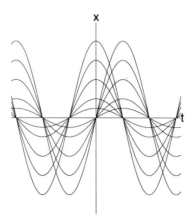

FIGURE 6.2.1. Selected solutions to $x'' = -x$.

The "general" solution to $x'' = -x$ is $x = A \sin t + B \cos t$, with two arbitrary constants, A and B. A natural idea might be to specify *boundary values* $x(t_a)$ and $x(t_b)$, rather than initial values $x(t_0)$ and $y(t_0)$. But sometimes such boundary values will not specify a unique solution (case (i)) ; other times they will (case (ii)). For instance, the following summary will be the result of Exercise 6.2#9, using the fact that all solutions are periodic with period 2π:

(i) If $(t_b - t_a) = k\pi$, where k can be any integer, negative or nonnegative, then specifying $x(t_a)$ and $x(t_b)$ will result in either

   (a) an *infinite number* of solutions,

$$\text{if } x(t_a) = \begin{cases} x(t_b) & \text{for } k \text{ even} \\ -x(t_b) & \text{for } k \text{ odd} \end{cases};$$

   (b) or *no* solutions,

$$\text{if } x(t_a) \neq |x(t_b)|$$

$$\text{if } x(t_a) = \begin{cases} -x(t_b) & \text{for } k \text{ even} \\ x(t_b) & \text{for } k \text{ odd}. \end{cases}$$

(ii) If $(t_b - t_a) \neq k\pi$, any pair of boundary values gives a unique solution.
▲

Specifying n parameters by *boundary* conditions is equivalent to choosing *two points*, $(t_a, x(t_a))$ and $(t_b, x(t_b))$, in the tx-plane. If in Example 6.2.5 $(t_b - t_a) \neq k\pi$, there will indeed be a unique solution between the two points; otherwise there will be no possible solution through those two points, or an infinite number of solutions.

Theorem 6.2.3, on the other hand, promises that if the n parameters are *initial* conditions, they *always* specify a unique solution. In Example 6.2.5, initial conditions are $x(t_0)$ and $x'(t_0) = y(t_0)$, corresponding in the tx-plane to *a point and the slope at that point*.

Example 6.2.5 illustrates that it is important to recognize that

> *an initial value problem is entirely different from a boundary value problem.*

Only in the initial value problem is uniqueness assured, with solutions guaranteed not to cross. In boundary value problems, specifying n conditions may produce one, zero, or infinitely many solutions.

Boundary value problems are in fact very complicated, and we shall postpone further discussion of them until Part III of this text.

6.3   Example: Sharks and Sardines

In this section we will study a famous example of a differential equation in $\mathbb{R}^2$, first written down and analyzed by Vito Volterra in the early 1920s; his book, *A Mathematical Theory of the Struggle for Life*, probably started mathematical ecology.

Volterra developed his theory when approached by Umberto d'Ancona, who was an official in the Italian bureau of fisheries employed in Trieste during the First World War. D'Ancona was puzzled by statistics he kept. Specifically, he observed that during the war, the proportion of the catch that consisted of sharks, skate, and other such predating and unappetizing fish increased markedly over what it had been before, and what it became later. The data are listed in Table 6.3.1, in the midst of many years of approximately 11% of catches consisting of predators:

TABLE 6.3.1. Proportion of predators in Italian fishing statistics.

Year	% of catch consisting of predators
1914	11.9%
1915	21.4%
1916	22.1%
1917	21.2%
1918	36.4%
1919	27.3%
1920	16.0%
1921	15.9%
1922	14.8%
1923	10.7%

He presented these data to Volterra, who came up with the following explanation (which we first considered in the Introduction to Part I of our text, but we shall repeat the arguments in this context):

Let $x(t)$ be the number of food fish (known generically as sardines) as a function of time, and $y(t)$ be the number of predators (known generically as sharks). We will write down a system of differential equations for x and y reflecting the following assumptions:

(a) The population of sardines is kept down exclusively by the sharks, i.e., is not close to the limits of its food supply.

(b) The population of sharks is at the limit of its food supply, and is kept in check by the lack of sardines.

Thus, if there were no sharks, x would obey the equation of exponential growth

$$x' = \frac{dx}{dt} = ax, \quad \text{for some fertility rate } a > 0.$$

Volterra argued that in the absence of sardines, y would similarly obey the equation of exponential decay

$$y' = \frac{dy}{dt} = -by, \quad \text{for some } b > 0.$$

[Actually, this is questionable. It seems more reasonable to think that in the absence of sardines or other food fish, y would obey something like the leaky bucket equation, since it would become zero in finite time. In fact, even that is perhaps optimistic (from the point of view of the sharks), since in fact they would presumably become extinct in one generation whatever their initial number, and a leaky bucket can take an arbitrarily long time to empty if it is big enough to begin with.]

In any case, the product $x(t)y(t)$ is proportional to the number of meetings of food fish with sharks, which are bad for food fish and good for sharks. This leads to the system of equations:

$$\begin{aligned} x' &= ax - cxy \\ y' &= -by + fxy, \end{aligned} \tag{7}$$

where a, b, c, f are > 0.

Innocent though this system may appear, no one seems to know how to solve it for $x(t)$ and $y(t)$. However, we still can analyze it fairly completely by its trajectories in the xy phase plane.

Dividing the second equation of (7) by the first, we see that

$$\frac{dy}{dx} = \frac{-by + fxy}{ax - cxy}. \tag{8}$$

This equation (8) is separable, leading to

$$\frac{a - cy}{y} dy = \frac{-b + fx}{x} dx,$$

which can be integrated to yield

$$|x|^b |y|^a e^{-(fx+cy)} = C. \qquad (9)$$

Note that the right-hand sides of equation (7) are continuously differentiable functions, which means that on any bounded region they satisfy a Lipschitz condition, so the solution to the equation through any point will be unique by Theorem 6.2.3. Therefore the set of trajectories defined by (9) forms a system of nonintersecting curves filling up the xy phase plane, and we can hope to draw them.

What we have shown in equation (9) is that these trajectories are *level curves* of the function

$$F(x, y) = |x|^b |y|^a e^{-(fx+cy)}. \qquad (10)$$

The fact that the differential equation (7) restricts the function F expressed in (10) to a constant value on a given trajectory represents some sort of a *conservation law* (to be discussed at greater length in Section 6.6), saying that you cannot have too many sharks without some number of sardines, and vice versa.

> *That is, every solution of the differential equation (7) in $\mathbb{R}^2$ has a trajectory in the phase plane that lies on a level curve of this function F. So to understand the trajectories (but not how a point moves along a trajectory as a function of time), all we need to do is understand the function that represents the conservation law.*

This is an important point to which we shall return in other examples and in Section 6.6.

For the model of sharks and sardines, we are only interested in the function F in the first quadrant where both populations are positive. We leave it as Exercise 6.3#2a to show that there the function F has a unique maximum at $x = b/f$, $y = a/c$, and that it decreases to zero when (x, y) approaches the axes or goes to infinity in any direction in the first quadrant. Note that in any case, it is clear from the differential equation (7) that $(b/f, a/c)$ is an *equilibrium* of the equation, since $x' = y' = 0$ there. Now level curves on a mountain are easy to imagine; we can expect our set of trajectories to be closed curves, which indeed the computer drawing confirms (see Figure 6.3.1).

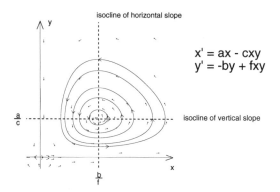

FIGURE 6.3.1. Phase plane trajectories for sharks (y) and sardines (x).

The placement of the arrowheads in Figure 6.3.1 can be made by observations on the original equation (7), as in the previous discussion in Section 6.1 about drawing phase portraits. For instance,

if $x = 0$, $x' = 0$ and y' is negative, y is decreasing;
if $y = 0$, $y' = 0$ and x' is positive, x is increasing.

All the first quadrant solutions to the differential equation (7) are periodic, because the trajectories cycle round and round, describing populations that oscillate (out of phase with each other). This is the key observation, because it allowed Volterra to speak of the *average* populations, $\bar{x}$ and $\bar{y}$, defined by the usual integral formulas

$$\bar{x} = \frac{1}{T} \int_0^T x(t)dt, \qquad \bar{y} = \frac{1}{T} \int_0^T y(t)dt$$

if the populations move on a trajectory of period T. Of course, you would expect the averages to depend on the trajectory, but from the second equation of (7), the computation

$$x(t) = \frac{y' + by}{fy}$$

gives

$$\bar{x} = \frac{1}{T} \int_0^T \frac{y'}{fy}dt + \frac{1}{T} \int_0^T \frac{b}{f}dt = \frac{1}{Tf}(\ln y(T) - \ln y(0)) + \frac{b}{f} = \frac{b}{f}. \quad (11)$$

The result (11) shows that $\bar{x}$ does not depend on the trajectory and is in fact, in this particular case, the x-coordinate of the equilibrium. You can confirm the steps of this computation in Exercise 6.3#1b. An identical computation shows that $\bar{y} = a/c$.

If you find all this far too clever, do not forget that Volterra was working before computers, and therefore was forced to look for such tricks. No doubt the mathematical model he arrived at was chosen partially because he had a trick up his sleeve that worked for it, and not only because the biology imposed it. As soon as we change the model slightly, no such tricks will work, and averages will have to be found numerically (and they will not be just constants). So in fact, in the battle between the genius Volterra without a computer and the ordinary student with one, the student probably comes out ahead.

WHAT DOES ALL THIS HAVE TO DO WITH THE FISHING?

It is fairly clear that the result of fishing is to replace in the differential equation a by $a - \varepsilon$ and b by $b + \delta$ for some $\varepsilon > 0$ and $\delta > 0$, since fishing simply subtracts from both populations a proportion of what is there. The difference in sign occurs because a is a rate of increase and b is a rate of decrease. Equation (7) now becomes

$$\begin{aligned} x' &= (a - \varepsilon)x - cxy \\ y' &= -(b + \delta)y + fxy. \end{aligned} \tag{12}$$

The new equilibrium occurs at

$$\overline{x} = \frac{b + \delta}{f}, \qquad \overline{y} = \frac{a - \varepsilon}{c},$$

and is presented in Figure 6.3.2.

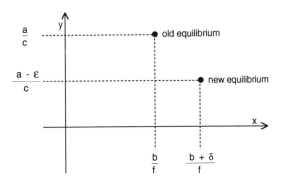

FIGURE 6.3.2. Revised phase plane for sharks and sardines *with fishing*.

As a result of the ε and δ, the equilibrium is moved down and to the right and yields the following amazing conclusion:

Fishing increases the number of food fish!

This is so remarkable that without corroboration one might be tempted to think it might mean that the model is worthless. The following example is offered as corroboration:

Example 6.3.1. Scale insects and ladybugs. In 1868, some acacia trees were imported from Australia and planted in California. Some insects of the species *Icerya purchasi*, better known as scale insects, were on them and promptly infested the orange trees. Scale insects suck the sap from the trees and, as they grow, they split their skins, which harden on the leaves and leave a white cottony cushion, hence the name. In any case, the damage they did to the trees was such as to nearly wipe out the citrus industry in California.

In Australia, the scale insect has a predator, *Rodolia cardinalis*, the ladybird beetle. An entomologist from the Department of Agriculture, Dr. Riley, imported some of these beetles, which promptly brought the population of scale insects under control (this was in 1889). Amazingly enough, it took only 18 months from the time 514 beetles were imported until they had all but wiped out the scale insects and were themselves starving in large numbers. Of course, they did not eliminate the scale insect, but just created a particularly large swing of an oscillation.

Shortly before World War II, DDT was discovered. The orange growers thought: "we almost got them with the beetle, now we will really get them with DDT," so they applied it to the orchards.

As a result, the scale insects became more numerous. In terms of Volterra's analysis, it is clear that this should happen: the scale insects are like food fish (the prey), the ladybird beetles are like the sharks (the predators), and the DDT is like the fishing, which destroys a proportion of both. Fishing increases the number of food fish, and DDT increases the number of scale insects. ▲

CRITICISM OF THE MODEL

It is easy to pick holes in Volterra's model. For one thing, it cannot be valid for large values of x, since exponential increase of a species cannot continue without limit. Also, without food, the population of sharks would in fact go to zero, not just decline exponentially (more probably, in reality, they would switch to an alternate food source).

A more serious question is whether perturbing terms change the qualitative description of the evolution of the system. Clearly they should; it seems highly unlikely that every initial condition will be on a periodic orbit, and far more likely that whatever the initial state, the system will settle down to some cycle (perhaps the equilibrium) independent of where it started.

In fact, we have a hint as to what the period ought to be. Such real ecological systems are forced by the seasons: almost all species have a fertility rate [the number a in equation (7)] that is not constant but varies with the time of year. So we might expect the population to settle down to a yearly cycle, or perhaps a cycle of several years.

Many numerical experiments have been run under these various assumptions, showing that even such simple mathematical models can have amazingly complicated behavior. There is still no coherent theory.

6.4 Higher Order Equations

Many differential equations encountered in applications involve higher order derivatives of the unknown function (or functions). For example, the differential equations of physics that involve *accelerations* are second order differential equations. We will see in this section that such a higher order equation is "equivalent," in a sense to be made precise, to a system of first order equations. There is a catch, of course; namely, that we need more unknown functions. Still, the theory for systems of first order equations is so simple, and the intuitive idea of what a differential equation means is so clear, that it is usually a good idea to replace a higher order equation by a system of first order equations.

This is especially true of the numerical theory: it is quite clear what it means to be blown around by a wind, even in several dimensions, and the idea of Euler's method is immediate. Consider, on the other hand, a differential equation such as

$$x'' = x' - x^2 + t.$$

As we shall see, an appropriate initial condition for such an equation is the specification of both x and x' at the initial time t_0. But what should you do from there? You do not have a wind to be blown by, and you do not know your velocity, only your acceleration (i.e., x''). Most people have little intuitive feel for acceleration, and much less yet for higher derivatives. Thus, replacing them by first derivatives is desirable.

Physicists have long known this; they speak of the *configuration space* of a system—its set of positions, on which motions are described by a system of second order differential equations (because of Newton's law $\mathbf{F} = m\mathbf{a}$, which expresses the accelerations in terms of the positions and the velocities). However, they soon replace the configuration space by the *phase space*— the set of states of the system. A *state* of the system is specified when you know both a position and a velocity for each component of the system. On the phase space, the motions are described by a system of first order equations. Of course, you have paid the price: there are now twice as many variables as before, both a position and a velocity where there was just a position. But what you have gained with phase space is that it gives the *only* "nice" picture, where the trajectories do not cross.

Not only does the replacement of a higher order equation by a system of first order equations give a better intuitive feel for the differential equation, and the possibility of clear phase space pictures of trajectories, but, as we shall see in Chapters 7–9, it allows the solution of a linear system (or the linearization of a nonlinear system) to proceed by simply manipulating a matrix of coefficients.

A differential equation of order n in one variable is an equation of the form

$$x^{(n)} = f(t, x, x', \dots, x^{(n-1)}),$$

where $x^{(i)}$ means the ith derivative of the function x, and f is a function defined in some region of $\mathbb{R} \times \mathbb{R}^n$. Of course, the case $n = 1$ is precisely what we considered in Volume I, Chapters 1–4.

The key idea to solution of an nth order differential equation is to introduce new variables representing successive derivatives, generalizing what we did in Example 6.0.1 when we changed $x'' + x = 0$ to $x' = y, y' = -x$.

Theorem 6.4.1. *The differential equation*

$$x^{(n)} = f(t, x, x', \dots, x^{(n-1)}) \tag{13}$$

is equivalent, if we set $x_0 \equiv x$, to the first order differential equation in $\mathbb{R}^n$

$$\mathbf{x}' = \begin{bmatrix} x' \\ x'' \\ x''' \\ \vdots \\ x^{(n)} \end{bmatrix} = \begin{bmatrix} x_0' \\ x_1' \\ x_2' \\ \vdots \\ x_{n-1}' \end{bmatrix} = \begin{bmatrix} x_1 \\ x_2 \\ x_3 \\ \vdots \\ f(t, x_0, x_1, \dots, x_{n-1}) \end{bmatrix} \equiv f(t, \mathbf{x}) \tag{14}$$

in the sense that a function $x = u(t)$ is a solution of the first equation (13) if and only if the n-dimensional vector function

$$\mathbf{u}(t) = \begin{bmatrix} u(t) \\ u'(t) \\ \vdots \\ u^{n-1}(t) \end{bmatrix} \quad \text{is a solution of the system (14).}$$

In other words, for an nth order differential equation, we set up n variables: $x_0 = x, x_i = x_{i-1}', \dots, x_{n-1} = x_{n-2}'$; then we solve the differential equation for $x^{(n)} = x_{n-1}'$ to obtain the nth first order differential equation of the system.

Proof. Just plug in the formulas. □

Example 6.4.2. The third order differential equation

$$x''' - 3x'' + tx' - x + t^2 = 0$$

is equivalent to the following system, with $x = x_0$:

$$\left. \begin{array}{l} x' = x_0' = x_1 \\ x'' = x_1' = x_2 \\ x''' = x_2' = 3x_2 - tx_1 + x_0 - t^2 \end{array} \right\} \quad \begin{array}{c} \text{nonautonomous system} \\ \text{of three first order} \\ \text{equations.} \quad \blacktriangle \end{array}$$

We hope that after all the buildup, reduction of a higher order differential equation to a system looks like a cheap trick. It is a cheap trick, but nevertheless important. For instance, we now have a uniqueness and existence theory for higher order differential equations, as well as various approximation algorithms (Euler, midpoint Euler, Runge–Kutta).

Theorem 6.4.3. *Let $f(t, \mathbf{x})$ be a function defined on some region U in $\mathbb{R} \times \mathbb{R}^n$ and satisfying a Lipschitz condition with respect to $\mathbf{x}$. Given any $t_0 \in \mathbb{R}$ and a vector $\mathbf{v} \in \mathbb{R}^n$, there exists a unique solution $u(t)$ of*

$$x^{(n)} = f(t, x, x', \dots, x^{(n-1)})$$

such that $\mathbf{u}(t_0) = \mathbf{v}$; that is,

$$\mathbf{u}(t_0) = \begin{bmatrix} u(t_0) \\ u'(t_0) \\ \vdots \\ u^{(n-1)}(t_0) \end{bmatrix} = \begin{bmatrix} v_0 \\ v_1 \\ \vdots \\ v_{n-1} \end{bmatrix}.$$

Proof. Combine Theorem 6.4.1 with Theorems 6.2.2 and 6.2.3. $\quad \square$

MOVING IN THE OTHER DIRECTION: TRADING DIMENSIONS FOR HIGHER ORDER

This is more delicate than the reverse process: let us see how to go from a first order equation in $\mathbb{R}^2$ to a second order equation in one variable.

Suppose

$$x' = f(x, y)$$
$$y' = g(x, y)$$

is a first order system in $\mathbb{R}^2$. Use the first equation to express y implicitly as a function of x and x', say $y = F(x, x')$, for some function $F(u, v)$ of two variables. Now differentiate this equation, to find

$$y' = \frac{\partial F}{\partial u}(x, x')x' + \frac{\partial F}{\partial v}(x, x')x''.$$

Substitute this expression for y' in the second equation and set the right-hand side equal to $g(x, y) = g(x, F(x, x'))$. This results in a second order equation solely in terms of x and its derivatives:

$$x'' = \frac{1}{(\partial F/\partial v)(x, x')} \left(\frac{\partial F}{\partial u}(x, x')x' - g(x, F(x, x')) \right).$$

Example 6.4.4. Consider the system of equations

$$x' = y + \sin x$$
$$y' = \cos(x + y).$$

The first equation gives $y = x' - \sin x$, with

$$y' = x'' - (\cos x)x' = \cos(x + x' - \sin x).$$

Substitution for y and y' in the second equation yields the rather fearsome differential equation

$$x'' = (\cos x)x' + \cos(x + x' - \sin x). \qquad \blacktriangle$$

The reader should appreciate that the process can be considerably more difficult: expressing the implicit function may be difficult or impossible in elementary terms; in higher dimensions, it requires the notorious Implicit Function Theorem from Chapter 13 (See Appendix T). The conditions for an implicit function to exist may be violated. Exercise 6.4#3 will allow you to explore this a bit further.

6.5 Mechanical Systems with One Degree of Freedom

Physics provides many beautiful examples of second order differential equations in a single variable, which we shall study as systems of two first order equations. We shall use the simple pendulum as an example to discuss several aspects of general theory.

Example 6.5.1. Consider the pendulum with motion restricted to the plane, with a bob on a string of length ℓ, as illustrated in Figure 6.5.1.

The arclength of the "sweep" of the pendulum is $\ell\theta$; the velocity of the bob is $\ell\theta'$, and the tangential component of its acceleration is $\ell\theta''$. The force on the pendulum bob in the direction of motion is a component (in the direction of sweep) of the gravitational force $m\mathbf{g}$. From Newton's Law

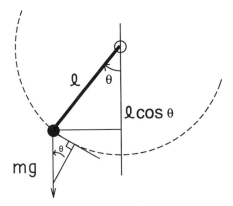

FIGURE 6.5.1. A pendulum in motion.

$\mathbf{F} = m\mathbf{a}$ we can write an equation that describes (in terms of θ) the behavior of the pendulum as moving opposite to:

$$m\ell\theta'' = -mg\sin\theta$$

or

$$\theta'' = -\left(\frac{g}{\ell}\right)\sin\theta, \tag{15a}$$

which we call the *equation of motion of the pendulum*. Solving the equation of motion for θ as a function of t describes exactly how the pendulum bob moves.

Equation (15a) can be replaced by the system

$$\theta' = y, \qquad y' = -\left(\frac{g}{\ell}\right)\sin\theta. \quad \blacktriangle \tag{15b}$$

In Example 6.5.1 there is only one variable, θ, that describes the motion of the pendulum, so we say it has only *one degree of freedom*.

POTENTIAL AND ENERGY

For a physical particle moving without friction in a manner involving only one variable (x, in general), a general theory of physics introduces a useful concept called *potential*. The potential $V(x)$ is defined by the following force equation:

$$mx'' = -\frac{dV(x)}{dx}. \tag{16a}$$

The potential function $V(x)$ may be a purely mathematical construction obtainable from integration with respect to x, but in physical systems it can represent *potential energy*.

In many mechanical systems we can calculate the potential energy $V(x)$ from physics: if the particle is under the influence of "constant" gravity, as on the surface of the earth, then

$V(x) = mgh$, where h is height above any arbitrary fixed level.

The differential equation (16a) for a one-parameter mechanical system can be replaced by the following system of first-order differential equations:

$$x' = m^{-1}y, \qquad y' = -\frac{dV}{dx}. \qquad (16b)$$

Note that, up to a constant multiple, y represents the *velocity* of the particle. Also note that equations (16a) and (16b) equate a derivative with respect to x with a second derivative with respect to t. Furthermore, since equations (16) depend only on the derivative of $V(x)$, not on $V(x)$ itself, the equations of motion do not change if a constant is added to V; consequently, the potential V is determined only up to an additive constant.

In order to form some intuitive idea of motion under a potential, you might think of a bead of mass m sliding without friction on a wire shaped to give the graph of $V(x)$, as shown in Figure 6.5.2. The bead has only one degree of freedom—it can only move back and forth along the wire. You can imagine letting the bead start from any point on the wire and then try to visualize how it will slosh up and down and back and forth according to the shape of the potential function $V(x)$.

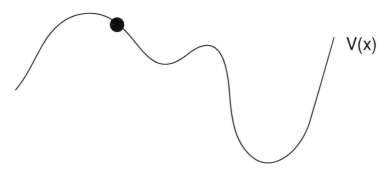

FIGURE 6.5.2. Bead and wire model for $x'' = -dV(x)/dx$.

This model of a bead on a wire is not quite accurate—at the end of this section in Example 6.5.6, we will properly analyze the bead-on-wire analogy for the pendulum—but it does give the proper qualitative features. In particular, you can see oscillations of x near the minima of V, exceptional solutions that tend (infinitely slowly) toward the maxima of V, and even more exceptional solutions that just sit at the maxima of V.

Although equation (16a) cannot generally be solved for $x(t)$ in terms of elementary functions, the mechanical system can nevertheless usually be understood. The key to this understanding is the following theorem:

Theorem 6.5.2. *For the differential equation $mx'' = -dV/dx$, the total energy function*

$$E(x, x') = \tfrac{1}{2} m(x')^2 + V(x)$$
$$= \text{kinetic energy} + \text{potential energy} \qquad (17a)$$

is constant along the trajectories of solutions in the xy phase plane; consequently, the mechanical system corresponding to this differential equation is called conservative.

Proof. This is a straightforward computation from (17a) of

$$\frac{d}{dt} E(x(t), x'(t)) = mx'x'' + \frac{dV}{dx} x',$$

which is zero by equation (15a). □

In terms of the alternative form (16b) of the mechanical system with one degree of freedom,

$$x' = m^{-1}y \qquad y' = -dV/dx, \qquad (16b, \text{ again})$$

Theorem 6.5.2 can be stated in terms of conserving

$$E(x, y) = \frac{1}{2} m^{-1} y^2 + V(x). \qquad (17b)$$

In the *xy-phase plane*, the point $[x(t), y(t)]$ of the system moves along a *trajectory*; Theorem 6.5.2 shows that these trajectories are *level curves* of the function E.

Example 6.5.3. Let us go back to the pendulum of Example 6.5.1, described by the second order differential equation, for $K = (g/\ell)$,

$$\theta'' = -K \sin \theta \qquad (15a, \text{ again})$$

and equivalent to the system

$$\theta' = y$$
$$y' = -K \sin \theta. \qquad (15b, \text{ again})$$

Equations (15) are of the proper mathematical form (16) for a mechanical system with one degree of freedom, so there exists a potential function $V(\theta)$. Combining equations (15a) and (16a) implies that

$$\frac{dV(\theta)}{d\theta} = K \sin \theta,$$

which can be integrated; so we know that $V(\theta) = -K \cos \theta$.

By Theorem 6.5.2, the trajectories of the solutions are level curves of the function

$$E(\theta, y) = \frac{y^2}{2} - K \cos \theta. \tag{18}$$

This function E is a mathematical total energy function, which is up to a constant multiple of the physical total energy, as you can confirm in Exercise 6.5#11.

The function $E(\theta, y)$ has a graph that looks like Figure 6.5.3.

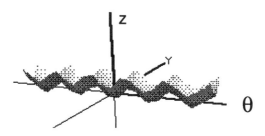

FIGURE 6.5.3. Graph of $E = y^2/2 - K \cos \theta$.

The level curves of E, along which the trajectories of the system lie, look like Figure 6.5.4 . The minima of E are at the points $(\theta, y) = (2k\pi, 0)$, for integer k; the saddle points of the surface E are at the points $((2k+1)\pi, 0)$.

Arrows on the curves in Figure 6.5.4 indicate in which direction the pendulum follows them:

$$y = \theta' \text{ decrees that } \begin{cases} \text{positive } y \text{ implies increasing } \theta \\ \text{negative } y \text{ implies decreasing } \theta. \end{cases}$$

This fact is confirmed by watching computer drawings evolve, using a program like *DiffEq, Phase Plane* in the *MacMath* package.

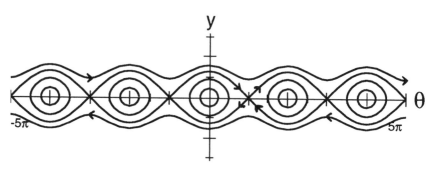

FIGURE 6.5.4. Trajectories for $\theta' = y$, $y' = -K \sin \theta$.

Look carefully at the different kinds of phase plane trajectories exhibited in Figure 6.5.4:

The *closed level curves* ($|E| < K$) correspond to back and forth oscillations of the pendulum.

The *points at the centers* of these closed level curves ($E = -K$) correspond to the *stable equilibrium* of the pendulum at rest at the bottom of its motion.

The *level curves above and beneath the closed curves* ($E > K$) correspond to motions in which the pendulum has enough energy to simply go full circle, round and round over the top.

The level curves joining the *saddles* ($E = K$) are those exceptional motions in which for $t \to \infty$ and for $t \to -\infty$ the pendulum tends to the *unstable equilibrium* at the very top of the swing. Note that the time required to go from one saddle to the next is infinite.   ▲

The solutions coming from or going to saddles are called *separatrices*, because they have exceptional behavior and separate the regions of generic behavior from each other. (You should be reminded of the exceptional solutions in antifunnels for equations in $\mathbb{R}^1$.) We will have much to say about saddles and separatrices in Chapter 8.

EFFECTS OF ADDING FRICTION

When we add to the system of the pendulum a *perturbation term* to represent friction, a physically more realistic situation, analysis changes. There will no longer be any potential or conservation of energy. Nevertheless, we can analyze the behavior in a similar manner.

Example 6.5.4. Consider the same pendulum of Example 6.5.3, but add a bit of friction. The motion of the pendulum bob will now obey an equation such as

$$\begin{aligned} \theta' &= y \\ y' &= -K \sin \theta - \varepsilon y, \end{aligned} \qquad (19)$$

where ε is a friction coefficient and we have assumed that the friction is proportional to the velocity. This system with friction is *not* of the form required by Theorem 6.5.2, because the equation in y' depends on y as well as on θ. In fact, the system with friction does not conserve total energy.

The function E is no longer constant on solutions, but we can easily see that it decreases. This corresponds to the intuitive idea that in a system in which energy is dissipated by friction, the global energy decreases. This fact is easy to compute from equation (17b):

$$\frac{dE}{dt} = \frac{d}{dt} \left[\frac{y(t)^2}{2} - K \cos \theta(t) \right]$$

$$= y(t)y'(t) - (-K\sin\theta(t))\,\theta'(t)$$
$$= y(t)\left[-K\sin\theta(t) - \varepsilon y(t)\right] + K\sin\theta(t)y(t) = -\varepsilon y(t)^2 \le 0.$$

The regions $\{(t,\theta) \mid E(t,\theta) \le C\}$ are now trapping regions: if a solution enters one of them, it can never get out again. (You should be reminded of funnels.) The phase plane will look like Figure 6.5.5 when ε is sufficiently small.

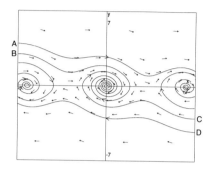

FIGURE 6.5.5. Four trajectories for $\theta' = y$, $y' = -\sin\theta - 0.2\,y$. Those starting at A and B begin with the pendulum bob to the left of center; those from C and D begin with the bob to the right.

We will see in Exercise 8.1#11 that the spiraling behavior disappears when the friction coefficient ε becomes large. When ε becomes zero, the picture becomes that of Figure 6.5.4.   ▲

We will see other examples of mechanical systems in Section 6.7.

Meanwhile, let us return to the intuitive visualization of behavior under the influence of a potential function, as suggested earlier in this section.

A BEAD SLIDING ON A WIRE: PRECISE ANALYSIS

At the beginning of this section, we claimed that a solution of

$$m\,x'' = -\frac{dV}{dx} \tag{16a, again}$$

behaves roughly like the x-coordinate of a bead of mass m sliding on a wire whose shape is the graph of V, in a constant gravitational field with $g = 1$. Note that we introduced this analogy only to give the reader an intuitive feel for the differential equation (16a): we think that most readers will know without any mathematics or physics how such a bead will move. But it is not a good method of analysis, as the detailed behavior of the bead is quite a bit more complicated than the original problem.

The differential equation describing the motion of the bead is as usual $\mathbf{F} = m\mathbf{a}$, but it is quite unpleasant to write the acceleration in terms of the x-coordinate of the bead. Instead, we will use the arclength s along the wire; in other words, we will record the position of the bead by recording how far along the wire it is (measuring from the point with x-coordinate 0, for instance). Clearly, if the motion is described by the function $s(t)$, then the acceleration is $s''(t)$. A force analysis (Exercise 6.5#12a) shows that at the point distance s along the wire, with x-coordinate $x(s)$, the force is

$$-m \frac{dV/dx}{\sqrt{1 + (dV/dx)^2}} = m s''. \tag{20}$$

Keep in mind that since $x = x(s)$, dV/dx will be expressed in terms of s.

Example 6.5.5. Suppose $V(x) = -\sqrt{\ell^2 - x^2}$; i.e., that the bead is constrained to remain on a wire of semi-circular shape, as shown in Figure 6.5.6.

Then, since $x = \ell \sin(\theta)$ and $\theta = s/\ell$,

$$x(s) = \ell \sin(s/\ell).$$

Furthermore, from the definition of $V(x)$ for this example,

$$\frac{dV}{dx} = \frac{x}{\sqrt{\ell^2 - x^2}} = \frac{\ell \sin(\theta)}{\ell \cos(\theta)} = \tan\left(\frac{s}{\ell}\right).$$

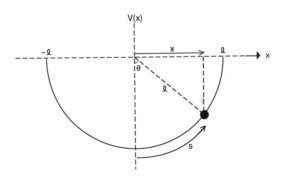

FIGURE 6.5.6. Bead on semi-circular wire.

Therefore, the expression (20) for the force becomes

$$ms'' = -m \frac{dV/dx}{\sqrt{1 + (dV/dx)^2}} = -m \frac{\tan(s/\ell)}{\sqrt{1 + \tan^2(s/\ell)}} = -m \sin(s/\ell).$$

Thus, the equation describing the motion of the bead is simply

$$s'' = -\sin(s/\ell). \quad \blacktriangle$$

Example 6.5.6. Note that if you set $\theta = s/\ell$ (a very reasonable definition), the equation $s'' = -\sin(s/\ell)$ of Example 6.5.5, describing the motion of a bead on a circular wire, becomes

$$\theta'' = -(1/\ell)\sin\theta,$$

the differential equation describing the pendulum of Examples 6.5.1 and 6.5.3. This should not be surprising: *there is no physical difference in constraining the bead to a circular path by attaching it to a lever arm or by putting it on a circular wire.* All we have done is rederived the equation of the pendulum in a far more complicated way.

To go back to the analogy, we are saying that the motion of the bob on a pendulum is *exactly* the same thing as motion of a bead constrained to move on a circular wire. This is shown in Figure 6.5.7 where the top half of the circular motion has been reflected to the sides (which allows us to think of the bead on its nth trip around the circle as running along the nth pattern).

The motion of the bob on the pendulum that we derived in Example 6.5.1 from $V(\theta) = K\cos\theta$ is *approximately* the same as the motion of a bead on a wire with the shape of the graph of $-\ell\cos(s/\ell)$, as shown in Figure 6.5.8.

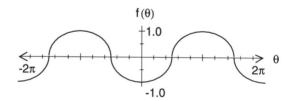

FIGURE 6.5.7. Exact bead-on-wire model.

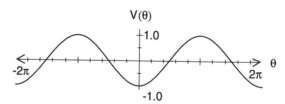

FIGURE 6.5.8. Approximate bead-on-wire model.

Notice that close to the minimum of potential, Figure 6.5.8 for approximate motion is very "similar" to Figure 6.5.7 for exact motion, but they are *not* identical.

Some motions are oscillations in the wells, others travel along the wires, and so forth. These statements should be intuitively clear; it should also be intuitively clear that the details of the motions will be different. ▲

The conclusion of Example 6.5.6 is also all that can be said in general. The differential equation describing the motion of the bead on the semi-circular wire is

$$s'' = \frac{dV/dx}{\sqrt{1 + (dV/dx)^2}},$$

which looks quite a bit like the differential equation describing the pendulum,

$$x'' = -V'(x),$$

in the sense that both have equilibria at corresponding points and of the same type.

For the function $s(x)$ that measures arclength along the wire from the point above zero to the point above x, we have

$$s(x) = \int_0^x \sqrt{1 + (f'(u))^2}\, du.$$

We want $V(s)$ to be "f measured with respect to s," i.e.,

$$V(s(x)) = f(x).$$

Then these last two equations can be combined to give

$$\frac{dV}{ds} = \frac{dV/dx}{ds/dx} = \frac{f'(x)}{\sqrt{1 + (f'(x))^2}}. \tag{21}$$

So, if you know $V(s)$, can you use equation (21) to find $f(x)$? It is not easy! But since the right-hand side of equation (21) is less than 1, the answer may be yes if $|dV/ds| < 1$.

In Exercises 6.5#12c, you can confirm that equation (21) works if $s = \ell\theta$ with ℓ fixed; that is, that

$$\frac{dV}{ds} = \frac{f'(\theta)}{\sqrt{1 + (f'(\theta))^2}} \quad \text{implies} \quad f(\theta) = -K\sqrt{1 - (2\theta/\pi)^2}.$$

Remark. One may wonder whether, given a function $V(x)$, there is some other function $W(x)$ such that the motion on the wire with the shape of the graph of W, that is, with equation

$$s'' = \frac{W'(x(s))}{\sqrt{1 + (W'(x(s)))^2}},$$

is precisely the same as motion under $x'' = -dV/dx$. Clearly, one requirement is $|dV/dx| < 1$. It turns out that this is the only requirement, as shown in Exercise 6.5#13.

6.6 Essential Size, Conservation Laws

We have seen in Section 6.4 that the *order* of a differential equation is sort of phony: it can be diminished at the expense of increasing the number of unknown variables.

Vice versa, the *dimension* is phony since it can usually be traded for higher order, as in Example 6.4.4.

Definition 6.6.1 (Essential size). We will call the *essential size* of a differential equation the smallest number n such that the differential equation can be transformed into an *autonomous, first order* equation in $\mathbb{R}^n$.

Example 6.6.2. The system

$$x'' = x^2 - y$$
$$y'' = y - x$$

has essential size four, because *four* equations are necessary to achieve a system of *first* order equations:

$$x' = w$$
$$y' = v$$
$$w' = x^2 - y$$
$$v' = y - x.$$

▲

Essential size as a measure of how hard it is to understand a differential equation is sometimes misleading.

Example 6.6.3. The system

$$x' = x + y$$
$$y' = y - x^2$$
$$z' = 1$$

looks, and *is*, three dimensional in xyz-space, but is really two independent systems—one of size two (in x' and y'), the other (in z') of size one. This system is no harder to understand than its parts.   ▲

On the other hand, a nonautonomous system can be turned into an autonomous system in one higher dimension; so a system may have larger essential size than it might first appear.

Example 6.6.4. The equation $x' = f(t, x)$ is nonautonomous but can be written as an autonomous system

$$x' = f(t, x)$$
$$t' = dt/dt = 1;$$

so the essential size of the original single first order equation is two. This *essential* size of two is why *DiffEq* draws in the plane, not the line.   ▲

Example 6.6.5. Consider the system

$$x' = x + y$$
$$y' = y - x^2$$
$$z' = 2 + x.$$

The first two equations form a system of essential size two. Then once the solution $x(t)$ has been found to this first system, the third equation becomes a new system, also of essential size two, because this third equation, with $x(t)$, is not autonomous.   ▲

The complexity of solutions to a differential equation is a function of the essential size: the greater the essential size, the greater the opportunity for the solutions to tangle. (Compare the examples of Section 6.1 for autonomous differential equations in $\mathbb{R}^2$ with those in earlier chapters for nonautonomous differential equations in $\mathbb{R}^1$.)

The name of the game in solving systems of differential equations is to reduce the essential size. This is done by finding *conservation laws*.

Definition 6.6.6 (Conservation Law). If a differential equation in $\mathbb{R}^n$ implies for some function $F(\mathbf{x})$ that $F'(\mathbf{x}) = \mathbf{0}$, then

$$F(\mathbf{x}) = \text{a constant} \qquad (22)$$

along the trajectories of the solutions, and equation (22) is called a *conservation law*.

The reason conservation laws are so important is that each one decreases the essential size. This is because

$$F(x_1, x_2, \dots, x_n) = C$$

will usually allow you to express some variable, say x_n, implicitly as a function of the others, $x_1, x_2, \dots, x_{n-1}$.

Example 6.6.7. In the mechanical system with one degree of freedom, as explored in Section 6.5,

$$m\, x'' = -\frac{dV}{dx} \tag{23}$$

is an equation in $\mathbb{R}^2$ that we want to solve for $x = u(t)$. Theorem 6.5.2 states as a conservation law that

$$E = \frac{m\, y^2}{2} + V(x) \tag{24}$$

is constant on the trajectories of the solutions. It is possible to solve equation (24) for $y = x'$:

$$x' = \sqrt{2(E - V(x))/m}, \tag{25}$$

which reduces the problem to an autonomous first order equation in $\mathbb{R}^1$. Thus, the conservation law (24) has reduced the essential size from *two* in equation (23) to *one* in equation (25).

Remark. You can observe that equation (25) *can* be reduced to computing integrals, but they tend to be unpleasant or impossible to integrate in elementary terms; furthermore, it can still be an unpleasant problem to solve for $x(t)$. Therefore, we did not explore this option in Section 6.5 but instead used the conservation law to get at the trajectories by the back door. That is, we used equation (24) to plot the trajectories *instead* of solving (25)! ▲

Example 6.6.8. A *central force* is some function $\mathbf{f}(r)$, depending only on distance from the origin, that acts in the radial direction $\mathbf{r} = x\mathbf{i} + y\mathbf{j} + z\mathbf{k}$, with $r = \|\mathbf{r}\|$. (Such a force will be explored in detail in Section 6.7. You might look at Figure 6.7.1.) A central force combined with Newton's law in $\mathbb{R}^3$ produces the following second order equation:

$$m\,\mathbf{a} = m\,\mathbf{r}'' = \|\mathbf{f}(r)\|\frac{\mathbf{r}}{r}, \tag{26}$$

of essential size six, because it represents a second order equation for each of the three components of $\mathbf{r}$: x, y, and z.

Differentiating the cross product $\mathbf{r} \times \mathbf{r}'$ gives [by several steps, including the use of equation (26), in Exercise 6.6#2a]

$$\frac{d}{dt}(\mathbf{r} \times \mathbf{r}') = \mathbf{r} \times \mathbf{r}'' = 0.$$

Consequently, we find as a conservation law the following fact:

$$\mathbf{r} \times \mathbf{r}' = \mathbf{M}, \quad \text{a constant vector.} \tag{27}$$

This vector $\mathbf{M}$ should be called the *angular momentum* vector, because its *magnitude* gives the angular momentum, as will be explained in Example

6.6.9. However, it is the direction of $\mathbf{M}$ that concerns us here. The constant direction of $\mathbf{M}$ ensures that $\mathbf{r}$ will never leave the *plane* spanned by $\mathbf{r}(t_0)$ and $\mathbf{r}'(t_0)$ at any particular time t_0. (In Exercise 6.6#2b, you can work through this derivation.) This means that the central force law (26) actually operates only in $\mathbb{R}^2$, so the second order equation now has essential size four. ▲

If the physical or mathematical system under consideration has *symmetries*, then it is possible to choose *suitable coordinates* so that the equations of motion will be independent of some of the coordinates, which will lead to conservation laws and reduce essential size. *This correspondence between symmetries and conservation laws is one of the guiding principles of physics.*

Example 6.6.9. We have shown in Example 6.6.8 that a central force law,

$$\mathbf{r}'' = f(r)\frac{\mathbf{r}}{r}, \tag{26, again}$$

operates in a *plane*. The obvious symmetries of a central force law will be most nicely described by *polar coordinates*, so we first describe the position vector

$$\mathbf{r}(t) = \begin{bmatrix} x(t) \\ y(t) \end{bmatrix}$$

in terms of two scalar functions $r(t)$ and $\theta(t)$, and we shall then find the differential equation that they will satisfy. This is a typical unpleasant change of variables and goes as follows:

$$\mathbf{r} = \begin{bmatrix} x \\ y \end{bmatrix} = \begin{bmatrix} r\cos\theta \\ r\sin\theta \end{bmatrix}$$

$$\mathbf{r}' = \begin{bmatrix} x \\ y \end{bmatrix}' = \begin{bmatrix} r'\cos\theta - r\theta'\sin\theta \\ r'\sin\theta + r\theta'\cos\theta \end{bmatrix}$$

$$\mathbf{r}'' = \begin{bmatrix} x \\ y \end{bmatrix}'' = \begin{bmatrix} (r'' - r(\theta')^2)\cos\theta - (2r'\theta' + r\theta'')\sin\theta \\ (r'' - r(\theta')^2)\sin\theta + (2r'\theta' + r\theta'')\cos\theta \end{bmatrix}$$

$$= (r'' - r(\theta')^2)\begin{bmatrix} \cos\theta \\ \sin\theta \end{bmatrix} + (2r'\theta' + r\theta'')\begin{bmatrix} -\sin\theta \\ \cos\theta \end{bmatrix}. \tag{28}$$

From the central force field (26), we know

$$\begin{bmatrix} x \\ y \end{bmatrix}'' = f(r)\begin{bmatrix} \cos\theta \\ \sin\theta \end{bmatrix}, \tag{29}$$

that is, that the force points in the direction of the origin. Since the vectors

$$\begin{bmatrix} \cos\theta \\ \sin\theta \end{bmatrix} \text{ and } \begin{bmatrix} -\sin\theta \\ \cos\theta \end{bmatrix}$$

are orthogonal, combining equations (28) and (29) leads to the system of equations

$$r'' - r(\theta')^2 = f(r) \tag{30a}$$

$$r\theta'' + 2r'\theta' = 0, \tag{30b}$$

which describes in polar coordinates the motion in a central force field.

Now we can observe that $(r^2\theta')' = r(r\theta'' + 2r'\theta') = 0$ by equation (30b), so that $r^2\theta'$ as a function of time is *constant*. This quantity $r^2\theta' = M$ is the *magnitude* of the *angular momentum* of the body around the z-axis, and the derivation above shows that *in any central force field, the angular momentum is preserved.* We have thus found another conservation law.

Remark. In Exercise 6.6#2c, you will show an alternate way to derive (30b) directly from

$$\mathbf{r} \times \mathbf{r}' = \mathbf{M}, \tag{27, again}$$

so that

$$\|\mathbf{M}\| = |r^2\theta'| \equiv M;$$

hence the angular momentum M is in fact the magnitude of $\mathbf{M}$. In fact, the conservation of angular momentum could have been derived directly from equation (27), but we needed the explicit derivatives of $\mathbf{r}$ to get equation (30a). That route also emphasizes the role that taking advantage of symmetry can play in finding a conservation law.

The conservation of angular momentum further reduces the essential size of the system. Since θ' can be expressed as a function of r,

$$\theta' = \frac{M}{r^2}, \tag{31}$$

equation (30a) becomes a *second* order equation in r alone (for which we have developed methods of analysis and will demonstrate them in Section 6.7). Once we know r, the conservation of angular momentum gives a *first* order equation (31) in θ, so the total essential size of the system has been reduced from four (at the end of Example 6.6.8) to three. That is, system (30) can now be replaced by

$$r'' - \frac{M^2}{r^3} = f(r) \quad \text{and} \quad \theta' = \frac{M}{r^2}. \quad \blacktriangle \tag{32}$$

6.7   The Two-Body Problem

This section is devoted to showing that Newton's law of gravitation implies Kepler's laws.

Johannes Kepler (1571–1630) was a German astronomer. He was not born to wealth and was employed, first as an assistant and later as a partner, by the Danish nobleman Tycho Brahe, who was rich and fascinated by astronomy. Kepler's job in Tycho's observatory was to observe the planet Mars. As a result of 14 years of observation, he came up with the following laws:

1. *The planets move on ellipses with the sun at one focus.*

2. *The radial segment from the sun to a planet sweeps out equal areas in equal time intervals.*

3. *The periods of rotation are proportional to the $\frac{3}{2}$ power of the semi-major axes of the ellipses.*

Even knowing that these laws are true, and using a calculator and a telescope, it is not clear what you should observe in order to confirm them. One cannot be but struck at the amazing genius it must have taken for a person, without a telescope or any accurate means of measurement, before the invention of analytic geometry, and without any way to determine the distances of any celestial bodies, to take the results of 14 years of observations and come up with such laws. If you consider that Tycho's observatory was on an island between Denmark and Sweden, which must be, next to Ithaca, the cloudiest place in the world, the records must have been spotty in the extreme.

By the time of Newton (1642–1724), these laws had been examined and largely accepted (although Galileo was tried by the church and convicted in 1633 for teaching the Copernican view that the earth turned around the sun).

We now come to another extraordinary display of genius. In a period of a year, Newton postulated the universal law of gravitation, saw that it gave a differential equation for the motions of the planets (inventing calculus in the process, not to mention the theory of differential equations), and solved the equation to the extent of showing that the postulate implied Kepler's laws. Nowadays, we hesitate to teach his solution in third year calculus because it is too difficult, even though it appeared as the focus of the first book on calculus.

All of this is intended to explain why we are including the solution here, even though it does not quite fit. Some might say (we are among them) that the work of Newton was one of the most important events in the history of humanity, the one that ushered in the scientific age. Considering that this work was largely the solution of a differential equation, it would be outrageous not to include it in a book on the subject, and at an early stage.

Given two bodies with masses m_1 and m_2 and positions $\mathbf{x}_1$ and $\mathbf{x}_2$, Newton's universal law of gravitation together with $\mathbf{F} = m\mathbf{a}$ gives the system

$$\mathbf{x}_1'' = \frac{Gm_2(\mathbf{x}_2 - \mathbf{x}_1)}{\|\mathbf{x}_2 - \mathbf{x}_1\|^3}$$

$$\mathbf{x}_2'' = \frac{Gm_1(\mathbf{x}_1 - \mathbf{x}_2)}{\|\mathbf{x}_1 - \mathbf{x}_2\|^3}.$$

(33)

This is a priori a differential equation in $\mathbb{R}^{12}$: there are three dimensions for each $\mathbf{x}_i$, for a total of six dimensions; but the differential equation for each of these is second order, so the system (33) is equivalent to a system of *twelve* first order equations. However, it is easy to reduce the dimension from $\mathbb{R}^{12}$ by taking advantage of conservation laws, which we shall proceed to do.

The *center of mass* is

$$\mathbf{X} = \frac{m_1\mathbf{x}_1 + m_2\mathbf{x}_2}{m_1 + m_2}$$

and it satisfies the differential equation $\mathbf{X}'' = 0$, so $\mathbf{X}(t)$ moves at *constant velocity* $\mathbf{X}'$ on some straight line, as you will prove in Exercise 6.7#1b. This is overall conservation of *linear momentum*, with $\mathbf{X} = \mathbf{a} + t\mathbf{b}$ and $\mathbf{X}' = \mathbf{b}$.

Remark. Simple though this observation may be, it is a typical application of the main method of solution for equations in mechanics:

> *You find some preserved quantity* (in this case linear momentum); *then setting it equal to a constant allows expression of some unknown functions in terms of others* (in this case the position of one body in terms of the position of the other).

In this manner we can now reduce the number of actual unknown quantities from twelve to six, so the problem is reduced from $\mathbb{R}^{12}$ to $\mathbb{R}^6$. One way to express this problem in $\mathbb{R}^6$ is as a *second* order equation for one position vector in $\mathbb{R}^3$. We shall use $\mathbf{r} \equiv \mathbf{x}_1 - \mathbf{X}$ to describe the position of the first body; the position of the second is then determined by the conservation of linear momentum described above.

In Exercise 6.7#1c you are asked to show that the vector $\mathbf{r}$ satisfies the differential equation

$$\mathbf{r}'' = -\frac{Gm_2^3}{(m_1 + m_2)^2} \frac{\mathbf{r}}{\|\mathbf{r}\|^3} = -K\frac{1}{\|\mathbf{r}\|^2}\frac{\mathbf{r}}{\|\mathbf{r}\|},$$

(34)

where $\frac{\mathbf{r}}{\|\mathbf{r}\|}$ is the unit vector in the $\mathbf{r}$ direction and $K = Gm_2^3/(m_1 + m_2)^2$ gathers together the constants. In this form (34), the system clearly satisfies an *inverse square law*.

Notice that if m_1 is very small in comparison with m_2, then equation (34) is nearly what you would get with the mass m_2 at rest at the origin. For instance, you could consider m_1 a planet and m_2 the sun, although in any case equation (34) is exact.

Equation (34) means that $\mathbf{r}$, the measurement of position with respect to the center of gravity, behaves as a point in a central force field. That is, the direction of the force is in the direction of the $\mathbf{r}$ vector, and the magnitude of the force depends only on $r = \|\mathbf{r}\|$, so

$$\mathbf{r}'' = f(r)\frac{\mathbf{r}}{r}, \qquad \text{(26, again)}$$

as in Examples 6.6.7 and 6.6.8.

The crucial thing is that in a central force field the force depends only on the magnitude r, and a picture of the force vectors will look like Figure 6.7.1.

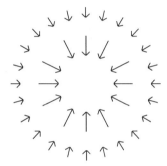

FIGURE 6.7.1. A particular central force field.

In Example 6.6.8, we analyzed motion in a central force field. Let us summarize what we found. First, the position vector $\mathbf{r}$ remains in the plane spanned by the initial position and the initial velocity. This followed from considering the *angular momentum vector*

$$\mathbf{r} \times \mathbf{r}' = \mathbf{M}, \text{ a constant vector,} \qquad \text{(27, again)}$$

and saying that its direction is constant. A second order equation in a plane is a first order equation in $\mathbb{R}^4$ rather than $\mathbb{R}^6$, so we have now reduced the essential size of the problem from six to four, which is a great improvement but not enough yet to solve the problem.

To further decrease the essential size, change variables so that the plane of the motion is the horizontal plane $\mathbb{R}^2 \subset \mathbb{R}^3$ and use polar coordinates in $\mathbb{R}^2$ to take advantage of the symmetries of the problem. Then the angular momentum vector points in the direction of the z-axis and has magnitude

$$r^2\theta' = \|\mathbf{M}\| \equiv M$$

as was shown in equation (30b) of Example 6.6.8.

Kepler's Second Law. *Conservation of angular momentum is equivalent to Kepler's second law*, because $r^2\theta'/2$ is the rate at which the vector $\mathbf{r}$ sweeps out area (Exercise 6.7#5a). So Kepler's second law is a way of stating conservation of angular momentum without mentioning derivatives. Note that Kepler's second law is valid in *any* central force field (26) and does not require the inverse square law (34).

As we saw in equation (32) of Example 6.6.8, the formula

$$\theta' = \frac{M}{r^2} \tag{35}$$

gives

$$r'' - r(\theta')^2 = f(r) \tag{36}$$

for *any* central force field. Finally, using the inverse square law (34), we find

$$r'' = -\frac{K}{r^2} + \frac{M^2}{r^3}, \tag{37}$$

which is a second order equation in r alone. We have brought the essential size down to two, and this we will be able to solve. Of course, when we have done so, we will have to go back to the angular momentum equation (35) to find θ, and finally use the polar coordinate formulas to find x and y.

Equation (37) is an equation for a mass moving in a one-dimensional force field: it can be written $r'' = -dW(r)/dr$, where

$$W(r) = -\frac{K}{r} + \frac{M^2}{2r^2}, \tag{38}$$

which is the equation of motion of a particle with one degree of freedom moving under a "potential." However, this potential W is fictitious rather than real (just a figment of the mathematics).

Equation (37) is equivalent, as we showed in Sections 6.4 and 6.5, to the system

$$r' = v \tag{39a}$$

$$v' = -\frac{K}{r^2} + \frac{M^2}{r^3}. \tag{39b}$$

We have seen in Theorem 6.5.2 that the energy function

$$E(r, v) = \frac{v^2}{2} + W(r) = \frac{v^2}{2} - \frac{K}{r} + \frac{M^2}{2r^2} \tag{40}$$

is constant along trajectories of motions in the rv-plane.

From equation (38), we can graph $W(r)$, as in Figure 6.7.2.

Then the level curves of E look like Figure 6.7.3, as you are asked to show in Exercise 6.7#3.

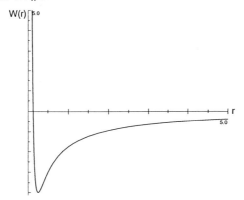

FIGURE 6.7.2. Mathematical "potential" $W(r)$.

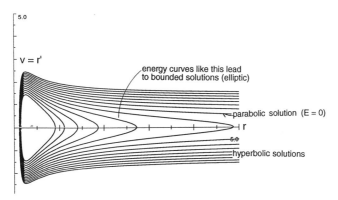

FIGURE 6.7.3. Level curves of $E = \frac{1}{2}v^2 + W(r)$.

We see in Figure 6.7.3 that orbits with $E < 0$ are bounded (and we shall show that they correspond to *elliptical* orbits in the xy-plane of configuration space), and that orbits with $E \geq 0$ are unbounded.

We are not able actually to solve these equations for r and v as functions of time, but here we will be able to solve for r as a function of θ. (This is as in Section 6.3, where we could not solve the Volterra equations $x' = ax - cxy$; $y' = -by + fxy$ for $x(t)$ or $y(t)$, but $dy/dx = y'/x'$ did happen to be integrable, so we were able to solve for trajectories in the xy-phase plane.)

Kepler's First Law. Suppose the motion considered has energy E and angular momentum M. Then putting together

$$\frac{d\theta}{dt} = \frac{M}{r^2} \qquad (35, \text{ again})$$

and

$$\frac{dr}{dt} = v = \sqrt{2\left(E + \frac{K}{r} - \frac{M^2}{2r^2}\right)}, \qquad \text{from (40)}$$

we get

$$\frac{d\theta}{dr} = \frac{d\theta}{dt} \Big/ \frac{dr}{dt} = \frac{M}{r^2\sqrt{2\left(E + K/r - M^2/2r^2\right)}}, \qquad (41)$$

as you can confirm in Exercise 6.7#6a.

Equation (41) can actually be integrated (and Newton did it, in the very first book on calculus!), yielding (Exercise 6.7#6b)

$$\theta = \text{arc cos} \frac{(M/r - K/M)}{\sqrt{2E + K^2/M^2}}. \qquad (42)$$

You can confirm the solution (42) by differentiating. Furthermore, now you *can* solve for r.

If you set $M^2/K = p$ and $\sqrt{1 + (2EM^2/K^2)} = e$, equation (42) becomes

$$r = \frac{p}{1 + e\cos\theta}, \qquad (43)$$

which is the equation of an *ellipse* with a focus at the origin, eccentricity e, and parameter p, as shown in Figure 6.7.4. (See Exercise 6.7#7.)

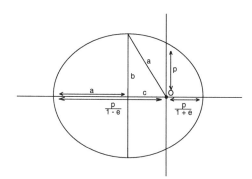

FIGURE 6.7.4. Ellipse.

Kepler's Third Law. From analytic geometry, recall that the area of the ellipse is πab, where a is the semi-major axis and b is the semi-minor axis. (Intuitively, you can see this as stretching a disk.) But we saw in the discussion of Kepler's Second Law, combined with equation (35), that $M/2$ is the speed at which area is swept out by the vector r, so that if an orbit has period T, then

$$2\pi ab = MT.$$

Now some elementary geometry on Figure 6.7.4 shows that

$$a = \frac{p}{1 - e^2} \quad \text{and} \quad b = \frac{p}{\sqrt{1 - e^2}},$$

so that

$$T = 2\pi a \frac{b}{M} = 2\pi a^{3/2} \frac{\sqrt{p}}{M} = \frac{2\pi a^{3/2}}{\sqrt{K}}.$$

Thus Kepler's Third Law is proved.

Summary. Having said so much in this section, we pause for an overview because many of the ideas are by no means limited to the example of the two-body problem:

1. The original problem concerns second order equations for two vectors with three components each, so there are $2 \times 2 \times 3 = 12$ dimensions.

2. A change of variable in terms of the *center of mass* of the physical system will allow us to use conservation of *linear momentum* to reduce the number of dimensions to six.

3. By another conservation law for the *direction* of the *angular momentum*, we can further notice that the actual motion would be in a plane rather than in a larger space, which further reduces the number of dimensions to four.

4. Next, conservation of the *magnitude* of the *angular momentum* allows us to once more reduce the number of actual dimensions to two.

5. Finally, conservation of *energy* allows us to solve the problem (but only for trajectories in the phase plane, not for the actual motion as a function of time).

6.8   Flows

We have been emphasizing solutions of differential equations as functions of time. *Flows* provide a language to emphasize the dependence on initial conditions. For instance, the value of a bank account depends on time, but we should remember that it also depends on the initial deposit.

Let $\mathbf{x}' = \mathbf{f}(\mathbf{x})$ be a differential equation in $\mathbb{R}^2$, and imagine drawing some shape in the plane (a rectangle, for instance, or a cat) and solving the differential equation for a *fixed amount of time*, starting at each point of this shape. The shape will move, and probably become distorted, as the following examples illustrate.

Example 6.8.1. For instance, the trajectories of the system of equations

$$\begin{bmatrix} x \\ y \end{bmatrix}' = \begin{bmatrix} x \\ -y \end{bmatrix}$$

are hyperbolas, and if you let the "normal" cat (in Figure 6.8.1) flow forward for time 1, you will find a cat like the "horizontally flattened" cat in the lower right, whereas if you let it flow backward, for time −1, you will get something like the "vertically squeezed" cat in the upper left.

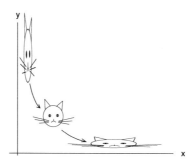

FIGURE 6.8.1. The flow of a region under $x' = x, y' = -y$. ▲

Examples 6.8.2.

A. $\dfrac{dx}{dt} = y$ \qquad\qquad B. $\dfrac{dx}{dt} = 1$ \qquad\qquad C. $\dfrac{dx}{dt} = 1$

$\quad\ \dfrac{dy}{dt} = -x$ \qquad\qquad\ \ \ $\dfrac{dy}{dt} = y$ \qquad\qquad\quad $\dfrac{dy}{dt} = y^2 - x$

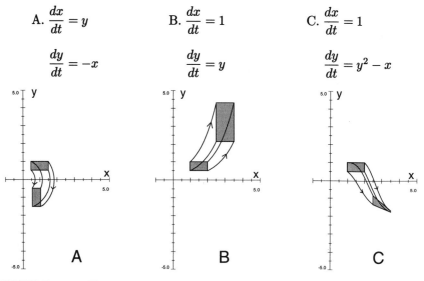

FIGURE 6.8.2. Flow in the xy-plane of the same rectangle for three different differential equations.

The pictures in Figure 6.8.2 were made simply by drawing 30 steps of an approximate solution with Euler's method from each of the four corners of the rectangle bounded by $x = 1, x = 2, y = 1/2$, and $y = 1$.

Notice that for equations A, the rectangle simply rotates in the plane; for equations B, the rectangle stretches vertically as it moves to the right; for equations C, this particular rectangle moves to the lower right but becomes considerably distorted. ▲

Of course, Examples 6.8.2 show only what is happening locally for a small rectangle in the phase plane. You should play around with Exercises 6.8#1–6.8#3 in order to explore what happens globally—what changes are caused by the flows of different rectangles for these same equations.

Now that you have begun to visualize a flow, we shall state a formal definition:

Definition 6.8.3. Let $\mathbf{f}$ be a vector field on a subset of $\mathbb{R}^n$. The flow of the autonomous differential equation $\mathbf{x}' = \mathbf{f}(\mathbf{x})$ is the function $\phi_{\mathbf{f}}(t, \mathbf{x})$ satisfying the conditions

1. $t \mapsto \phi_{\mathbf{f}}(t, \mathbf{x})$ is a solution of the differential equation, as a function of t for each fixed $\mathbf{x}$,

2. $\phi_{\mathbf{f}}(0, \mathbf{x}) = \mathbf{x}$.

In other words, $\phi_{\mathbf{f}}(t, \mathbf{x})$ is the position at time t of the solution with initial position $\mathbf{x}$ at time 0.

As we have defined it, the flow of a differential equation exists even if the equation is not autonomous, but the definition is not natural if the equation is time dependent. It only describes the solutions that start at time 0. For a time-dependent differential equation, we would have to let the "flow" depend on two "times," an initial and a final time. If you start an autonomous equation at the same point later, the solution describes the same trajectory later; the motion depends only on the "flowing time," not on the particular initial and final times.

The procedure to write a closed formula for a flow, when it is possible, is to

1. Solve the differential equation.

2. Evaluate the constant of integration at $t = 0$.

3. Set $\phi_{\mathbf{f}}(t, \mathbf{x})$ equal to the solution with that value of the constant.

The flow $\phi_{\mathbf{f}}(t, \mathbf{x})$ of a vector field $\mathbf{f}(\mathbf{x})$ will usually not be defined for all $(t, \mathbf{x})$, because the solution starting at time 0 at $\mathbf{x}$ will not always be defined at time t. This occurs for equations C of Example 6.8.2, for instance, but it is quite difficult to visualize the domain of $\phi_{\mathbf{f}}$ in $\mathbb{R}^3$. Here is a simpler example.

Example 6.8.4. For the differential equation $x' = x^2$, the flow is given by

$$\phi_{\mathbf{f}}(t, \mathbf{x}) = \frac{x}{1 - tx} \quad \text{when} \quad |tx| < 1.$$

In this case, the domain of the flow is the region between the two branches of the hyperbola (Figure 6.8.3).

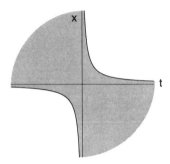

FIGURE 6.8.3. The domain for the flow of $x' = x^2$. ▲

You are asked in Exercise 6.8#5 to verify the assertions of Example 6.8.4, and in Exercises 6.8#6 and 6.8#7 to consider several more examples.

Examples 6.8.5. For equation A of Example 6.8.2,

$$\phi_{\mathbf{f}}\left(t, \begin{bmatrix} x \\ y \end{bmatrix}\right) = \begin{bmatrix} \cos(t)x + \sin(t)y \\ -\sin(t)x + \cos(t)y \end{bmatrix}.$$

For equations B of Example 6.8.2,

$$\phi_{\mathbf{f}}\left(t, \begin{bmatrix} x \\ y \end{bmatrix}\right) = \begin{bmatrix} x + t \\ ye^t \end{bmatrix}.$$

For equations C of Example 6.8.2, there is no closed formula for the flow, because the equation $x' = x^2 - t$ cannot be solved in closed form (the main example of Part I of this text). ▲

The following properties of flows follow immediately from the definition.

Proposition 6.8.6. *The flow of an autonomous differential equation* $\mathbf{x}' = \mathbf{f}(\mathbf{x})$ *satisfies the following whenever the indicated flows are defined:*

$$\phi_{\mathbf{f}}(t_1, \phi_{\mathbf{f}}(t_0, \mathbf{x})) = \phi_{\mathbf{f}}(t_0 + t_1, \mathbf{x}).$$

Theorem 6.8.7 (Continuity of Flows). *If* $f(x)$ *is a Lipschitz vector field, then* $\phi_f(t, x)$ *is a continuous function of both variables, i.e., as a function of the pair* (t, x), *wherever it is defined.*

Proof. The fundamental inequality gives a slightly more precise result than the statement above. Choose x_0 and t_0 such that the flow is defined at (t_0, x_0); suppose that K is a Lipschitz constant for the differential equation $x' = f(x)$ valid in a neighborhood U of the set

$$\{\phi_f(t, x_0), 0 \le t \le t_0\}.$$

Further, suppose that, on U, $\|f(x)\| \le M$ for some $M \in \mathbb{R}$, which will be possible if U is chosen sufficiently small.

Then, by the Fundamental Inequality 6.2.2,

$$\|\phi_f(t_1, x_1) - \phi_f(t_0, x_0)\| \le M|t_1 - t_0| + \|x_0 - x_1\|e^{K|t_0|} \qquad (44)$$

so long as (t_1, x_1) is sufficiently close to (t_0, x_0) that $\phi(t, x_1)$ stays in U for t between 0 and t_1. Clearly, the right-hand side of equation (44) can be made arbitrarily small by choosing (t_1, x_1) sufficiently close to (t_0, x_0). □

In Chapters 8 and 8*, it will be useful to have this formal notation for flows.

Chapter 6   Exercises

Exercises 6.1 Graphical Representation

6.1#1. Hand sketch phase plane trajectories for the following systems by finding the isoclines of horizontal and vertical slope, then finding the general direction of trajectories in the regions between (left or right, up or down). Use one color for the information from $dx/dt = 0$, and another for the information from $dy/dt = 0$. Use a third color to mark the resultant directions NE, NW, SE, or SW; draw sample trajectories in a fourth color.

Some of the pitfalls encountered in first attempts at sketching phase planes are highlighted by the following:

(i) An isocline of *horizontal* slope usually separates *vertical* behaviors, i.e., regions where trajectories have positive slopes from regions where slopes are negative. [There are exceptions, for instance, see part (c).] Similarly, an isocline of *vertical* slope usually separates *horizontal* behaviors.

(ii) Equilibria occur where an isocline of horizontal slopes meets an isocline of vertical slopes. Where two isoclines of horizontal (resp. vertical) slope meet, nothing happens except that you are doubly sure the slope is horizontal (resp. vertical).

(iii) Solutions cannot actually cross or meet, except at a point where the Existence and Uniqueness Theorem (which you will meet as Theorem 6.2.3) fails. You will observe solutions approaching or coming from an equilibrium, but they will not actually start or end there because if they were *at* the equilibrium, they would just stay there!

Try the following exercises to illustrate these points:

(a) $\begin{array}{l} dx/dt = x \\ dy/dt = x - y \end{array}$

(d) $\begin{array}{l} dx/dt = x \\ dy/dt = x^2 - y^2 \end{array}$

(b) $\begin{array}{l} dx/dt = x \\ dy/dt = -(x - y) \end{array}$

(e) $\begin{array}{l} dx/dt = y \\ dy/dt = x^2 - y^2 \end{array}$

(c) $\begin{array}{l} dx/dt = x \\ dy/dt = (x - y)^2 \end{array}$

(f) $\begin{array}{l} dx/dt = y + x^2 - 1 \\ dy/dt = x^2 - y^2 \end{array}$

6.1#2. As in the previous exercise, hand sketch phase plane trajectories for the following systems:

(a) $\begin{array}{l} dx/dt = 1 - x \\ dy/dt = x^2 - y \end{array}$

(f) $\begin{array}{l} dx/dt = 3x - 2xy \\ dy/dt = 2y - y^2 \end{array}$

(b)° $\begin{array}{l} dx/dt = x + 2 - y \\ dy/dt = x^2 - y \end{array}$

(g)° $\begin{array}{l} dx/dt = x - (1/4)x^2 - xy \\ dy/dt = 2y - y^2 - xy \end{array}$

(c) $\begin{array}{l} dx/dt = y - \sin x \\ dy/dt = x/4 - y \end{array}$

(h) $\begin{array}{l} dx/dt = 2x - x^2 - xy \\ dy/dt = y - y^2 - (1/4)xy \end{array}$

(d) $\begin{array}{l} dx/dt = y - \sin x \\ dy/dt = x/4 + y \end{array}$

(i) $\begin{array}{l} dx/dt = x - 4x^2 - xy \\ dy/dt = 2y - y^2 - 3xy \end{array}$

(e) $\begin{array}{l} dx/dt = y(x + 1) \\ dy/dt = (3 - y)x \end{array}$

Compare the results of part (b) with Example 6.1.5. Parts (g), (h), and (i) are three of the four cases you will find in Exercise 6.3#3.

6.1#3. Express the equation $x'' = \cos t$ as a system of first order equations. Then use *DiffEq, 3D Views* from *MacMath* or a similar computer graphics program to draw the following sample solutions in xyt, xt, yt, and xy views:

Start solutions at $x_0 = 0$, $y_0 = 0$, 1, 2, 3, π, $\pi/2$, $3\pi/2$, 2π, etc. You can solve this particular second order equation just by integrating; compare with your pictorial results.

6.1#4. Find phase portraits with a computer program, such as *DiffEq, Phase Plane* from *MacMath*, for the following systems. Print them and draw on your graphs (by hand if necessary) the isoclines of horizontal and vertical slopes, to verify that the trajectories indeed cross these isoclines with the proper slopes.

(a) $\quad \begin{aligned} dx/dt &= y^2 - 1 \\ dy/dt &= x^2 + 2x \end{aligned}$

(b) $\quad \begin{aligned} dx/dt &= y(y - 1)(y + 1) \\ dy/dt &= \sin(x + y) \end{aligned}$

(c)° $\quad \begin{aligned} dx/dt &= [(x - 2)^2 + y^2 - 1][x^2 + y^2 - 9] \\ dy/dt &= (x - 1)^2 + y^2 - 4 \end{aligned}$

6.1#5. Experiment in $\mathbb{R}^3$ with the Lorenz strange attractor of Example 6.1.6. As in Figure 6.1.9, make a printout from *DiffEq, 3D Views* in Mac-Math showing seven views: xyt, xy, xz, yz, xt, yt, and zt. Experiment with what changes in the graphs when you make a slight change in the initial conditions. Make observations and at least one conjecture. *Proving* a conjecture may be too much to ask at this point, but see how far you can get.

6.1#6. Another strange attractor in $\mathbb{R}^3$, attributed to O.E. Rossler (1979; see References), is derived from a physical situation of a dripping faucet. It is similar to the Lorenz attractor, but simpler. The following set of equations provides an example for computer experimentation.

$$\begin{aligned} dx/dt &= -(y + z) \\ dy/dt &= x + 0.2\,y \\ dz/dt &= 0.2 + z(x - 5.7). \end{aligned}$$

Exercises 6.2 Theorems

6.2#1°. Prove Theorem 6.2.2. Notice that if $\mathbf{u}_1(t)$ and $\mathbf{u}_2(t)$ are continuous piecewise linear functions, then $\|\mathbf{u}_1(t) - \mathbf{u}_2(t)\|$ is not a piecewise linear function (give an example). Why is the proof still correct?

6.2#2. Prove Theorem 6.2.3.

6.2#3. Prove Corollary 6.2.4.

6.2#4. Justify the formulas in Section 6.2 for the various numerical methods.

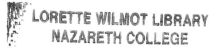

6.2#5. Let $f(t, \mathbf{x})$ be defined in the region $I \times S$, where I is the interval $t_0 \leq t \leq t_1$ and $S = \{\mathbf{x} \mid a_i \leq x_i \leq b_i\}$. Suppose the solution $\mathbf{u}(t)$ with $\mathbf{u}(t_0) = \mathbf{x}_0$ is a mapping $\mathbf{u} : I \to S$ and that the Euler approximations $\mathbf{u}_h(t)$ also map I to S for $h \leq h_0$. Then find a bound for $\|\mathbf{u}_h(t) - \mathbf{u}(t)\|$ of the function $\|\mathbf{u}_h(t) - \mathbf{u}(t)\| \leq Ch$ for all $t \in I$, where C should be expressed in terms of $|f|$, $\left|\frac{\partial f}{\partial t}\right|$, $\left|\frac{\partial f}{\partial x_i}\right|$ for $i = 1, \ldots, n$.

Hint: Look at Theorem 4.5.2.

6.2#6.

(a) For Euler's method, find stepsize h such that a solution of $x' = y$, $y' = -x$ starting at $(0, 1)$ is guaranteed to be accurate to three decimal places at $t = \pi$.

(b) For Euler's method, find h such that a solution of $x'' = x^2 - t$ starting at $(0, 0)$ is guaranteed to be accurate to three decimal places at $t = 1$.

6.2#7. For the following differential equations with $x(0) = 1, y(0) = 0$, and stepsize 0.1, approximate by hand the first two steps by (i) the Euler method, (ii) the midpoint Euler method, and (iii) the Runge–Kutta method:

(a) $x'' + x = 0$

(b) $\begin{aligned} dx/dt &= y \\ dy/dt &= -2x \end{aligned}$ (You'll meet this in Section 6.3.)

(c) $\begin{aligned} d\theta/dt &= y \\ dy/dt &= -k \sin\theta \end{aligned}$ (You'll meet this in Section 6.5.)

What differences does the method make in the drawings of trajectories?

6.2#8°.

(a) Show that along the solution of $dx/dt = y$, $dy/dt = -x$, the function $F(x, y) = x^2 + y^2$ is constant. What are the trajectories of the solutions in the phase plane?

(b) Show by hand calculation that if you solve the same system of differential equations using Euler's method, the approximate solution will spiral out to infinity. Using a moderately large stepsize, hand sketch the approximate trajectory in the phase plane. Confirm with a computer printout, checking especially that your hand sketch goes in the proper direction.

(c) What happens if you use Runge–Kutta? (It depends on the stepsize h, and to see how requires quite a bit of computation, as in Chapter 5.4 of Part I. We do not ask for an analytic treatment here, but rather to see how far you can get with what is more easily available.) Since the trajectories in *MacMath's DiffEq, Phase Plane* are by default calculated by Runge–Kutta, you can experiment with the computer to see what happens, using different stepsizes, trying both small and large h values.

(d) What happens if you use midpoint Euler? Again, you can experiment with the computer, then write a theoretical support for your observations.

6.2#9. For Example 6.2.5, with $dx/dt = y$, $dy/dt = -x$, show exactly

(a) when a pair of *boundary conditions* $x(t_a)$ and $x(t_b)$ will *not* specify a unique solution (resulting in either an infinite number of solutions, or no solution at all);

(b) when a pair of boundary conditions *will* specify a unique solution.

Hint: Use the fact that all the solutions are periodic with period 2π.

Exercises 6.3 Sharks and Sardines

6.3#1. Complete the steps of the following calculations in Section 6.3:

(a) from (7) to (9) to show that $F(x, y)$ in (10) is constant along a solution,

(b) to verify equation (11).

6.3#2°. Referring back to the sharks and sardines example of Section 6.3, consider the function F described by equation (10):

$$F(x, y) = |x|^b \, |y|^a \, e^{-(fx+cy)}.$$

(a) Show that this function F has a unique maximum at $x = b/f$, $y = a/c$, and that it decreases to zero when (x, y) approaches the axes or goes to infinity in either direction.

(b) What might this mean ecologically?

6.3#3. Let $x(t)$ and $y(t)$ represent the populations of two species, both competing for the same food supply.

(a) Explain (interpret) how the following model represents such a system, for positive constants a_1, a_2, b_1, b_2, c_1, c_2; that is, by factoring out a common factor, consider each population rate of change as a multiple of the population size, e.g., $(a_1 - b_1x - c_1y)x$. Then identify which terms are caused by crowding, which by competition, and which rates are independent of both.

$$dx/dt = a_1 x - b_1 x^2 - c_1 xy$$
$$dy/dt = a_2 y - b_2 y^2 - c_2 xy.$$

(b) Show that in the first quadrant of the xy phase plane the isocline of horizontal slope is a straight line, the isocline of vertical slope is another straight line, and that both of these lines have negative slope. Argue why the following sketches show the four possible cases for relative position of these isoclines:

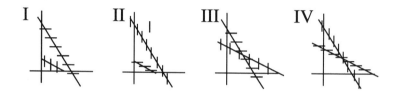

(c) If you consider dy/dx as $(dy/dt)/(dx/dt)$, you have a first order differential equation in x and y. The lines (isoclines) of horizontal and vertical slopes act as fences in the xy-plane. For each case, I–IV, mark each region determined by the isoclines with the general direction (right or left, up or down) of the xy slope, as determined by whether dx/dt and dy/dt, respectively, are positive or negative. Then draw trajectories that match these directions and those of the isoclines. Identify and label "funnels" and "antifunnels" *in the direction of increasing t*.

(d) This is a matching exercise: tell which of the above cases in (b) and (c) represents each of the following possibilities:

A. The two species tend to an equilibrium where both survive.

B. Species x surely becomes extinct.

C. Species y surely becomes extinct.

D. In almost all cases, one of the two species will become extinct, but which one depends on the initial condition. Which initial conditions provide an exception?

(e) Make computer examples for each of the four cases; three of them occurred in Exercise 6.1#2(g), (h), (i). Find values (nonzero) of the parameters (a_i, b_i, c_i) that will satisfy the axes intercepts for each of the four possibilities; for each, provide a computer printout from *MacMath's DiffEq, Phase Plane* showing trajectories in the xy phase plane. Draw arrows to indicate the direction of increasing t on the trajectories for both of these graphs. (Remember that trajectories in the phase plane do not necessarily move from left to right.)

If possible, make a printout from *DiffEq, 3D Views*. Use the xyt view to clarify your choice of trajectory directions in the phase plane; explain exactly how the xt and yt views relate to the phase plane trajectories.

6.3#4. For the system of differential equations,

$$\frac{dx}{dt} = 0.2 - 0.09\,y, \qquad \frac{dy}{dt} = -0.04\,xy$$

(a) Sketch the xy phase plane with isoclines for horizontal and vertical slopes. Add an arrow in each region bounded by these isoclines and the axes to show the general direction (up or down, left or right) for the trajectories.

(b) Then, with the aid of the drawing below, showing the separatrices of the saddle, make a clean sketch with these isoclines and a representative set of trajectories, adding arrows to show the direction of increasing t . Add arrowheads also for the separatrices.

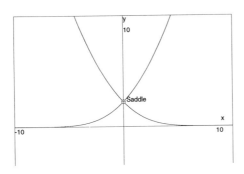

(c) This model represents warfare between conventional troops x (*in thousands*) and guerilla troops y (*in hundreds*), with t measured *in days*. The conventional force sends in reinforcements and suffers significant losses only at the hands of the guerillas. The guerillas are confined to a single forest area and remain essentially invisible, so the chances of hitting one decreases as their number decreases; furthermore, the guerillas are cut off from reinforcements.

    (i) Show how the equations account for reinforcements and combat losses by labeling each term.

(ii) Tell how many men per day are sent as reinforcements to the conventional force.

(iii) If at the beginning of the campaign the conventional force numbers 2000 and the guerilla force 600, use the graph to tell which side wins and why (mathematically!).

Exercises 6.4 Higher Order Equations

6.4#1. Express each of the following differential equations as a system of first order differential equations:

(a) $x'' + 3x' + 5x = 0$

(b) $x'' + 3tx' = t$

(c) $x''' - tx'' + x' - 5x + t^2 = 0$

(d) $x'' - xx' = 0$.

6.4#2°. A classical method for solving second order differential equations with either the dependent or independent variable missing [i.e., either x or t is missing from $f(x'', x', x, t) = 0$] is to make a substitution $y = dx/dt$ which results in the original equation being expressable as a first order equation in either

$$y \text{ and } t, \text{ if } f(x'', x', t) = f(y', y, t) = 0 ,$$

or

$$y \text{ and } x, \text{ if } f(x'', x', x) = 0, \text{ using}$$

$$x'' = \frac{dy}{dx}\frac{dx}{dt} = y\frac{dy}{dx},$$

so you end up with $f(dy/dx, y, x) = 0$.

In either case, this new equation is a first order equation in y; once it is solved for y (as a function of t in the first case, of x in the second), you can get to $x = u(t)$ by solving another first order equation, $dx/dt = y$, which comes from the original substitution.

You can apply this method to actually solve parts (a), (b), and (d) of Exercise 6.4#1. Take note of the difference in character for each equation. That of part (a) is autonomous (no explicit dependence on t) and linear, with constant coefficients, and will be solved by the general algebraic methods of Chapter 7. If you make the substitution $y = dx/dt$, you find a first order equation that can be separated in the variables (x, v) if you set $y = xv$, so that $dy/dx = xdv/dx + v$ (see Part I, Exercise 2 misc#2 about "homogeneous" first order equations).

The equation of part (b) is a second order nonhomogeneous linear equation, with nonconstant coefficients. The methods of Chapter 7 do not quite apply, but the substitution $y = dx/dt$ makes it a *first* order nonhomogeneous linear equation that is solvable (by the methods of Chapter 2, Sections 3 and 4).

The equation of part (d) is not linear, so it would *not* be solvable by the methods of Chapter 7; thus, this classical substitution can be a helpful trick.

6.4#3. For the system

$$dx/dt = 3x - y$$
$$dy/dt = x + 2y$$

(a) Replace by a second order equation in x.

(b) Replace by a second order equation in y. Does the result surprise you?

(c) Prove that for *any* system of linear first order equations with constant coefficients,

$$dx/dt = ax + by$$
$$dy/dt = cx + dy$$

the second order equations in x and y are always exactly the same.

(d) Show that, despite the results of (c), the *solutions* to (a) and (b) are *not* exactly the same. Explain. Hint: Show what happens with initial conditions $x(0) = 0, y(0) = 1$.

6.4#4. Replace, where possible, the following systems of first order differential equations by a single equation of higher order.

(a) $\quad dx/dt = 3x - y^2$
$ dy/dt = x + 2y$

(b) $\quad dx/dt = 3x - y^2$
$ dy/dt = x^2 + 2y$

(c) $\quad dx/dt = y + \cos(x + y)$
$ dy/dt = x - y \sin x$

(d) $\quad dx/dt = x$
$ dy/dt = y \sin x$

Exercises 6.5 Mechanical Systems with One Degree of Freedom

6.5#1. Consider a particle with mass 1 moving on a line under the potential $V(x) = x^4 - 1$.

(a) Show that the particle obeys the differential equation

$$x'' = -4x^3,$$

and write this second order equation as a system of two first order differential equations.

(b) Write down the energy function and show that the energy is constant along solutions.

(c) Sketch some level curves of the energy function. One nice way is to use *MacMath's Analyzer* program to sketch the two y functions (one positive, one negative) for each value of E.

6.5#2. Consider a particle of mass m moving on a line under the potential $V(x) = 2x^3 - 3x^2$.

(a) Show that the particle obeys the differential equation

$$mx'' = -6x^2 + 6x,$$

and write this second order equation as a system of two first order differential equations.

(b) Sketch the level curves of the energy function in the phase plane; find and describe the equilibria of the motion.

6.5#3°. Consider a particle with mass $m = 1$, which moves with one degree of freedom under the potential

$$V(x) = x^4 - x^2.$$

(a) From a sketch of the graph of $V(x)$ [or of $2V(x)$], sketch in the phase plane different level curves for the total energy. Find the equilibria. Mark at least three different level curves corresponding to $E(x, y) = C$, where C is negative, 0, and positive, respectively. Put arrows on the curves to indicate in which direction the particle follows them. (To make it clearer, use different colors for different trajectories). Describe the various types of motion for the particle.

(b) What happens if a small friction term is added on, proportional to the velocity? Sketch the corresponding drawing in the phase plane and describe the equilibria and the motions.

6.5#4. For a spring with no friction, the potential energy is

$$U(x) = \frac{1}{2}kx^2, \qquad k > 0.$$

Find the differential equation describing the motions of a bob of mass m at the end of such a spring and draw the trajectories in phase space. Explain why the trajectories look the way they do.

6.5#5. A mass m is suspended from a spring with spring constant k, as shown.

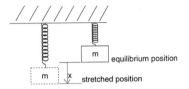

Show that if we measure the position of the mass relative to the stretched equilibrium position, we can rewrite the downward force,

$$F = mg - kx = mx'',$$

without the mg term and thus the force is dependent on position alone.

6.5#6. The force exerted by a spring on a frictionless cart at its end is often modeled by $F = -kx$ where x is the displacement of the cart.

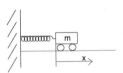

Suppose that a better model is $F = -k(x)x$, where $k(x) = e^{x^2}$. Let the mass of the cart be m.

(a) Sketch the phase portrait of this system.

(b) If the cart is held still at a displacement of $l > \frac{1}{2}$ and then released, what will its velocity be the first time its displacement is $\frac{1}{2}$?

6.5#7°. Find a function $V(x, y)$ that is constant along solutions for the system

$$\frac{dx}{dt} = -(1 - x^2), \qquad \frac{dy}{dt} = 3x^2 - 2x^2y.$$

(Warning: This is truly an exercise in integration!)

6.5#8. Figures 6.5.3 and 6.5.4 show a saddle point at $(\pi, 0)$. Use an approximation of $\sin \theta$ to show that this is indeed a saddle point of $E = y^2/2 - k \cos \theta$. Draw the graph of the trajectories near $(\pi, 0)$.

6.5#9. Consider the differential equation

$$x'' = \alpha x - x^3.$$

(a) Turn this equation into a system of two first order equations

$$dx/dt = f(x, y)$$
$$dy/dt = g(x, y)$$

and state precisely what the relation is between the two.

(b) Find a function $E(x, y)$ which is constant on the trajectories.

(c) What do the level curves of E look like for $\alpha > 0$, $\alpha = 0$, $\alpha < 0$?

(d) Classify the solutions of the original second order equation in the three cases of part (c).

6.5#10°. Consider the dynamical system with one degree of freedom

$$x'' = -\frac{dV}{dx},$$

where $V(x) = \sin x + \alpha x^2$ with $\alpha > 0$ a parameter.

(a) Turn the equation into a system of first order equations.

(b) Find a conserved quantity for this system. Plot approximately the level curves of this conserved quantity, for $\alpha = 0.01$, $\alpha = 0.1$, $\alpha = 0.5$, $\alpha = 1$.

(c) Describe the equilibria of the system as functions of α.

6.5#11. Confirm that the mathematical total energy function (18),

$$E(\theta, y) = \frac{y^2}{2} - K \cos \theta, \tag{18}$$

is within a constant multiple of the physical total energy, kinetic energy plus potential energy.

6.5#12. Consider the case of a bead constrained to move along a semicircular wire.

(a) Show that if the bead is a distance s along the wire (from the bottom), with x-coordinate $x(s)$, the force ms'' is

$$-m \frac{V'(x(s))}{\sqrt{1+(V'(x(s)))^2}}. \tag{20}$$

(b) Confirm all the steps of the derivation in Example 6.5.5 to show that

$$s'' = -\sin(s/\ell).$$

(c) Confirm that in this case, with $s = \ell\theta$, that

$$\frac{dv}{ds} = \frac{f'(\theta)}{\sqrt{1+(f'(\theta))^2}} \quad \text{implies} \quad f(\theta) = -K\sqrt{1-(2\theta/\pi)^2}.$$

6.5#13. Consider the final remark of Section 6.5, that one may wonder whether, given a function $V(x)$, there is some other function $W(x)$ such that the motion on the wire with the shape of the graph of W, that is with equation

$$s'' = \frac{W'(x(s))}{\sqrt{1+(W'(x(s)))^2}},$$

is precisely the same as motion under $x'' = -dV/dx$. Clearly, one requirement is $|dV/dx| < 1$. Show that this is the only requirement.

Exercises 6.6 Essential Size, Conservation Laws

6.6#1. Find the essential size of the following systems:

(a) $x' = x, y' = 2y + x^2$

(b) $x' = xz, y' = 2y + x^2, z' = y$

(c) $x' = x, y' = 2y, z' = x^2$

6.6#2°.

(a) In Example 6.6.8, fill in all the steps (there are several) that lead from equations (26) to (27).

(b) In Example 6.6.8, verify the statement that $\mathbf{r}$ will never leave the plane spanned by $\mathbf{r}(t_0)$ and $\mathbf{r}'(t_0)$.

(c) In Example 6.6.9, verify the statement that the magnitude of $\mathbf{r} \times \mathbf{r}' = \mathbf{M}$, which is conserved, is in fact the angular momentum.

Exercises 6.7 Two-Body Problem

6.7#1. For two bodies moving under Newton's Universal Law of Gravitation,

$$\mathbf{x}_1'' = \frac{Gm_2(\mathbf{x}_2 - \mathbf{x}_1)}{\|\mathbf{x}_2 - \mathbf{x}_1\|^3}$$

$$\mathbf{x}_2'' = \frac{Gm_1(\mathbf{x}_1 - \mathbf{x}_2)}{\|\mathbf{x}_1 - \mathbf{x}_2\|^3}$$

(33)

(a) Show that linear momentum is a preserved quantity.

(b) Using $m_1 = 1 \times 10^5$, $m_2 = 2 \times 10^5$ and the initial conditions

$$\mathbf{x}_1(t_0) = \begin{bmatrix} 1 \\ 2 \\ -3 \end{bmatrix}, \qquad \mathbf{x}_1'(t_0) = \begin{bmatrix} -2 \\ -1 \\ 2 \end{bmatrix},$$

$$\mathbf{x}_2(t_0) = \begin{bmatrix} -4 \\ -1 \\ 0 \end{bmatrix}, \qquad \mathbf{x}_2'(t_0) = \begin{bmatrix} 0 \\ 0 \\ 0 \end{bmatrix}, \qquad t_0 = 1,$$

show that the position vector

$$\mathbf{X} = \frac{m_1\mathbf{x}_1 + m_2\mathbf{x}_2}{m_1 + m_2}$$

for the center of mass can be written as $\mathbf{X}(t) = \mathbf{a} + t\,\mathbf{b}$, hence showing that the center of mass moves at constant velocity in a straight line.

(c) After translation to the center of mass, $\mathbf{X}$, show that $\mathbf{r} = \mathbf{x}_1 - \mathbf{X}$ satisfies

$$\mathbf{r}'' = -K\frac{\mathbf{r}}{\|\mathbf{r}\|^3}.$$

(d) We could set $\mathbf{r}_1 = \mathbf{x}_1 - \mathbf{X}$ and $\mathbf{r}_2 = \mathbf{x}_2 - \mathbf{X}$. Show that then

$$\mathbf{r}_1 = \frac{-m_2}{m_1 + m_2}(\mathbf{x}_2 - \mathbf{x}_1)$$

$$\mathbf{r}_2 = \frac{m_1}{m_1 + m_2}(\mathbf{x}_2 - \mathbf{x}_1).$$

6.7#2°. For three bodies we would have three second order differential equations in $\mathbf{x}_1, \mathbf{x}_2, \mathbf{x}_3$. The center of mass is defined by

$$\mathbf{X} = \frac{m_1\mathbf{x}_1 + m_2\mathbf{x}_2 + m_3\mathbf{x}_3}{m_1 + m_2 + m_3}.$$

(a) Show the conservation of linear momentum for three bodies:

$$m_1\mathbf{x}_1'' + m_2\mathbf{x}_2'' + m_3\mathbf{x}_3'' = \mathbf{0}.$$

(b) How does the center of mass move?

6.7#3. Show that in the two-body problem the level curves of the total energy E look like Figure 6.7.3, using the graph of the "potential" $W(r)$ given in Figure 6.7.2.

6.7#4.

(a)° Study the motion under a central force

$$-\frac{k}{r^3}\frac{\mathbf{r}}{r},$$

instead of the gravitational force

$$-\frac{k}{r^2}\frac{\mathbf{r}}{r}$$

of the two-body problem of Section 6.7.

Write down the equations similar to (35) through (39) in Section 6.7. Make the drawings similar to Figures 6.7.2 and 6.7.3. Explain why there is *no* bounded motion.

(b) More generally, if the central force is

$$-\frac{k}{r^\alpha}\frac{\mathbf{r}}{r},$$

for which α can we have bounded motions?

6.7#5.

(a) Prove that Kepler's Second Law is equivalent to conservation of angular momentum. Note that both of these statements are valid in *any* central force field, not just one with an inverse square law such as the gravitational field.

(b) Use *MacMath's Planets* program to illustrate your examples.

6.7#6.

(a) Confirm equation (41) in the derivation of Kepler's First Law by integration of

$$\frac{d\theta}{dr} = \frac{M}{r^2\sqrt{2\left(E + K/r - M^2/2r^2\right)}}. \tag{41}$$

(b) Confirm by integration of (41) that

$$\theta = \text{arc } \cos \frac{(M/r - K/M)}{\sqrt{2E + K^2/M^2}}. \qquad (42)$$

6.7#7. Show that

$$r = \frac{p}{1 + e \cos t}$$

is the equation of an ellipse, as claimed in the proof of Kepler's First Law. Recall that an ellipse is the set of points such that the sum of the distances to the foci is constant ($= 2a$).

6.7#8. The constant of gravity G is $(6.673 \pm 0.003) \times 10^{-8} \text{cm}^3/(g \sec^2)$. Note, incidentally, that it is not known with great precision.

(a) If a body with small mass m_1 rotates around one with large mass m_2 with period T and semi-major axis a, give a formula for m_2 in terms of G, T, and a. How sensitive is this formula to the assumption that one mass is negligible? For instance, how wrong is it if $m_1 = m_2/4$?

(b) Assume that the mass of the earth is negligible with respect to the mass of the sun. Given that the distance from the earth to the sun is approximately 1.496×10^{13}cm, and that the earth revolves around the sun in 1 year (1 year $= 3.156 \times 10^7$ sec), what is the mass of the sun?

(c) Assume that the mass of the moon is negligible with respect to that of the earth. Knowing that the distance of the moon to the earth is approximately 3.85×10^{10}cm, roughly what is the mass of the earth?

6.7#9. A recent practical application of the two-body problem was calculation of the mass of Pluto, from the 1978 discovery that Pluto has a moon:

(a) Pluto was discovered in 1930, as a result of perturbations in the orbit of Neptune. In 1978, astronomers J. Christy and R. Harrington discovered that Pluto has a satellite, which they called Charon. Charon appears, from the earth, to have an orbit with a radius of about 20,000 kilometers about Pluto, and a period of about 6 days and 10 hours. If the orbit is circular, what is the mass of Pluto?

Remark. We have asked some astronomers why the orbit is assumed to be circular. They answer that a very eccentric orbit leads to large tidal effects, which would presumably either break up the satellite, or at least absorb enough energy to bring it to a more circular orbit.

(b) The number found in part (a) is about 100 times too small to explain the irregularities in the orbit of Neptune. Assume rather that the orbit of Charon is very elliptical, with the long axis pointed toward the earth. How eccentric would the orbit need to be in order to bring the mass up to the level required to explain the irregularities?

6.7#10. We are collecting exercises derived from playing with *MacMath's Planets* program and welcome more examples.

Exercises 6.8 Flows

6.8#1. Use a phase portrait program like *MacMath's DiffEq Phase Plane* to explore the flow of *different* rectangles for equations A of Example 6.8.2. For example, some possibilities are shown in the following diagram for various positions of a rectangle like the original one (darkest) of the example:

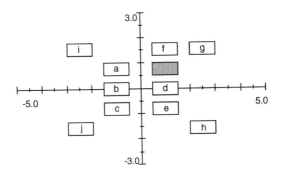

(i) Show the flow for the solid rectangles symmetric to the original rectangle.

(ii) Explain the similarities and differences in the results.

(iii) Predict what will happen to the dotted rectangles and verify with the computer.

(iv) Conjecture and explain what happens geometrically to the flow in general, globally, for this differential equation. One way to do this is to imagine the phase plane filled with little squares and think about what will happen to them all as you take a step in t.

6.8#2. Repeat Exercise 6.8#1 for equations B of Example 6.8.2.

6.8#3. Repeat Exercise 6.8#1 for equations C of Example 6.8.2. Note that the behavior of this flow is far less obviously predictable than those of equations A and B; this is because the dependence on initial condition really rules this flow! You should find several different behaviors to explain in this example. This system C is equivalent to the famous one-dimensional equation $dx/dt = x^2 - t$ of Part I of this text. Does that help to explain your results?

6.8#4. For each of the cases A, B,and C of Example 6.8.2, draw or compute and print the direction field with a number of trajectories. Then explain what additional information you get from the flow that you do not get from the direction field.

6.8#5. Verify that the formula and the domain in Example 6.8.4 are correct.

6.8#6°. Give a formula and the appropriate domain for the flow of the following equations:

$$x' = x^3, \quad x' = x^3 - x.$$

6.8#7. For the following differential equation, an explicit formula for the flow cannot be given in elementary terms. Sketch the domain of the flow, and give a formula for the domain using definite integrals.

$$x' = x^2 - \sin(x).$$

6.8#8. This exercise is quite a bit more challenging than the previous ones. Try to sketch, in $\mathbb{R}^3$, the domain of the flow from equations C of Example 6.8.1. In particular, describe how this domain is related to the exceptional solution of the equation. How does this domain intersect the plane $x = a$ for a small (say $a = -5$), or for a large (say $a = 5$)?

7

Systems of Differential Equations

The only large and important class of systems of differential equations that has been extensively analyzed is that of *linear* differential equations, which we begin to study in this chapter. We will give examples in Section 7.1 and the formal (simple) side of the theory in Section 7.2. The more substantive theoretical results (for linear equations with varying coefficients) are deferred to Chapter 12 in Part III.

The class of linear differential equations for which we can in general find *explicit* solutions is restricted to those *with constant coefficients*. The remainder of this chapter will be devoted to these: how to compute and analyze their solutions. Chapters 8 and 9 deal with important facts that extend from linear to nonlinear differential equations; Chapters 10 and 11 (in Part III) will work out interesting classes of examples.

As we will see, the theory for constant coefficients is largely linear algebra. When you need to review basic linear algebra operations, consult Appendix L, Sections L1–L6. The related computer programs for finding eigenvalues and eigenvectors are elaborated in Appendices L7 and L8.

7.1 Linear Differential Equations in General

A *linear differential equation* for a vector function $\mathbf{x}(t)$ is one that can be written as

$$\mathbf{x}' = A(t)\mathbf{x} \tag{1}$$

or, more generally,

$$\mathbf{x}' = A(t)\mathbf{x} + \mathbf{g}(t), \tag{2}$$

where A is a matrix and $\mathbf{g}$ is a vector, both of which can be filled with functions of t. The word "linear" means with respect to the dependent variable $\mathbf{x}$; the entries of the matrix function $A(t)$ and the vector function $\mathbf{g}(t)$ need *not* be linear functions of the independent variable t.

Example 7.1.1. The system

$$x' = y$$
$$y' = -x,$$

which was introduced in Examples 6.1.2, 6.1.3, and 6.2.5 and is one of the basic linear equation examples that you will see often, can be written

$$\begin{bmatrix} x \\ y \end{bmatrix}' = \begin{bmatrix} 0 & 1 \\ -1 & 0 \end{bmatrix} \begin{bmatrix} x \\ y \end{bmatrix}, \quad \text{or}$$

$$\mathbf{x}' \quad = \quad A \quad \mathbf{x}. \qquad \blacktriangle$$

Example 7.1.2. The system

$$x' = t^2 x - y$$
$$y' = (\ln t)x + e^t$$

is also linear in x and y and can be written

$$\begin{bmatrix} x \\ y \end{bmatrix}' = \begin{bmatrix} t^2 & -1 \\ \ln t & 0 \end{bmatrix} \begin{bmatrix} x \\ y \end{bmatrix} + \begin{bmatrix} 0 \\ e^t \end{bmatrix}, \quad \text{or}$$

$$\mathbf{x}' \quad = \quad A(t) \quad \mathbf{x} \quad + \quad \mathbf{g}(t).$$

where the matrix A as well as the vector $\mathbf{g}$ depends on t, but their entries are in fact *not* linear functions of t. $\quad \blacktriangle$

Example 7.1.3. The system of equations

$$x' = y + xy$$
$$y' = x + y^2$$

is *not* linear. It can be written

$$\begin{bmatrix} x \\ y \end{bmatrix}' = \begin{bmatrix} 0 & (1+x) \\ 1 & y \end{bmatrix} \begin{bmatrix} x \\ y \end{bmatrix}$$

or as

$$\begin{bmatrix} x \\ y \end{bmatrix}' = \begin{bmatrix} y & 1 \\ 1 & y \end{bmatrix} \begin{bmatrix} x \\ y \end{bmatrix},$$

but however you write it, the matrix A must contain entries in the dependent variables x and y. $\quad \blacktriangle$

The case of linear differential equations where x is one-dimensional has been extensively treated in Chapter 2. We present here two archetypal examples in *two* dimensions:

Example 7.1.4. The damped harmonic oscillator. Suppose a mass m is suspended from a spring with restoring force F and damping constant k, as shown in Figure 7.1.1. If x denotes the displacement of the mass from equilibrium and y its velocity, then it obeys the system of differential equations

$$x' = y$$
$$my' = -Fx - ky, \tag{3}$$

which can be written alternatively as

$$\begin{bmatrix} x \\ y \end{bmatrix}' = \begin{bmatrix} 0 & 1 \\ -F/m & -k/m \end{bmatrix} \begin{bmatrix} x \\ y \end{bmatrix}. \tag{4}$$

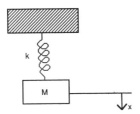

FIGURE 7.1.1. Mass on spring.

You may be more familiar with the same problem in the form of a single second order differential equation,

$$mx'' + kx' + Fx = 0, \tag{5}$$

which you can obtain by differentiating the first equation of system (3) and substituting $x'' = y'$ and $x' = y$ into the second equation of the system.
▲

Exercise 7.1–7.2#5 elaborates on Example 7.1.4 for the case of varying rather than constant mass (a water bucket that leaks).

Example 7.1.5. The driven RLC circuit. Consider the electrical circuit of Figure 7.1.2, consisting of a resistor with resistance R, a capacitor with capacitance C, an inductor with inductance L, and a generator which furnishes a potential drop $f(t)$.

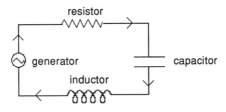

FIGURE 7.1.2. Electric circuit.

Then, as you have probably seen in physics or engineering courses (and as we shall show in Chapter 10 of Part III), the voltage drop $V(t)$ across the capacitor and the current $I(t)$ through the inductor satisfy the system of equations

$$CV' = I$$
$$LI' = -V - RI + f(t),$$

which can be written alternatively as

$$\begin{bmatrix} V \\ I \end{bmatrix}' = \begin{bmatrix} 0 & 1/C \\ -1/L & -R/L \end{bmatrix} \begin{bmatrix} V \\ I \end{bmatrix} + \begin{bmatrix} 0 \\ f(t)/L \end{bmatrix}. \tag{6}$$

Again, you may be more familiar with the equation in second order form:

$$LQ'' + RQ' + (1/C)Q = f(t), \tag{7}$$

where $Q = CV$ is the charge on the capacitor. ▲

Notice that the two equations (5) and (7) of Examples 7.1.4 and 7.1.5 are almost the same; in fact the equation describing a forced damped harmonic oscillator *would* be precisely the same as the equation describing the driven RLC circuit. These models will be greatly expanded in Chapters 11 and 10 of Part III, respectively.

Equations like (5) and (7) appear constantly in physics and engineering, both in their own right and as approximations to other more complicated equations.

NONLINEAR EQUATIONS

Certainly not all systems of differential equations are linear. Nearly every example of such a system that was discussed in Chapter 6, such as

   i. the *predator–prey model:* $\begin{aligned} x' &= ax - bxy \\ y' &= -cy + dxy, \end{aligned}$

   ii. the *simple pendulum:* $\theta'' = -K \sin \theta,$

   iii. the derivation of *Kepler's First Law:* $r'' = -K/r + M^2/r^3,$

is *not* linear, because terms like xy, $\sin \theta$, and r^m (for $m \neq 1$) are not linear in the variables

$$\mathbf{x} = \begin{bmatrix} x \\ y \end{bmatrix}, \quad \begin{bmatrix} \theta \\ \theta' \end{bmatrix}, \quad \text{or} \quad \begin{bmatrix} r \\ r' \end{bmatrix},$$

respectively, that underlie each of these examples.

HIGHER ORDER DIFFERENTIAL EQUATIONS

Very often those systems of differential equations that *are* linear appear in the form of *higher order* linear differential equations [as defined in the Introduction, equation (5)], like this section's equations (5) and (7), rather than as *systems* of first order differential equations. However, the method of Section 6.4, translated into the language of matrices, easily changes a higher order equation into a system of first order equations.

Example 7.1.6. The equation $x''' - 3x'' + tx' - x + t^2 = 0$, of Example 6.1.6 in Section 6.4, can be written as

$$\mathbf{x}' = \begin{bmatrix} 0 & 1 & 0 \\ 0 & 0 & 1 \\ 1 & -t & 3 \end{bmatrix} \mathbf{x} + \begin{bmatrix} 0 \\ 0 \\ -t^2 \end{bmatrix},$$

where the desired solution is $x = u(t) = x_1(t)$. [Recall that $x_2(t) = u'(t)$ and $x_3(t) = u''(t)$.]   ▲

Example 7.1.6 illustrates the following procedure for converting an nth order linear differential equation to a system of first order linear differential equations in matrix form:

(1) Solve the original nth order equation for $x^{(n)}$.

(2) The bottom line of the matrix $A(t)$ is formed by the coefficients of $x, x', x'', \ldots, x^{(n-1)}$, *in that order*; the bottom entry in the vector $\mathbf{g}(t)$ is the extra function of t.

(3) Above the bottom line of the matrix, all entries are zeroes except in positions just above the main diagonal, where they are ones.

If you have already run into higher order linear differential equations such as (5) and (7), then you are probably also familiar with the cookbook method for their solution:

Recipe 7.1.7. *To solve the equation*

$$\alpha x'' + \beta x' + \gamma x = 0, \tag{8}$$

substitute $u(t) = e^{mt}$ into equation (8), giving $(\alpha m^2 + \beta m + \gamma)e^{mt} = 0$. Since e^{mt} does not vanish, we get the characteristic equation

$$\alpha m^2 + \beta m + \gamma = 0. \tag{9}$$

There are three cases to consider:

(a) $\beta^2 - 4\alpha\gamma > 0$, so the quadratic equation (9) has two real roots m_1 and m_2. Then the general solution is

$$u(t) = C_1 e^{m_1 t} + C_2 e^{m_2 t}.$$

(b) $\beta^2 - 4\alpha\gamma < 0$, so the quadratic equation (9) has two complex roots $\mu \pm i\sigma$. Then the general solution is

$$u(t) = e^{\mu t}(C_1 \cos \sigma t + C_2 \sin \sigma t).$$

(c) $\beta^2 - 4\alpha\gamma = 0$, so the quadratic equation (9) has a single double root m. Then the general solution is

$$u(t) = e^{mt}(C_1 + C_2 t).$$

Most of the remainder of this chapter is devoted to understanding this recipe and its generalizations beyond two-dimensional systems of linear differential equations.

7.2 Linearity and Superposition Principles

In exact analogy with the one-dimensional linear differential equations that we studied in Sections 2.2 and 2.3 in Part I, a linear equation of the form

$$\mathbf{x}' = A(t)\mathbf{x} + \mathbf{g}(t)$$

is called *homogeneous* if $\mathbf{g}(t) = 0$ and *nonhomogeneous* if $\mathbf{g}(t) \neq 0$. In the latter case, the equation $\mathbf{x}' = A(t)\mathbf{x}$ is called the *associated homogeneous* equation.

There are a certain number of important *algebraic properties* of linear differential equations, concerning sums of solutions and so forth. They take a bit of time to state, but they are very easy (which does not mean they are not important).

These properties may remind you of similar statements about linear (non-differential) equations and their associated homogeneous equations. This is not surprising: they are somehow the essence of *linearity*. At the end of this section, we will give the general framework, which includes both.

Let $A(t)$ be an $n \times n$ matrix with entries continuous functions of t.

Theorem 7.2.1 (Superposition of solutions for homogeneous equations). *If $\mathbf{u}_1(t)$ and $\mathbf{u}_2(t)$ are two solutions of the homogeneous linear differential equation $\mathbf{x}' = A(t)\mathbf{x}$, then any linear combination of those solutions*

$$\mathbf{u}(t) = C_1 \mathbf{u}_1(t) + C_2 \mathbf{u}_2(t),$$

for any two numbers C_1 and C_2, is also a solution.

Proof. Just plug $\mathbf{u}(t) = C_1\mathbf{u}_1(t) + C_2\mathbf{u}_2(t)$ into $\mathbf{x}' = A(t)\mathbf{x}$ and rearrange terms.   □

Example 7.2.2. Consider $x'' + 3x' + 2x = 0$. You can obtain from Recipe 7.1.7 that two solutions are $x = u_1(t) = e^{-t}$ and $x = u_2(t) = e^{-2t}$ and that

$$u(t) = C_1 e^{-t} + C_2 e^{-2t} \tag{10}$$

is also a solution, or, if you prefer to work with a system of first order equations, by the algorithm of Section 6.4,

$$\begin{bmatrix} x \\ y \end{bmatrix}' = \begin{bmatrix} 0 & 1 \\ -2 & -3 \end{bmatrix} \begin{bmatrix} x \\ y \end{bmatrix}.$$

Two solutions are $\mathbf{u}_1(t) = \begin{bmatrix} e^{-t} \\ -e^{-t} \end{bmatrix}$ and $\mathbf{u}_2(t) = \begin{bmatrix} e^{-2t} \\ -2e^{-2t} \end{bmatrix}$, as you can confirm by substitution, and so also is

$$\mathbf{u}(t) = C_1\mathbf{u}_1(t) + C_2\mathbf{u}_2(t) = C_1 \begin{bmatrix} e^{-t} \\ -e^{-t} \end{bmatrix} + C_2 \begin{bmatrix} e^{-2t} \\ -2e^{-2t} \end{bmatrix}.$$

Note that the top line of the vector solution matches the expression you got for $x = u(t)$ in (10). But it is not obvious that this includes *all* of the solutions.   ▲

In order to show that indeed we have *all* solutions for a linear differential equation like Example 7.2.2, we shall wait until Theorem 7.4.12 when we will have a formula showing the existence of solutions. You can show in Exercise 7.2#6 that linear differential equations with constant coefficients automatically satisfy a Lipschitz condition, so from the Fundamental Inequality 6.2.2 you will also have uniqueness.

Theorem 7.2.3 (General solution of nonhomogeneous equations).
If $\mathbf{u}_p(t)$ is any (particular) solution of a nonhomogeneous differential equation

$$\mathbf{x}' = A(t)\mathbf{x} + \mathbf{g}(t), \tag{11}$$

then a vector function $\mathbf{u}(t)$ is a solution of (11) if and only if $\mathbf{u}(t)$ can be written

$$\mathbf{u}(t) = \mathbf{u}_p(t) + \mathbf{u}_h(t),$$

where $\mathbf{u}_h(t)$ is a solution of the associated homogeneous equation

$$\mathbf{x}' = A(t)\mathbf{x}.$$

Proof. We show that $\mathbf{u}(t) = \mathbf{u}_p(t) + \mathbf{u}_h(t)$ is a solution by plugging in:

$$(\mathbf{u}_p(t) + \mathbf{u}_h(t))' = \underbrace{A(t)\mathbf{u}_p(t)}_{\mathbf{g}(t)} + \underbrace{A(t)\mathbf{u}_h(t)}_{\mathbf{0}}.$$

Then we show that the solution must be of form $\mathbf{u}(t) = \mathbf{u}_p(t) + \mathbf{u}_h(t)$ by proving that $\mathbf{u}(t) - \mathbf{u}_p(t)$ satisfies the homogeneous equation $A(t)\mathbf{x} = 0$:

$$
\begin{aligned}
(\mathbf{u}(t) - \mathbf{u}_p(t))' &= \mathbf{u}'(t) - \mathbf{u}'_p(t) \\
&= A(t)\mathbf{u}(t) + \mathbf{g}(t) - (A(t)\mathbf{u}_p(t) + \mathbf{g}(t)) \\
&= A(t)(\mathbf{u}(t) - \mathbf{u}_p(t)). \qquad \square
\end{aligned}
$$

As with the linear equations in one dimension discussed in Sections 2.2 and 2.3 in Part I, the usual ways of finding a particular solution $\mathbf{u}_p(t)$ to a nonhomogeneous equation are the methods of undetermined coefficients (an educated guess) and variation of parameters. We shall use the first of these here and discuss the second for higher dimensional systems of linear equations later, in Section 7.7.

Example 7.2.4. Consider

$$
\begin{bmatrix} x \\ y \end{bmatrix}' = \begin{bmatrix} 0 & 1 \\ -2 & -3 \end{bmatrix} \begin{bmatrix} x \\ y \end{bmatrix} + \begin{bmatrix} 0 \\ \sin t \end{bmatrix}.
$$

A particular solution to the entire nonhomogeneous equation is

$$
\mathbf{u}_p(t) = \begin{bmatrix} 0.1 \sin t - 0.3 \cos t \\ 0.1 \cos t + 0.3 \sin t \end{bmatrix}
$$

as you can confirm, and a solution to the associated homogeneous equation was found in Example 7.2.2. Therefore,

$$
\mathbf{u}(t) = \begin{bmatrix} 0.1 \sin t - 0.3 \cos t \\ 0.1 \cos t + 0.3 \sin t \end{bmatrix} + C_1 \begin{bmatrix} e^{-t} \\ -e^{-t} \end{bmatrix} + C_2 \begin{bmatrix} e^{-2t} \\ -2e^{-2t} \end{bmatrix}
$$

for any C_1 and C_2 is a solution to the original equation. ▲

Example 7.2.4 is the system that represents $x'' + 3x' + 2x = \sin t$, so, as in Example 7.2.2, you can confirm that a solution to this second order equation is read from the top line of the vector equation:

$$
u(t) = 0.1 \sin t - 0.3 \cos t + C_1 e^{-t} + C_2 e^{-2t}.
$$

Theorems 7.2.1 and 7.2.3 are both special cases of a more inclusive theorem:

Theorem 7.2.5 (Superposition of solutions for nonhomogeneous equations). *If $\mathbf{u}_1(t)$ and $\mathbf{u}_2(t)$ are solutions of two nonhomogeneous linear differential equations*

$$\mathbf{x}' = A(t)\mathbf{x} + \mathbf{g}_1(t) \quad and \quad \mathbf{x}' = A(t)\mathbf{x} + \mathbf{g}_2(t),$$

respectively, with the same associated homogeneous equation $\mathbf{x}' = A(t)\mathbf{x}$, then

$$\mathbf{u}(t) = \mathbf{u}_1(t) + \mathbf{u}_2(t)$$

is a solution of the equation $\mathbf{x}' = A(t)\mathbf{x} + \mathbf{g}_1(t) + \mathbf{g}_2(t)$.

Proof. Just plug in. □

Example 7.2.6. Consider

$$\begin{bmatrix} x \\ y \end{bmatrix}' = \begin{bmatrix} 0 & 1 \\ -2 & -3 \end{bmatrix} \begin{bmatrix} x \\ y \end{bmatrix} + \begin{bmatrix} 0 \\ \sin t + e^t \end{bmatrix}.$$

A particular solution to the nonhomogeneous equation

$$\begin{bmatrix} x \\ y \end{bmatrix}' = \begin{bmatrix} 0 & 1 \\ -2 & -3 \end{bmatrix} \begin{bmatrix} x \\ y \end{bmatrix} + \begin{bmatrix} 0 \\ e^t \end{bmatrix}$$

is, as you can confirm, $\mathbf{u}_2(t) = \begin{bmatrix} \frac{1}{6} e^t \\ \frac{1}{6} e^t \end{bmatrix}$; in Example 7.2.4, we found that the solution to the nonhomogeneous equation

$$\begin{bmatrix} x \\ y \end{bmatrix}' = \begin{bmatrix} 0 & 1 \\ -2 & -3 \end{bmatrix} \begin{bmatrix} x \\ y \end{bmatrix} + \begin{bmatrix} 0 \\ \sin t \end{bmatrix}$$

is

$$\mathbf{u}_p(t) = \begin{bmatrix} 0.1 \sin t - 0.3 \cos t \\ 0.1 \cos t + 0.3 \sin t \end{bmatrix}.$$

Therefore,

$$\mathbf{u}(t) = \begin{bmatrix} 0.1 \sin t - 0.3 \cos t + (1/6)e^t \\ 0.1 \cos t + 0.3 \sin t + (1/6)e^t \end{bmatrix} + C_1 \begin{bmatrix} e^{-t} \\ -e^{-t} \end{bmatrix} + C_2 \begin{bmatrix} e^{-2t} \\ -2e^{-2t} \end{bmatrix}$$

is, for any C_1 and C_2, a solution to the whole thing. ▲

In the language of linear algebra, the above results may be restated as follows:

> Theorem 7.2.1 says that the set of solutions of a *homogeneous* linear differential equation forms a *vector space*.

Theorem 7.2.3 says that the set of solutions of a nonhomo-geneous linear differential equation is an *affine space*, i.e., the translate of the vector space of solutions of the associated ho-mogeneous equation by any solution of the nonhomogeneous one.

Now we will relate these statements to the more standard statements from linear algebraic equations. Even if you have never seen these state-ments exactly, they should be clear.

Let M be a matrix (not necessarily square).

Statement A. Any linear combination of solutions of $M\mathbf{x} = 0$ is a solution.

Statement B. If $\mathbf{x}_p$ is a solution of $M\mathbf{x} = \mathbf{a}$, then a vector $\mathbf{x}$ is another solution if and only if $\mathbf{x} = \mathbf{x}_p + \mathbf{x}_c$ with $\mathbf{x}_c$ a solution of the associated homogeneous equation $M\mathbf{x} = 0$.

Statement C. If $\mathbf{x}_1$ is a solution of $M\mathbf{x} = \mathbf{a}_1$ and $\mathbf{x}_2$ is a solution of $M\mathbf{x} = \mathbf{a}_2$, then $\mathbf{x}_1 + \mathbf{x}_2$ is a solution of $M\mathbf{x} = \mathbf{a}_1 + \mathbf{a}_2$.

Clearly these statements are parallel to Theorems 7.2.1, 7.2.3, and 7.2.5. The explanation of this fact is *linearity*.

Let V and W be vector spaces and $T: V \to W$ a linear transformation. Then we have the following three linearity statements:

Statement 1. Any linear combination of solutions of $T\mathbf{v} = 0$ is a solution.

Statement 2. If $\mathbf{v}_p$ is a solution of $T\mathbf{v} = \mathbf{a}$, then a vector $\mathbf{v}$ is another solution if and only if $\mathbf{v} = \mathbf{v}_p + \mathbf{v}_c$ with $\mathbf{v}_c$ a solution of the associated homogeneous equation $T\mathbf{v} = 0$.

Statement 3. If $\mathbf{v}_1$ is a solution of $T\mathbf{v} = \mathbf{a}_1$ and $\mathbf{v}_2$ is a solution of $T\mathbf{v} = \mathbf{a}_2$, then $\mathbf{v}_1 + \mathbf{v}_2$ is a solution of $T\mathbf{v} = \mathbf{a}_1 + \mathbf{a}_2$.

Statements A, B, and C about systems of linear algebraic equations are the special case of the linearity statements 1, 2, and 3, where T is the linear transformation $\mathbb{R}^n \to \mathbb{R}^m$ represented by the matrix M.

How about the differential equations of Theorems 7.2.1, 7.2.3, and 7.2.5? What are the vector spaces and the linear transformation in that case? Well, the answer is a bit more unpleasant.

The space V is the *vector space of continuously differentiable vector functions* $\mathbf{f}(t)$ defined for $t \in I$, where I is the interval on which $A(t)$ and all the $\mathbf{g}(t)$ are defined.

W is the space of continuous vector functions on the same in-terval I.

The transformation $T: \mathbf{f} \mapsto \mathbf{f}'(t) - A\mathbf{f}$.

Of course, the spaces V and W are infinite dimensional, and so are a bit unfamiliar, but all the results above are so formal that the nature of the vector space involved is irrelevant.

Such a linear transformation $T: \mathbf{f} \mapsto \mathbf{f}'(t) - A\mathbf{f}$ is called a *differential operator* (more accurately, a vector first order differential operator). Of course, the granddaddy of all differential operators is the derivative D, and it has close relatives $D^2, D^3, \ldots$ (the second, third, ... derivatives). Another way of writing the differential equation

$$ax'' + bx' + cx = 0$$

is

$$(aD^2 + bD + c)x = 0.$$

Following this method of notation, all of the theory of linear differential equations can be rewritten in terms of differential operators. We will not do this, although we will use the language when it seems convenient.

7.3 Linear Differential Equations with Constant Coefficients; Eigenvectors and Decoupling

There is *no formula* for solving linear differential equations if the coefficients are not constant. In fact, it can be proved that the innocent looking equation $x'' - tx = 0$ has no solutions, other than the zero solution, that can be written in terms of the elementary functions, or their indefinite integrals ad infinitum. (Reference: Theorem 6.6, p. 43 of Kaplansky's sixty-seven page book on *Differential Algebra*.)

In fact, the only substantial class of systems of equations that can be solved in elementary terms is the class of *linear differential equations with constant coefficients*. As a result, and also because of their intrinsic importance, these equations have traditionally been the bread and butter of differential equation courses. Ours will not be different; they will play a central part in our development.

The following theorem is really the most important result about linear differential equations with constant coefficients. It says that the component of a solution *in the direction of an eigenvector* evolves independently of everything else and obeys a linear differential equation in *one* variable, about which we know just about everything from Chapter 2 in Part I. This phenomenon of being able to look at components independently is called decoupling.

Accordingly, the theory of linear differential equations reduces (almost) to the algebraic theory of *eigenvalues* and *eigenvectors*. This theory is fairly simple, but when it comes to actually computing eigenvalues, the going gets stickier. You can review the essentials of eigenvalues and eigenvectors in Appendix L6; actually finding them will most conveniently be done by the computer using the QR method described in Appendix L7 or Jacobi's method, described in Appendix L8.

Theorem 7.3.1 (Evolution in the direction of an eigenvector). *Let A be an $n \times n$ matrix, and suppose $\mathbf{v}$ is an eigenvector of A, with $A\mathbf{v} = \lambda\mathbf{v}$. Then*

$$\mathbf{u}(t) = \alpha e^{\lambda(t-t_0)}\mathbf{v} \tag{12}$$

is the solution of $\mathbf{x}' = A\mathbf{x}$ with initial condition $\mathbf{u}(t_0) = \alpha\mathbf{v}$.

Proof. As always, when an explicit solution of a differential equation is proposed, it is a straightforward matter of differentiating it and plugging the derivative into the equation to check whether or not it is correct. Let us compute

$$\frac{d}{dt}(\mathbf{u}(t)) = \lambda\alpha\, e^{\lambda(t-t_0)}\mathbf{v} = \alpha\, e^{\lambda(t-t_0)}(\lambda\mathbf{v}) = \alpha\, e^{\lambda(t-t_0)}A\mathbf{v}$$

$$= A(\alpha\, e^{\lambda(t-t_0)}\mathbf{v}) = A\mathbf{u}(t).$$

It is easy to verify that the initial condition is satisfied. □

Do not be fooled by the extreme simplicity of the above derivation. It does not have to be hard to be important. That easy computation is an important service of mathematics to applied science.

Theorem 7.3.2. *Let A be an $n \times n$ matrix of constants, and suppose $\mathbf{v}_1, \ldots, \mathbf{v}_m$ are eigenvectors of A, with $A\mathbf{v}_i = \lambda_i\mathbf{v}_i$. Then for the differential equation $\mathbf{x}' = A\mathbf{x}$, if the initial condition is a linear combination of these $\mathbf{v}_i$'s,*

$$\mathbf{u}(t_0) = \sum_{i=1}^{m} a_i\mathbf{v}_i,$$

the solution is

$$\mathbf{u}(t) = \sum_{i=1}^{m} a_i\, e^{\lambda_i(t-t_0)}\mathbf{v}_i. \tag{13}$$

Proof. This follows from Theorems 7.2.1 and 7.3.1. □

Theorem 7.3.2 is of particular interest when the $\mathbf{v}_i$'s form a *basis* of $\mathbb{R}^n$ or $\mathbb{C}^n$, for in that case formula (13) gives *all* solutions to $\mathbf{x}' = A\mathbf{x}$, since *any* initial condition can be stated as a linear combination of eigenvectors. In real life, this is the typical situation: in general, an $n \times n$ matrix will have n distinct eigenvalues and therefore provide an *eigenbasis* (a basis of eigenvectors) of $\mathbb{R}^n$.

Theorem 7.3.3. *Let A be an $n \times n$ matrix, and suppose $\mathbf{v}_1, \ldots, \mathbf{v}_n$ form an eigenbasis of $\mathbb{R}^n$, with $A\mathbf{v}_i = \lambda_i \mathbf{v}_i$. Then for the differential equation $\mathbf{x}' = A\mathbf{x}$, the initial condition can be written*

$$\mathbf{u}(t_0) = \sum_{i=1}^{n} a_i \mathbf{v}_i,$$

and the solution is

$$\mathbf{u}(t) = \sum_{i=1}^{n} a_i \, e^{\lambda_i (t - t_0)} \mathbf{v}_i. \tag{14}$$

The proof has been covered in the paragraph preceding the theorem.

CHANGE OF BASIS

Theorem 7.3.3 permits another point of view. Its key point is that

> in a *basis of eigenvectors*, the coordinates *evolve independently.*

This statement is equivalent, in the language of change of basis, to the statement that

> in a *basis of eigenvectors*, a system of linear differential equations with constant coefficients will *decouple.*

We shall now restate and rederive Theorem 7.3.3 in this language of change of basis.

Suppose that $\mathbf{v}_1, \ldots, \mathbf{v}_n$ is a basis of $\mathbb{R}^n$ (or C^n) made up of eigenvectors of A with corresponding eigenvalues $\lambda_1, \ldots, \lambda_n$, as above. Let P be the *change of basis matrix*, as described in Appendix L2, so that

$$P = \begin{bmatrix} | & | & & | \\ \mathbf{v}_1, & \mathbf{v}_2, & \ldots, & \mathbf{v}_n \\ | & | & & | \end{bmatrix}, \quad \text{and} \quad P^{-1}AP = \begin{bmatrix} \lambda_1 & & \\ & \ddots & 0 \\ 0 & & \\ & & \lambda_n \end{bmatrix},$$

the *diagonal* matrix with the *eigenvalues* λ_i along the diagonal. This is why, when the eigenvectors of A form a basis for $\mathbb{R}^n$, the matrix is said to be *diagonalizable.*

Theorem 7.3.4 (Decoupling $\mathbf{x}' = A\mathbf{x}$ by change of basis). *Consider the differential equation $\mathbf{x}' = A\mathbf{x}$ in $\mathbb{R}^n$. If $\mathbf{v}_1, \ldots, \mathbf{v}_n$ are eigenvectors of A forming a basis, with P as the change of basis matrix composed of the $\mathbf{v}_i$ as columns, and $\tilde{\mathbf{x}} \equiv P^{-1}\mathbf{x}$, then*

> i. *each entry $\tilde{x}_i$ of $\tilde{\mathbf{x}}$ satisfies the scalar differential equation*

$$\tilde{x}_i' = \lambda_i \tilde{x}_i;$$

ii. *the solution* $\mathbf{x} = \mathbf{u}(t)$ *to the original differential equation can be written in terms of the initial condition* $\mathbf{x}(t_0)$ *as follows*:

$$\mathbf{x}(t) = P\tilde{\mathbf{x}}(t) = P \begin{bmatrix} e^{\lambda_1(t-t_0)} & & 0 \\ & \ddots & \\ 0 & & e^{\lambda_n(t-t_0)} \end{bmatrix} P^{-1}\mathbf{x}(t_0). \qquad (15)$$

Proof. If we define $\tilde{\mathbf{x}} \equiv P^{-1}\mathbf{x}$, then the entries of $\tilde{\mathbf{x}}$ are the coordinates of $\mathbf{x}$ with respect to the basis $\mathbf{v}_1, \ldots, \mathbf{v}_n$ (see Theorem L2.22 in Appendix L2). Now we are ready to prove the theorem:

i. Multiply the equation $\mathbf{x}' = A\mathbf{x}$ by P^{-1} on the left and introduce a factor of $PP^{-1} = I$ between A and $\mathbf{x}$, giving

$$P^{-1}\mathbf{x}' = (P^{-1}AP)(P^{-1}\mathbf{x}),$$

or

$$\tilde{\mathbf{x}}' = (P^{-1}AP)\tilde{\mathbf{x}}.$$

Since $P^{-1}AP$ is diagonal, this last equation breaks up into

$$\tilde{x}_1' = \lambda_1 \tilde{x}_1 , \; \ldots \; , \; \tilde{x}_n' = \lambda_n \tilde{x}_n, \qquad (16)$$

a completely decoupled set of ordinary first order linear differential equations.

ii. Each of the set (16) of decoupled equations can be explicitly solved as in Section 2.2 in Part I, giving

$$\tilde{x}_i(t) = e^{\lambda_i(t-t_0)}\tilde{x}_i(t_0).$$

Since $\tilde{\mathbf{x}}(t) = P^{-1}\mathbf{x}(t)$, then $\tilde{\mathbf{x}}(t_0) = P^{-1}\mathbf{x}(t_0)$, and putting all these solutions together in vector form gives equation (15). $\quad \square$

Remember that Theorem 7.3.4 is simply a restatement of the message of Theorem 7.3.3. We shall solve a sample linear differential equation using both theorems.

Example 7.3.5.

$$\begin{bmatrix} x \\ y \\ z \end{bmatrix}' = \begin{bmatrix} 0 & 1 & 0 \\ 0 & 0 & 1 \\ -5 & 5 & 1 \end{bmatrix} \begin{bmatrix} x \\ y \\ z \end{bmatrix}.$$

To solve this equation, we first find eigenvalues and eigenvectors. The characteristic polynomial is

$$-\lambda^3 + \lambda^2 + 5\lambda - 5,$$

so the eigenvalues are 1 and $\pm\sqrt{5}$, and corresponding eigenvectors are

$$\begin{bmatrix} 1 \\ 1 \\ 1 \end{bmatrix}, \begin{bmatrix} 1 \\ \sqrt{5} \\ 5 \end{bmatrix}, \begin{bmatrix} 1 \\ -\sqrt{5} \\ 5 \end{bmatrix}, \text{ with } P = \begin{bmatrix} 1 & 1 & 1 \\ 1 & \sqrt{5} & -\sqrt{5} \\ 1 & 5 & 5 \end{bmatrix}.$$

i. *Solution by straight eigenvectors.* The general solution, by Theorem 7.3.3, is

$$a e^t \begin{bmatrix} 1 \\ 1 \\ 1 \end{bmatrix} + b e^{\sqrt{5}t} \begin{bmatrix} 1 \\ \sqrt{5} \\ 5 \end{bmatrix} + c e^{-\sqrt{5}t} \begin{bmatrix} 1 \\ -\sqrt{5} \\ 5 \end{bmatrix}. \tag{17}$$

Suppose we know the initial condition as $\mathbf{x}(0) = \begin{bmatrix} \alpha \\ \beta \\ \gamma \end{bmatrix}$. Then in order to find a, b, and c in solution (17), we must solve the system of linear algebraic equations

$$\begin{bmatrix} \alpha \\ \beta \\ \gamma \end{bmatrix} = P \begin{bmatrix} a \\ b \\ c \end{bmatrix}. \tag{18}$$

ii. *Solution by change of basis.* Theorem 7.3.4 gives the following, in terms of equation (15):

$$\mathbf{x}(t) = P \begin{bmatrix} e^t & 0 & 0 \\ 0 & e^{\sqrt{5}t} & 0 \\ 0 & 0 & e^{-\sqrt{5}t} \end{bmatrix} P^{-1} \begin{bmatrix} \alpha \\ \beta \\ \gamma \end{bmatrix}, \tag{19}$$

which you can confirm (Exercise 7.3#9) gives the same result as equations (17) and (18). ▲

It is interesting to compare the amount of work necessary to solve Example 7.3.5 by the two theorems. In both cases, almost all the work is the same—you have to find all the eigenvalues and eigenvectors. But at the end, the first method, with equation (17), requires solving a (nondifferential) system of equations (18); the second method given by equation (19) requires computing the matrix inverse of P.

The first method is easier if you are solving just for a single set of initial conditions. But if you are going to be solving the problem for a number of different values of α, β, and γ, the alternative second method gives an actual formula which will save a great deal of computation.

COMPLEX EIGENVALUES AND EIGENVECTORS

Up to this point, we have been working as if everything were in real numbers. Actually, there is no difference if the eigenvalues are complex:

Example 7.3.6. We shall solve the basic equation of Example 7.1.1,

$$\begin{bmatrix} x \\ y \end{bmatrix}' = \begin{bmatrix} 0 & -1 \\ 1 & 0 \end{bmatrix} \begin{bmatrix} x \\ y \end{bmatrix}, \text{ with initial condition } \begin{bmatrix} x_0 \\ y_0 \end{bmatrix} \text{ at } t = 0.$$

First we find the eigenvalues and eigenvectors. The characteristic polynomial is $\lambda^2 + 1$, so the eigenvalues are $\pm i$ and corresponding eigenvectors are

$$\begin{bmatrix} 1 \\ -i \end{bmatrix} \text{ and } \begin{bmatrix} 1 \\ i \end{bmatrix}.$$

The matrices P and P^{-1} are

$$P = \begin{bmatrix} 1 & 1 \\ -i & i \end{bmatrix}, \qquad P^{-1} = \begin{bmatrix} 1/2 & i/2 \\ 1/2 & -i/2 \end{bmatrix},$$

and we can observe that indeed

$$P^{-1}AP = \begin{bmatrix} 1/2 & i/2 \\ 1/2 & -i/2 \end{bmatrix} \begin{bmatrix} 0 & -1 \\ 1 & 0 \end{bmatrix} \begin{bmatrix} 1 & 1 \\ -i & i \end{bmatrix} = \begin{bmatrix} i & 0 \\ 0 & -i \end{bmatrix},$$

the diagonal matrix of eigenvalues.

By the change of basis solution (12) of Theorem 7.3.4, we get

$$\begin{bmatrix} x(t) \\ y(t) \end{bmatrix} = \begin{bmatrix} 1 & 1 \\ -i & i \end{bmatrix} \begin{bmatrix} e^{it} & 0 \\ 0 & e^{-it} \end{bmatrix} \begin{bmatrix} 1/2 & i/2 \\ 1/2 & -i/2 \end{bmatrix} \begin{bmatrix} x_0 \\ y_0 \end{bmatrix}$$

$$= \begin{bmatrix} 1/2(e^{it} + e^{-it}) & i/2(e^{it} - e^{-it}) \\ -i/2(e^{it} - e^{-it}) & 1/2(e^{it} + e^{-it}) \end{bmatrix} \begin{bmatrix} x_0 \\ y_0 \end{bmatrix} = \begin{bmatrix} \cos t & -\sin t \\ \sin t & \cos t \end{bmatrix} \begin{bmatrix} x_0 \\ y_0 \end{bmatrix}.$$

▲

Thus, Example 7.3.6 shows that computations with complex eigenvalues *will* yield, if painfully, terms with trigonometric functions in them.

FINDING EIGENVALUES AND EIGENVECTORS

The method we have just explained in this section is the most important for solving $\mathbf{x}' = A\mathbf{x}$; the main task is to find the eigenvectors and eigenvalues of A. In principle, you could try to do this by computing the characteristic polynomial of A, finding the eigenvalues as its roots, and solving for the eigenvectors by solving linear equations.

It turns out that as soon as the matrix A is at all large, this method is quite impractical. First, computing the characteristic polynomial is quite laborious, even if something smarter than the formula for the determinant is used. Above, we spoke rather glibly about "finding the roots" of the polynomial, but it is not obvious how to do this (and in fact it is quite hard).

So we turn to the computer, but if we were to work directly with finding the roots of the characteristic polynomial, we would find that in the computation all precision is lost, and the coefficients of the characteristic polynomials are known only approximately. In addition, the errors in the coefficients may lead to much larger errors in the roots. (See Exercise 7.3#10.)

We have two other computer approaches, the *QR method* and *Jacobi's method*, that alleviate some of these problems (although finding eigenvectors and eigenvalues really is hard, and nothing will make it quite simple). Appendices L7 and L8 respectively describe the QR method (and its simpler cousin, the power method), and Jacobi's method (which only works for symmetric matrices).

7.4 Linear Differential Equations with Constant Coefficients; Exponentials of Matrices

The eigenvector method of Section 7.3 has some limitations. If there are not enough eigenvectors to form a basis of $\mathbb{R}^n$ or $\mathbb{C}^n$, then there is an alternative approach.

Just as, from Section 2.2 in Part I, the solution of $x' = ax$ with $u(0) = x_0$ is

$$u(t) = e^{ta}x_0,$$

we shall see (in Theorem 7.4.12) that the general solution to $\mathbf{x}' = A\mathbf{x}$ with $\mathbf{u}(0) = \mathbf{x}_0$ is

$$\mathbf{u}(t) = e^{tA}\mathbf{x}_0. \tag{20}$$

This requires the *exponentials of matrices*, to which we now proceed. However, the reader should not think this is a panacea—it just shifts the problem from solving differential equations to computing exponentials.

Definition 7.4.1 (Exponentials of matrices). If A is a *square* $n \times n$ matrix, we define

$$e^A \equiv I + A + \frac{A^2}{2!} + \frac{A^3}{3!} + \cdots = \sum_{m=0}^{\infty} \frac{A^m}{m!}. \tag{21}$$

First let us give a few examples.

Example 7.4.2.

$$e^{tI} = I + tI + \frac{t^2 I^2}{2!} + \frac{t^3 I^3}{3!} + \cdots = e^t I, \tag{22}$$

since $I = I^2 = I^3 = \cdots$. In particular, if $t = 0$,

$$e^{(0)I} = I = e^{(0)A}. \qquad \blacktriangle \tag{23}$$

Example 7.4.2 is a special case of Example 7.4.3:

Example 7.4.3. If A is a diagonal matrix with diagonal entries $\lambda_1, \ldots, \lambda_n$, then e^{tA} is also diagonal, with entries $e^{t\lambda_1}, \ldots, e^{t\lambda_n}$ along the diagonal. ▲

Example 7.4.3 is important, because it shows the main way in which exponentials are actually computed, as we shall see in Theorem 7.4.13.

The next example is less familiar but actually easier. It illustrates a phenomenon that occurs for exponentials of matrices but not for exponentials of numbers; that is, if there exist polynomials in the exponent, then e^{tA} may have polynomial terms and the series may be finite.

Example 7.4.4.

$$e^{\begin{bmatrix} 0 & t \\ 0 & 0 \end{bmatrix}} = \begin{bmatrix} 1 & 0 \\ 0 & 1 \end{bmatrix} + \begin{bmatrix} 0 & t \\ 0 & 0 \end{bmatrix} + \frac{1}{2}\begin{bmatrix} 0 & 0 \\ 0 & 0 \end{bmatrix} + \cdots = \begin{bmatrix} 1 & t \\ 0 & 1 \end{bmatrix}$$

because, as you can confirm, $\begin{bmatrix} 0 & t \\ 0 & 0 \end{bmatrix}^m = \begin{bmatrix} 0 & 0 \\ 0 & 0 \end{bmatrix}$, for $m \geq 2$. ▲

When real matrices have complex eigenvalues, sines and cosines make their way into the exponential. This is hardly surprising, since $e^{i\theta} = \cos\theta + i\sin\theta$, but please observe that complex numbers are never mentioned in the next example:

Example 7.4.5.

$$e^{t\begin{bmatrix} 0 & -1 \\ 1 & 0 \end{bmatrix}} = \begin{bmatrix} 1 & 0 \\ 0 & 1 \end{bmatrix} + t\begin{bmatrix} 0 & -1 \\ 1 & 0 \end{bmatrix} + \frac{1}{2!}t^2\begin{bmatrix} -1 & 0 \\ 0 & -1 \end{bmatrix} + \frac{1}{3!}t^3\begin{bmatrix} 0 & 1 \\ -1 & 0 \end{bmatrix} + \cdots$$
$$= \begin{bmatrix} \cos t & -\sin t \\ \sin t & \cos t \end{bmatrix}$$

from the power series for sine and cosine. Note that this calculation indeed gives the solution for the differential equation of Example 7.3.6 that you could expect from equation (20). ▲

Remark. Since e^{tA} is a matrix, the order of multiplication must be $(e^{tA})(x_0)$, i.e., (matrix) (vector); the opposite order makes no sense. Moreover, since e^{tA} is a matrix function of t, it can be *differentiated* with respect to t: the derivative of a matrix function $M(t)$ of t is another matrix $M'(t)$ of the same size, where each element $M'(t)_{ij}$ is the derivative of the corresponding element $M(t)_{ij}$.

PROPERTIES OF e^A

Formula (21) defining e^A for an $n \times n$ matrix A is exactly the same as the power series for the exponential of a number; we may hope it will have similar properties. This is true up to a point, as we shall show in the following properties:

Property 7.4.6 (Convergence). *The series defining e^A converges for every A.*

Proof. If $\sup_{i,j} |Ai, j| \leq k$ and the matrix A is $n \times n$, then it is can be shown by induction that

$$|(A^m)_{i,j}| \leq n^{m-1} k^m,$$

because from the formula for multiplication of matrices, each multiplication is the sum of n terms. Therefore,

$$\left| \sum_{m=0}^{\infty} \frac{(A^m)_{i,j}}{m!} \right| \leq \frac{1}{n} \sum_{m=0}^{\infty} \frac{(nk)^m}{m!} = \frac{e^{nk}}{n},$$

and we see that the series does converge by the comparison test.   □

The fundamental property of the exponential function is $e^{a+b} = e^a e^b$. So we might hope that for *any* two $n \times n$ matrices A and B, we would have $e^{A+B} = e^A e^B$. Unfortunately, this is *false* in general, and this is the reason why systems are more difficult than scalar equations. If the two matrices *commute*, which means that the order of multiplication does not matter, then the hoped-for result is true:

Property 7.4.7 (Addition formula). *If A and B are two $n \times n$ matrices that commute, so that $AB = BA$, then*

$$e^A e^B = e^{A+B}. \tag{24}$$

Proof. Let us look at the two series developments, carefully keeping track of the order of multiplication (*not* just using the binomial expansion formulas):

i. $\begin{aligned} e^{(A+B)} &= I + (A + B) + \frac{1}{2}(A + B)^2 + \frac{1}{6}(A + B)^3 + \cdots \\ &= I + (A + B) + \frac{1}{2}(A^2 + AB + BA + B^2) \\ &\quad + \frac{1}{6}(A^3 + A^2B + ABA + AB^2 + BA^2 + BAB + B^2A + B^3) \\ &\quad + \cdots \end{aligned}$

ii. $e^A e^B = \left(I + A + \dfrac{1}{2}A^2 + \dfrac{1}{6}A^3 + \cdots \right)\left(I + B + \dfrac{1}{2}B^2 + \dfrac{1}{6}B^3 + \cdots \right)$

$\qquad = I + (A + B) + \dfrac{1}{2}(A^2 + 2AB + B^2)$

$\qquad\quad + \dfrac{1}{6}(A^3 + 3A^2B + 3AB^2 + B^3) + \cdots .$

Proceed until it is clear that if $AB = BA$, the series are equal. $\square$

The most important thing to remember about Property 7.4.7 is the necessity, which you see in the proof, that the matrices commute. In general (when $AB \neq BA$),

$$e^A e^B \neq e^{A+B}.$$

Example 7.4.8. Let $A = t\begin{bmatrix} 0 & -1 \\ 0 & 0 \end{bmatrix}$ and $B = t\begin{bmatrix} 0 & 0 \\ 1 & 0 \end{bmatrix}$. You can confirm that $AB \neq BA$ and that $A^2 = B^2 = 0$. We shall show here that $e^A e^B \neq e^{A+B}$:

$$e^A e^B = e^{t\begin{bmatrix} 0 & -1 \\ 0 & 0 \end{bmatrix}} e^{t\begin{bmatrix} 0 & 0 \\ 1 & 0 \end{bmatrix}} = \left(I + t\begin{bmatrix} 0 & -1 \\ 0 & 0 \end{bmatrix} \right) \cdot \left(I + t\begin{bmatrix} 0 & 0 \\ 1 & 0 \end{bmatrix} \right)$$

$$= I + t\begin{bmatrix} 0 & -1 \\ 1 & 0 \end{bmatrix} + t^2\begin{bmatrix} -1 & 0 \\ 0 & 0 \end{bmatrix} = \begin{bmatrix} 1 - t^2 & -t \\ t & 1 \end{bmatrix}. \qquad (25)$$

From Example 7.4.5,

$$e^{A+B} = e^{t\begin{bmatrix} 0 & -1 \\ 1 & 0 \end{bmatrix}} = \begin{bmatrix} \cos t & -\sin t \\ \sin t & \cos t \end{bmatrix}. \qquad (26)$$

Clearly, these results, (25) and (26), do not match (although we do get the same first terms in the power series!), so $e^A e^B \neq e^{A+B}$ in this case, where $AB \neq BA$. ▲

Property 7.4.9 (Differentiation). *We have the following formula for the derivative of e^{tA}:*

$$\frac{d}{dt}(e^{tA}) = Ae^{tA}. \qquad (27)$$

Proof. By the definition of derivative,

$$\frac{d}{dt}e^{tA} = \lim_{h \to 0} \left(\frac{1}{h} \right) (e^{(t+h)A} - e^{tA}).$$

Certainly tA and hA commute, so $e^{(t+h)A} = e^{hA}e^{tA}$, and we see that

$$\frac{d}{dt}e^{tA} = \lim_{h \to 0} \left(\frac{1}{h} \right) (e^{hA} - I)e^{tA}.$$

Now $e^{hA} - I = hA + \frac{1}{2}h^2A^2 + \cdots$, so $\lim_{h\to 0}(1/h)(e^{hA} - I) = A$ and

$$\frac{d}{dt}e^{tA} = A e^{tA}. \qquad \square$$

Remarks. For any square matrix A, the matrices A and e^{tA} commute, so in the previous formula, the factor A can be written on the right or the left.

In Exercise 7.4#4 you can show that the product and chain rules for differentiation also hold for expressions involving exponentials of matrices.

Example 7.4.10.

$$\frac{d}{dt}e^{t\begin{bmatrix} 0 & -1 \\ 1 & 0 \end{bmatrix}} = \begin{bmatrix} 0 & -1 \\ 1 & 0 \end{bmatrix} e^{t\begin{bmatrix} 0 & -1 \\ 1 & 0 \end{bmatrix}}$$

$$= \begin{bmatrix} 0 & -1 \\ 1 & 0 \end{bmatrix}\begin{bmatrix} \cos t & -\sin t \\ \sin t & \cos t \end{bmatrix} = \begin{bmatrix} -\sin t & -\cos t \\ \cos t & -\sin t \end{bmatrix},$$

as you might have expected from differentiating both sides of the last line of Examples 7.4.5 and 7.4.8. ▲

Property 7.4.11 (Change of basis). *If M is an invertible matrix, then*

$$e^{M^{-1}AM} = M^{-1}e^A M. \qquad (28)$$

Proof. This follows from the power series (21),

$$e^{M^{-1}AM} \equiv I + M^{-1}AM + \frac{(M^{-1}AM)^2}{2!} + \frac{(M^{-1}AM)^3}{3!} + \cdots$$

$$\equiv I + M^{-1}AM + \frac{M^{-1}A^2M}{2!} + \frac{M^{-1}A^3M}{3!} + \cdots$$

term by term, by observing that

$$(M^{-1}AM)^m = (M^{-1}A\underbrace{M)(M^{-1}}_{I}AM)\cdots(M^{-1}AM) = M^{-1}A^m M. \qquad \square$$

all pairs in the middle cancel by $MM^{-1} = I$

We are now ready to talk about differential equations again; Property 7.4.9 is precisely what it takes to prove the following major theorem:

Theorem 7.4.12 (Solving $x' = Ax$ by exponentiation). *The solution* $u(t)$ *of $x' = Ax$ with $u(t_0) = x_0$ is*

$$u(t) = e^{(t-t_0)A}x_0. \qquad (29)$$

Proof. Just differentiate (using Property 7.4.9) and check. □

So, we have a "formula" for the solution to a homogeneous linear differential equation. Unfortunately, Theorem 7.4.12 suffers from the standard defect of formulas: it says very little by itself. We want to know the answers to such questions as whether solutions go to 0 or to ∞, and it is unclear that formula (29) says anything about this. *We have just shifted the problem from solving the differential equation to computing the exponential.*

However, *if* we have a basis of eigenvectors, the exponential reduces to something familiar. Property 7.4.11 implies that an eigenbasis for A is also an eigenbasis for e^A.

Theorem 7.4.13 (Computing e^{tA} by eigenvectors). *Let A be an $n \times n$ matrix and let $\mathbf{v}_1, \ldots, \mathbf{v}_n$ be an eigenbasis for A with corresponding eigenvalues $\lambda_1, \ldots, \lambda_n$. Let P be the change of basis matrix (i.e., the $\mathbf{v}_i$ are the columns of P); then the exponential e^{tA} can be computed as follows:*

$$
e^{tA} = P \begin{bmatrix} e^{t\lambda_1} & & 0 \\ & \ddots & \\ 0 & & e^{t\lambda_n} \end{bmatrix} P^{-1}.
$$

Proof. This is just a restatement of Property 7.4.11 and Example 7.3.4, as you can show in Exercise 7.4#5. □

So, in the diagonalizable case where you do have a *basis* of eigenvectors, Theorems 7.4.12 and 7.4.13 give the same result as Theorem 7.3.4, the eigenvalue/eigenvector method that leads quickly to an easy-to-interpret answer. Note, however, that the method of exponentiating matrices is in a different basis and usually does not lead to anything recognizable as the same thing. (See Exercise 7.4#6.)

Even if there is no basis of eigenvectors, you can fall back on the exponential approach, working directly from the power series using Theorem 7.4.12 and Definition 7.4.1. Only sometimes will this result in anything you can calculate. Calculating the exponential of a matrix can be quite complicated, requiring generalized eigenvectors or Jordan Canonical Form if you want explicit formulas. Although, compared to nonlinear problems, this is still "simple," if you try to put this into practice, you will come to understand the expletive that "hell is undiagonalizable matrices."

Before leaving this subsection, we will make one more algebraic point. We go back and note that Example 7.4.5 is also an example of the following result:

Theorem 7.4.14. *If A is an antisymmetric real matrix (or more generally an anti-Hermitian matrix), then e^{tA} is orthogonal (or, more generally, unitary).*

Proof. As defined in Appendix L4, antisymmetric means that $A^T = -A$; orthogonal means that the column vectors of the matrix are orthonormal, which in turn means that $A^T = A^{-1}$. We will treat the real case; the complex case is Exercise 7.4#7. Let us *differentiate*:

$$\frac{d}{dt}\left[(e^{tA})(e^{tA})^T\right] = (e^{tA})A(e^{tA})^T + (e^{tA})A^T(e^{tA})^T$$
$$= (e^{tA})\underbrace{(A + A^T)}_{= 0}(e^{tA})^T = 0.$$

Thus, $(e^{tA})(e^{tA})^T$ is constant, and at time $t = 0$ it is the identity, so it is always the identity. Hence, e^{tA} is indeed orthogonal. $\square$

Notice that all of the theory of Section 7.4, as well as that of Section 7.3, is perfectly true with *complex numbers* as eigenvalues or entries in eigenvectors or the matrix A itself. In particular, the power series of Definition 7.4.1 makes sense for complex-valued matrices; we shall need this fact.

We shall examine in Section 7.5 the case of 2×2 matrices, where the mathematics can, surprisingly enough, be done directly from the power series of the exponentials.

7.5 Two by Two Matrices and the Bifurcation Diagram

Garrett Birkhoff says that according to his father, George Birkhoff, the most famous American scholar of differential equations:

> *If you understand linear equations,*
> *you almost understand differential equations;*
>
> *If you understand linear equations of degree 2,*
> *you almost understand all linear equations;*
>
> *Finally, you almost understand linear equations of degree 2*
> *if you understand those with constant coefficients.*

We find the first of these claims overoptimistic; the second and third, on the other hand, are certainly true. Second order linear differential equations with constant coefficients are ubiquitous. For that reason, we devote this section to them.

Second order linear equations with constant coefficients lead to the simplest system of the form $\mathbf{x}' = A\mathbf{x}$, with 2×2 matrices A:

$$\begin{bmatrix} x \\ y \end{bmatrix}' = \begin{bmatrix} a & b \\ c & d \end{bmatrix}\begin{bmatrix} x \\ y \end{bmatrix} = A\begin{bmatrix} x \\ y \end{bmatrix}.$$

In this section, we will consider what the trajectories for this differential equation look like in the xy-plane, when A is a real matrix. In principle, we already could figure out the answer from Sections 7.3 or 7.4, at least if the eigenvalues of A are distinct, but we want to look at what those sections say in more detail in this particular case.

Note that the only equilibrium point is at $(0,0)$. There are three cases of possible eigenvalues for a, b, c, and d real:

(1) real and distinct eigenvalues,

(2) complex eigenvalues (necessarily distinct since they are complex conjugate),

(3) a double eigenvalue (necessarily real).

You can confirm in Exercise 7.5#5 that with $\operatorname{tr} A = a + d$ and $\det A = ad - bc$ the characteristic equation is $\lambda^2 - (\operatorname{tr} A)\lambda + (\det A) = 0$, with *discriminant*

$$\Delta = (\operatorname{tr} A)^2 - 4\det A,$$

and that cases (1), (2), and (3) correspond respectively to $\Delta > 0$, $\Delta < 0$, and $\Delta = 0$.

1. *Real (nonzero, distinct) eigenvalues.* If $\Delta > 0$, there are two real distinct eigenvalues λ_1 and λ_2 and the corresponding eigenvectors $\mathbf{v}_1$ and $\mathbf{v}_2$ form a basis of $\mathbb{R}^2$. The general solution of $\mathbf{x}' = A\mathbf{x}$ can in that case be written, by Theorem 7.3.4, as

$$\mathbf{x}(t) = C_1 e^{\lambda_1 t}\mathbf{v}_1 + C_2 e^{\lambda_2 t}\mathbf{v}_2,$$

and the behavior depends on the signs of λ_1 and λ_2. We discuss here the results for distinct and nonzero real eigenvalues; those for the case with a double eigenvalue are discussed below in subsection 3, and those for the case where an eigenvalue is zero are discussed in subsection 4.

i. *Node source.* If $0 < \lambda_1 < \lambda_2$, then all trajectories tend to ∞ as $t \to \infty$ and to 0 as $t \to -\infty$. (See Figure 7.5.1.) Moreover, all trajectories leave $(0,0)$ tangentially to $\mathbf{v}_1$ except those that are multiples of $\mathbf{v}_2$ (Exercise 7.5#6a).

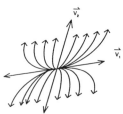

FIGURE 7.5.1. Node source.

ii. *Saddle.* If $\lambda_1 < 0 < \lambda_2$, then the trajectories with $C_1 = 0$ go from 0 to ∞ on the line in the direction of positive and negative multiples of $\mathbf{v}_2$, and those with $C_2 = 0$ go from ∞ to 0 on the line in the direction of multiples of $\mathbf{v}_1$; other trajectories are superpositions of these motions (Exercise 7.5#6b). (See Figure 7.5.2.)

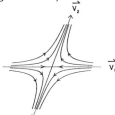

FIGURE 7.5.2. Saddle.

iii. *Node sink.* If $\lambda_1 < \lambda_2 < 0$, then all trajectories tend to 0 as $t \to \infty$ and to ∞ as $t \to -\infty$. (See Figure 7.5.3.) Moreover, all trajectories approach $(0,0)$ tangentially to $\mathbf{v}_2$ except those that are multiples of $\mathbf{v}_1$ (Exercise 7.5#6c).

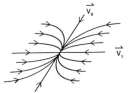

FIGURE 7.5.3. Node sink.

Remark. A node sink is exactly the same thing as a node source with time going backward.

2. *Complex eigenvalues.* Suppose the two eigenvalues of A are $\alpha \pm i\beta$, with $\beta > 0$. Of course, there exists a basis $\mathbf{v}_1$, $\mathbf{v}_2$ of eigenvectors, but the vectors are in $\mathbb{C}^2$, and it is not quite clear how to use them to describe the behavior of trajectories in the real phase plane. The following trick gives a way:

Observe first that if $\mathbf{v}$ is an eigenvector for $\lambda_1 = \alpha + i\beta$, then $\overline{\mathbf{v}}$ is an eigenvector for $\lambda_2 = \overline{\lambda}_1 = \alpha - i\beta$. Let us write $\mathbf{v} = \mathbf{w}_1 - i\mathbf{w}_2$, where $\mathbf{w}_1$ and $\mathbf{w}_2$ are real. Neither $\mathbf{w}_1$ or $\mathbf{w}_2$ are eigenvectors, but they are linearly independent (Exercise 7.5#7) form a basis of $\mathbb{R}^2$; it turns out that in that basis A has a rather pleasant form. Indeed,

$$A\mathbf{w}_1 + iA\mathbf{w}_2 = A(\mathbf{w}_1 - i\mathbf{w}_2) = A\mathbf{v} = \lambda_1\mathbf{v} = (\alpha + i\beta)(\mathbf{w}_1 - i\mathbf{w}_2)$$
$$= (\alpha\mathbf{w}_1 + \beta\mathbf{w}_2) + i(\alpha\mathbf{w}_2 - \beta\mathbf{w}_1),$$

and identifying real and imaginary parts, we find

$$A\mathbf{w}_1 = \alpha\mathbf{w}_1 + \beta\mathbf{w}_2$$
$$A\mathbf{w}_2 = \beta\mathbf{w}_1 - \alpha\mathbf{w}_2,$$

i.e., in the basis $\mathbf{w}_1$, $\mathbf{w}_2$, the linear transformation A has the matrix

$$A_1 = \begin{bmatrix} \alpha & -\beta \\ \beta & \alpha \end{bmatrix}.$$

Alternately, let $P = [\mathbf{w}_1, \mathbf{w}_2]$, with $A_1 = P^{-1}AP$.

It is easy to compute e^{tA_1}, because

$$A_1 = \alpha I + \beta \begin{bmatrix} 0 & -1 \\ 1 & 0 \end{bmatrix},$$

and the two matrices in this sum commute since the identity I commutes with everything. By Property 7.4.7 and a calculation as in Example 7.3.6, we see that

$$e^{tA_1} = e^{t\alpha} \begin{bmatrix} \cos \beta t & -\sin \beta t \\ \sin \beta t & \cos \beta t \end{bmatrix}.$$

Now it is quite easy to see what the trajectory of a point looks like. It simultaneously turns at speed β radians per unit time, under the influence of the rotation matrix above, and approaches or leaves 0, under the influence of $e^{t\alpha}$, depending on whether $\alpha < 0$ or $\alpha > 0$. This leads to the following classification:

i. *Spiral sink.* If $\alpha < 0$, the trajectories spiral in to $(0,0)$. (See Figure 7.5.4.)

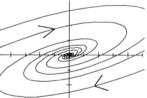

FIGURE 7.5.4. Spiral sink.

ii. *Center.* If $\alpha = 0$, the trajectories turn periodically on an ellipse and are called "centers," because each ellipse is centered at $(0,0)$. (See Figure 7.5.5.)

FIGURE 7.5.5. Center.

iii. *Spiral source.* If $\alpha > 0$, the trajectories spiral out to ∞. (See Figure 7.5.6.)

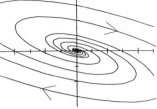

FIGURE 7.5.6. Spiral source.

The direction of rotation of the spiral trajectories depends on the sign of c, the lower left entry of the matrix A: if $c > 0$, the rotation is *counterclockwise*; if $c < 0$, the rotation is *clockwise*. (You will prove this fact in Exercise 7.5#8.)

Centers are exceptional, but they turn up often in differential equations describing conservative mechanical systems as discussed in Chapter 6.5. Exercises 7.5#9, and especially 8.1#12, explain why this will occur.

THE TRACE/DETERMINANT PLANE; BIFURCATION DIAGRAM

The information we have assembled for the behavior of trajectories under real or complex eigenvalues leads us to begin Figure 7.5.7, a picture of

the plane determined by $\operatorname{tr} A$ *and* $\det A$.

This plane (which is *not* an xy phase plane) merely helps us study how $\operatorname{tr} A$ and $\det A$ and the relationship between them determines the behavior of trajectories to $\mathbf{x}' = A\mathbf{x}$.

The $(\operatorname{tr} A, \det A)$-plane is divided by the axes $\det A = 0$ and $\operatorname{tr} A = 0$ and by the parabola

$$(\operatorname{tr} A)^2/4 = \det A,$$

the dividing locus between the cases $\Delta > 0$ and $\Delta < 0$. We shall examine more closely all of these divisions.

The *sketches* in each region of the $(\operatorname{tr} A, \det A)$-plane of Figure 7.5.7 are just insets showing a typical drawing (in the xy phase plane) for the *trajectories* to $x' = Ax$, for values of A corresponding to that region. The conventional labels for each type of behavior accompany each of these insets.

We still need to understand the remaining possibilities; i.e., what occurs along boundaries of the regions already discussed:

3. along the parabola where $\Delta = 0$,

96   7. Systems of Linear Differential Equations

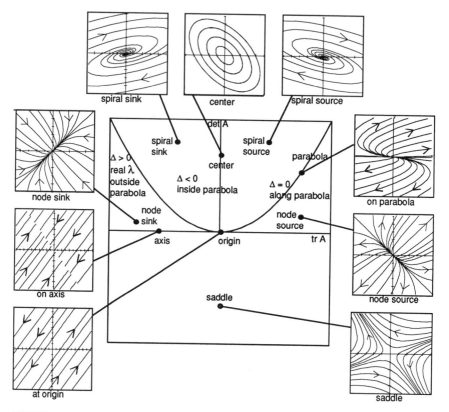

FIGURE 7.5.7. Bifurcation diagram in $(\operatorname{tr} A, \det A)$-plane for the two-dimensional
equation $\mathbf{x}' = A\mathbf{x}$.

and the division within the case of real eigenvalues between nodes and
saddles:

4. along the horizontal axis $\det A = 0$.

3. *Double eigenvalue.* Along the parabola, $\Delta = 0$. If there is only one
(double) eigenvalue λ, let $\mathbf{v}_1 = \begin{bmatrix} 1 \\ 0 \end{bmatrix}$ be an eigenvector, and let $\mathbf{v}_2$ be any
vector linearly independent of $\mathbf{v}_1$. Then $\mathbf{v}_1$ and $\mathbf{v}_2$ form a basis, and in that
basis the linear transformation A has matrix

$$A_1 = \begin{bmatrix} \lambda & \alpha \\ 0 & \lambda \end{bmatrix}$$

for some appropriate number α. You can see this by noting that the first
column is correct since $A\mathbf{v}_1 = \lambda\mathbf{v}_1$ and that, regardless of the entries in the

second column, the matrix A_1 will be upper triangular. By Theorem L6.8, A has the eigenvalues (i.e., the two λ's) along the diagonal.

It is easy to compute e^{tA_1}, since $A_1 = \lambda I + \alpha \begin{bmatrix} 0 & 1 \\ 0 & 0 \end{bmatrix}$, and the two matrices commute. So by Property 7.4.7 of the exponential and Example 7.4.4, we see that

$$e^{tA_1} = e^{t\lambda} \begin{bmatrix} 1 & \alpha t \\ 0 & 1 \end{bmatrix},$$

and there are two quite different cases: $\alpha \neq 0$ and $\alpha = 0$. Note that the first corresponds to there being only one eigenvector, and the second to all vectors being eigenvectors. Both of these possibilities are allowed, since there is one (double) root of the characteristic polynomial. However, the first possibility is *usual*, and the second *exceptional*. Indeed, the space of 2×2 matrices is four-dimensional, and those whose characteristic polynomial has a double root are defined by the one equation $(\operatorname{tr} A)^2 = 4 \det A$, so they form a three-dimensional locus. However, if there is a single eigenvalue and two linearly independent eigenvectors, the matrix is simply a multiple of the identity, and the set of such is one-dimensional (given by the multiple in question).

i. In the (exceptional) case of infinitely many eigenvectors, the drawing in the phase plane looks like Figure 7.5.8, as should be clear.

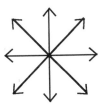 

FIGURE 7.5.8. Phase plane behavior with double eigenvalue at zero.

ii. In the (usual) case of only one eigenvector, the motion of a point under e^{tA_1} is a superposition of the part due to the exponential, i.e., exponential attraction or repulsion depending on whether $\lambda < 0$ or $\lambda > 0$, and motion under the matrix, which is simply motion at constant speed α along the horizontal line through the initial condition. Note that if you go back to the initial variables, this part becomes motion at constant speed along the line parallel to the eigenvector $\mathbf{v}_1$ and through the initial point.

So if $\lambda = 0$, we find a "shearing" motion, and the drawing in the phase plane appears as in Figure 7.5.9.

FIGURE 7.5.9. Phase plane behavior with single eigenvector and zero eigenvalue.

If $\lambda \neq 0$, then the exponential term superposes its effect on the above linear motion, leading to Figure 7.5.10.

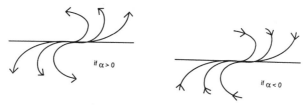

FIGURE 7.5.10.Phase plane behavior with single eigenvector and nonzero eigenvalue.

4. *A zero eigenvalue.* The remaining boundary between regions we have studied in the $(\operatorname{tr} A, \det A)$-plane is along the horizontal axis, where $\det A = 0$; the origin has been included in the preceding discussion of the parabola; elsewhere along this axis one of the eigenvalues must be zero (as a result of Theorems L5.2 and L6.8). These are the *degenerate* cases, because $\mathbf{x}' = A\mathbf{x}$ vanishes on some *line*, which becomes a line of equilibria. The trajectories appear as in Figure 7.5.11.

FIGURE 7.5.11. Phase plane behavior with zero eigenvalue.

But with the slightest perturbation, this whole line of equilibria disappears, and you get either saddles, sinks, centers, or nodes (depending on $\operatorname{tr} A$ and the perturbation).

We have now described phase plane behavior for every region on the bifurcation diagram of Figure 7.5.7, in the $(\operatorname{tr} A, \det A)$-plane.

It may be difficult to imagine how to continuously go from the motions with eigenvectors (nodes) across the parabola to the spiraling motions; what happens in general is that the eigenvectors make a smaller and smaller angle, until they coincide and disappear. The sequence of drawings in Figure 7.5.12 illustrates the change.

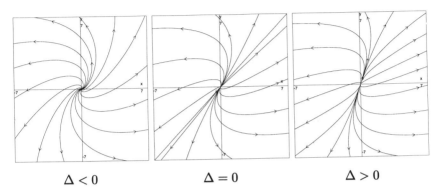

$$\Delta < 0 \qquad\qquad \Delta = 0 \qquad\qquad \Delta > 0$$

FIGURE 7.5.12. How a spiral becomes a node as the matrix changes to cross the parabola in the trace/determinant plane for $x' = \alpha x - y$, $y' = x + y$.

Example 7.5.1. Consider the system

$$\begin{bmatrix} x \\ y \end{bmatrix}' = \begin{bmatrix} 0 & 1 \\ -1 & \alpha \end{bmatrix} \begin{bmatrix} x \\ y \end{bmatrix}.$$

and observe how the choice of α affects the trajectories. Even before finding the eigenvalues, we can observe the following:

$(0,0)$ is the only equilibrium;
$\operatorname{tr} A = \alpha$;
$\det A = 1$;
$\Delta = \alpha^2 - 4 = 0$ for $\alpha = -2$ or 2;
c is negative, so spirals are clockwise.

Therefore, we can list how the phase plane behavior depends on α:

$$\begin{array}{ll} \alpha < -2, & \text{node sink,} \\ -2 < \alpha < 0 & \text{spiral sink,} \\ \alpha = 0 & \text{center,} \\ 0 < \alpha < 2 & \text{spiral source,} \\ \alpha > 2 & \text{node source.} \end{array}$$

For $\alpha = \pm 2$, A is not a multiple of the identity, so the trajectories appear respectively as in Figure 7.5.13.

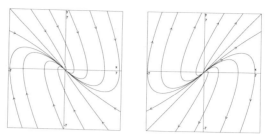

FIGURE 7.5.13.

As α increases from $-\infty$ to $+\infty$, the trajectories change as you cross the $(\operatorname{tr} A, \det A)$-plane from left to right, along the line $\det A = 1$, as shown in Figure 7.5.14.

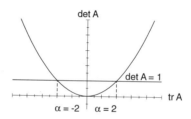

FIGURE 7.5.14.

Some insight into the role of α and the way the eigenvectors fit into the diagram can be gained from solving the characteristic equation

$$\lambda^2 - \alpha\lambda + 1 = 0, \qquad \lambda_{1,2} = \frac{\alpha \pm \sqrt{\alpha^2 - 4}}{2}.$$

Because of the fortuitous form of the matrix in this example, we can read the eigenvectors directly from the definition of eigenvalue and eigenvector: $A\mathbf{v} = \lambda\mathbf{v}$.

$$\begin{bmatrix} 0 & 1 \\ -1 & \alpha \end{bmatrix} \begin{bmatrix} x \\ y \end{bmatrix} = \lambda \begin{bmatrix} x \\ y \end{bmatrix},$$

so the first equation of the multiplication tells us that the eigenvectors have form $y = \lambda x$, and we have

$$y = [(\alpha \pm \sqrt{\alpha^2 - 4})/2]x.$$

You can show in Exercise 7.5#12 that this eigenvector equation is equivalent to

$$y^2 - \alpha xy + x^2 = 0,$$

which is the equation of a *surface* in α xy-space, a third degree equation in x, y, and α.

This surface is called the *Whitney umbrella* (after Hassler Whitney, a contemporary mathematician at the Institute for Advanced Study), and it looks like Figure 7.5.15.

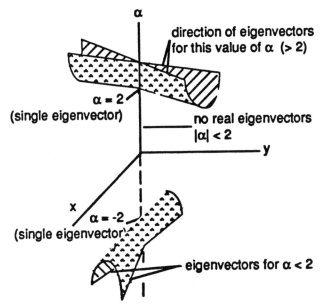

FIGURE 7.5.15. Whitney umbrella, showing the position of the eigenvectors in the xy phase plane according to the value of the parameter α. ▲

7.6   Eigenvalues and Global Behavior

The most important thing to observe about the bifurcation diagram of Section 7.5, which is perhaps a surprising observation, is that

the behavior of solutions is controlled by the signs of the *real* parts of the eigenvalues (in particular, by the eigenvalues themselves if they are real).

You can easily check, from looking at the diagram, that the origin is a *sink* if both eigenvalues have negative real part;

a *source* if both eigenvalues have positive real part;

a *saddle* if the eigenvalues have opposite sign

(in which case they are necessarily real).

In this section we will generalize this classification to arbitrary dimension.

SINKS AND SOURCES

The following theorem provides a criterion under which all solutions of a linear differential equation with constant coefficients will tend to 0. Typically, this is the kind of equation we will meet when describing dynamical systems with friction, or electrical systems with resistance. In many real systems, such as the electrical distribution system of a geographical region, it is essential to make sure that the conditions of this theorem are satisfied, to guarantee stability of the system.

Theorem 7.6.1. *If A is an $n \times n$ matrix and all the eigenvalues of A have negative real parts, then all solutions $\mathbf{x}(t)$ of $\mathbf{x}' = A\mathbf{x}$ satisfy*

$$\lim_{t \to \infty} \mathbf{x}(t) \to 0.$$

Remarks.

i. The origin is called a *sink* when Theorem 7.6.1 holds.

ii. There is an exactly analogous statement when all eigenvalues have a *positive* real part, turning time backward; then the origin is called a *source*, and $\lim_{t \to \infty} \mathbf{x}(t) \to 0$.

iii. The converse of Theorem 7.6.1 is also correct (Exercise 7.6#2).

Proof. The theorem is easy to prove *if the matrix A is diagonalizable over $\mathbb{C}$*: just switch to a basis of eigenvectors, and all coefficients decrease exponentially. The (exceptional) case where A is not diagonalizable is, unfortunately, a good deal harder. However, the following argument works in general:

Look at all the eigenvalues and take $-\Lambda$ to be the largest real part of an eigenvalue; that is, let

$$\Lambda = -\sup\{\operatorname{Re} \lambda_i \mid \lambda_i \text{ is an eigenvalue of } A\}.$$

From Appendix L, we know

Theorem L6.19. *For any ε, we can find a complex basis $\mathbf{v}_1, \ldots, \mathbf{v}_n$ such that in that basis, the matrix of the linear transformation A is upper-triangular, with the eigenvalues along the diagonal, and all off-diagonal terms of absolute value less than ε.*

That is, let P be the change of basis matrix $[\mathbf{v}_1, \ldots, \mathbf{v}_n]$; then $A_1 = P^{-1}AP$ is upper-triangular as described.

Write $A_1 = B + C$, where B is the diagonal part, and C is the strictly upper-triangular part. Please note that A_1, B, and C are usually complex matrices, and so we must use complex inner products (Appendices L3 and L4).

Lemma. *For $\mathbf{y} \in \mathbb{R}^n$, and the ε of Theorem L6.9, we have the inequalities*

$$\operatorname{Re}(B\mathbf{y} \cdot \mathbf{y}) \leq -\Lambda \|\mathbf{y}\|^2 \quad and \quad |\operatorname{Re}(C\mathbf{y} \cdot \mathbf{y})| \leq n\varepsilon \|\mathbf{y}\|^2.$$

Proof of Lemma. For the first inequality, since B is diagonal with the eigenvalues λ_i along the diagonal, we see

$$\operatorname{Re}(B\mathbf{y} \cdot \mathbf{y}) = \operatorname{Re}\left(\sum \lambda_i y_i \overline{y}_i\right) = \sum \operatorname{Re}(\lambda_i)|y_i|^2 \leq -\Lambda \|\mathbf{y}\|^2.$$

For the second, observe that $\|C\| \leq n\varepsilon$, so using Schwarz's inequality (Theorem L3.3) and then Exercise 7.1-7.2#7, we find

$$|\operatorname{Re}(C\mathbf{y} \cdot \mathbf{y})| \leq |C\mathbf{y} \cdot \mathbf{y}| \leq \|C\mathbf{y}\| \, \|\mathbf{y}\| \leq \|C\| \, \|\mathbf{y}\|^2 \leq n\varepsilon \|\mathbf{y}\|^2. \qquad \square$$

Remark. In Exercise 7.6#3 you show the sharper inequality $\|C\| \leq \sqrt{n(n-1)/2}\,\varepsilon$.

Proof of Theorem, continued. Set $\mathbf{y} = P\mathbf{x}$. Then $\mathbf{y}$ obeys the differential equation $\mathbf{y}' = A_1\mathbf{y}$, so

$$\frac{d}{dt}(\|\mathbf{y}(t)\|^2) = 2\operatorname{Re}(\mathbf{y}' \cdot \mathbf{y}) = 2\operatorname{Re}(A_1\mathbf{y} \cdot \mathbf{y}) = 2(\operatorname{Re}(B\mathbf{y} \cdot \mathbf{y}) + \operatorname{Re}(C\mathbf{y} \cdot \mathbf{y}))$$
$$\leq 2(-\Lambda + n\varepsilon)\|\mathbf{y}\|^2.$$

This means that $\|\mathbf{y}(t)\|^2$ is a lower fence for the differential equation

$$\mathbf{u}' = 2(-\Lambda + n\varepsilon)\mathbf{u},$$

with solution $\mathbf{u}(t) = e^{2(-\Lambda + n\varepsilon)t}\mathbf{u}(0)$. If ε is chosen so small that $n\varepsilon - \Lambda$ is negative, then since

$$\|\mathbf{y}(t)\|^2 \leq e^{2(-\Lambda + n\varepsilon)t}\|\mathbf{y}(0)\|^2,$$

we see that $\mathbf{y}(t) \to 0$ as $t \to \infty$. $\quad \square$

Theorem 7.6.1 describes solutions of the differential equation $\mathbf{x}' = A\mathbf{x}$ if all the eigenvalues of A have a negative real part. Two points should be noted:

(i) All solutions decrease exponentially, but *the basis with respect to which lengths must be measured in order for this to occur may be pretty distorted.*

(ii) The exponential rate at which solutions decrease may not be exactly the largest real part of an eigenvalue.

Example 7.6.2. illustrates both (i) and (ii). It shows a distorted basis, and its solutions decrease like $C\,te^{-t}$, although this does go to zero as $t \to \infty$ since the decreasing exponential wins over the increasing polynomial; still, it does not decrease quite as fast as e^{-t}. This explains why the rate of decrease we will get is only $e^{t(n\varepsilon-\Lambda)}$.

Example 7.6.2. Consider the differential equation

$$x' = -x + 100\,y$$
$$y' = -y.$$

From Section 7.5 we know all about the solutions to this equation; its solution with $x(0) = 0$, $y(0) = 1$ is

$$x(t) = 100\,te^{-t}$$
$$y(t) = e^{-t}.$$

At time 0, it has length 1 (with respect to the standard basis), and at time 1, it has length $\sqrt{10001}/e = 36.789\ldots$. The length of the solution certainly does not appear to be decreasing, and it is not if it is measured with respect to the standard basis. If, however, you measure lengths with respect to a new basis

$$\mathbf{v}_1 = \mathbf{e}_1, \qquad \mathbf{v}_2 = \mathbf{e}_2(\varepsilon/100),$$

you will find that all solutions decrease if $|\varepsilon| \leq 2$ (Exercise 7.6#4).

Such a new basis is quite distorted: one basis vector is at least 50 times longer than the other. ▲

SADDLES

What happens if the real parts of eigenvalues are both positive and negative? Well, if no eigenvalues have a real part 0, then the space $\mathbb{R}^n$ breaks up into two subspaces, V_+ and V_-, with the property that $A(V_+) = V_+$ and $A(V_-) = V_-$, so that in each of these subspaces Theorem 7.6.1. applies. In the usual case where A is diagonalizable, just take V_+ to be the span of the eigenvectors with eigenvalues having a positive real part, and V_- the span of the eigenvectors corresponding to eigenvalues with a negative real part. The general case is, as usual, harder and follows almost immediately from Theorem L6.12.

Theorem 7.6.3. *Let A be an $n \times n$ matrix, with no eigenvalues having a real part 0. Let p_A be the characteristic polynomial of A, and write $p_A = p_- p_+$, where the roots of p_- are all the roots of p_A that have a negative real part, and the roots of p_+ are all the roots of p_A that have a positive real part. Let $V_\pm = \ker(p_\pm(A))$. Then for the differential equation $\mathbf{x}' = A\mathbf{x}$,*

(a) *Any solution $\mathbf{x}(t)$ can be written $\mathbf{x}(t) = \mathbf{x}_-(t) + \mathbf{x}_+(t)$ with $\mathbf{x}_-(t)$ and $\mathbf{x}_+(t)$ in V_- and V_+, respectively;*

(b) *$\mathbf{x}_-(t)$ tends to $\mathbf{0}$ as $t \to +\infty$ and $\mathbf{x}_+(t)$ tends to $\mathbf{0}$ as $t \to -\infty$.*

Proof. By Theorem L6.12, any vector $\mathbf{x} \in \mathbb{R}^n$ can be written uniquely as

$$\mathbf{x} = \mathbf{x}_- + \mathbf{x}_+, \text{ with } \mathbf{x}_- \in V_- \text{ and } \mathbf{x}_+ \in V_+.$$

This follows from the fact that p_- and p_+ are relatively prime since they have no common roots, hence no common factors.

Suppose that the initial value $\mathbf{x}(0) = \mathbf{x}_-(0) + \mathbf{x}_+(0)$ of a solution is decomposed as above. Since $A(V_\pm) \subset V_\pm$, the solution $\mathbf{x}_\pm(t)$ with initial condition $\mathbf{x}_\pm(0)$ stays in $V_\pm$ for all t, and in this subspace obeys a differential equation with constant coefficients and all eigenvalues negative or positive respectively. This follows from the fact that the characteristic polynomial of A restricted to $V_\pm$ is $p_\pm$. Finally, we use the fact that $\mathbf{x}(t) = \mathbf{x}_-(t) + \mathbf{x}_+(t)$ is just the superposition of solutions, Theorem 7.2.1. $\square$

The technique used in the proof of Theorem 7.6.3 can be extended considerably to isolate the behavior coming from any particular eigenvalue or from other groupings besides the one into roots with positive and negative real parts.

A ZOOLOGICAL DESCRIPTION OF LINEAR DIFFERENTIAL EQUATIONS IN $\mathbb{R}^3$

According to the development so far, we see that linear differential equations with constant coefficients in $\mathbb{R}^3$ will fall into the following main classes:

(i) those where all three eigenvalues a have negative real part (*sinks*);

(ii) those with two eigenvalues having a negative real part and one positive eigenvalue (*saddles* of type 2,1);

(iii) those with one negative eigenvalue and two eigenvalues having a positive real part (*saddles* of type 1,2);

(iv) those where all three eigenvalues have a positive real part (*sources*).

In the first case, all solutions tend to **0** as $t \to \infty$ and to infinity as $t \to -\infty$; in the second case, there is a plane of solutions tending to **0** as $t \to +\infty$ and a line of solutions tending to **0** as $t \to -\infty$, with all other solutions superpositions of these; the third and fourth cases are like the second and first, respectively, with time running backward.

There are various exceptional equations that do not fit into the description above, namely,

(v) those with a pair of purely imaginary eigenvalues (*centers*),

(vi) those with a zero eigenvalue.

7.7   Nonhomogeneous Linear Equations

Very frequently, the linear differential equations to be solved are not simply

$$\mathbf{x}' = A\mathbf{x},$$

which correspond to free physical systems, but rather

$$\mathbf{x}' = A\mathbf{x} + \mathbf{g}(t),$$

which correspond to driven physical systems.

The question of finding a particular solution $\mathbf{u}_p(t)$ for the nonhomogeneous equation of the driven system can be attacked again as in the one dimensional case of Sections 2.2 and 2.3 of Part I.

There is a formula for solving such systems: the *variation of parameters formula*, which we will see in the second part of this section. Although of great importance, variation of parameters generally leads to unpleasant computations in practical cases. For special cases of the "driving function" $\mathbf{g}(t)$, there is an alternative method for finding a particular solution $\mathbf{u}_p(t)$ which is usually much simpler: the *method of undetermined coefficients*. This method really comes into its own when the driving function $\mathbf{g}(t)$ is given as an infinite superposition of such special driving functions; this occurs, in particular, if the function is given as a Fourier series, a Fourier transform, or a Laplace transform (see the third subsection below, on Limitations and Extensions of the Method of Undetermined Coefficients).

METHOD OF UNDETERMINED COEFFICIENTS

The special cases for which the method of undetermined coefficients is appropriate occur when the entries of $\mathbf{g}(t)$ are linear combinations of functions of the form $t^k e^{at}$; these include polynomials (the case $a = 0$), plain exponentials (the case $k = 0$), and functions such as $t^k \sin \omega t$ and $t^k \cos \omega t$, because of the formulas

$$\sin \omega t = \frac{1}{(2i)}(e^{i\omega t} - e^{-i\omega t}), \quad \cos \omega t = \frac{1}{2}(e^{i\omega t} + e^{-i\omega t}).$$

Using the superposition principle, Theorem 7.2.5, we see that we may consider also the case where $\mathbf{g}(t)$ is of the form $\mathbf{p}(t)e^{at}$, where $\mathbf{p}(t)$ is a vector of polynomial functions. In that case, try setting

$$\mathbf{x}(t) = \mathbf{q}(t)\,e^{at},$$

where $\mathbf{q}(t)$ is an unknown vector of polynomial functions of the same degree as $\mathbf{p}(t)$. Substituting into the equation and equating like terms, you will be led to a system of linear (nondifferential) equations for the coefficients of $\mathbf{q}$ that will in fact have the same number of equations as variables. If the vector $\mathbf{g}(t)$ contains terms in sine and cosine, it may be simpler to stick to sines and cosines rather than passing to complex exponentials.

Example 7.7.1. Solve the differential equation

$$\begin{bmatrix} x \\ y \end{bmatrix}' = \begin{bmatrix} -2 & -1 \\ 5 & 4 \end{bmatrix} \begin{bmatrix} x \\ y \end{bmatrix} + \begin{bmatrix} 0 \\ e^{2t} \end{bmatrix}. \tag{30}$$

There are two steps to solving this differential equation:

(i) Solve the associated homogeneous equation

$$\begin{bmatrix} x \\ y \end{bmatrix}' = \begin{bmatrix} -2 & -1 \\ 5 & 4 \end{bmatrix} \begin{bmatrix} x \\ y \end{bmatrix}.$$

Using the techniques discussed in Sections 7.2 and 7.3, that solution is

$$\begin{bmatrix} x \\ y \end{bmatrix} = C_1\,e^{3t} \begin{bmatrix} 1 \\ -5 \end{bmatrix} + C_2\,e^{-t} \begin{bmatrix} 1 \\ -1 \end{bmatrix}. \tag{31}$$

(ii) Solve the nonhomogeneous equation (30) with

$$g(t) = \begin{bmatrix} 0 \\ e^{2t} \end{bmatrix} = \begin{bmatrix} 0 \\ 1 \end{bmatrix} e^{2t}.$$

That is, try substituting

$$\begin{bmatrix} x \\ y \end{bmatrix} = \mathbf{q}(t)e^{2t} \quad \text{with} \quad \mathbf{q}(t) = \begin{bmatrix} a \\ b \end{bmatrix}.$$

By solving the resulting linear equations for the coefficients a and b, you will find $a = \frac{1}{3}$, $b = -\frac{4}{3}$. The final solution is

$$\begin{bmatrix} x \\ y \end{bmatrix} = C_1\,e^{3t} \begin{bmatrix} 1 \\ -5 \end{bmatrix} + C_2\,e^{-t} \begin{bmatrix} 1 \\ -1 \end{bmatrix} + e^{2t} \begin{bmatrix} 1/3 \\ -4/3 \end{bmatrix}. \quad \blacktriangle$$

Example 7.7.2. Solve the differential equation

$$\begin{bmatrix} x \\ y \end{bmatrix}' = \begin{bmatrix} -2 & -1 \\ 5 & 4 \end{bmatrix} \begin{bmatrix} x \\ y \end{bmatrix} + \begin{bmatrix} 0 \\ e^t + te^t \end{bmatrix}. \tag{32}$$

(i) Solve the associated homogeneous equation as in Example 7.7.1; the solution is equation (31).

(ii) Using undetermined coefficients, solve the nonhomogeneous equation (32) with

$$\mathbf{g}(t) = \begin{bmatrix} 0 \\ 1+t \end{bmatrix} e^t \quad \text{and} \quad \mathbf{x}(t) = \begin{bmatrix} a+tb \\ c+td \end{bmatrix} e^t.$$

Using this equation for $\mathbf{x}$, you can substitute into the original equation, which yields $a = \frac{1}{4}$, $b = \frac{1}{4}$, $c = -1$, and $d = -\frac{3}{4}$. The final solution is

$$\mathbf{x}(t) = C_1 e^{3t} \begin{bmatrix} 1 \\ 5 \end{bmatrix} + C_2 e^{-t} \begin{bmatrix} 1 \\ 1 \end{bmatrix} + e^t \begin{bmatrix} \frac{1}{4} + \frac{1}{4}t \\ -1 - \frac{3}{4}t \end{bmatrix}. \quad \blacktriangle$$

RESONANCE

Even with the nonhomogeneous term $\mathbf{g}(t)$ of the proper form, the method of undetermined coefficients does not always work. Let us see why, in the simplest case: assume the driving term $\mathbf{g}(t)$ is of the form $\mathbf{G}\,e^{\alpha t}$, where $\mathbf{G}$ is a constant vector. Then, if we substitute $\mathbf{x}(t) = \mathbf{Q}\,e^{\alpha t}$, we find

$$\mathbf{x}'(t) = \alpha \mathbf{Q}\,e^{\alpha t} = A\mathbf{Q}\,e^{\alpha t} + \mathbf{G}\,e^{\alpha t},$$

which leads to

$$(\alpha I - A)\mathbf{Q} = \mathbf{G}.$$

This is the system of linear equations mentioned above, which can usually be solved *only if the matrix* $(\alpha I - A)$ *is invertible*:

$$\mathbf{Q} = (\alpha I - A)^{-1}\mathbf{G}. \tag{33}$$

The *inverse* $(\alpha I - A)^{-1}$ *only exists if* α *is not an eigenvalue of* A. In general, you can expect trouble with undetermined coefficients if the driving term is an exponential term with the coefficient of the exponent equal to one of the eigenvalues.

The way around this problem is fairly simple, just as in Part I, Chapter 2: it can be shown that if instead of trying for a solution of the form $\mathbf{Q}\,e^{\alpha t}$, you had given yourself a bit more leeway and tried a solution of the form $\mathbf{P}(t)e^{\alpha t}$, with $\mathbf{P}$ a vector of polynomials of degree equal to the multiplicity of α as a root of the characteristic polynomial of A, then the undetermined coefficients procedure does work.

Example 7.7.3. Solve the differential equation

$$\begin{bmatrix} x \\ y \end{bmatrix}' = \begin{bmatrix} -2 & -1 \\ 5 & 4 \end{bmatrix} \begin{bmatrix} x \\ y \end{bmatrix} + \begin{bmatrix} 0 \\ e^{-t} \end{bmatrix}. \tag{34}$$

This will not work as well as Examples 7.7.1 and 7.7.2, because -1 is an eigenvalue of $\begin{bmatrix} -2 & -1 \\ 5 & 4 \end{bmatrix}$. To solve the nonhomogeneous equation (34), try

$$\begin{bmatrix} x \\ y \end{bmatrix} = \begin{bmatrix} a + bt \\ c + dt \end{bmatrix} e^{-t}.$$

Substitution yields $a = -c - \frac{1}{4}$, $b = \frac{1}{4}$, $c = c$, $d = -\frac{1}{4}$.

Combining these results with the solution (31) to the homogeneous equation, the final solution to equation (34) is

$$\begin{bmatrix} x \\ y \end{bmatrix} = C_1 e^{3t} \begin{bmatrix} 1 \\ -5 \end{bmatrix} + C_2 e^{-t} \begin{bmatrix} 1 \\ -1 \end{bmatrix} + e^{-t} \begin{bmatrix} -c - \frac{1}{4} + \frac{1}{4}t \\ c - \frac{1}{4}t \end{bmatrix}.$$

It looks as if this solution depends on three arbitrary constants, but in fact c and C_2 play the same role: the freedom in the solution to the inhomogeneous equation is a solution to the associated homogeneous equation. ▲

This comment about degrees of freedom is not special to this example: if undetermined coefficients lead to a family of solutions depending on a parameter, then this freedom will always correspond to adding a solution of the associated homogeneous equation, since the difference of two solutions of the inhomogeneous equation is a solution of the associated homogeneous equation.

The phenomenon in which the driving term includes $e^{\lambda t}$ for some eigenvalue λ of the matrix A is called *resonance*, and it is very important in applications, either because resonance is wanted (for instance, to amplify a radio signal at a particular frequency) or to be avoided (for instance, in a building buffeted by winds or earthquakes).

In many cases of practical importance, such as conservative mechanical systems without friction or electrical systems without resistors, all the eigenvalues of the corresponding matrices describing the system are purely imaginary. If the system is driven with a harmonically oscillating term $e^{i\omega t}$ (or $\sin \omega t$) and if $i\omega$ is not one of the eigenvalues of A, then you can expect a response that is also harmonically oscillating, i.e., bounded. But if the frequency ω of the driving term is an eigenvalue of A, then the response will be a polynomial times an oscillating term and will grow without bound.

This makes it sound as if there is a dichotomy: resonance or no resonance. But if you drive a system *near* resonance, the response becomes large. For instance, the equation

$$x'' + x = \sin \omega t$$

with general solution

$$x(t) = \frac{1}{1 - \omega^2} \sin \omega t + a \sin t + b \cos t$$

has no periodic solution when $\omega = \pm 1$, and when ω is close to ± 1, the solutions all exhibit large oscillations. Note that the first order system associated to this equation is

$$\begin{bmatrix} x \\ y \end{bmatrix}' = \begin{bmatrix} 0 & 1 \\ -1 & 0 \end{bmatrix} \begin{bmatrix} x \\ y \end{bmatrix} + \begin{bmatrix} 0 \\ \sin \omega t \end{bmatrix}$$

with eigenvalues $\pm i$, which corresponds to $\omega = \pm 1$. Exploration of the behavior of this system is the topic of Exercise 7.7#8.

LIMITATIONS AND EXTENSIONS OF THE METHOD OF UNDETERMINED COEFFICIENTS

A more serious limitation of the method of undetermined coefficients is that the driving terms are restricted; if the driving term were

$$1/(1+t^2) \quad \text{or} \quad \sin(t^2) \quad \text{or} \quad e^{t^2},$$

the method simply does not seem to work. Actually, it is so convenient that people have tried really hard to make it work anyway, and they have succeeded. The key point is that the driving term does not have to be *one* term of the form $p(t)\,e^{at}$; it can be a sum of such terms. When you start looking at sums of such terms, it is amazing how many functions you can approximate.

In fact, if you are willing to allow infinite superpositions, i.e., to consider driving terms of the form

$$\mathbf{g}(t) = \sum \mathbf{a}_n\, e^{int} \qquad \text{(Fourier series)},$$

or continuous superpositions, for instance,

$$\mathbf{g}(t) = \int \mathbf{a}(s)\, e^{ist} ds \qquad \text{(Fourier transforms)}$$

or

$$\mathbf{g}(t) = \int \mathbf{a}(s)\, e^{-st} ds \qquad \text{(Laplace transforms)},$$

you can get practically any function $\mathbf{g}$ at all. The first form is suitable for periodic driving forces, the second for functions that decrease at both $+\infty$ and $-\infty$, and the third for functions that decrease at $+\infty$ but are not controlled near $-\infty$.

Example 7.7.4. Suppose you want to solve

$$\begin{bmatrix} x \\ y \end{bmatrix}' = \begin{bmatrix} -2 & -1 \\ 5 & 4 \end{bmatrix} \begin{bmatrix} 0 \\ \sin(t)/t \end{bmatrix}. \qquad (35)$$

A priori, this seems like something that cannot be solved by undetermined coefficients. But observe that

$$\frac{\sin(t)}{t} = \frac{1}{2} \int_{-1}^{1} e^{ist} ds,$$

as you can check by computing the integral. If we try to solve this equation by undetermined coefficients, i.e., if we set

$$x(t) = \int_{-1}^{1} a(s) e^{ist} ds,$$

$$y(t) = \int_{-1}^{1} b(s) e^{ist} ds,$$

and substitute into the equation, we find

$$\int_{-1}^{1} is\, a(s) e^{ist} ds = \int_{-1}^{1} -2a(s) e^{ist} ds - \int_{-1}^{1} b(s) e^{ist} ds,$$

$$\int_{-1}^{1} is\, b(s) e^{ist} ds = \int_{-1}^{1} 5a(s) e^{ist} ds + \int_{-1}^{1} 4b(s) e^{ist} ds + \frac{1}{2} \int_{-1}^{1} e^{ist} ds.$$

Combining integrands and factoring out e^{ist}, you end up with the system of linear algebraic equations

$$(is + 2)a(s) + \qquad\quad b(s) = 0,$$
$$-5a(s) + (is - 4)b(s) = \tfrac{1}{2}.$$

This system of equations can be straightforwardly solved, to yield

$$a(s) = \frac{1}{2(s^2 + 2is + 3)},$$

$$b(s) = \frac{-is - 2}{2(s^2 + 2is + 3)}.$$

We now know a particular solution in the form

$$x(t) = \frac{1}{2} \int_{-1}^{1} \frac{1}{s^2 + 2is + 3} e^{ist} ds,$$

$$y(t) = \frac{1}{2} \int_{-1}^{1} \frac{-is - 2}{s^2 + 2is + 3} e^{ist} ds.$$

These are not integrals that can be computed in elementary form, but that they are still easy to evaluate numerically, and their properties are easy to understand. The final solution is

$$\begin{bmatrix} x \\ y \end{bmatrix} = C_1 e^{3t} \begin{bmatrix} 1 \\ -5 \end{bmatrix} + C_2 e^{-t} \begin{bmatrix} 1 \\ -1 \end{bmatrix} + \frac{1}{2} \begin{bmatrix} \int_{-1}^{1} \frac{1}{s^2 + 2is + 3} e^{ist} ds \\ \int_{-1}^{1} \frac{-is - 2}{s^2 + 2is + 3} e^{ist} ds \end{bmatrix}. \quad \blacktriangle$$

Exercises 7.7#9 and 7.7#10 will give further examples of this kind of solution.

METHOD OF VARIATION OF PARAMETERS

There is also an explicit formula for solving nonhomogeneous equations: the formula that goes under the name of *variation of parameters*. The name comes from one possible derivation, which we do not find particularly enlightening and will not repeat. Most students are repelled by its apparent complication, and one must admit that the integrals that it leads to do tend to be unpleasant.

Theorem 7.7.5. *The solution* $\mathbf{u}(t)$ *to* $\mathbf{x}' = A\mathbf{x} + \mathbf{g}(t)$ *with* $\mathbf{u}(t_0) = \mathbf{x}_0$ *is*

$$\mathbf{u}(t) = e^{(t-t_0)A}\mathbf{x}_0 + \int_{t_0}^{t} e^{(t-s)A}\mathbf{g}(s)ds.$$

We have already, in Section 2 of Part I, given an explanation of what the various terms in this formula mean in one dimension, to which the reader should return: it makes the formula *obvious*. A brief statement appears in Appendix T.

We will come back to Theorem 7.7.5 in Chapter 12 in Part III (it turns out that the formula is not restricted to the case of constant coefficients). Here, we will simply prove by an easy but unmotivated computation that it is correct.

Proof. Clearly, $\mathbf{u}(t_0) = \mathbf{x}_0$. Differentiating $\mathbf{u}(t)$, we find

$$\frac{d}{dt}\mathbf{u}(t) = Ae^{(t-t_0)A}\mathbf{x}_0 + \mathbf{g}(t) + A\int_{t_0}^{t} e^{(t-s)A}\mathbf{g}(s)ds = A\mathbf{u}(t) + \mathbf{g}(t),$$

where the $\mathbf{g}(t)$ is the derivative of the integral with respect to the upper limit of integration, and the integral term is the derivative with respect to the term t which appears in the exponent. □

Example 7.7.6. Solve the differential equation

$$\begin{bmatrix} x \\ y \end{bmatrix}' = \begin{bmatrix} 3 & -4 \\ 1 & -1 \end{bmatrix}\begin{bmatrix} x \\ y \end{bmatrix} + \begin{bmatrix} 2 \\ 1 \end{bmatrix}.$$

Using variation of parameters, the solution is

$$\mathbf{u}(t) = e^{(t-t_0)A}\mathbf{x}_0 + \int_{t_0}^{t} e^{(t-s)A}\begin{bmatrix} 2 \\ 1 \end{bmatrix}ds$$

and you can confirm that

$$e^{tA} = \begin{bmatrix} e^t + 2te^t & -4te^t \\ te^t & e^t - 2te^t \end{bmatrix};$$

so

$$\mathbf{u}(t) = \begin{bmatrix} e^t + 2te^t & -4te^t \\ te^t & e^t - 2te^t \end{bmatrix} \mathbf{x}_0 + \int_{t_0}^t \begin{bmatrix} 2e^{t-s} \\ e^{t-s} \end{bmatrix} ds.$$

The last term is a result of exceedingly good fortune in the matrix multiplication, which you can confirm in Exercises 7.7#11. The resulting final solution, after integrating this last term is

$$\mathbf{u}(t) = \begin{bmatrix} e^t + 2te^t & -4te^t \\ te^t & e^t - 2te^t \end{bmatrix} \mathbf{x}_0 + \begin{bmatrix} 2e^{t-t_0} - 2 \\ e^{t-t_0} - 1 \end{bmatrix}.$$

The reader is invited to check that this agrees with the solution you could find by undetermined coefficients. ▲

Chapter 7   Exercises

Exercises 7.1-7.2 Introduction to Linear Systems

7.1-7.2#1. In order to become more familiar with matrix notation:

(i) Find $A(t)$ and $\mathbf{g}(t)$ in order to write each of the following systems in the form $\mathbf{x}' = A(t)\mathbf{x} + \mathbf{g}(t)$:

(a)° $\begin{aligned} x' &= ty + t \\ y' &= x + y \end{aligned}$

(b)° $\begin{aligned} x' &= ty + t + z \\ y' &= x + y \\ z' &= 2x \end{aligned}$

(ii) Turn the following second order equations into a system of first order differential equations, by finding $A(t)$ and $\mathbf{g}(t)$ as above:

(c)° $x'' - 3x' + 5x = t^2$ (d)° $x'' - x\cos t = e^t$.

7.1-7.2#2°. Solve the following differential equations by the cookbook method 7.1.7:

(a) $x'' - 4x' + x = 0$.

(b) $x'' - 4x' + 8x = 0$.

(c) $x'' + 12x' + 36x = 0$, with $x(0) = -2$ and $x'(0) = -3$.

(d) $x'' - x' - x = 0$, with $x(0) = x'(0) = 1$.

7.1-7.2#3°. Solve the following differential equations by turning each into a system of first order differential equations (setting $y = x'$). Initial conditions are given.

(a) $3x'' - 6x' + 4x = 0$ with $\begin{bmatrix} x(0) \\ y(0) \end{bmatrix} = \begin{bmatrix} 0 \\ 1 \end{bmatrix}$.

(b) $2x'' - 2x' + x = 0$ with $\begin{bmatrix} x(0) \\ y(0) \end{bmatrix} = \begin{bmatrix} 1 \\ 2 \end{bmatrix}$.

7.1-7.2#4°. The following equations are of a class called Euler equations, which are discussed in detail in Chapter 12 (in Part III) . For now it suffices to say that there exist solutions of the form $x = t^\alpha$, so you can substitute that expression in the differential equations, as in the cookbook method, Recipe 7.1.7, and evaluate the resulting expression in α:

(a) Solve $t^2 x'' + 4tx' + 2x = 0$.

(b) Determine the domain of definition for the equation in part (a).

(c) Solve $t^2 x'' - x = 0$.

(d) Discuss the domain of definition for the equation in part (c); give a basis for the solutions with positive t and a basis for the solutions with negative t.

7.1-7.2#5. Consider a mass $m(t)$, depending on t, suspended from a spring as in Example 7.1.1.

(a) Write a system of differential equations describing the motion of the mass.

(b) Now suppose that the mass is a bucket filled with water and that it is punctured, leaking at some rate *independent of the motion*. Verify as in Example 4.1.1, the leaky bucket, that for a cylindrical bucket with radius r, the height h of the water varies as $(t - t_e)^2$, for $0 \le t \le t_e =$ the time at which the bucket empties, so that

$$m(t) = \rho \pi r^2 h + m_0 = Q(t - t_e)^2 + m_0,$$

with Q representing an appropriate constant expression and m_0 representing the mass of the empty bucket.

(c) Show that in case (b), the matrix system becomes, for $0 \le t \le t_e$,

$$\begin{bmatrix} x \\ y \end{bmatrix}' = \begin{bmatrix} 0 & 1 \\ -F/m(t) & -k/m(t) \end{bmatrix} \begin{bmatrix} x \\ y \end{bmatrix}$$

$$= \left[\begin{array}{cc} 0 & 1 \\ \frac{-F}{Q(t-t_e)^2+m_0} & \frac{-k}{Q(t-t_e)^2+m_0} \end{array} \right] \left[\begin{array}{c} x \\ y \end{array} \right],$$

which is still a linear differential equation in $\left[\begin{array}{c} x \\ y \end{array} \right]$, regardless of the exceedingly *non*linear appearance of t's within the matrix.

7.1-7.2#6. Consider again the leaky bucket suspended by a spring as in Exercise 7.1-7.2 #5, but now assume that the leaking *depends on the velocity y*. Show that in this case the matrix equation becomes

$$\left[\begin{array}{c} x \\ y \end{array} \right]' = \left[\begin{array}{cc} 0 & 1 \\ -F/m(y) & -k/m(y) \end{array} \right] \left[\begin{array}{c} x \\ y \end{array} \right],$$

which is *not* linear in $\left[\begin{array}{c} x \\ y \end{array} \right]$.

7.1-7.2#7°. This is an exercise in the use of Schwarz's inequality, Theorem L3.3, in order to get uniqueness of solutions to linear equations. Given a square matrix A, denote

$$\|A\| = \sqrt{\sum_{i,j} |a_{i,j}|^2}.$$

(a) Show that for any square matrix A and any vector $\mathbf{x}$,

$$\|A\mathbf{x}\| \le \|A\| \, \|\mathbf{x}\|.$$

(b) What does it mean to say that $\mathbf{x}' = A\mathbf{x}$ satisfies a Lipschitz condition on $\mathbb{R} \times \mathbb{R}^n$?

(c) Show that every linear differential equation with constant coefficients does satisfy a Lipschitz condition on $\mathbb{R} \times \mathbb{R}^n$.

(d) Find a Lipschitz constant for

$$\mathbf{x}' = \left[\begin{array}{cc} 0 & 1 \\ 1 & 2 \end{array} \right] \mathbf{x}.$$

Exercises 7.3 Eigenvectors and Decoupling

7.3#1.

(i)° Find the eigenvalues and eigenvectors of the following matrices:

(a) $\left[\begin{array}{cc} 5 & -6 \\ 3 & -4 \end{array} \right]$ (e) $\left[\begin{array}{ccc} 1 & 4 & 0 \\ 3 & 2 & 1 \\ 3 & -3 & 2 \end{array} \right]$

(b) $\begin{bmatrix} 2 & -1 \\ 0 & 2 \end{bmatrix}$

(f) $\begin{bmatrix} 1 & 4 & 0 \\ 3 & 2 & 1 \\ -1 & -3 & 2 \end{bmatrix}$

(c) $\begin{bmatrix} -2 & 1 \\ -1 & -2 \end{bmatrix}$

(g) $\begin{bmatrix} 1 & 4 & 0 \\ 3 & 2 & 1 \\ -13 & -3 & 2 \end{bmatrix}$

(d) $\begin{bmatrix} 1 & 0 & 0 \\ -1 & 2 & 0 \\ 1 & 0 & 2 \end{bmatrix}$

(h) $\begin{bmatrix} 0 & 2 & 0 \\ 0 & 0 & 3 \\ 1 & 0 & 0 \end{bmatrix}$

(ii) Write the general solution in cases (a) and (c) to the differential equation $\mathbf{x}' = A\mathbf{x}$ and give the vectors $\mathbf{x}(0)$ which tend to 0 as $t \to \infty$ (initial condition).

7.3#2. Solve the following differential equations and draw in both cases the trajectory in the xy plane:

(a) $\begin{bmatrix} x \\ y \end{bmatrix}' = \begin{bmatrix} 2 & 1 \\ 1 & 1 \end{bmatrix} \begin{bmatrix} x \\ y \end{bmatrix}$ with initial condition $\begin{bmatrix} x(0) \\ y(0) \end{bmatrix} = \begin{bmatrix} 1 \\ 1 \end{bmatrix}$,

(b) $\begin{bmatrix} x \\ y \end{bmatrix}' = \begin{bmatrix} 2 & 1 \\ 1 & 4 \end{bmatrix} \begin{bmatrix} x \\ y \end{bmatrix}$ with initial conditions $\begin{bmatrix} x(0) \\ y(0) \end{bmatrix} = \begin{bmatrix} 2 \\ 0 \end{bmatrix}$.

7.3#3°. Solve the differential equation

$$\begin{bmatrix} x \\ y \end{bmatrix}' = \begin{bmatrix} 2 & 10 \\ -1 & 4 \end{bmatrix} \begin{bmatrix} x \\ y \end{bmatrix},$$

(a) with initial condition $\begin{bmatrix} x(0) \\ y(0) \end{bmatrix} = \begin{bmatrix} 1 \\ 0 \end{bmatrix}$,

(b) with initial condition $\begin{bmatrix} x(0) \\ y(0) \end{bmatrix} = \begin{bmatrix} 0 \\ 1 \end{bmatrix}$.

(c) Draw the two trajectories from parts (a) and (b) in the xy phase plane. Write down the complete solution to the equation.

7.3#4°.

(a) Solve the differential equation

$$\begin{bmatrix} x \\ y \\ z \end{bmatrix}' = \begin{bmatrix} 0 & 1 & 0 \\ 0 & 0 & 1 \\ 6 & 7 & 0 \end{bmatrix} \begin{bmatrix} x \\ y \\ z \end{bmatrix} \quad \text{with} \quad \begin{bmatrix} x(0) \\ y(0) \\ z(0) \end{bmatrix} = \begin{bmatrix} 3 \\ 0 \\ 14 \end{bmatrix}.$$

(b) Solve the differential equation

$$\begin{bmatrix} x \\ y \\ z \end{bmatrix}' = \begin{bmatrix} 1 & 0 & 0 \\ -2 & 5 & 4 \\ 1 & -3 & -3 \end{bmatrix} \begin{bmatrix} x \\ y \\ z \end{bmatrix}.$$

7.3#5°. For the following differential equations, find a basis for $\mathbb{R}^2$ such that vectors expressed with respect to that basis decouple the differential equations:

(a) $\begin{aligned} x' &= y \\ y' &= x + y \end{aligned}$

(b) $\begin{aligned} x' &= 2x - y \\ y' &= 3y \end{aligned}$

7.3#6°. Consider the differential equation $\mathbf{x}' = A\mathbf{x}$, where

$$A = \begin{bmatrix} 0 & 2 & -2 \\ -2 & 0 & 1 \\ 2 & -1 & 0 \end{bmatrix}.$$

(a) Find the eigenvalues of A. Find a basis of $\mathbb{C}^3$ made of eigenvectors of A.

(b) Write the general solution.

(c) What is the solution if $\mathbf{x}(0) = \begin{bmatrix} 1 \\ 1 \\ 0 \end{bmatrix}$?

(d) Describe the solutions geometrically.

7.3#7. Find a 2×2 matrix with eigenvalues 1 and 3, and corresponding eigenvectors

$$\begin{bmatrix} 1 \\ 1 \end{bmatrix}, \quad \begin{bmatrix} 1 \\ 2 \end{bmatrix}.$$

7.3#8°. For what values of a, b, and c are the following matrices diagonalizable?

(a) $\begin{bmatrix} 0 & a \\ -1 & 2 \end{bmatrix}$

(c) $\begin{bmatrix} 0 & c & 0 \\ 1 & 0 & c \\ c & 1 & 0 \end{bmatrix}$

(b) $\begin{bmatrix} 1 & 0 & 0 \\ 1 & 1 & 1 \\ 0 & 0 & b \end{bmatrix}$

(d) $\begin{bmatrix} 1 & 1 & a \\ 0 & 2 & b \\ c & 0 & 2 \end{bmatrix}$

7.3#9. Confirm in Example 7.3.5 that equation (19) gives the same result as equations (17) and (18).

7.3#10. This exercise illustrates how a small error or change in the coefficients of the characteristic polynomial can lead to surprisingly larger errors in its roots. In each case, compare the roots of the first polynomial with the roots of the second:

(a) $\lambda^2 - 2\lambda + 1$; $\lambda^2 - 2\lambda + 0.99$.

(b) λ^4; $\lambda^4 - 0.01$

(c) λ^4; $\lambda^4 - 0.01\lambda - 0.01$.

Exercises 7.4 Exponentials of Matrices

7.4#1. Compute

(a) $e^{t\begin{bmatrix} 1 & 4 \\ -2 & 3 \end{bmatrix}}$

(d) $e^{t\begin{bmatrix} 2 & 10 \\ -1 & 4 \end{bmatrix}}$

(b) $e^{t\begin{bmatrix} 2 & 1 \\ 1 & 1 \end{bmatrix}}$

(e) $e^{t\begin{bmatrix} 0 & 1 & 0 \\ 0 & 0 & 0 \\ -4 & 4 & 1 \end{bmatrix}}$

(c) $e^{t\begin{bmatrix} 2 & 1 \\ 1 & 4 \end{bmatrix}}$

7.4#2. Compute the exponentials of the following matrices:

(a) $\begin{bmatrix} 5 & -6 \\ 3 & -4 \end{bmatrix}$

(d) $\begin{bmatrix} 1 & 0 & 0 \\ -1 & 2 & 0 \\ 1 & 0 & 2 \end{bmatrix}$

(b) $\begin{bmatrix} 2 & -1 \\ 0 & 2 \end{bmatrix}$

(e) $\begin{bmatrix} 0 & 2 & 0 \\ 0 & 0 & 3 \\ 1 & 0 & 0 \end{bmatrix}$

(c) $\begin{bmatrix} -2 & 1 \\ -1 & -2 \end{bmatrix}$

7.4#3. Let $A = \begin{bmatrix} 0 & 1 & 0 \\ 0 & 0 & 1 \\ -4 & 4 & 1 \end{bmatrix}$.

(a) Find a matrix P such that $P^{-1}AP = \begin{bmatrix} \lambda_1 & 0 & 0 \\ 0 & \lambda_2 & 0 \\ 0 & 0 & \lambda_3 \end{bmatrix}$.

(b) Compute e^{tA}.

(c) Solve $\begin{bmatrix} x \\ y \\ z \end{bmatrix}' = A \begin{bmatrix} x \\ y \\ z \end{bmatrix}$ with initial condition

$$\begin{bmatrix} x(0) \\ y(0) \\ z(0) \end{bmatrix} = \begin{bmatrix} 1 \\ 1 \\ 1 \end{bmatrix}.$$

(d) It is easy to show that $\mathbf{x}(t)$ satisfies a third order differential equation with constant coefficients. Find it, and solve it by the cookbook Recipe 7.1.7.

7.4#4. Show that the ordinary product and chain rules for differentiation also hold for expressions involving exponentials of matrices.

7.4#5. Prove Theorem 7.4.13.

7.4#6°. Apply the method of exponentiating matrices to $x'' + 3x' + 2x = 0$, the equation of Example 7.2.2. You will observe that, in the standard basis in which this method operates, the answer will be quite unrecognizable as being the same as that found in the basis of eigenvectors. The moral should be obvious: use eigenvectors whenever you can! That is, use eigenvectors whenever you have a basis of them.

7.4#7. Prove the complex case of Proposition 7.4.14.

7.4#8. Verify Examples 7.4.2,3,4.

7.4#9. Verify Examples 7.4.8 and 7.4.10. For each one, tell what initial conditions lead to solutions that tend to 0 as $t \to \infty$.

7.4#10. For the following second order differential equations:

(i) Translate the problem into a system of equations.

(ii) Write the system in the form $\mathbf{x}' = A\mathbf{x}$.

(iii) Find the solution of $\mathbf{x}' = A\mathbf{x}$ by calculating e^{tA}.

(iv) State the answer to the original problem.

(a) $x'' + 12x' + 36x = 0$, with $x(0) = -2$ and $x'(0) = -3$.
(b) $x'' - 4x' + x = 0$.
(c) $x'' - 4x' + 8x = 0$.

You can compare your results with Exercises 7.1-7.2#2a,b,c.

7.4#11°. Let A be a 2×2 matrix, with $\operatorname{tr} A = 0$.

(a) Show that $A^2 = -(\det A)I$.

(b) Write the power series for e^{tA} as

$$e^{tA} = \alpha(t)I + \beta(t)A,$$

where $\alpha(t)$ and $\beta(t)$ are scalar power series.

(c) Express $\alpha(t)$ and $\beta(t)$ as known functions of t.

Remark: You will need to separate the case $\det A > 0$, which leads to hyperbolic functions, and $\det A < 0$, which leads to trigonometric functions.

7.4#12. Let A be any 2×2 matrix, and let $B = A - \frac{\operatorname{tr} A}{2}I$.

(a) Show that $\operatorname{tr} B = 0$.

(b) Compute $\det B$ in terms of $\det A$ and $\operatorname{tr} A$.

(c) Using that $A = B + \frac{\operatorname{tr} A}{2}I$, and that B and I commute, write e^{tA} in terms of $\det B$ and $\operatorname{tr} A$.

Exercises 7.5 Two by Two Matrices and the Bifurcation Diagram

7.5#1. Solve each of the following differential equations and, in each case, draw the track of the solution in the xy-plane.

(a) $\begin{bmatrix} x \\ y \end{bmatrix}' = \begin{bmatrix} 1 & 4 \\ -2 & 3 \end{bmatrix} \begin{bmatrix} x \\ y \end{bmatrix}$ with $\begin{bmatrix} x(0) \\ y(0) \end{bmatrix} = \begin{bmatrix} 1 \\ 1 \end{bmatrix}$

(b) $\begin{bmatrix} x \\ y \end{bmatrix}' = \begin{bmatrix} 2 & 1 \\ 1 & 1 \end{bmatrix} \begin{bmatrix} x \\ y \end{bmatrix}$ with $\begin{bmatrix} x(0) \\ y(0) \end{bmatrix} = \begin{bmatrix} 0 \\ 1 \end{bmatrix}$

(c) $\begin{bmatrix} x \\ y \end{bmatrix}' = \begin{bmatrix} 2 & 1 \\ 1 & 4 \end{bmatrix} \begin{bmatrix} x \\ y \end{bmatrix}$ with $\begin{bmatrix} x(0) \\ y(0) \end{bmatrix} = \begin{bmatrix} 2 \\ 0 \end{bmatrix}.$

7.5#2. Consider the linear system of differential equations $\mathbf{x}' = A\mathbf{x}$ given by

$$\begin{aligned} x' &= x - 2y \\ y' &= x + y. \end{aligned} \tag{i}$$

(a) Compute e^{tA} (by Theorem 7.4.13).

(b) Find the solution with initial condition $(x, y) = (-6, 0)$. Make a drawing (by hand or computer).

(c) Consider the related single differential equation

$$\frac{dy}{dx} = \frac{x+y}{x-2y} \qquad \text{(ii)}$$

Explain the similarities and differences between studying (i) and (ii).

7.5#3°. Show that if $\operatorname{tr} A = 0$, the equation

$$\frac{dy}{dx} = \frac{dy/dt}{dx/dt}$$

is *exact* and can be solved analytically.

7.5#4. Consider the linear system of differential equations

$$\mathbf{x}' = A\mathbf{x}, \quad \text{where } A = \begin{bmatrix} -1 & 2 \\ -3 & 4 \end{bmatrix}.$$

(a) Determine the eigenvalues and eigenvectors of the matrix A, and write down the general solution in terms of the eigenvalues and eigenvectors. Explain the behavior of the solutions when $t \to -\infty$.

(b) Determine the solution with initial condition $\mathbf{x}(0) = \begin{bmatrix} 0 \\ 1 \end{bmatrix}$.

7.5#5. Show that for a 2×2 matrix, $A = \begin{bmatrix} a & b \\ c & d \end{bmatrix}$, the characteristic polynomial can be written $\lambda^2 - (\operatorname{tr} A)\lambda + (\det A)$, with discriminant $\Delta = (\operatorname{tr} A)^2 - 4(\det A)$. Show that $\Delta > 0$ leads to real, distinct eigenvalues, $\Delta < 0$ leads to complex eigenvalues, and that $\Delta = 0$ means a double eigenvalue.

7.5#6. Confirm the text statements about the phase plane behavior with real distinct nonzero eigenvalues as follows:

(a) If both eigenvalues are positive, show that all trajectories leave 0 tangentially to the eigenvector associated with the smaller eigenvalue, except for those leaving along a multiple of the eigenvector associated with the larger eigenvalue; hence the phase plane portrait is a node source.

(b) If the eigenvalues are of opposite sign, show that the trajectories move away from 0 along the eigenvector associated with the positive eigenvalue and toward 0 along the eigenvector associated with the negative eigenvalue; hence the phase plane portrait is a saddle.

(c) If both eigenvalues are negative, show that all trajectories approach 0 tangentially to the eigenvector associated with the smaller (in absolute value) eigenvalue, except for those approaching along a multiple of the eigenvector associated with the larger (in absolute value) eigenvalue; hence the phase plane portrait is a node sink.

7.5#7°. If for a real matrix A, λ is a nonreal eigenvalue, and $\mathbf{v} = \mathbf{w}_1 + i\mathbf{w}_2$, where $\mathbf{w}_1$ and $\mathbf{w}_2$ are real, is a corresponding eigenvector, then show that $\mathbf{w}_1$ and $\mathbf{w}_2$ are linearly independent.

7.5#8°. Consider the differential equation $\mathbf{x}' = A\mathbf{x}$, where $A = \begin{bmatrix} a & b \\ c & d \end{bmatrix}$ is a real matrix. Suppose that the trajectories spiral; show that the direction of spiral trajectories depends on the sign of c. Show that if $c > 0$, the rotation is counterclockwise; if $c < 0$, the rotation is clockwise.

7.5#9. Suppose that $x(t)$ describes the position of a particle moving in one dimension, under the potential $V(x)$. If $V(x) = ax^2$, show that the differential equation that $x(t)$ obeys is linear. If $a > 0$, show that it has a center at 0, and if $a < 0$, show that it has a saddle.

7.5#10. If you look at solutions of the system $x' = 0.2x - 0.04y$, $y' = x + 0.2y$ on the computer, they spiral completely around one or two times before leaving the screen. Computer drawn solutions of $x' = 0.96x - 0.04y$, $y' = x + y$, on the other hand, seem to twist a little and then travel out along almost straight lines. Will they ever make one revolution? If not, what is the boundary between spiral sources whose solutions make one or more revolutions and those whose solutions do not?

7.5#11. Show with computer and hand drawings how one can move smoothly from saddles to having one zero eigenvalue [i.e., $\det(A) = 0$] to node sources. Show how one can move smoothly from saddles to centers without going through the zero matrix.

7.5#12. In Example 7.5.1 verify the calculations leading from the eigenvector equation to the equation of a surface in $xy\alpha$-space. That is, show that

$$y = \frac{\alpha \pm \sqrt{\alpha^2 - 4}}{2} x$$

implies that $y^2 - \alpha xy + x^2 = 0$.

7.5#13. Consider the linear system of differential equations

$$\begin{bmatrix} x \\ y \end{bmatrix}' = \begin{bmatrix} 0 & -1 \\ 10 + \alpha & \alpha \end{bmatrix} \begin{bmatrix} x \\ y \end{bmatrix},$$

where the matrix A_α depends on the parameter α.

(a) Each matrix determines a point $(\operatorname{tr} A_\alpha, \det A_\alpha)$ in the $(\operatorname{tr} A) - (\det A)$-plane. Draw the curve in the $(\operatorname{tr} A) - (\det A)$-plane made up of these points when α increases from $-\infty$ to $+\infty$.

(b) For what values of α (if any) is $(0,0)$ a saddle, a node sink, a spiral sink, a center, a spiral source, or a node source?

(c) Make computer drawings of trajectories in the phase plane for $\alpha = -15, -10, -7, -3, 0, 3, 15$. Show directions on trajectories by arrows. Draw directions for real eigenvectors (in color) whenever possible.

(d) Draw in the (x, y, z)-space the set of all real eigenvectors, i.e., in the plane at height α draw the real eigenvectors of A_α (if any).

7.5#14. Consider the system of differential equations

$$\begin{bmatrix} x \\ y \end{bmatrix}' = \begin{bmatrix} 0 & 1+\alpha \\ -1 & \alpha \end{bmatrix} \begin{bmatrix} x \\ y \end{bmatrix}$$

depending on the parameter α.

(a) Classify the differential equation according to values of α.

(b) Draw the phase plane for $\alpha = -2, -1, -\frac{1}{2}, 0, \frac{1}{2}, 5$.

(c) Draw, in the xyz-space, the set of all eigenvectors (i.e., in the plane at height α, draw the eigenvectors of $\begin{bmatrix} 0 & 1+\alpha \\ -1 & \alpha \end{bmatrix}$).

7.5#15. Investigate the system

$$\begin{bmatrix} x \\ y \end{bmatrix}' = \begin{bmatrix} y \\ -4x^3 + 2x \end{bmatrix} + (x^4 - x^2 - \alpha) \begin{bmatrix} 4x^3 - 2x \\ y \end{bmatrix}.$$

Draw the phase plane for $\alpha = -\frac{1}{2}, -\frac{1}{4}, -\frac{1}{8}, 0, \frac{1}{2}$. Can you understand why the solutions look the way they do?

7.5#16. Show how the variable α affects the trajectories for different values of α for

$$\begin{bmatrix} x \\ y \end{bmatrix}' = \begin{bmatrix} \alpha & -1 \\ 1 & 1 \end{bmatrix} \begin{bmatrix} x \\ y \end{bmatrix}.$$

7.5#17. Consider the Euler equation

$$x'' + \frac{\alpha}{t} x' + \frac{\beta}{t^2} x = 0$$

defined for $0 < t < \infty$.

(a) Find the general solution.

(b) For what values of (α, β) is the space of solutions that tend to 0 as $t \to 0$ of dimension zero, one, and two, respectively?

(c) For what values of (α, β) are there any solutions at all that tend to a limit other than zero or infinity as $t \to 0$?

Exercises 7.6 Eigenvalues and Global Behavior

7.6#1.

(a) Consider the differential equations $\mathbf{x}' = A\mathbf{x}$, where A is the matrices

$$
\text{(i)} \begin{bmatrix} 0 & 1 & 0 \\ 0 & 0 & 1 \\ \alpha & -1 & 1 \end{bmatrix}
\qquad
\text{(ii)} \begin{bmatrix} 0 & 1 & 0 & 0 \\ 0 & 0 & 1 & 0 \\ 0 & 0 & 0 & 1 \\ \alpha & -1 & 1 & 1 \end{bmatrix}
$$

and α is a parameter. For what values of α does the type of the eigenvalues change, for instance, from real to complex, or from positive to negative?

(b) For each of the intervals where the type does not change, describe the solutions of the differential equation. (You may need *Analyzer*, or *Eigenfinder*, or both.)

7.6#2. Prove the converse of Theorem 7.6.1. That is, prove the following:

Theorem 7.6.1*. *Consider* $\mathbf{x}' = A\mathbf{x}$ *where A is an $n \times n$ matrix. If all solutions $\mathbf{x}(t)$ satisfy*

$$\lim_{t \to \infty} \mathbf{x}(t) \to \mathbf{0},$$

then all the eigenvalues of A have negative real parts.

7.6#3. In the proof of Theorem 7.6.1 we proved a lemma showing that $\|C\| \le n\varepsilon$. Prove the sharper limit that $\|C\| \le \sqrt{n(n-1)/2}\,\varepsilon$.

7.6#4. Consider the differential equation of Example 7.6.2:

$$
\begin{aligned}
x' &= -x + 100y \\
y' &= -y.
\end{aligned}
$$

Show that if you measure lengths with respect to a new basis

$$\mathbf{v}_1 = \mathbf{e}_1, \qquad \mathbf{v}_2 = \mathbf{e}_2(\varepsilon/100),$$

then all solutions decrease if $|\varepsilon| \le 2$.

7.6#5°. Consider the linear system of differential equations

$$
\begin{bmatrix} x \\ y \end{bmatrix}' = \begin{bmatrix} -1 & 1 \\ \alpha & \alpha \end{bmatrix} \begin{bmatrix} x \\ y \end{bmatrix}
$$

depending on the parameter $\alpha \in \mathbb{R}$.

(a) Explain why $(x, y) = (0, 0)$ is a singular point for the differential equation for any α. For what values of α (if any) is $(0,0)$ a saddle point, a node sink, a spiral sink, a center, a spiral source, or a node source?

(b) Determine the eigenvalues and eigenvectors of A_α for $\alpha = 1$, and sketch the trajectories in the phase plane.

7.6#6. Consider the differential equation

$$\begin{bmatrix} x \\ y \\ z \end{bmatrix}' = \begin{bmatrix} 0 & 1 & 0 \\ 0 & 0 & 1 \\ -\alpha & 1 & 0 \end{bmatrix} \begin{bmatrix} x \\ y \\ z \end{bmatrix}.$$

(a) Find all values of α where the behavior of the differential equation bifurcates.

(b) For each of these values of α, and for one value of α in each of the intervals that they bound, describe and sketch the solutions of the differential equation.

7.6#7. Consider the differential equation

$$\begin{bmatrix} x \\ y \\ z \end{bmatrix}' = \begin{bmatrix} 0 & 1 & 0 \\ 0 & 0 & 1 \\ 1 & -(\alpha+1) & (\alpha+1) \end{bmatrix} \begin{bmatrix} x \\ y \\ z \end{bmatrix}.$$

(a) Find the characteristic polynomial of the matrix and find its roots (as functions of α).

(b) At what values of α does the behavior of the trajectories change completely?

(c) In each of the intervals of the α-line bounded by the bifurcation values, what is the dimension of the space of initial conditions x_0 such that the solution with that initial condition gives a solution that tends to 0 as $t \to +\infty$ and as $t \to -\infty$?

(d) Sketch, up to linear change of variables, the solutions of the differential equation in each of the intervals above.

7.6#8. Consider the system of differential equations

$$\begin{bmatrix} x \\ y \end{bmatrix}' = \begin{bmatrix} \sqrt{3}\cos\alpha & \sin\alpha \\ -\sqrt{3} & 0 \end{bmatrix} \begin{bmatrix} x \\ y \end{bmatrix}, \qquad 0 \le \alpha \le 2\pi.$$

(a) Find the values of α for which the equation above bifurcates and classify the equation in the intervals bounded by these values.

(b) Solve the equation for $\alpha = 0$, $\pi/6$, $\pi/2$, $3\pi/2$.

7.6#9. Consider the differential equation

$$\begin{bmatrix} x \\ y \\ z \end{bmatrix}' = \begin{bmatrix} 0 & 1 & 0 \\ 0 & 0 & 1 \\ \alpha\beta & (-\alpha+\beta+\alpha\beta) & (\alpha+\beta+1) \end{bmatrix} \begin{bmatrix} x \\ y \\ z \end{bmatrix},$$

depending on two parameters, α and β.

(a) Find the eigenvalues of this matrix. (It is easier than you might think.)

(b) For the four values $(\pm 1, \pm 1)$ of (α, β), describe the solutions of the differential equation.

(c) Describe how the dimensions of the spaces of initial conditions leading to solutions that tend to 0 as $t \to \pm\infty$ depend on (α, β).

(d) For what values of (α, β) does the matrix admit a basis of eigenvectors? (Try to compute the dimension of the space of eigenvectors associated to any eigenvalue.)

7.5#17. Consider the Euler equation

$$x'' + \frac{\alpha}{t}x' + \frac{\beta}{t^2}x = 0$$

defined for $0 < t < \infty$.

(a) Find the general solution.

(b) For what values of (α, β) is the space of solutions that tend to 0 as $t \to 0$ of dimension zero, one, and two, respectively?

(c) For what values of (α, β) are there any solutions at all that tend to a limit other than zero or infinity as $t \to 0$?

Exercises 7.7 Nonhomogeneous Linear Equations

7.7#1. Solve the following differential equations by the method of undetermined coefficients:

(a) $\mathbf{x}' = \begin{bmatrix} 0 & 3 & 5 \\ 3 & 0 & 2 \\ 2 & 0 & 1 \end{bmatrix} \mathbf{x} + \begin{bmatrix} e^t \\ \cos t \\ 0 \end{bmatrix}$

(b) $\mathbf{x}' = \begin{bmatrix} 3 & 3 \\ 3 & 4 \end{bmatrix} \mathbf{x} + \begin{bmatrix} \cos t \\ \sin t \end{bmatrix}$

(c) $\mathbf{x}' = \begin{bmatrix} 1 & -2 \\ 3 & 6 \end{bmatrix} \mathbf{x} + \begin{bmatrix} e^t + e^{3t} \\ 0 \end{bmatrix}$

(d) $\mathbf{x}' = \begin{bmatrix} 3 & 5 \\ 3 & 4 \end{bmatrix} \mathbf{x} + \begin{bmatrix} te^t \\ 0 \end{bmatrix}$

7.7#2°.

(a) Show that trying $\begin{bmatrix} a + ct \\ b + dt \end{bmatrix} e^t$ as a solution of

$$\mathbf{x}' = \begin{bmatrix} 2 & 1 \\ -5 & -4 \end{bmatrix} \mathbf{x} + \begin{bmatrix} te^t \\ 0 \end{bmatrix}$$

will not work.

(b) Try setting $Q(t)$ as a quadratic instead of a linear function and solve the equation.

7.7#3. Find the general solutions for the following differential equations.

(a) $x'' + 2x' + x = e^{-t}$

(b) $x'' + 3x' + 2x = e^{-t}$

7.7#4°.

(a) Find the general solutions of the differential equations

$$x'' + x' + x = \sin \omega t \quad \text{and} \quad x'' + 2x = \sin \omega t,$$

using the method of undetermined coefficients.

(b) Among the solutions for the first equation, which one is the *steady-state solution*? What is its amplitude as a function of ω?

(c) For the second equation, there should be a value ω_0 of ω for which you could not solve by the method of undetermined coefficients. Solve the equation by variation of parameters for that value ω_0.

7.7#5. Show that $u(t) = t + 1$ is a solution to $x'' - 4x' + x = t - 3$. Find the complete solution to the equation.

7.7#6. Use variation of parameters to solve the differential equation

$$\begin{bmatrix} x \\ y \end{bmatrix}' = \begin{bmatrix} 0 & 1 \\ -1 & 0 \end{bmatrix} \begin{bmatrix} x \\ y \end{bmatrix} + \begin{bmatrix} 0 \\ \tan t \end{bmatrix}$$

with $x(0) = y(0) = 0$.

7.7#7.

(a) Write down the general solution to the second order system of differential equations

$$x'' = Ax, \quad \text{where } A = \begin{bmatrix} -\omega_1^2 & 0 \\ 0 & -\omega_2^2 \end{bmatrix}.$$

(b) Let $\mathbf{g}(t) = \begin{bmatrix} \cos \omega t \\ \sin \omega t \end{bmatrix}$ and assume $\omega \neq \omega_i$, $i = 1, 2$. Determine $c_1, c_2 \in \mathbb{C}$, such that

$$\mathbf{h}(t) = \begin{bmatrix} c_1 e^{i\omega t} \\ c_2 e^{i\omega t} \end{bmatrix}$$

and $\operatorname{Re} \mathbf{h}(t) = \mathbf{g}(t)$.

(c) Use the method of undetermined coefficients to find a solution

$$\mathbf{x}(t) = \begin{bmatrix} \xi_1 e^{i\omega t} \\ \xi_2 e^{i\omega t} \end{bmatrix}$$

to the differential equation $\mathbf{x}'' = A\mathbf{x} + \mathbf{h}(t)$. Determine a real solution to the differential equation $\mathbf{x}'' = A\mathbf{x} + \mathbf{g}(t)$.

7.7#8. Explore with *MacMath's DiffEq Phase Plane* or *DiffEq 3D Views* the driven harmonic oscillator $x'' + x = \sin \omega t$. Try values of ω near but on either side of the natural frequency ($\omega = 1$). Print or sketch your results.

What aspects of the phase plane behavior change? How do they change? Describe what these changes imply for a physical pendulum.

7.7#9°. The square wave $g(t)$, which is periodic of period 2π, and which on $(-\pi, \pi]$ is given by

$$g(t) = \begin{cases} +1 & \text{if } t \geq 0 \\ -1 & \text{if } t < 0, \end{cases}$$

is given by the Fourier series

$$g(t) = \frac{1}{\pi} \left(\sin t + \frac{1}{3} \sin 3t + \frac{1}{5} \sin 5t + \cdots \right) = \frac{1}{\pi} \sum_{n=1}^{\infty} \frac{\sin(2n-1)t}{2n-1}.$$

This is something you may already know, or can take on faith (if your faith needs bolstering, try plotting the sum of the first five terms).

Solve the differential equation $x'' + \omega^2 x = g(t)$ by "undetermined coefficients", by setting $x(t) = \sum a_n \sin nt$ and considering the a_n as coefficients to be determined.

For what values of ω does this scheme work? Can you solve the equation in the other cases also.

7.7#10. Observe that

$$\int_0^\infty e^{-st} ds = \frac{1}{t}.$$

Find a solution of

$$x'' + x = \frac{1}{t}.$$

by "undetermined coefficients," by writing

$$x(t) = \int_0^\infty f(s)e^{-st}ds,$$

where $f(s)$ is the undetermined coefficient, and plugging into the equation.

7.7#11. Confirm the calculations of the solution to Example 7.7.6.

8

Systems of Nonlinear Differential Equations

The general autonomous differential equation on $\mathbb{R}^n$ is

$$\mathbf{x}' = \mathbf{f}(\mathbf{x}) = \begin{bmatrix} f_1(\mathbf{x}) \\ \vdots \\ f_n(\mathbf{x}) \end{bmatrix}, \tag{1}$$

where $\mathbf{f}$ should be thought of as a vector field on an open subset of $\mathbb{R}^n$. It describes the evolution of innumerable actual systems, and even the two-dimensional case

$$\begin{aligned} x' &= f(x,y) \\ y' &= g(x,y) \end{aligned} \tag{2}$$

has a great many applications.

In Chapter 7 we studied the case where $\mathbf{f}$ is linear; in this chapter, we will concentrate on nonlinear equations. Of course, even when $n = 2$, we cannot expect the sort of detailed description that we found in the linear case. After all, the linear case depends on only four parameters, whereas the nonlinear case depends on infinitely many, since f and g are both arbitrary functions, elements of an infinite-dimensional vector space. [Actually, f and g are not quite arbitrary; they must satisfy a Lipschitz condition for equation (1) to have unique solutions through every point and, in practice, will usually be required to be twice continuously differentiable.]

Still, we will get a surprising amount of information about nonlinear vector fields from our analysis of the linear case (and also from iteration in one dimension, in Part I, Chapter 5).

We will see in Sections 8.1–8.3 that near the *zeroes* of the vector field [the points $\mathbf{x}_0$ where $\mathbf{f}(\mathbf{x}_0) = 0$], the behavior of the differential equation is usually controlled by the linear approximation to the vector field at that point. The linearized problem is precisely what we studied in Chapter 7. Then in Sections 8.4–8.6 we will stand back from the zeroes and see what we can say about the global behavior of nonlinear systems in dimension two.

The two-dimensional system (2) in $\mathbb{R}^2$ is most amenable to analysis. The functions f and g do not depend explicitly on time, thus making system (2) autonomous. We know from Section 6.2 that it makes sense to examine the trajectories of the solutions in the xy phase plane, because for an autonomous system, these trajectories will usually not cross (except where the Existence and Uniqueness Theorem 6.2.3 fails).

Curves in the plane which cannot intersect are so strongly constrained that "chaos" cannot appear. In dimension three, however, trajectories have every opportunity to knot, link, and tangle. This is discussed informally in Sections 8.7 and 8.8, but actual proofs are difficult and are reserved for Chapter 8*, where we can take sufficient detours to explain the difficult points. Chapter 8* consists of many discussions extending the topics introduced here in Chapter 8.

One more topic that belongs here in the introduction to nonlinear differential equations in $\mathbb{R}^n$ is that of symmetries and volume-preserving equations, which appears as Section 8.6 (but does not have a companion section in Chapter 8*).

Chapter 8 can be extended by various alternative routes:

- After reading Sections 8.7 and 8.8, you may go directly to Chapter 9.

- After reading Sections 8.7 and 8.8, you may read Chapter 8* and then go on to Chapter 9.

- You can alternate sections of Chapters 8 and 8* as follows: 8.1, 8*.1, 8.2, 8*.2, and so on through 8.5, 8*.5, and finally 8*.6. In this case you will find it enlightening to read 8.7 and 8.8 before 8*.5 and 8*.6.

A flowchart is given in the Preface to clarify these suggestions.

In this chapter, we shall consider mainly nonlinear autonomous systems in $\mathbb{R}^2$, the xy-plane. Occasionally, both statements and proofs are no harder in $\mathbb{R}^n$, and there we will do things in more generality.

8.1   Zeroes of Vector Fields and Their Linearization

LOCAL AND GLOBAL BEHAVIOR OF A VECTOR FIELD IN THE PLANE

Example 8.1.1. Consider the differential equation

$$x' = x - y^2 + a + bxy$$
$$y' = 0.2y - x + x^3 \tag{3}$$

for the values $a = 1.28$ and $b = 1.4$. The drawing in the phase plane is shown in Figure 8.1.1.   ▲

We see in Example 8.1.1 that the trajectories just form systems of curves in the plane, more or less parallel on a small scale, except at the six points (and there are only six) where both x' and y' are zero. At these points the vector field vanishes, so they are called the *zeroes of the vector field* (sometimes *singular points* or *singularities*). Because the zeroes are the points where the system is at equilibrium, they are also sometimes called *equilibria*, especially for physical systems.

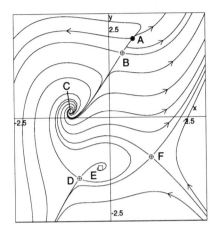

FIGURE 8.1.1. The phase plane for equation (3). Observe the six points where the vector field vanishes, all within the indicated rectangle.

Definition 8.1.2. Let $\mathbf{x}' = \mathbf{f}(\mathbf{x})$ be a differential equation, defined in some region U of $\mathbb{R}^n$. Then a point $\mathbf{x}_0$ where $\mathbf{f}(\mathbf{x}_0) = \mathbf{0}$ is called a *zero* or *equilibrium point* of the differential equation.

In particular, when $n = 2$, a point (x_0, y_0) is a zero of the differential equation

$$x' = f(x, y)$$
$$y' = g(x, y)$$

if $f(x_0, y_0) = g(x_0, y_0) = 0$.

Near the zeroes, the solutions form patterns that should be familiar: they look like the drawings we obtained in the linear case.

More precisely, point A looks like a node source, C looks like a spiral source, E looks like a spiral sink, and B, D, and F look like saddles. We shall now give a rough idea of why this is true by showing how to construct a linearized equation at each zero.

LINEARIZATION: LOCAL BEHAVIOR NEAR SINGULARITIES

If f and g are twice continuously differentiable near a zero (x_0, y_0), it is natural to expand them in a Taylor polynomial around this point:

$$f(x, y) = \left.\frac{\partial f}{\partial x}\right|_{(x_0, y_0)} (x - x_0) + \left.\frac{\partial f}{\partial y}\right|_{(x_0, y_0)} (y - y_0) + P(x - x_0, y - y_0)$$

$$g(x,y) = \frac{\partial g}{\partial x}\bigg|_{(x_0,y_0)} (x - x_0) + \frac{\partial g}{\partial y}\bigg|_{(x_0,y_0)} (y - y_0) + Q(x - x_0, y - y_0)$$

where $P(x - x_0, y - y_0)$ and $Q(x - x_0, y - y_0)$ start with terms that are at least quadratic in $(x - x_0)$ and $(y - y_0)$. One may expect that near a zero of the vector field, P and Q will be negligible compared with the linear terms of the Taylor series; of course, this cannot be true if the linear terms vanish identically, and we will see that there are other exceptions.

In order to study the behavior of a system near its zeroes, change the coordinates to indicate displacement from equilibrium:

$$\xi = x - x_0, \qquad \eta = y - y_0,$$

and take only the first order terms of the Taylor polynomials. The original nonlinear differential equation

$$\begin{bmatrix} x \\ y \end{bmatrix}' = \begin{bmatrix} f(x,y) \\ g(x,y) \end{bmatrix} \tag{2}$$

is approximated by

$$\begin{bmatrix} \xi \\ \eta \end{bmatrix}' = \begin{bmatrix} \frac{\partial f}{\partial x}\big|_{(x_0,y_0)} & \frac{\partial f}{\partial y}\big|_{(x_0,y_0)} \\ \frac{\partial g}{\partial x}\big|_{(x_0,y_0)} & \frac{\partial g}{\partial y}\big|_{(x_0,y_0)} \end{bmatrix} \begin{bmatrix} \xi \\ \eta \end{bmatrix}, \tag{4}$$

which is called the *linearization* of the system (2) near (x_0, y_0) .

Example 8.1.3. For the differential equation of the damped pendulum,

$$\begin{aligned} x' &= y \\ y' &= -\sin x - y, \end{aligned} \tag{5}$$

the points $(0,0)$ and $(\pi, 0)$ are zeroes of the vector field.

The linearization at $(0,0)$ is

$$\begin{bmatrix} \xi \\ \eta \end{bmatrix}' = \begin{bmatrix} 0 & 1 \\ -1 & -1 \end{bmatrix} \begin{bmatrix} \xi \\ \eta \end{bmatrix}, \tag{6}$$

with parameters $(-1, 1)$ in the trace-determinant plane. Thus, the linearization at $(0,0)$ is a spiral sink, as shown in Figure 8.1.2.

The linearization at $(\pi, 0)$ is

$$\begin{bmatrix} \xi \\ \eta \end{bmatrix}' = \begin{bmatrix} 0 & 1 \\ 1 & -1 \end{bmatrix} \begin{bmatrix} \xi \\ \eta \end{bmatrix}, \tag{7}$$

with parameters $(-1, -1)$ in the trace-determinant plane. Thus, the linearization at $(\pi, 0)$ is a saddle, as also shown in Figure 8.1.2.

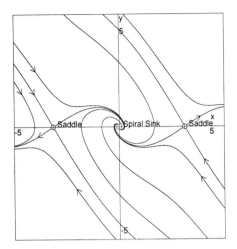

FIGURE 8.1.2. The phase plane for equation (5). The zero at $(0,0)$ is a spiral sink; the zeroes at $(\pm\pi, 0)$ are saddles.   ▲

Example 8.1.4. We return to our opening Example 8.1.1.

$$\begin{aligned} x' &= x - y^2 + a + bxy \\ y' &= 0.2y - x + x^3 \end{aligned} \tag{3}$$

for the values $a = 1.28$ and $b = 1.4$.

For the four zeroes at A, B, C, and E we have numerically computed the linearization, and Figure 8.1.3 on the next page shows side-by-side blowups of the phase plane for the nonlinear equation (3) and the associated linearizations.   ▲

Let us now formalize these notions of what is happening at the zeroes or equilibria of a nonlinear differential equation.

Definition 8.1.5. Let

$$\mathbf{f}(\mathbf{x}) = \begin{bmatrix} f_1(\mathbf{x}) \\ \vdots \\ f_n(\mathbf{x}) \end{bmatrix}$$

be a twice-differentiable vector field on an open subset U of $\mathbb{R}^n$, and let $\mathbf{x}_0$ be a zero of $\mathbf{f}$. The *linearization* of the differential equation $\mathbf{x}' = \mathbf{f}(\mathbf{x})$ at $\mathbf{x}_0$ is the linear differential equation

$$\boldsymbol{\xi}' = A\boldsymbol{\xi}, \tag{8}$$

where

$$\boldsymbol{\xi} = \mathbf{x} - \mathbf{x}_0$$

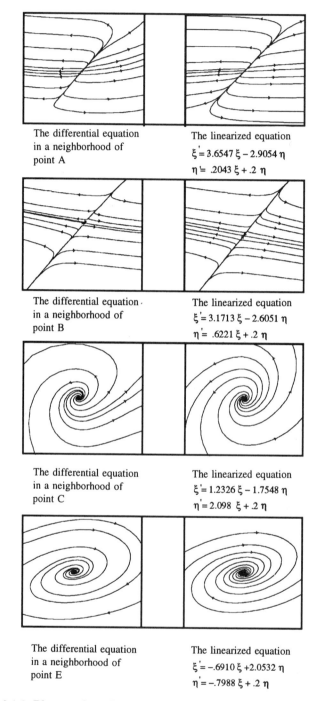

The differential equation in a neighborhood of point A

The linearized equation
$$\xi' = 3.6547\,\xi - 2.9054\,\eta$$
$$\eta' = .2043\,\xi + .2\,\eta$$

The differential equation in a neighborhood of point B

The linearized equation
$$\xi' = 3.1713\,\xi - 2.6051\,\eta$$
$$\eta' = .6221\,\xi + .2\,\eta$$

The differential equation in a neighborhood of point C

The linearized equation
$$\xi' = 1.2326\,\xi - 1.7548\,\eta$$
$$\eta' = 2.098\,\xi + .2\,\eta$$

The differential equation in a neighborhood of point E

The linearized equation
$$\xi' = -.6910\,\xi + 2.0532\,\eta$$
$$\eta' = -.7988\,\xi + .2\,\eta$$

FIGURE 8.1.3. Blowups from Figure 8.1.1, with associated linearizations.

and

$$A = \begin{bmatrix} \frac{\partial f_1}{\partial x_1}\big|_{\mathbf{x}_0} & \cdots & \frac{\partial f_1}{\partial x_n}\big|_{\mathbf{x}_0} \\ \vdots & & \vdots \\ \frac{\partial f_n}{\partial x_1}\big|_{\mathbf{x}_0} & \cdots & \frac{\partial f_n}{\partial x_n}\big|_{\mathbf{x}_0} \end{bmatrix}.$$

By Taylor's Theorem, we can write

$$\mathbf{f}(\mathbf{x}) = A(\mathbf{x} - \mathbf{x}_0) + \mathbf{h}(\mathbf{x} - \mathbf{x}_0),$$

where $\mathbf{h}(\boldsymbol{\xi})$ starts with quadratic terms, so $\boldsymbol{\xi}(t)$ "should" resemble $\mathbf{x}(t) - \mathbf{x}_0$.

Principle 8.1.6. *In general, near a zero or equilibrium point* (x_0, y_0), *the solutions of a nonlinear system of differential equations* $\mathbf{x}' = \mathbf{f}(\mathbf{x})$ *look like those of its linearization*

$$\xi' = \begin{bmatrix} \frac{\partial f_1}{\partial x_1}\bigg|_{\mathbf{x}_0} & \cdots & \frac{\partial f_1}{\partial x_n}\bigg|_{\mathbf{x}_0} \\ \vdots & & \vdots \\ \frac{\partial f_n}{\partial x_1}\bigg|_{\mathbf{x}_0} & \cdots & \frac{\partial f_n}{\partial x_n}\bigg|_{\mathbf{x}_0} \end{bmatrix} \xi.$$

We call this a principle, rather than a theorem, because the term "look like" is undefined. It will remain undefined, because it means rather different things depending on whether the linearized equation has a saddle or a source or whatever. What we will do in Sections 8.1 and 8.2 is to pick out from the linearized equation the properties of particular interest and show that they go over to the nonlinear setting.

Furthermore, let us elaborate on the caveat "in general":

Principle 8.1.6 is false when the matrix A has any purely imaginary eigenvalues or, in particular, if any eigenvalues are 0.

In the plane, this corresponds to the two following cases:

(i) *If* $\operatorname{tr} A = 0$ *and* $\det A > 0$ (implying "centers" for the linearization), we shall show by example that almost anything can happen. For differential equations with the same linearization, we get very different results in the nonlinear cases.

Example 8.1.7. Consider the differential equation

$$\begin{bmatrix} x \\ y \end{bmatrix}' = \begin{bmatrix} -y \\ x \end{bmatrix} + \varepsilon(x^2 + y^2) \begin{bmatrix} x \\ y \end{bmatrix}. \tag{9}$$

About $(0,0)$ the linearized equation is simply

$$\begin{bmatrix} x \\ y \end{bmatrix}' = \begin{bmatrix} -y \\ x \end{bmatrix} = \begin{bmatrix} 0 & -1 \\ 1 & 0 \end{bmatrix} \begin{bmatrix} x \\ y \end{bmatrix}, \tag{10}$$

so $\operatorname{tr} A = 0$, and the trajectories of solutions to the linearized equation (10) are circles. However, the solutions of the nonlinear system (9) spiral out if $\varepsilon > 0$ and spiral in if $\varepsilon < 0$. ▲

We can make far more complicated examples; for instance, see Example 8.4.3, which has infinitely many periodic solutions.

(ii) *If* $\det A = 0$, the linearization is degenerate and the linearized vector field vanishes identically on a line (or even all of $\mathbb{R}^2$). Adding on an arbitrarily small perturbing term can destroy this feature, so again the appearance of the perturbed vector field can differ essentially from the appearance of the linearization.

Example 8.1.8. Consider the system of equations

$$\begin{aligned} x' &= x^2 \\ y' &= -y. \end{aligned} \tag{11}$$

Exercise 8.1#4 asks you to find the linearization, to show that an eigenvalue of the linearized equation is 0, and to show that the equation does *not* behave like its linearization. ▲

Remark. In Chapter 9, we shall examine general cases and show that Example 8.1.7 leads to "Hopf bifurcation," and Example 8.1.8 leads to "saddle-node bifurcation."

In the plane, if all eigenvalues have a nonzero real part, Principle 8.1.6 means what it says. There is a unified way of making this statement: By making a smooth (nonlinear) change of variables near the equilibrium point you can make the equation into its linearization; you can even require that the derivative of the change of variables be the identity at the zero. A graphic way of saying this is that if the space of trajectories were drawn on a piece of rubber paper, then you could stretch and distort the rubber sheet so that the trajectories of the original equation will fit precisely on those of the linearization; you can even keep your thumb down on the equilibrium point, thereby fixing the point and the tangent vectors, and prevent the rubber sheet from moving or turning around that point.

The proof of the result in that form is quite delicate, so we will take a different approach. We will try to isolate the salient features of each type of zero of a vector field and show that these features carry over to the nonlinear equation near the zero.

GLOBAL BEHAVIOR OF SOLUTIONS

Look back at Figure 8.1.1. Solutions tend either to the sink E or to ∞, except for the zeroes and the exceptional solutions which tend to the saddles. Since such solutions separate generic forms of behavior, they are called *separatrices*; their existence and basic properties are covered in Section 8.3. In $\mathbb{R}^2$, it will be true that almost all solutions have a generic form of behavior and that the classes with a given behavior are separated by *exceptional solutions*. Figure 8.1.4 shows all the zeroes of the vector field, but only the trajectories that are the separatrices of the saddles; consider how these separatrices divide Figure 8.1.4 into regions of generic behavior.

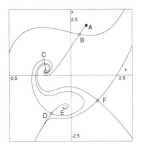

FIGURE 8.1.4. The zeroes and separatrices for Examples 8.1.1 and 8.1.4. Point E is a sink and points A and C are sources.

Figure 8.1.5 shows how the separatrices of the saddles divide the plane respectively into the *basin* of the sink (the points that tend to the sink as $t \to \infty$) and the points that tend to ∞ as $t \to \infty$. The second picture gives the "basins," called *cobasins*, of the sources (the points that tend to the sources as $t \to -\infty$).

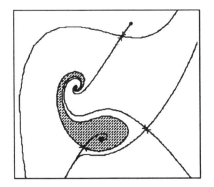
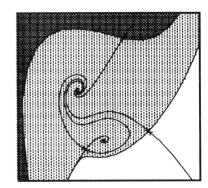

FIGURE 8.1.5. Basins and cobasins for Examples 8.1.1 and 8.1.4.

Figures 8.1.1–8.1.5 illustrating sinks, sources, and saddles show the generic *local* forms of behavior. For nonlinear differential equations in $\mathbb{R}^2$, there is also a *global* generic form of behavior: *limit cycles*, which do not appear in the linear case. Solutions can be attracted to stable limit cycles as well as to sinks; classes of solutions can be separated by unstable limit cycles as well as separatrices.

Example 8.1.9. Consider the differential equation

$$\begin{bmatrix} x \\ y \end{bmatrix}' = [\, y(2 - x^2)y - x \,],$$

which will be discussed in more detail in Example 8.4.2.

As is shown in Figure 8.1.6, this example has an attractive and stable limit cycle, with a node source in the center.

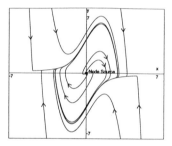

FIGURE 8.1.6. A limit cycle is attracting all solutions except the constant solution at the origin.   ▲

An unstable limit cycle is repelling, looking like Figure 8.1.6 with all the arrows reversed and a sink at the center. From the forward point of view, an unstable limit cycle acts as a separatrix; if time runs backward, it acts as an attractor.

There is a theorem that allows you to locate limit cycles, whether stable or unstable, if you can guess where they ought to be. This result is called the Poincaré–Bendixson Theorem, and will be discussed in Section 8.5. We will also see in that section how hard it can be to actually prove that there is only one limit cycle.

SUMMARY

The general strategy for understanding an autonomous system in the plane will be to

(1) Locate the zeroes of the vector field and analyze them.

(2) Draw the separatrices with arrows by starting at each saddle in the direction of the eigenvectors and then "follow your nose" through the vector field.

(3) Locate the stable and unstable limit cycles if you can. (Using a phase plane computer program is the best bet; see *MacMath.*)

(4) Figure out from the separatrices and limit cycles the boundaries of the basins and cobasins of the sinks, sources, and limit cycles.

If you are able to do all this, you can claim to *understand the differential equation.* In reality, locating the zeroes is hard: it involves solving a system of nonlinear (nondifferential) equations, and there is no very good way to do that. The *DiffEq Phase Plane* computer program in *MacMath* will zoom in on zeroes (singularities) one at a time if you give it a good enough initial guess, but it gives no guarantee that you have found them all.

Once you have located a zero, analyzing it is routine. (The computer, however, had to be taught to differentiate):

(1) Set new coordinates (ξ, η) to indicate displacement from that particular equilibrium.

(2) Replace each term in the original nonlinear differential equation by the linear term in its Taylor polynomial (expanded about the equilibrium point).

(3) Calculate the trace and determinant of the linearization to determine the behavior of trajectories at that equilibrium. (See Figure 7.5.7, the bifurcation diagram.)

Locating limit cycles is even more difficult, and impossible without a computer. However, with computer graphics phase portraits, you can, in fact, usually locate and analyze limit cycles.

We shall now proceed to discuss in detail each of the possible types of zeroes of the vector field (Sections 8.2 and 8.3) and limit cycles (Section 8.4).

8.2 Sources Are Sources and Sinks Are Sinks

The object of this section is well described by the title: we wish to show that if a differential equation has a singularity, and if the linearization there is a sink, then the nonlinear differential equation also has a sink. In this case, it is not much easier to deal with the two-dimensional case than the general case, so we will begin work in $\mathbb{R}^n$.

First we need to know what a nonlinear sink is (and its opposite, a source).

Definition 8.2.1. Let $\mathbf{x}' = \mathbf{f}(\mathbf{x})$ be an autonomous differential equation in $\mathbb{R}^n$ and let $\mathbf{x}_0$ be a zero of the vector field.

Then $\mathbf{x}_0$ is a *sink* if there is a neighborhood U of $\mathbf{x}_0$ such that any solution $\mathbf{u}(t)$ with $\mathbf{u}(t_0) \in U$ remains in U for $t \geq t_0$, and $\lim_{t \to \infty} \mathbf{u}(t) = \mathbf{x}_0$.

The point $\mathbf{x}_0$ is a *source* if there is a neighborhood U of $\mathbf{x}_0$ such that any solution $\mathbf{u}(t)$ with $\mathbf{u}(t_0) \in U$ was in U for all $t \leq t_0$ and $\lim_{t \to -\infty} \mathbf{u}(t) = \mathbf{x}_0$.

Remark. The limit condition is needed to exclude centers.

Theorem 8.2.2. *If at a zero of an autonomous differential equation the linearization is a sink or a source, then the zero is itself a sink or a source. Furthermore, all solutions sufficiently close to the zero tend to it exponentially fast as $t \to \infty$ for a sink or as $t \to -\infty$ for a source.*

Proof. We will treat only the case of sinks; the source case is similar. We may change coordinates so that the zero under consideration is at the origin and the differential equation can be written

$$\mathbf{x}' = A\mathbf{x} + \mathbf{q}(\mathbf{x}), \tag{12}$$

where $\mathbf{q}$ starts with at least quadratic terms, so that for some constants C and R, we have

$$\|\mathbf{q}(\mathbf{x})\| \leq C\|\mathbf{x}\|^2 \quad \text{for } \|\mathbf{x}\| < R. \tag{13}$$

The proof will consist of Lemma 8.2.3 and Lemma 8.2.4. The idea is straightforward: we already know that the linearization is a sink, and we need to know that the nonlinear term $\mathbf{q}$ is sufficiently small near the zero of the differential equation and negligible by comparison with the linear term.

To deal with the linear term, we cannot quite use the argument of Theorem 7.6.1. In Chapter 7, we used Theorem L6.9, which provided us with a *complex* basis with respect to which solutions have decreasing norm. Of course, a real matrix can always be considered as a complex matrix, so the use of a complex basis was justified so long as we were only using linear equations.

In our present case, the function $\mathbf{q}(\mathbf{x})$ may not be defined for complex values of $\mathbf{x}$, and although we could extend it in some way, we prefer to stick to real bases.

Lemma 8.2.3. *Let A be a real $n \times n$ matrix whose eigenvalues have a negative real part, so that if*

$$-\Lambda = \sup\{\operatorname{Re} \lambda | \lambda \text{ is an eigenvalue of } A\},$$

then $\Lambda > 0$. Then for any C with $0 < C < \Lambda$, there exists a basis of $\mathbb{R}^n$ in which $A\mathbf{x} \cdot \mathbf{x} < -C\|\mathbf{x}\|^2$ for every $\mathbf{x} \in \mathbb{R}^n$.

Proof. Theorem L6.19 asserts: If A has k pairs of conjugate nonreal eigenvalues, then there exists a basis $\mathbf{w}_1, \ldots, \mathbf{w}_n$ of $\mathbb{R}^n$ such that for any ε, in that basis, A is k-pseudo-upper-triangular with terms of absolute value $< \varepsilon$ above the pseudo-diagonal. Exercise 8.2–8.3#3 asks you to prove that

$$A\mathbf{x} \cdot \mathbf{x} \le (-\Lambda + n\varepsilon)\|\mathbf{x}\|^2,$$

where $-\Lambda$ is the largest real part of an eigenvalue. Choose $\varepsilon < (\Lambda - C)/n$, and the lemma follows. $\square$

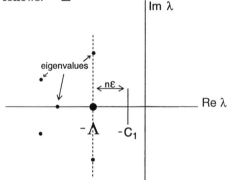

FIGURE 8.2.1. Relative positions of various constants.

Now we can deal with the nonlinear case, referring to Figure 8.2.1.

Lemma 8.2.4. Let A be a real matrix satisfying $A\mathbf{x} \cdot \mathbf{x} \le -C_1\|\mathbf{x}\|^2$, and let $\mathbf{q}(\mathbf{x})$ satisfy $\|\mathbf{q}(\mathbf{x})\| \le C_2\|\mathbf{x}\|^2$ for $\|\mathbf{x}\| \le R$. Let $\rho = \min(R, C_1/2C_2)$. Then any solution of $\mathbf{x}' = A\mathbf{x} + \mathbf{q}(\mathbf{x})$ with $\|\mathbf{x}(0)\| \le \rho$ satisfies

$$\|\mathbf{x}(t)\|^2 \le \|\mathbf{x}(0)\|^2 e^{-C_1 t}$$

for $t \ge 0$.

Proof. This is a "fence" result, as illustrated in Figure 8.2.2: If $\mathbf{x}(t)$ is a solution to $\mathbf{x}' = A\mathbf{x} + \mathbf{q}(\mathbf{x})$, then

$$d/dt(\|\mathbf{x}(t)\|^2) = 2\mathbf{x}'(t) \cdot \mathbf{x}(t) = 2(A\mathbf{x}(t)) \cdot \mathbf{x}(t) + 2\mathbf{q}(\mathbf{x}(t)) \cdot \mathbf{x}(t) \\ \le -2C_1\|\mathbf{x}(t)\|^2 + 2C_2\|\mathbf{x}(t)\|^3 \tag{14}$$

whenever $\|\mathbf{x}(t)\| < R$. Our definition of ρ then gives, when $\|\mathbf{x}(t)\| \le \rho \le C_1/(2C_2)$,

$$d/dt(\|\mathbf{x}(t)\|^2) \le -C_1\|\mathbf{x}(t)\|^2. \tag{15}$$

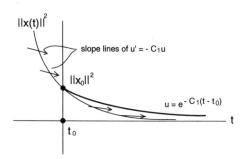

FIGURE 8.2.2. The fence described in the proof of Lemma 8.2.4.

Equation (15) says that $\|\mathbf{x}(t)\|^2$ is a lower fence for the differential equation $u' = -C_1 u$, so

$$\|\mathbf{x}(t)\|^2 \leq \|\mathbf{x}(0)\|^2 e^{-C_1 t}$$

if $\|\mathbf{x}(0)\| \leq \rho$.

Hence, $\|\mathbf{x}\| \leq \rho$ for all $t > 0$, which justifies the use of (15).   □

This completes the proof of Theorem 8.2.2.   □

Remark. Unlike the case of linear equations, it is quite possible for a zero to be a sink without the linearization being a sink, as shown by the following:

Example 8.2.5. Consider the differential equation

$$\begin{bmatrix} x \\ y \end{bmatrix}' = \begin{bmatrix} 0 & -1 \\ 1 & 0 \end{bmatrix} \begin{bmatrix} x \\ y \end{bmatrix} \begin{bmatrix} x \\ y \end{bmatrix} - (x^2 + y^2) \begin{bmatrix} x \\ y \end{bmatrix} \qquad (16)$$

whose linearization at $(0,0)$ is a center, but whose phase portrait looks like Figure 8.2.3.

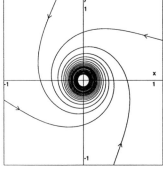

FIGURE 8.2.3. Phase plane for equation (16).

Let $u(t) = x(t)^2 + y(t)^2$; then

$$u'(t) = 2x(t)x'(t) + 2y(t)y'(t) = 2x(-y - ux) + 2y(x - uy)$$
$$= -2u(x(t)^2 + y(t)^2) = -2(u(t))^2,$$

and $u(t) = 1/(2t - C)$ is the general solution of this new differential equation in u, representing distance from the origin of a solution to equation (16). In particular, this information about u tells us that solutions to the original system (16) go to $(0,0)$ as $t \to \infty$, although much more slowly than exponentially. ▲

Sinks like that in Example 8.2.5 will be called *weak sinks*. They will be major actors in Chapter 9, under Hopf bifurcations.

LIAPUNOV FUNCTIONS

There is a way to deal simultaneously with zeroes of vector fields for which the linearization is a sink, and at least some of those for which the linearization is not a sink but which are nevertheless (weak) sinks.

Observe that in Example 8.2.5, the quantity $d/dt\|\mathbf{x}(t)\|^2$ is strictly negative when $\mathbf{x}(t) \neq 0$. The proof of Theorem 8.2.4 used the same principle: the equation

$$\frac{d}{dt}\|\mathbf{x}(t)\|^2 \le -C_1\|\mathbf{x}(t)\|^2 \tag{15}$$

says that with respect to an appropriate basis, the function $\|\mathbf{x}(t)\|^2$ decreases with time (and at a particular rate). In both cases, it is handy to define a *Liapunov function*. Then the fact that the origin is a sink will follow from Theorem 8.2.7.

This special type of function is named for Aleksandr Mikhailovich Liapunov, a Russian mathematician from the late nineteenth and early twentieth centuries, who specialized in the stability of systems, particularly those due to rotating heavy fluids.

Definition 8.2.6. A continuously differentiable function $F(\mathbf{x})$ defined in a neighborhood U of a zero $\mathbf{x}_0$ of the vector field $\mathbf{f}$ is a *Liapunov function* for the differential equation $\mathbf{x}' = \mathbf{f}(\mathbf{x})$ on U if

(1) F has its *unique* minimum in U at $\mathbf{x}_0$,

(2) $\nabla F(\mathbf{x}) \cdot \mathbf{f}(\mathbf{x}) < 0$ when $\mathbf{x}(t) \in U$ and $\mathbf{x}(t) \neq \mathbf{x}_0$.

Note that you do not need to solve the differential equation $\mathbf{x}' = \mathbf{f}(\mathbf{x})$ to check that a function is a Liapunov function. The second condition above, by the chain rule, can be written as $\frac{d}{dt}F(\mathbf{x}(t)) < 0$, which means that F decreases along solutions.

In the plane, this condition means that all solutions of the differential equation cross the level curves of F from the outside to the inside, as in Figure 8.2.4.

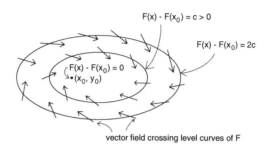

vector field crossing level curves of F

FIGURE 8.2.4. Two level curves of a Liapunov function in a direction field for a differential equation.

Theorem 8.2.7. *If a zero of a differential equation admits a Liapunov function, then the point is a sink.*

The proof we will give for Theorem 8.2.7 is rather nonconstructive, and this cannot be avoided unless we make some assumption about how fast $\mathbf{f}$ is decreasing along solutions. In Exercise 8.2–8.3#8, we will propose some alternative proofs under stronger hypotheses (of course, the proof of Theorem 8.2.2 or Lemma 8.2.4 is already such an alternative proof). We will write the proof in the plane; it is not really harder in higher dimensions, but the geometric language of "level curves of F'" becomes more intimidating.

Proof of Theorem 8.2.7. Suppose $F(\mathbf{x}_0) = 0$. [If not, use $F(\mathbf{x}) - F(\mathbf{x}_0)$ instead of F throughout this proof.]

Choose a circle C in U centered at $\mathbf{x}_0$; the function F must assume a minimum $M > 0$ on this circle C. Let V be the region inside C where $F < M$; this is a new neighborhood of $\mathbf{x}_0$, smaller than U, and we will show that every solution which enters V is attracted to $\mathbf{x}_0$.

Suppose that $\mathbf{u}(t_0) \in V$. Since the function $F(\mathbf{u}(t))$ is decreasing, it can never increase to M for $t > t_0$. Hence, the solution will stay in V for all $t > t_0$.

Remark. What makes this proof unpleasant is that we cannot speak of $\lim_{t \to \infty}(\mathbf{u}(t))$ without some precautions: $\mathbf{u}(t)$ might accumulate on a limit cycle or some more complicated object and not have a limiting value. We *can* say that for any sequence $t_i \to \infty$, there is a subsequence t_{i_j} such that

$$\lim_{i \to \infty} \mathbf{u}(t_{i_j})$$

exists. This is simply the statement "any sequence in a compact set has a convergent subsequence," which you should take on faith if you do not know it already. However, you should also be a little wary—this is where the nonconstructivity is hidden. The "simple" statement above gives no hint how to find the subsequence.

Proof, continued. Take any well-spaced sequence of times tending to ∞, for instance, the integers. Since the sequence $F(\mathbf{u}(n))$ has a limit, the sequence of differences

$$F(\mathbf{u}(n+1)) - F(\mathbf{u}(n))$$

tends to 0. By the Mean Value Theorem, there exists some number $s_n \in [n, n+1]$ such that

$$\frac{d}{dt}F(\mathbf{u}(s_n)) = F(\mathbf{u}(n+1)) - F(\mathbf{u}(n)) \tag{17}$$

tends to 0.

Extracting a subsequence of s_n, we may suppose that $\mathbf{u}(s_n)$ converges to some point $\mathbf{u}_\infty$. But if $\mathbf{u}_\infty \neq \mathbf{x}_0$,

$$\frac{d}{dt}F(\mathbf{u}_\infty) < 0.$$

This contradicts equation (17), so $\mathbf{u}(s_n)$ tends to $\mathbf{x}_0$ as $n \to \infty$.

This is not quite enough to prove the theorem; it just shows that there is a sequence of times at which the solution tends to $\mathbf{x}_0$. It does show, however, that $\lim_{t\to\infty} F(\mathbf{u}(t)) = 0$, since the function $F(\mathbf{u}(t))$ is decreasing and certainly has a limit. And this is good enough. If $\mathbf{u}(t)$ does not converge to $\mathbf{x}_0$, then there is a sequence t_i tending to ∞ such that $\mathbf{u}(t_i)$ does converge in V to some point $\mathbf{v}_\infty \neq \mathbf{x}_0$. Then $F(\mathbf{u}(t_i))$ tends to $F(\mathbf{v}_\infty) \neq 0$, and this is a contradiction.   □

Example 8.2.8. Consider the second order equation

$$x'' + f(x)x' + g(x) = 0, \tag{18}$$

with $f(x) > 0$ and $x, g(x)$ of the same sign for $x \neq 0$. This equation is called Liénard's equation and should make you think of the damped harmonic oscillator; the condition $f(x) > 0$ says that the friction is positive, and the condition $xg(x) > 0$ says that the force really *restores*. Of course, the ordinary damped harmonic oscillator equation is a special case of Liénard's equation, but in general the equation is nonlinear, with the friction and the restoring force depending on the position.

It is not really surprising, in view of the signs, that the origin should be a sink, and it is.

Replace the second order equation (18) by the equivalent system

$$\begin{bmatrix} x \\ y \end{bmatrix}' = \begin{bmatrix} y \\ -f(x)y - g(x) \end{bmatrix} = \mathbf{f}(\mathbf{x}).$$

Set $G(x) = \int_0^x g(u)du$ and consider

$$F(x, y) = \frac{y^2}{2} + G(x).$$

Note first that F does have an absolute minimum at $(0,0)$, since G has an absolute minimum at 0 [that is why we assumed $xg(x) > 0$], and, second, that

$$\nabla F(x, y) \cdot \mathbf{f}(x, y) = -y^2 f(x) < 0$$

when $(x, y) \neq (0, 0)$.

If we make the stronger assumption that $G(x) \to \infty$ as $x \to \infty$, we get the stronger statement that $(0, 0)$ attracts the entire plane. Actually, this is already proved in the proof of Theorem 8.2.7. In that proof, there is at the beginning a seemingly fussy argument to get a compact region V which the vector field always enters; we then showed that all solutions that enter such a region are attracted to the minimum of F.

For this example, every point (x_1, y_1) is in the region $F(x, y) \leq C$ if $C > F(x_1, y_1)$; the hypothesis on G is exactly what is needed to see that this region is compact.

Exercise 8.2–8.3#7 provides an example to show what can go wrong if the hypothesis is removed.   ▲

SINKS AND SOURCES IN THE PLANE

We now know that if the linearization of a vector field $\mathbf{f}$ at a zero $\mathbf{x}_0$ of $\mathbf{f}$ is a sink or a source, then the differential equation $\mathbf{x}' = \mathbf{f}(\mathbf{x})$ also has a sink or a source at $\mathbf{x}_0$. One might like more precise information, at least for vector fields in the plane, where the linear equations are completely classified. For instance, if the linearization spirals, then do the solutions of the nonlinear differential equation spiral also? This is indeed true, as we will now show.

The case of spiral sinks. A typical spiral sink is shown in Figure 8.2.5.

FIGURE 8.2.5. The nonlinear sink E from Figure 8.1.1.

Definition 8.2.9. If $\mathbf{f}$ is a vector field in $\mathbb{R}^2$ and $\mathbf{x}_0$ is a zero of $\mathbf{f}$ at which the linearization of $\mathbf{x}' = \mathbf{f}(\mathbf{x})$ is a spiral sink, we will say that $\mathbf{x}_0$ is a *spiral sink of the differential equation* $\mathbf{x}' = \mathbf{f}(\mathbf{x})$.

This definition is justified by the following result.

Theorem 8.2.10. *If $\mathbf{x}_0$ is a spiral sink for $\mathbf{x}' = \mathbf{f}(\mathbf{x})$, then the solutions starting sufficiently close to $\mathbf{x}_0$ spiral around $\mathbf{x}_0$.*

Proof. We may change coordinates to put $\mathbf{x}_0$ at the origin, and if we use the real and imaginary parts of an eigenvector as a basis, then we saw in Section 7.5 that the differential equation will be of the form

$$\mathbf{x}' = \begin{bmatrix} a & -b \\ b & a \end{bmatrix} \mathbf{x} + \mathbf{q}(\mathbf{x}). \tag{19}$$

We would like to know that the solutions spiral. Let $\theta(t)$ be the polar angle of a solution $(x(t), y(t))$. Since $\theta = \arctan(y/x)$, we have

$$\theta' = \frac{xy' - x'y}{x^2 + y^2} = \frac{x(bx + ay + Q(x,y)) - y(ax - by + P(x,y))}{x^2 + y^2}$$

$$= b + \frac{xQ(x,y) - yP(x,y)}{x^2 + y^2}, \tag{20}$$

where P and Q are the coordinate functions of $\mathbf{q}$.

We saw in Theorem 8.2.2 that if $(x(0), y(0))$ is sufficiently small, then $\|(x(t), y(t))\|$ will be (exponentially) small as $t \to \infty$. Furthermore, since in expression (20) for θ' the terms other than b are at least cubic in the numerator and quadratic in the denominator, they give something arbitrarily small for (x, y) near $(0, 0)$. We see that for any ε, if $(x(0), y(0))$

is chosen sufficiently small, then $\theta'(t)$ has constant sign for $t > 0$, and $|\theta'(t)| > |b| - \varepsilon$ for $t > 0$. It immediately follows that the solution turns infinitely many times around 0. $\square$

In Exercises 8.2–8.3#9 and #10, we will show that Theorem 8.2.10 is in some sense the best possible: if the linearization of a differential equation at a zero of the vector field is a sink with equal eigenvalues, then the solutions do not spiral.

The case of node sinks. A typical node sink is shown in Figure 8.2.6.

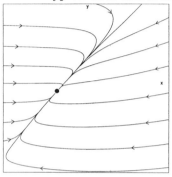

FIGURE 8.2.6. A nonlinear node sink. (This is the opposite of the node source A in Figure 8.1.1.)

Theorem 8.2.11. *A differential equation in $\mathbb{R}^2$ has a node sink at a singular point if its linearization at that point has a node sink. This means that:*

(i) *All solutions which start sufficiently close to the sink approach the sink tangentially to an eigendirection as $t \to \infty$.*

(ii) *Precisely two trajectories approach tangentially to the eigendirection corresponding to the eigenvalue having smaller absolute value, from opposite sides; all others approach tangentially to the eigendirection with eigenvalue having larger absolute value.*

Proof. We have already seen in Theorem 8.2.2 that the singular point is a sink if the linearization is a sink, so we only need to show the statements about directions of trajectories. In the case of node sinks, the linearization has two linearly independent eigenvectors. The proof of Theorem 8.2.11 is much easier if we begin our analysis

- using coordinates that place the node equilibrium (x_0, y_0) at the origin, so that $\xi = x$ and $\eta = y$,

- using the eigenvectors as a basis, which places them along the axes.

In that case the differential equation is, for P and Q starting with quadratic terms,

$$x' = -\mu x + P(x, y)$$
$$y' = -\nu y + Q(x, y)$$
(21)

where $-\mu$ and $-\nu$ are the distinct negative eigenvalues, and we will assume that $-\mu < -\nu < 0$, so the picture looks like Figure 8.2.8; that is, for every trajectory not on the x-axis, the x-coordinate approaches 0 faster than the y-coordinate; thus, such a trajectory ends up tangent to the y-axis.

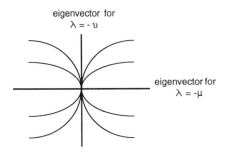

FIGURE 8.2.7. Phase portrait for $x' = -\mu x, y' = -\nu y$ with $-\mu < -\nu < 0$.

For the linearized equation, all solutions approach the origin tangentially to the y-axis except those whose trajectory is on the x-axis. We would like to know that for the nonlinear equation there is also a single trajectory tangent to the x-axis and that all the other solutions tend to the sink tangentially to the y-axis. This kind of statement should remind you of the *funnels* and *antifunnels* of Part I; indeed, the proofs are just clever choices of fences so that the funnel and antifunnel theorems (with minor modifications) apply (See Appendix T concerning Chapters 1 and 4 in Part I).

Funnels and antifunnels apply to differential equations in one variable, not to systems; however, the trajectories of solutions to the system lie along graphs of solutions to

$$\frac{dy}{dx} = \frac{\nu y - Q(x, y)}{\mu x - P(x, y)} = \psi(x, y)$$
(22)

thinking of y as a function of x, and of

$$\frac{dx}{dy} = \frac{\mu x - P(x, y)}{\nu y - Q(x, y)} = \phi(x, y)$$
(23)

thinking of x as a function of y.

You do need to worry about the curves along which the denominators vanish, as in equations (22) and (23) there is no differential equation there and, in particular, none of the theorems about funnels or antifunnels are true.

However, the curve of the equation $\mu x = P(x, y)$ where equation (22) fails to be defined is a twice-differentiable curve tangent to the y-axis, and similarly the curve of the equation $\nu y = Q(x, y)$ where equation (23) fails to be defined is a twice-differentiable curve tangent to the x-axis. As we will see, our choice of antifunnels will always exclude the curves along which the differential equations fail to be defined.

To prove Theorem 8.2.11, we will define three regions in the xy-plane, called U, V, and W, bounded by curves as shown in Figure 8.2.8, so that

(a) The differential equation (23) is defined in U, and U is a *funnel* for (23), so that all solutions that enter it stay in it as y increases to 0; in particular, they approach $(0, 0)$ tangentially to the y-axis.

(b) The differential equation (22) is defined in V, and V is an *antifunnel* for (22) with the uniqueness condition, so there exists a unique trajectory staying in V as x increases to 0.

(c) W is also an *antifunnel* for (22), so the unique trajectory specified in (b) approaches $(0, 0)$ tangentially to the x-axis.

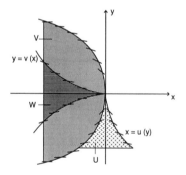

FIGURE 8.2.8. Funnels and antifunnels for system (21).

The most delicate part of the proof is the uniqueness in (b), because the solutions in this antifunnel are in fact converging, but not quite fast enough to stay between the fences.

It should be clear that similar statements will hold if x and y are positive; we chose the regions above because it is easier to think of x and y increasing.

Proof of (a). *The region U is a funnel, forcing solutions to approach 0 tangentially to the y-axis.*

Choose c with $1 < c < 2$ and $c < \mu/\nu$ and let $x = u(y) = |y|^c$.

Since $c > 1$, this curve does indeed approach $(0,0)$ tangentially to the y-axis. However, since $c < 2$, it approaches the y-axis more slowly than any twice-differentiable curve that is tangent to the y-axis. In particular, equation (22) is defined in the region V. Of course, equation (23) is defined in U, since ϕ is infinite only on a curve tangent to the x-axis.

Because y is negative in the region considered, we have $u'(y) = -c|y|^{c-1}$ and $|y|/y = -1$. This means that by factoring νy out of the denominator (Reference: Part I, Appendix A on Asymptotic Development), we can write

$$\phi(|y|^c, y) = \frac{\mu|y|^c - P(|y|^c, y)}{\nu y - Q(|y|^c, y)}$$

$$= \left(-\frac{\mu}{\nu}|y|^{c-1} - \frac{P(|y|^c, y)}{\nu y} \right) \left(\frac{1}{1 - Q(|y|^c, y)/\nu y} \right)$$

$$= \left(-\frac{\mu}{\nu}|y|^{c-1} - \frac{P(|y|^c, y)}{\nu y} \right) \left(1 + \frac{Q(|y|^c, y)}{\nu y} + \cdots \right),$$

and the leading term in y is $-(\mu/\nu)|y|^{c-1}$. Note that this used the requirement $c < 2$, since $\frac{P(|y|^c, y)}{\nu y}$ might contain linear terms in y, contributed by terms in y^2 of P.

Since $c < \mu/\nu$, we have

$$u'(y) = -c|y|^{c-1} > \phi(u(y), y)$$

for $y < 0$ and $|y|$ sufficiently small. Thus, part (a) is proved.

A similar computation with $u_1(y) = -u(y)$ shows that U is a funnel and, at the same time, that V is an antifunnel, which is the first part of the next step, (b).

Proof of (b). *There is a unique trajectory in the antifunnel V.*

We first need to find a function $K(x)$ such that in V we have

$$\frac{\partial \psi}{\partial y}(x, y) \geq K(x);$$

i.e., K gives a lower bound for the rate at which solutions can converge in V.

We compute from (22), using the fact that $(\partial P/\partial y)(0,0) = 0$, that

$$\frac{\partial \psi}{\partial y}(x, y) = \left(\frac{\nu}{\mu x} - \frac{\partial Q/\partial y}{\mu x} \right) \left(\frac{1}{1 - P(x, y)/\mu x} \right)$$

$$= \left(\frac{\nu}{\mu x} - \frac{\partial Q/\partial y}{\mu x} \right) \left(1 + \frac{P(x, y)}{\mu x} + \cdots \right).$$

Since $|y| \leq |x|^{1/c}$ in V, $\nu/\mu x$ is the leading term, and we can take

$$K(x) = \frac{\nu/\mu + \varepsilon}{x}$$

with an arbitrarily small ε if we choose $|x_0|$ sufficiently small. Note how far we are from being able to apply the easy uniqueness criterion $K(x) > 0$; our $K(x)$ goes to $-\infty$ as x increases to 0. However, the harder antifunnel Theorem 4.7.4 still applies.

Let us repeat the argument: If $\gamma_1(x)$ and $\gamma_2(x)$ are solutions in the antifunnel with $\gamma_1 > \gamma_2$, let $\eta(x) = \gamma_1(x) - \gamma_2(x)$. Then $\eta'(x) \geq K(x)\eta(x)$, so that

$$\eta(x) \geq \eta(x_0) e^{\int_{x_0}^x K(s)ds} = \eta(x_0) \left| \frac{x}{x_0} \right|^{(\nu/\mu)+\varepsilon}.$$

How does $|x|^{(\nu/\mu)+\varepsilon}$ compare to $|x|^{1/c}$? Since, from the beginning of (i), $c < \mu/\nu$, then $1/c > \nu/\mu$, and if ε is chosen sufficiently small, we have $1/c > \nu/\mu + \varepsilon$.

The smaller the power of $|x|$, the higher the graph of the function near 0, and we see that the function $\eta(x)$ is larger than the distance $2|x|^{1/c}$ between the top and the bottom of the antifunnel, so only one solution can pass through the origin. This proves uniqueness.

Proof of (c). *The exceptional solution is tangent to the x-axis.*

Let $v(x) = |x|^\delta$ for any δ with $1 < \delta < 2$. Then, since x is negative, $v'(x) = -\delta|x|^{\delta-1}$, and

$$\psi(x, v(x)) = \frac{\nu|x|^\delta - Q(x, |x|^\delta)}{\mu x - P(x, |x|^\delta)}$$

$$= \left(-\frac{\nu}{\mu}|x|^{\delta-1} - \frac{Q(x, |x|^\delta)}{\mu x} \right) \frac{1}{(1 + P(x, |x|^\delta)/\mu x)}$$

$$= \left(-\frac{\nu}{\mu}|x|^{\delta-1} - \frac{Q(x, |x|^\delta)}{\mu x} \right) \left(1 - \frac{P(x, |x|^\delta)}{\mu x} + \cdots \right).$$

The leading term is $-(\nu/\mu)|x|^{\delta-1}$, and since $\nu/\mu < 1 < \delta$, we have

$$v'(x) < \psi(x, v(x)).$$

A similar argument about $v_1 = -v$ shows that W is an antifunnel. The unique solution in the antifunnel V must also lie in W, which forces it to be tangent to the x-axis.

The proof of Theorem 8.2.11 is now complete.   □

The following example will explain why noninteger exponents were necessary in the proof of the preceding theorem.

Example 8.2.12. Consider the system

$$x' = -2x + y^2$$
$$y' = -y.$$

The associated linear system has solutions whose trajectories lie on parabolas $x = Cy^2$; we will see that the nonlinear equation is different in a small but annoying way.

This particular nonlinear system can easily be solved: the second equation alone yields

$$y(t) = C_1 e^{-t},$$

and then the first equation becomes

$$x' = -2x + C_1^2 e^{-2t},$$

which can be solved by variation of parameters, giving

$$x(t) = e^{-2t} \left(\int_0^t e^{2s} C_1^2 e^{-2s} ds + C_2 \right)$$
$$= e^{-2t} \left(C_1^2 t + C_2 \right).$$

To find the trajectories in the xy phase plane, we need to eliminate t from the equations for $x(t)$ and $y(t)$. From the equation for $y(t)$, we get

$$t = -\log|y/C_1|,$$

which if inserted in the equation for $x(t)$ gives

$$x = -y^2 \log|y| + C_3 y^2,$$

where $C_3 = (C_2/C_1^2 + \log|C_1|)$.

The important thing to realize about these trajectories for the nonlinear system is that they do not approach the y-axis as fast as the parabolas that we found for the linear equation. Any attempt to fence in the solutions approaching $(0,0)$ tangentially to the y-axis by parabolas is going to fail, and we will have to use a funnel with a wider mouth.   ▲

8.3 Saddles

Definition 8.3.1. An autonomous differential equation $x' = f(x)$ in $\mathbb{R}^2$ has a *saddle* at a zero x_0 of f if its linearization there has a saddle.

Theorem 8.3.2. *If an autonomous differential equation $x' = f(x)$ has a saddle at x_0, then there are precisely* *two trajectories that tend to $\mathbf{x}_0$ as $t \rightarrow \infty$, that together with $\mathbf{x}_0$ form a smooth curve tangent at $\mathbf{x}_0$ to the line of eigenvectors with negative eigenvalue (the stable direction),*

and precisely

two trajectories that tend to $\mathbf{x}_0$ as $t \rightarrow -\infty$, that together with $\mathbf{x}_0$ form a smooth curve tangent at $\mathbf{x}_0$ to the line of eigenvectors with positive eigenvalue (the unstable direction).

As we have seen in Figures 8.1.4 and 8.1.5, the trajectories of Theorem 8.3.2, called separatrices, typically separate the regions of generic behavior, and are, as such, the most important objects to understand about a vector field in the plane. A typical saddle is shown with its separatrices in Figure 8.3.1.

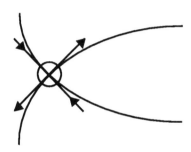

FIGURE 8.3.1. A saddle with eigenvectors and separatrices.

Proof of Theorem 8.3.2. The proof is similar to that for Theorem 8.2.11. Here also we

(i) use coordinates that place the saddle equilibrium (x_0, y_0) at the origin, so that $\xi = x$ and $\eta = y$,

(ii) use the eigenvectors as a basis, which places them along the axes.

In that case, the nonlinear differential equation is

$$\begin{bmatrix} x \\ y \end{bmatrix} = \begin{bmatrix} -\mu & 0 \\ 0 & \nu \end{bmatrix} \begin{bmatrix} x \\ y \end{bmatrix} + \begin{bmatrix} P(x,y) \\ Q(x,y) \end{bmatrix}$$

or

$$\begin{aligned} x' &= -\mu x + P(x,y) \\ y' &= \nu y + Q(x,y), \end{aligned} \tag{24}$$

with $-\mu$ and ν the negative and positive eigenvalues, respectively, and P and Q starting with quadratic terms.

The fact that the separatrices of a saddle divide the vector field into regions of generic behavior should remind you of *antifunnels*, and indeed antifunnels are used in the proof. As before, recall that funnels and antifunnels apply to differential equations in one variable, not to systems; however, the trajectories of solutions to system (24) follow the graphs of solutions to the first order differential equation

$$\frac{dy}{dx} = \frac{\nu y + Q(x, y)}{-\mu x + P(x, y)}. \tag{25}$$

Equation (25) is less satisfactory than equation (24): it is not a differential equation wherever $\mu x = P(x, y)$, so that the denominator vanishes, and in particular, none of the theorems about funnels and antifunnels are true when this happens. However, the curve $\mu x = P(x, y)$ is an (implicitly defined) twice-differentiable curve tangent to the y-axis (Exercise 8.2–8.3#11), and our choice of antifunnels will completely avoid curves tangent to the y-axis at the origin.

Consider in the left half of the (x, y)-plane the shaded region U, bounded above and below by curves of equation

$$y = \pm \gamma x^2 \quad \text{for some } x_0 < x < 0 \text{ and } \gamma > 0,$$

and the lighter region V for which $|y| < |x|$, as shown in Figure 8.3.2.

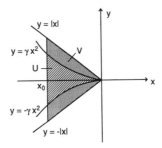

FIGURE 8.3.2.

Lemma 8.3.3. *For γ sufficiently large and x_0 sufficiently small, the regions U and V described above are both antifunnels for equation (25); therefore, these regions contain a unique solution defined for $-x_0 < x < 0$, and this unique solution is tangent to the x-axis.*

Proof of Lemma. (i) U *is an antifunnel:* first we show that $y(x) = \gamma x^2$ is a lower fence. If $y(x) = \gamma x^2$, then $y'(x) = 2\gamma x < 0$. On the other hand, the solutions to the differential equation (25) have slope $\psi(x, y(x))$

$$\psi(x, y(x)) = \frac{\nu y + Q(x,y)}{-\mu x} \left(\frac{1}{1 - \frac{P(x,y)}{\mu x}} \right)$$

$$= \left(-\frac{\nu \gamma x^2}{\mu x} - \frac{Q(x,y)}{\mu x} \right) \left(1 + \frac{P(x,y)}{\mu x} + \left(\frac{P(x,y)}{\mu x} \right)^2 + \cdots \right)$$
(26)

(Reference: Part I, Appendix A on Asymptotic Development) and we need
to locate the leading terms. If Q has a term in x^2, then both terms in the
first parentheses are linear in x. Otherwise, the first term dominates the
second for sufficiently small x_0 when $-x_0 < x < 0$. But if Q has such a term
in x^2, then there are no y's in that term and its coefficient is independent
of γ. So by taking γ sufficiently large and a correspondingly small x_0,
we can guarantee that the first term dominates and, in particular, that
$\psi(x, y(x)) > 0$. Therefore, $y = \gamma x^2$ is a lower fence. Similarly, $y = -\gamma x^2$ is
an upper fence, so U is an antifunnel.

(ii) V *is an antifunnel:* This proof is left as Exercise 8.2–8.3#12.

(iii) *Unique solution in the antifunnel:* In order to use the Antifunnel
Theorem 1.4.5 that promises a unique solution in V, we need first that V
is a *narrowing* antifunnel, which it is as x increases to 0, and second that
$\partial \psi / \partial y > 0$ in V. If you differentiate the quotient (25) for $\psi(x, y)$, you get

$$\frac{\partial \psi}{\partial y} = \left(-\frac{\nu}{\mu x} - \frac{\partial Q / \partial y}{\mu x} \right) \left(1 + \frac{P}{\mu x} + \cdots \right) - \frac{(\nu y + Q)(\partial P / \partial y)}{(\mu x - P)^2}. \quad (27)$$

As $x \uparrow 0$, the only term of (27) that approaches infinity is $-\nu/(\mu x)$,
and this leading term is positive in V. [$(\partial Q / \partial y)$ starts with a linear term,
so $(\partial Q / \partial y)/(\mu x)$ is at most a constant; P starts with a quadratic term,
so $P/\mu x$ starts with at most a linear term that approaches zero.] So if
$(x, y) \in V$ and (x, y) is sufficiently close to the origin, $\partial \psi / \partial y > 0$.

Therefore, in this narrowing antifunnel V, as we approach the origin,
there is a unique solution to the differential equation (25). The origin is
the equilibrium point for the nonlinear system whose trajectories lie along
the solutions to that differential equation; therefore, the unique solution is
a separatrix of the nonlinear system.

(iv) *Tangency of separatrix:* This separatrix is tangent to the x-axis, the
eigenvector with the negative eigenvalue: The unique solution in V is also
the unique solution in U, by the same arguments given in part (iii), so it
is squeezed in between two parabolas tangent to the axis at the origin.

Thus, the lemma is proved. $\square$

Proof of Theorem 8.3.2, continued. All the statements about saddles
follow from Lemma 8.3.2, applied four times, with the necessary changes

of variables for the four directions in which the saddle can be approached. We need the first antifunnel U to control the direction of the separatrix; we need the second antifunnel V, which is broader, to capture all solutions approaching the equilibrium point from the left diagonal quadrant, so as to avoid in the final argument any "no man's land" between the four rotated versions of U, shown in Figure 8.3.3, where we would not know how the solutions behave.

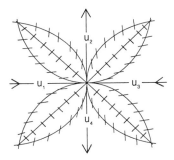

FIGURE 8.3.3. Four versions of U with four approaches to x_0, y_0.

We have now proved the theorem. □

8.4   Limit Cycles

The simplest thing a solution to a differential equation can do is to be attracted to an equilibrium. In the preceding three sections we have examined equilibria in some detail. The next simplest is for the system to undergo periodic motion, or for the phase plane trajectory to be a *cycle*. This behavior cannot occur for autonomous systems in $\mathbb{R}^1$, since a circle cannot be embedded in a line.

In the plane, cycles exist and are relatively common; for instance, the solutions of the harmonic oscillator equation

$$\begin{bmatrix} x \\ y \end{bmatrix}' = \begin{bmatrix} y \\ -x \end{bmatrix} \tag{28}$$

are cycles.

For nonlinear equations, cycles usually behave quite differently from those associated to linear equations: they are usually isolated and control the behavior of nearby solutions in the same sense that equilibria control the behavior of nearby solutions. This is perhaps the main way in which nonlinear equations differ from linear equations, at least in the plane.

Example 8.4.1. Consider the system of differential equations

$$\begin{bmatrix} x \\ y \end{bmatrix}' = \begin{bmatrix} y \\ -x \end{bmatrix} + \alpha(1 - x^2 - y^2)\begin{bmatrix} x \\ y \end{bmatrix}. \tag{29}$$

In the phase plane, this differential equation produces Figure 8.4.1.

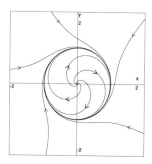

FIGURE 8.4.1. Trajectories for equation (29) with $\alpha = 1$.

Certainly it appears that all solutions other than the constant solution at the origin are attracted to a solution whose trajectory is the unit circle. This is, in fact, quite easy to prove. This differential equation has been carefully cooked up to be easy to translate (Exercise 8.4#1a) into *polar coordinates,* as follows:

$$r^2 = x^2 + y^2, \qquad \text{so} \qquad 2rr' = 2xx' + 2yy';$$

$$\theta = \arctan\frac{y}{x}, \qquad \text{so} \qquad \theta' = \frac{xy' - yx'}{x^2 + y^2}.$$

Now substituting x' and y' from (29) gives (Exercise 8.4#1b)

$$\begin{aligned} r' &= \alpha(1 - r^2)r \\ \theta' &= -1. \end{aligned} \tag{39}$$

It is fairly easy to solve the first equation of (30), but even easier to analyze it from fields of slopes in the rt-plane (Exercise 8.4#1b): For $\alpha > 0$, all solutions for $r > 0$ are attracted to the constant solution $r = 1$; for $\alpha < 0$, the solution $r = 1$ is repelling, and if $0 \le r(0) < 1$, the solution tends to 0, whereas if $1 < r(0)$, then the solution tends to infinity.   ▲

Example 8.4.1 shows a *stable* (or attracting) limit cycle when $\alpha > 0$ and an *unstable* (or repelling) limit cycle when $\alpha < 0$. Stability means that the cycle attracts nearby solutions as $t \to +\infty$; instability means that the cycle attracts nearby solutions as $t \to -\infty$.

In the plane, these are the only kinds of limit cycles that appear "generically" (there are other "exceptional ones," which we will examine in Section 9.4). In higher dimensions, there are many other generic possibilities, which we will consider briefly in Chapter 13 in Part III.

Considering that the existence of limit cycles is a perfectly common form of behavior for a differential equation in the plane, it is amazing how rarely they come up when differential equations are entered at random. Limit cycles do come up rather often, however, in differential equations describing real phenomena. Moreover, as we will see in Chapter 9, limit cycles are essential to understanding how phase plane drawings change their qualitative appearance as a parameter changes.

Here are a few further examples of limit cycles:

Example 8.4.2. Van der Pol's equation,

$$x'' + (x^2 - 1)x' + x = 0, \tag{31}$$

describes a nonlinear electrical circuit that we will study further in Chapter 10 in Part III. It can be turned into a system in the standard way:

$$\begin{aligned} x' &= y \\ y' &= (1 - x^2)y - x, \end{aligned} \tag{32}$$

and the associated phase plane drawing appears as in Figure 8.4.2.

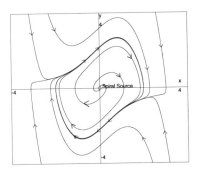

FIGURE 8.4.2. Trajectories for Van der Pol's equation.

There certainly seems to be a limit cycle in Figure 8.4.2, and this phenomenon can be understood as follows. We can think of the differential equation (31) as describing a damped harmonic oscillator, with friction $x^2 - 1$. But $x^2 - 1$ is only positive for $|x| > 1$; in other words, the oscillator is being *driven* if the displacement x is small, and *damped* or slowed if it is large. It is not so surprising that it settles down to some particular oscillation. ▲

It is quite possible to produce differential equations with several limit cycles. The following is an example:

Example 8.4.3. Consider the equation

$$\begin{bmatrix} x \\ y \end{bmatrix}' = \begin{bmatrix} y \\ -x \end{bmatrix} + (x^2 + y^2)^2 \sin \frac{1}{x^2 + y^2} \begin{bmatrix} x \\ y \end{bmatrix},$$

which yields Figure 8.4.3.

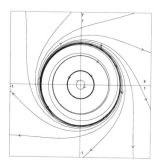

FIGURE 8.4.3. Several limit cycles for Example 8.4.3.

In Exercises 8.4#3, you can prove that this equation has infinitely many periodic solutions, with different trajectories. ▲

There are many open problems about limit cycles. For instance, it is unknown how many limit cycles a differential equation

$$x' = P(x, y)$$
$$y' = Q(x, y)$$

with quadratic polynomials P and Q can have, or in what configurations such cycles can lie.

There is one major theorem, the Poincaré–Bendixson theorem, that guarantees the existence of limit cycles. This is the subject of the next section.

8.5   The Poincaré–Bendixson Theorem

A major theorem that guarantees the existence of limit cycles is due to Henri Poincaré (French) and I. Bendixson (Swedish), in the early twentieth century. To see a statement and a proof, you should read Section 8*.5, which does not require the previous sections of Chapter 8*. Here we will give a useful and simpler special case of the Poincaré–Bendixson Theorem that can be stated as follows:

Theorem 8.5.1 (Poincaré–Bendixson Theorem, Annulus Form).
Let $0 < \alpha(\theta) < \beta(\theta)$ be two periodic functions of period 2π, continuous and piecewise differentiable. Let U be the "annular" region in the plane given in polar coordinates by

$$U = \{(r, \theta) \mid \alpha(\theta) \leq r \leq \beta(\theta)\}$$

and let

$$\mathbf{f}(x, y) = \begin{bmatrix} f_1(x, y) \\ f_2(x, y) \end{bmatrix} \tag{33}$$

be a vector field on U, pointing into U along both components of the boundary. If

$$\mathbf{f}(x, y) \cdot \begin{bmatrix} -y \\ x \end{bmatrix} \neq 0 \tag{34}$$

in U, then U must contain a limit cycle of $\mathbf{x}' = \mathbf{f}(x, y)$.

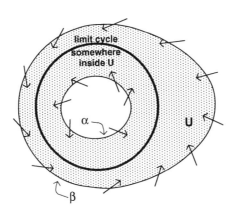

FIGURE 8.5.1. Annulus for Poincaré–Bendixson Theorem.

The hypothesis (34) means that the vector field curls around the origin, never pointing in the radial direction. Since the scalar product does not vanish, it must be positive or negative, according to whether the vector field turns counterclockwise or clockwise. Without loss of generality, we will assume that the scalar product (34) is positive.

Proof of Theorem 8.5.1. We will consider U first as a wraparound funnel, and it is easy to associate to U a genuine funnel: consider the region

$$V = \{(r, \theta) \mid \alpha(\theta) \leq r \leq \beta(\theta)\}$$

in the (θ, r)-plane, as shown in Figure 8.5.2.

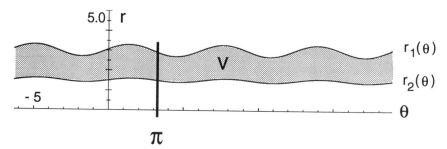

FIGURE 8.5.2. Unwrapping the annulus in the θr-plane.

The differential equation can be written in polar coordinates

$$
\begin{aligned}
r' &= g_1(r, \theta) \\
\theta' &= g_2(r, \theta),
\end{aligned}
\tag{35}
$$

as in Example 8.4.1, by writing

$$r^2 = x^2 + y^2, \quad \text{so} \quad rr' = xx' + yy'$$

$$\theta = \arctan \tfrac{y}{x}, \quad \text{so} \quad \theta' = (y'x - yx')x/r^2.$$

Substituting $x = r\cos\theta$, $y = r\sin\theta$ leads to

$$r' = \cos\theta f_1(r\cos\theta, r\sin\theta) + \sin\theta f_2(r\cos\theta, r\sin\theta) = g_1(r, \theta)$$

$$\theta' = \frac{\cos\theta f_2(r\cos\theta, r\sin\theta) - \sin\theta f_1(r\cos\theta, r\sin\theta)}{r} = g_2(r, \theta)$$

and our hypothesis (34) implies that $g_2(r, \theta) > 0$ in the region V. Then the trajectories will coincide, in the (θ, r)-plane, with the solutions of the equation

$$\frac{dr}{d\theta} = \frac{g_1(r, \theta)}{g_2(r, \theta)} = G(r, \theta). \tag{36}$$

Our hypothesis imply that V is a funnel for equation (36). Moreover, the function G is periodic of period 2π with respect to θ, so this sort of equation is just what was discussed at length in Part I, Section 5.5, including the period mapping $P : J \to \mathbb{R}$, which associates to $r \in J$ the value of the solution to equation (36) which passes through $(0, r)$ at $\theta = 2\pi$. The interval J is the maximal interval on which P is defined; in our case, $J = [\alpha(0), \beta(0)]$ since V is a funnel, and P maps J into itself.

Any continuous mapping from a closed and bounded interval into itself must have a fixed point (Exercise 8.5#1). The solution through this fixed point is, of course, periodic as a function of θ, and the solution to $\mathbf{x}' = \mathbf{f}(x, y)$ [as given by Equation (33)] with corresponding trajectory is clearly a cycle. $\square$

There is a fairly obvious condition which implies that the cycle in U is *unique*: the region V is a funnel for equation (36), but it is also a backward antifunnel. Suppose now that, in addition, $dG/dr < 0$, so that since G is periodic, the derivative $dG/dr < -C$ for some constant $C > 0$. Then the region V, viewed as a backward antifunnel, satisfies the uniqueness condition of Theorem 4.7.5, so that the periodic solution to (36) is unique, and correspondingly, the equation (33) has a unique cycle in U.

The condition $dG/dr < 0$ is usually quite difficult to verify, since G is usually a complicated function, but there is a more natural condition which leads to the same conclusion: if the divergence

$$\operatorname{div} \mathbf{f} = \frac{\partial f_1}{\partial x} + \frac{\partial f_2}{\partial y}$$

is strictly negative throughout U, then there is a unique cycle in U that attracts every solution in U. This follows from the fact that the divergence of a vector field $\mathbf{f}$ in the plane measures the extent to which its flow $\phi_{\mathbf{f}}$ contracts or dilates areas. If there were two limit cycles, the region between them would have an area preserved by the flow, which contradicts negative divergence. This will be covered in Section 8.6, and the uniqueness criterion when the divergence is negative is the object of Exercise 8.6#7.

For examples to illustrate the Poincaré–Bendixson Theorem, we go back to the limit cycles we have just discovered in Section 8.4.

Example 8.5.2. Consider the system of differential equations from Example 8.4.1:

$$\begin{bmatrix} x \\ y \end{bmatrix} = \begin{bmatrix} y \\ -x \end{bmatrix} + \alpha(1 - x^2 - y^2) \begin{bmatrix} x \\ y \end{bmatrix}. \tag{29}$$

We return to the case when $\alpha = 1$, where we already know the unit circle is a limit cycle; we see that an annulus formed by a circle of radius $r \geq 1$ and a circle of radius $r \leq 1$ will satisfy the conditions of the Poincaré–Bendixson Theorem 8.5.1, as shown in Figure 8.5.3.

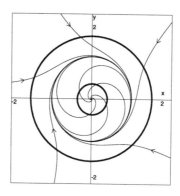

FIGURE 8.5.3. A Poincaré–Bendixson annulus for the limit cycle of $x' = y + x(1 - x^2 - y^2)$, $y' = -x + y(1 - x^2 - y^2)$. ▲

However, the sad fact about the Poincaré–Bendixson Theorem 8.5.1 is that it does not tell us how to *find* a suitable annulus in order to know where to look for a limit cycle. That is often quite difficult, even in a case where we already can see a limit cycle.

Example 8.5.3. Returning to the Van der Pol equation (31) or (32) of Example 8.4.2, it is far less clear how to define an annulus around the limit cycle—give it a try before you read on!

A helpful approach to defining an appropriate annulus is to use a dense grid of slope marks (an option in the *MacMath* program *DiffEq, Phase Plane*). Figure 8.5.4 shows one graphical effort to successfully wander through the morass, always crossing the vectors in the proper direction.

Figure 8.5.4 is a more organized annulus for the same problem, for which actual equations can be written. We leave the writing of the equations as Exercise 8.5#4, giving the hint that the nonstraight curves drawn on the figure are the isoclines for zero slope in the vector field.

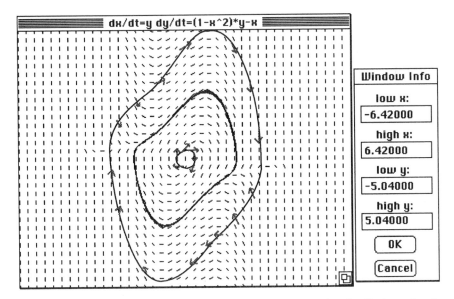

FIGURE 8.5.4. A sketched Poincaré–Bendixson annulus for the limit cycle of $x'' + (x^2 - 1)x' + x = 0$.

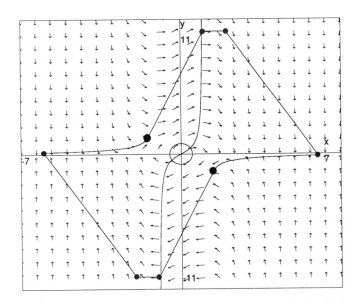

FIGURE 8.5.5. A well-defined Poincaré–Bendixson annulus for the limit cycle of $x'' + (x^2 - 1)x' + x = 0$. ▲

8.6 Symmetries and Volume-Preserving Equations

As you can see in Section 8*.6, for a "general" vector field

- the zeroes are isolated;

- the zeroes are sinks, saddles or sources;

- all solutions are attracted forward by sinks and attractive limit cycles, and backwards by sources and repelling limit cycles, except for separatrices and limit cycles themselves.

However, differential equations that you study will often not be "general" in this sense and will display quite different behavior. This is usually for one or both of the two following reasons:

(i) The vector field may have *symmetries*. See Example 8.6.1.

(ii) The vector field may be *area-preserving*. See Example 8.6.2.

Example 8.6.1. Consider the system

$$x' = y + x^2 + xy$$
$$y' = -x + xy + y^2.$$

The computer shows the picture in Figure 8.6.1 for the phase plane.

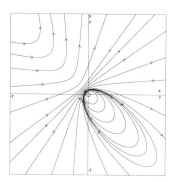

FIGURE 8.6.1. Phase plane for $x' = y + x^2 + xy$, $y' = -x + xy + y^2$.

It appears in Figure 8.6.1 that there is a *continuous* family of closed curves surrounding the origin. Why should they be there? A careful look at the picture appears to show that the diagram is *symmetric* around the line $x + y = 0$. This is, in fact, true and easy to show, as follows:

The symmetry around the line $x + y = 0$ is given by $(x, y) \mapsto (-y, -x)$. One way of expressing the symmetry is to say that if $(u(t), v(t))$ is a solution of the differential equation, then $(-v(-t), -u(-t))$ is also a solution, as you can verify. This solution follows the symmetric trajectory, but with time reversed.

The symmetry explains the observed behavior as follows: if a solution $(x(t), y(t))$ crosses the line $x + y = 0$ at some time $t_0 - h$, and again at some time t_0, then between time t_0 and time $t_0 + h$ we will have

$$(x(t_0 + s), y(t_0 + s)) = (-y(t_0 - s), -x(t_0 - s)),$$

since this is a solution of the differential equation with the correct initial condition. But this solution returns to the point where we started, giving rise to a periodic solution. Thus, there will be a continuous family of periodic solutions for this equation. ▲

Example 8.6.2. Consider the system of differential equations

$$x' = 2\cos(x + 2y) - x\sin(xy)$$
$$y' = -\cos(x + 2y) + y\sin(xy).$$

The computer shows the picture in Figure 8.6.2 for the phase plane.

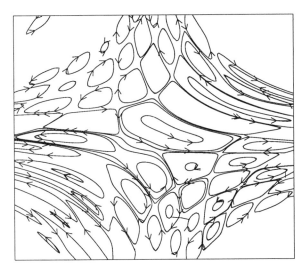

FIGURE 8.6.2. Phase plane for $x' = 2\cos(x + 2y) - x\sin(xy)$, $y' = -\cos(x + 2y) + y\sin(xy)$.

In Figure 8.6.2, we see lots of things that are not expected: families of periodic solutions and separatrices of saddles leading to other saddles. Yet, there does not seem to be any symmetry to explain these unusual phenomena as in the previous example. ▲

The differential equation of Example 8.6.2 is special because it is *area-preserving*. This is an important notion in itself, which we now explain, using the notion of *flows* from Section 6.8.

Definition 8.6.3. The differential equation $\mathbf{x}' = \mathbf{f}(\mathbf{x})$ in $\mathbb{R}^n$, with flow $\phi_{\mathbf{f}}$ is *volume-preserving* (area-preserving if $n = 2$) if for any subset U of $\mathbb{R}^n$ and any t, the n-dimensional volume

$$n\text{-volume}(\phi_{\mathbf{f}}(t, U)) = n\text{-volume}(U).$$

This is obviously an important property that a differential equation might have, but it is not obvious that there is any easy way to check whether the equation has the property or not. As Corollary 8.6.6 will show, nothing could be simpler. Recall from calculus, or simply define,

$$\operatorname{div} \mathbf{f} = \frac{\partial f_1}{\partial x_1} + \ldots + \frac{\partial f_n}{\partial x_n}.$$

Example 8.6.4. The divergence of the vector field in Example 8.6.1 is

$$\frac{\partial}{\partial x}\left(y + x^2 + xy\right) + \frac{\partial}{\partial y}\left(-x + xy + y^2\right) = 2x + y + x + 2y = 3(x + y).$$

The divergence of the vector field in Example 8.6.1 is

$$\frac{\partial}{\partial x}\left(2\cos(x + 2y) - x\sin(xy)\right) + \frac{\partial}{\partial y}\left(-\cos(x + 2y) + y\sin(xy)\right) = 0. \quad \blacktriangle$$

The role of the divergence is largely explained by the following theorem:

Theorem 8.6.5. *Given a differential equation $\mathbf{x}' = \mathbf{f}(\mathbf{x})$, the divergence of $\mathbf{f}$ measures the rate of change of volumes under the flow $\phi_{\mathbf{f}}$; i.e., for any $U \subset \mathbb{R}^n$, we have*

$$\frac{d}{dt}\operatorname{vol}(\phi_{\mathbf{f}}(t, U))\Big|_{t=0} = \int \ldots \int_U \operatorname{div} \mathbf{f}\, dx_1 \ldots dx_n.$$

Proof. Using the notation of asymptotic expansion (Part I, Appendix A), we will write

$$\phi_{\mathbf{f}}(t, \mathbf{x}) = \begin{bmatrix} u_1(t, \mathbf{x}) \\ \vdots \\ u_n(t, \mathbf{x}) \end{bmatrix} = \begin{bmatrix} u_1(t, (x_1, \ldots, x_n)) \\ \vdots \\ u_n(t, (x_1, \ldots, x_n)) \end{bmatrix},$$

where $u_i(t, \mathbf{x}) = x_i + t f_i(t, \mathbf{x}) + o(|t|)$ as $t \to 0$, since it is a coordinate of a solution to the differential equation. Note: The notation $o(|t|)$ means, since $t \to 0$, terms of strictly higher order than $|t|$.

We start with a formula from Appendix L5:

$$\text{vol}(\phi_{\mathbf{f}}(t, U)) = \int \cdots \int_U \det \begin{bmatrix} \partial u_1/\partial x_1 & \cdots & \partial u_1/\partial x_n \\ & \cdots & \\ \partial u_n/\partial x_1 & \cdots & \partial u_n/\partial x_n \end{bmatrix} dx_1 \ldots dx_n.$$

This gives

$$\frac{d}{dt}\text{vol}(\phi_{\mathbf{f}}(t, U)) = \int \cdots \int_U \frac{d}{dt}\det \begin{bmatrix} \partial u_1/\partial x_1 & \cdots & \partial u_1/\partial x_n \\ & \cdots & \\ \partial u_n/\partial x_1 & \cdots & \partial u_n/\partial x_n \end{bmatrix} dx_1 \ldots dx_n.$$

We can also use the fact that

$$\frac{\partial u_i}{\partial x_j}(t, x) = \begin{cases} t\partial f_i/\partial x_j(t, x) + o(|t|) & \text{if } i \neq j \\ 1 + t\partial f_i/\partial x_i(t, x) + o(|t|) & \text{if } i = j, \end{cases}$$

which gives, for the determinant above,

$$\det \begin{bmatrix} \partial u_1/\partial x_1 & \cdots & \partial u_1/\partial x_n \\ & \cdots & \\ \partial u_n/\partial x_1 & \cdots & \partial u_n/\partial x_n \end{bmatrix} = 1 + t\,\text{div}\,\mathbf{f} + o(|t|).$$

Differentiating this result with respect to t and evaluating at $t = 0$ yields the desired result.   □

Corollary 8.6.6. *The differential equation* $\mathbf{x}' = \mathbf{f}(\mathbf{x})$ *is volume-preserving if and only if* $\text{div}\,\mathbf{f} = 0$.

Proof. This is immediate from Theorem 8.6.5; the only problem is that the theorem only computes the derivative at $t = 0$. But the following formula repairs this difficulty:

$$\frac{d}{dt}\text{vol}(\phi_{\mathbf{f}}(t, V))\Big|_{t=s} = \frac{d}{dt}(\text{vol}(\phi_{\mathbf{f}}(t, \phi_{\mathbf{f}}(s, V))))\Big|_{t=0}.$$

Since the theorem holds for any domain V, in particular, $V = \phi_{\mathbf{f}}(s, U)$, this proves the derivative of the volume is zero, so it is constant if the divergence of the vector field vanishes.   □

Of course, a volume-preserving vector field cannot have any sources or sinks, since a neighborhood of a sink would be mapped strictly inside itself under the flow, or a neighborhood of a source would be mapped strictly outside itself—in either case, its volume would not be preserved.

This can also be seen from the linearization: if $\mathbf{u}' = A\mathbf{u}$ is the linearization of the differential equation $\mathbf{x}' = \mathbf{f}(\mathbf{x})$ at a zero $\mathbf{x}_0$, then Exercise 8.6#3 asks you to show

$$\text{div}\,\mathbf{f}(\mathbf{x}_0) = \text{tr}\,A.$$

If the vector field is volume-preserving, we must have $\text{tr}\,A = 0$; in dimension $n = 2$, this implies that all zeroes must be centers or (rather special) saddles. In fact, from Theorem 8.6.5 we see that

- the flow is volume-expanding if tr $A > 0$;

- the flow is volume-shrinking if tr $A < 0$.

This goes a long way toward explaining why sources satisfy tr $A > 0$ and sinks satisfy tr $A < 0$.

If all this volume-preserving stuff seems a little mysterious, it may come as a surprise that in the two-dimensional case, the subject was essentially covered way back in Part I, Chapter 2.6. More specifically, the system of equations

$$x' = f(x, y)$$
$$y' = g(x, y) \tag{37}$$

is area-preserving if and only if the associated equation

$$\frac{dy}{dx} = \frac{g}{f} \tag{38}$$

is *exact*. We saw in Section 2.6 in Part I on exact differential equations how to construct on any rectangle where both f and g are defined a function $F(x, y)$ which is constant on the solutions of (38), i.e., on the trajectories of (37). This means you can get actual *equations* for the trajectories, relating x and y (usually implicitly).

Example 8.6.7. In Example 8.6.2, the function $F(x, y) = \sin(x + 2y) + \cos(xy)$ is constant on the trajectories.

This essentially completely explains the observed behavior. Such a function F will have maxima, minima, saddle points, and, perhaps, more complicated extrema. In any case, near a maximum or a minimum the trajectories form families of closed curves, and a trajectory through a saddle point has no choice but to go to another at the same level, or come back to itself, or go off to infinity. ▲

This (almost) completely demystifies area-preserving differential equations in $\mathbb{R}^2$: they are all obtained by taking a function $F(x, y)$ and setting

$$x' = \frac{\partial F}{\partial y}$$
$$y' = -\frac{\partial F}{\partial x}.$$

Such equations are called *Hamiltonian*; in two dimensions their theory is simple: the function F is constant on trajectories, which completely determines them. These Hamiltonian equations are central in physics. In higher dimensions, one function alone does not determine a curve, and as we will see in Chapter 11 (in Part III), Hamiltonian equations are very elaborate in higher dimensions.

For an area-preserving differential equation, we can reduce its solution to integrating and inverting functions, at least in principle:

$$x' = f(x, y)$$
$$y' = g(x, y).$$

(1) First find (as explained in Section 2.6 in Part I on exact differential equations) the function F such that

$$\frac{\partial F}{\partial y} = f, \qquad \frac{\partial F}{\partial x} = -g.$$

(This function F will be called the *Hamiltonian function* in Chapter 11 in Part III.)

(2) Next consider the equation $F(x, y) = C_1$ as expressing y implicitly as a function of x, say $y = \phi(x)$. Note: Seeing or understanding the curve $y = \phi(x)$ may be difficult in practice.

(3) Now the first equation becomes separable, $dx/(f(x, \phi(x))) = dt$, and if it is possible to integrate $dx/(f(x, \phi(x)))$, you can find an expression $G(x) = t + C_2$. The function G will still need to be inverted to express x as a function of t.

The "almost" inserted in the sentence about demystification comes from the fact that if the equation is defined on a domain with holes, the function F may fail to be defined in the entire domain. This difficulty is important, but we will not discuss it here.

Example 8.6.8. Consider the differential equation from Section 6.5,

$$x' = y$$
$$y' = -\sin x,$$

describing the pendulum, or, more generally,

$$x' = y$$
$$y' = -\frac{\partial V(x, y)}{\partial x}.$$

These equations often had continuous families of closed trajectories, and saddle connections. We can now clearly see why:

(i) They are symmetric with respect to the y-axis, which you can see either directly or from the obvious symmetry of the Hamiltonian function when $V(x)$ is symmetric.

(ii) They are area-preserving, and, in fact, the Hamiltonian function is

$$F(x, y) = y^2/2 + V(x),$$

also called the total energy. Note the very important fact that conservation of energy and conservation of area are the same thing in this case! ▲

You might wonder if just by looking at the phase plane of a differential equation, you can tell whether the equation is volume-preserving. In the large (globally), you may get hints that this is so from the phase plane presence of continuous families of periodic solutions and of centers. In the small, if you see sinks or sources, you can be sure it is not area-preserving. But locally, and away from zeroes, you *cannot*: in the absence of equilibria, any pattern of trajectories is locally compatible with the property of being area-preserving.

More specifically, given any vector field $\vec{\xi}$ on a region of the plane with no holes, which furthermore does not vanish in that region, then there is a function $f(x, y) > 0$ such that the vector field $f\vec{\xi}$ is area-preserving. This means that locally (away from zeroes) you can adjust the *length* of the vectors forming the vector field to make it area-preserving.

Such a function, which acts as a multiplier on the vectors, is classically called an *integrating factor* and is a standard fixture of elementary courses on differential equations. In most cases, it is just as hard to find an integrating factor as to solve the equation, so we will not expand on this topic.

8.7   Chaos in Higher Dimensions

The theory throughout Chapter 8, especially in Sections 8.5 and 8.6, is quite specific to the plane. Although there are generalizations to higher dimensions, they are much weaker, either because the hypotheses are seldom satisfied or because the conclusions do not describe the flow with any precision. In this section, we want to give some examples that bring out the difficulties.

Example 8.7.1. We already encountered the Lorenz equation in Example 6.1.6. Recall it as the following rather innocent looking autonomous system in $\mathbb{R}^3$:

$$\begin{aligned}
x' &= \sigma(y - x) \\
y' &= \rho x - y - xz \\
z' &= -\beta z + xy.
\end{aligned} \tag{17}$$

Lorenz produced these equations as a reduction to three dimensions of a set of partial differential equations describing fluid convection; in this case, the parameters σ, ρ, and β are all positive. We will only consider the equation in this range.

Solutions of the Lorenz equation for the "traditional parameter values" $\sigma = 10, \beta = 8/3$, and $\rho = 28$ are represented in Figure 8.7.1. It is much easier to get a feel for this figure by seeing it move on a computer screen (or a video) than by looking at static pictures.

-25.000 < x < 35.000	**dx/dt = 10*(y-x)**
-30.000 < y < 35.000	**dy/dt = 28*x-y-x*z**
-10.000 < z < 70.000	**dz/dt = -(8/3)*z+x*y**

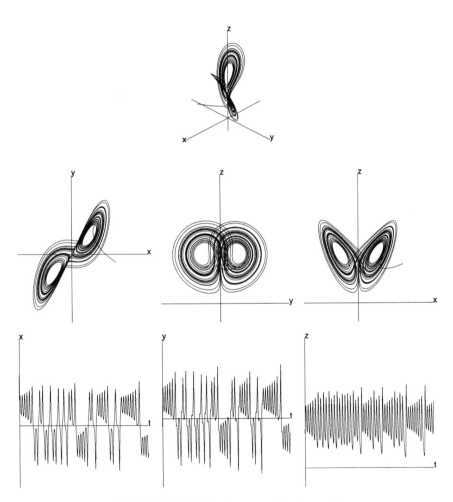

FIGURE 8.7.1. The Lorenz attractor.   ▲

In Example 8.7.1, the reduction to three dimensions is drastic, and there does not appear to be any close relation between the solutions of (17) and the original meteorological problem. Despite this, the literature about the Lorenz equations is very large, with at least one entire book [C. Sparrow, *The Lorenz Equations: Bifurcations, Chaos, and Strange Attractors*] and many papers devoted to its solutions, but without any complete description in sight. We will largely follow the description in Guckenheimer and Holmes [*Nonlinear Oscillations, Dynamical Systems, and Bifurcations of Vector Fields*].

We will now give a way of thinking about these solutions by a geometric picture of the Lorenz attractor proposed by R. F. Williams in *The Structure of Lorenz Attractors*.

Cut out three semi-circles of cardboard, marked as in Figure 8.7.2, and glue them according to the pattern of Figure 8.7.3.

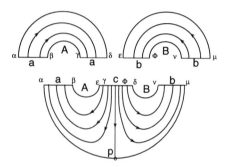

FIGURE 8.7.2. Pattern for cutouts.

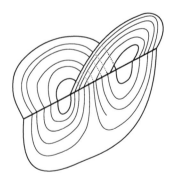

FIGURE 8.7.3. How to glue together cutouts of Figure 8.7.2.

You can now draw curves on this model by following the marked curves around; they can be followed (forward) forever unless they end at p. They turn a certain number of times around hole A, then around hole B, etc. The details of the behavior depend in a very sensitive way on the lengths a, b, and c.

The actual behaviors of the solutions to the Lorenz equations (17) behave much the same way: they turn around the zero A for a while, then around B, then around A again, etc.

No one has managed to prove that the behavior of this cardboard model really reflects the differential equation, but extensive computer experimentation backs up the view that this is how the solutions behave.

We will discuss one other phenomenon that appears in three dimensions. Although it is rather easier to understand than the Lorenz equation, it has perplexing properties. The discussion will necessarily be a bit vague: a precise definition of a section mapping is given in Section 8*.4, but we hope we can convey the idea here without the technicalities.

CYCLES AND THE SECTION MAPPING IDEA

Take a cycle and choose a transversal to the cycle (see Figure 8.7.4): in the plane, this transversal will be a line segment; in space, it will be a piece of a plane. If you start in the transversal sufficiently close to the cycle and solve the differential equation, you will stay close to the cycle and hence will come around and intersect the transversal again. This defines a mapping from the transversal to itself, called the *section mapping*, which has a fixed point at the intersection of the cycle with the transversal. To be accurate about things, you need to worry about the domain, etc., but we will leave all that to Chapter 8*.

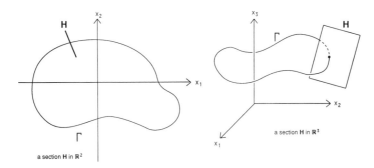

FIGURE 8.7.4. Transversals to cycles used for section mapping.

In the plane, cycles are usually attracting or repelling. This means that the section mapping will usually have an attractive or a repelling fixed point. This is not surprising; after all, the derivative of the section mapping at the fixed point is some number, and it is usually either bigger or smaller than one.

This sort of fixed point for one-dimensional iteration has been discussed in Part I, Chapter 5, and will be thoroughly discussed for two dimensions in Part III, Chapter 13. For now an informal introduction should suffice.

In three dimensions, there is another generic possibility, beyond being attracting or repelling. The derivative of the section mapping is now a 2×2 matrix and will usually have two eigenvalues. It seems intuitively likely that if they both have absolute values smaller than 1, the fixed point will be attracting, and if they both have absolute values greaterr than 1, the fixed point will be repelling. But what if one is bigger than 1 and the other smaller?

When this happens, we will see in Chapter 13 (Part III) that the mapping has a saddle: there is a curve through the origin (in the section) that is attracted to the origin under the section mapping, and another that is attracted under the inverse of the section mapping as shown in Figure 8.7.5. The situation is very similar to that of Section 8.3. As before, we will call these curves the stable and unstable separatrices of the origin (for the section mapping).

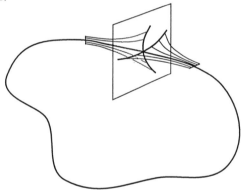

FIGURE 8.7.5. Stable and unstable separatrices in section mapping, and the surfaces they generate.

These "section mapping separatrices" are *not solutions of the differential equation*: they are contained in a transverse section. A solution through a point of the stable separatrix spirals toward the cycle in forward time; a solution through the unstable separatrix spirals toward the cycle in backward time. The solutions through the stable separatrix form a surface, which deserves to be called the *stable separatrix of the cycle*, and, similarly, the solutions through the unstable separatrix form a surface called the *unstable separatrix of the cycle*. An idea of these surfaces is sketched in Figure 8.7.6.

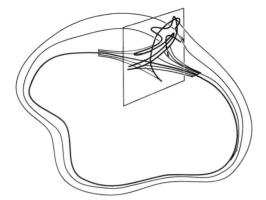

FIGURE 8.7.6. If the separatrices are extended, they may intersect.

Exercise 8.7#5 asks you to compute these separatrices explicitly in some simple cases.

Amazing things happen when the stable and unstable separatrices intersect. You can alternately think of the curves intersecting in the section, or the solution through such an intersection point, which spirals toward the cycle both backward and forward. These two ways of seeing the intersection are pictured in Figure 8.7.7.

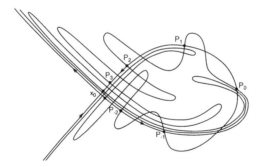

FIGURE 8.7.7. Intersections and tangling of stable and unstable separatrices in the section plane.

As Figure 8.7.7 shows, such a solution is now forced to spiral away from the cycle, wander around for awhile, and then spiral back toward the solutions. It may sound as if this gives great opportunities for tangling, and it does. In fact, it is not easy to believe quite how much complication must occur any time such a *homoclinic trajectory* exists.

Once one point p_0 is on both curves, all the forward and backward images

$$\ldots, p_{-2}, p_{-1}, p_0, p_1, p_2, \ldots$$

must be on both curves, so they must intersect infinitely often.

Now think of the piece of the unstable separatrix going through the points $p_n, p_{n+1}, \ldots$ for large n. This sequence of points is approaching the fixed point along the stable separatrix, and we claim that the unstable manifold must oscillate wildly as it approaches itself (without ever intersecting itself). Indeed, the section mapping is expanding in the direction of the unstable separatrix, mapping each oscillation into a bigger one.

For one thing, the oscillations will follow longer and longer parts of the unstable separatrix, until they start following the oscillations themselves, forming oscillations on the oscillations, etc.

For another, the stable separatrix has to go into the same kind of oscillations as it approaches itself along the unstable separatrix. This implies a whole network of new intersections of the stable and unstable separatrix.

One can show that when this happens, there must be infinitely many different periodic solutions, with periods tending to infinity. These solutions will mostly be knotted, and linked with each other, forming a remarkable mess that has come to be called the "homoclinic tangle," as shown in Figure 8.7.8. There are many, more complicated things than these in the homoclinic tangle.

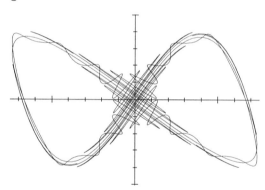

FIGURE 8.7.8. A more complete picture of the homoclinic tangle.

ORDER AND STRUCTURAL STABILITY

We have so far given the impression that structural stability comes only when order reigns and solutions have simple behavior: they are usually attracted to sinks or limit cycles, and the others are also easy to understand.

Two extraordinary results from the 1950s showed that this idea, however plausible, is wrong. It is quite possible for very chaotic systems to be structurally stable.

Smale, in his study of a class of mappings which have come to be called "Smale horseshoes," showed that some mappings (not differential equations) in the plane could simultaneously have infinitely many periodic points, orbits that are dense in a Cantor set, and be structurally stable. Moreover, he showed that if a mapping of the plane has a periodic point, the stable and unstable separatrices of which intersect transversally, then inevitably horseshoe behavior will occur. It is because of his results that we know that if a differential equation in $\mathbb{R}^3$ has a cycle, with stable and unstable separatrices which intersect transversally, then there must be infinitely many cycles and all sorts of other complications. But this does *not* contradict structural stability. Neither does it imply it, of course. We will describe and discuss the Smale horseshoe in Chapter 13 (Part III).

Anosov, in his study of the geodesic flow on closed surfaces of negative curvature, discovered that these are also structurally stable.

The geodesic flow is the differential equation, whose solutions are the geodesics on the surface, parametrized by arclength. To specify a solution, you give a point and a direction, and the point starts out in that direction and goes as straight as it can, at constant speed. Since we needed an initial velocity (the direction) as well as an initial position, we see that this is a second order differential equation on the surface, which can be turned in the standard way into a first order differential equation on the space of unit tangent vectors to the surface. This space of unit tangent vectors is three dimensional: there is a surface's worth of points at which to attach them, and a circle's worth of directions for each point.

A surface of positive curvature focuses geodesics, whereas negative curvature makes them spread out and diverge. It is hard to imagine anything more chaotic than geodesic flow on manifolds of negative curvature; for instance, there are infinitely many closed geodesics, each of which is a cycle in the space of unit vectors, which has stable and unstable separatrices that intersect transversally. In this case, it is fairly easy to imagine what these stable and unstable separatrices are: they are the geodesics that spiral toward the closed geodesic forward and backward.

Moreover, these separatrices intersect transversely: there are geodesics which spiral towards the closed geodesics both forwards and backwards. So there are infinitely many Smale horseshoes, all interrelated in some complicated way. Somehow, the whole geodesic flow is maximally disordered and tangled. But precisely this maximal disorder leads to structural stability. We will discuss geodesics in Chapter 11 (Part III), but will not be able to approach Anosov's theorem.

8.8   Structural Stability

A reader who plays with the phase plane program will observe that, in general, the phase portrait of a differential equation $\mathbf{x}' = \mathbf{f}(\mathbf{x})$ in the plane changes only slightly when the vector field $\mathbf{f}$ is slightly modified:

The zeroes of the modified equation are close to the zeroes of the origi-
nal equation; they are still sinks, sources, and saddles as they were for
the original equation; the basins of the sinks are bounded by the same
separatrices, and so forth.

Example 8.8.1.

(a) $\begin{bmatrix} x \\ y \end{bmatrix}' = \begin{bmatrix} y - x^2 + 3.7 \\ x - y^2 + 3.4 \end{bmatrix}$

(b) $\begin{bmatrix} x \\ y \end{bmatrix}' = \begin{bmatrix} y - x^2 + 2.2 \\ x - y^2 + 4.4 \end{bmatrix}$

(c) $\begin{bmatrix} x \\ y \end{bmatrix}' = \begin{bmatrix} y - x^2 + 5.9 \sin x \\ x - y^2 + 3.6 \end{bmatrix}$

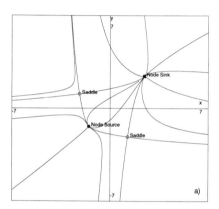

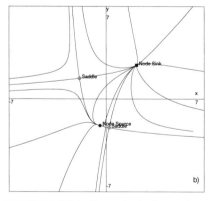

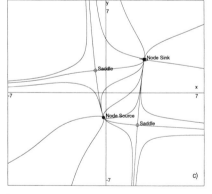

FIGURE 8.8.1. Perturbed phase planes, showing how the key structures perse-
vere. ▲

In Chapter 8*, we will state and give a complete proof of two theorems which make this precise. However, these results are quite difficult and
technical, and the content of the main statements (not the proofs) can be

conveyed without the heavy apparatus introduced in Section 8*.1. In this section, we will try to say enough to allow the reader to go directly to Chapter 9, which is considerably easier than Chapter 8*; bypassing Chapter 8* is probably the right course to follow in a first reading.

Let us pin down more precisely what we expect the reader exploring phase planes to observe. We claim that, in general, in $\mathbb{R}^2$,

(1) all zeroes of **f** are sinks, sources or saddles;

(2) all cycles are attracting or repelling;

(3) all solutions tend as $t \to +\infty$ to sinks or to attractive limit cycles, except the saddles, their stable separatrices, and the unstable limit cycles;

(4) all solutions tend as $t \to -\infty$ to sources or to repelling limit cycles, except the saddles, their unstable separatrices, and the stable limit cycles.

With some minor technical caveats (*linearly* attracting, etc.), we will call a vector field in the plane which satisfies conditions (1)–(4) *structurally stable*.

Let us try to see what must happen in order for one of these conditions to fail. At least one fact is clear: there can be a zero of **f** at which the linearization is neither a sink, a source, or a saddle. This can happen in two rather different ways: one eigenvalue of the linearization can be 0 or there can be a pair of purely imaginary eigenvalues. We discussed this in Section 7.5.

There is, of course, something worse that might happen: both eigenvalues might be zero. The reader should see that this is "more exceptional" than the previous two possibilities, as we shall elaborate by considering the "codimension" of a bifurcation.

THE CODIMENSION OF A BIFURCATION

If you take a 2×2 matrix "at random" (for instance, the linearization of **f** at some zero), you do not expect any eigenvalues to be 0 or purely imaginary. But if you have a one-parameter family of 2×2 matrices, the determinant is a continuous function of the parameter, and if it takes on both positive and negative values, it will have to vanish somewhere. Thus, we expect to see the eigenvalue 0 in one-parameter families. Similarly, the trace will have to vanish somewhere, which leads to purely imaginary eigenvalues.

Both a zero eigenvalue and a pair of purely imaginary eigenvalues are accidents which occur generically in one-parameter families: we say these are bifurcations of codimension 1.

On the other hand, we would not expect both the determinant and the trace to vanish for the same value of the parameter. But if we have a matrix that depends on two parameters, it is hard to see how we could avoid some point mapping to $(0,0)$ in the trace-determinant plane.

A double zero eigenvalue is an accident that occurs generically in two-parameter families but not in one-parameter families: we say this is a codimension 2 bifurcation.

This sort of analysis can be continued: we might look not just at the linearization at a zero but at higher terms of the Taylor polynomial and see how many of these vanish, finding more and more degenerate zeroes of vector fields, which occur generically in higher dimensional families, and near which the differential equation behaves more and more bizarrely. There is a small industry of investigating more and more degenerate zeroes of vector fields, which we will only touch on in Chapter 9.

But the fact that "accidents" can be more or less exceptional, according to the number of parameters needed to make them occur generically, is a very important idea. Without some assumption about differential equations being "general," or belonging to "general" one-parameter and two-parameter families, all kinds of exceptional behavior can occur, and there do not appear to be any straightforward results.

Another accident that might occur is that a limit cycle might fail to be attracting or repelling: it might, for instance, be attracting on one side and repelling on the other as shown in Figure 8.8.2.

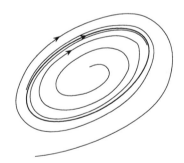

FIGURE 8.8.2. A semi-stable limit cycle.

There is actually only one other accident that can occur: the unstable separatrix of a saddle may also be the stable separatrix of another saddle, or even of the same saddle. The first case is called a *heteroclinic* saddle connection, the second a *homoclinic* saddle connection. See Figure 8.8.3.

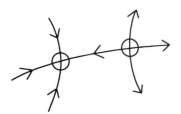
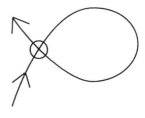

FIGURE 8.8.3. A heteroclinic and a homoclinic saddle connection.

Again, one can imagine more complicated things, such as the simultaneous saddle connections shown in Figure 8.8.4.

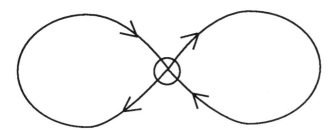

FIGURE 8.4.4. Two simultaneous saddle connections.

The reader should convince herself or himself that both of the accidents seen in Figure 8.8.3 occur generically in one-parameter families, but that the accident in Figure 8.8.4 occurs generically only in two-dimensional families; it is "more exceptional."

Example 8.8.2. Consider a one-parameter family of vector fields $\mathbf{f}_\alpha$ with a saddle, such that for two values of the parameters, the separatrices intersect a line segment I as shown in Figure 8.8.5.

Consider the (signed) distance $d(\alpha)$ from the intersection A_s of the stable separatrix with I to the intersection A_u of I with the unstable separatrix. The distance $d(\alpha)$ is a real-valued continuous function of α, which is positive for one value of the parameter and negative for another, so it must vanish somewhere.

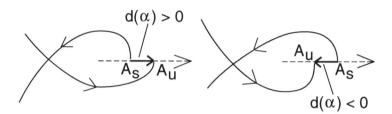

FIGURE 8.8.5. How separatrices intersect a transverse line segment.

It is hard to see how we could avoid such things in one-parameter families; on the other hand, near a vector field as shown in Figure 8.8.4, there are two analogous distances, and in a one-parameter family one would not expect them to vanish simultaneously. ▲

There is not any really obvious way to see that this is the end of the list of accidents, yet it is, and the result can be summarized as follows, to be proved and elaborated in Chapter 8*:

Proposition 8.8.3. *If all zeroes are sinks, sources, or saddles, all cycles are attracting or repelling, and there are no saddle connections, then the vector field is structurally stable.*

The reader who wants to see a proof will have to look at Section 8*.5 (which can be read without reading the earlier sections of Chapter 8*), and Section 8*.6 (which requires everything which comes before). But it is rather hard to imagine how Proposition 8.8.3 could be wrong, and so we ask you temporarily to take it on faith.

You can move directly to Chapter 9 on bifurcations without loss of any prerequisites. But we encourage you to at least come back to Chapter 8* later.

Chapter 8   Exercises

Exercises 8.1   Zeroes; Linearization

8.1#1. For each of the following differential equations

(a)° $\begin{bmatrix} x \\ y \end{bmatrix}' = \begin{bmatrix} y \\ -\sin(x) - 3y \end{bmatrix}$ (c) $\begin{bmatrix} x \\ y \end{bmatrix}' = \begin{bmatrix} x^2 + y^2 - 25 \\ xy - 12 \end{bmatrix}$

(b) $\begin{bmatrix} x \\ y \end{bmatrix}' = \begin{bmatrix} y^2 - 1 \\ x^2 + y^2 - 2 \end{bmatrix}$ (d) $\begin{bmatrix} x \\ y \end{bmatrix}' = \begin{bmatrix} \sin(\pi(x - y)) \\ xy - 1 \end{bmatrix}$

(i) Locate the singularities and write out the linearization about each of them.

(ii) Find the eigenvalues and eigenvectors of the linearization matrix for each of the singularities. This can be done by hand or with *Eigen-Finder* in *MacMath*.

(iii) With *MacMath's DiffEq, Phase Plane* or other phase plane computer software, make a phase portrait for the equation, locating all the singularities. Print three copies.

(iv) On the first phase portrait, draw the eigenvectors at each singularity where they are appropriate, marking the direction of motion by arrows according to the sign of the corresponding eigenvalues. You should be able to see exactly how the phase portrait trajectories are related to these eigenvalues and eigenvectors.

(v) On the second phase portrait, shade in all the basins of the sinks.

(vi) On the third phase portrait, shade in all the cobasins of the sources.

8.1#2. For each of the following differential equations

(a) $\begin{bmatrix} x \\ y \\ z \end{bmatrix}' = \begin{bmatrix} y \\ z \\ x^2 - yz - 1 \end{bmatrix}$ (b) $\begin{bmatrix} x \\ y \\ z \end{bmatrix}' = \begin{bmatrix} y + z \\ x^2 - 2y \\ x + y \end{bmatrix}$

(i) Locate the singularities and write out the linearization about each of them.

(ii) Find the eigenvalues and eigenvectors of the linearization matrix for each of the singularities. This can be done very quickly with *Eigen-Finder* in *MacMath*.

(iii) Using *MacMath's DiffEq, 3DViews*, draw a phase portrait for the equation, locating all the singularities. The xyz view will look like spaghetti, but you may find more "order" in the xy, yz, and xz views. Print out a copy of these graphs.

(iv) On each of the two-dimensional graphs printed in (iii), draw the eigenvectors at each singularity where they are appropriate, marking the direction of motion by arrows according to the sign of the corresponding eigenvalues. (That is, draw the appropriate projection of each eigenvector, which means ignoring the relevant third component.) You should gain some insight as to how the phase portrait trajectories are related to these eigenvalues and eigenvectors.

8.1#3. For each of the following differential equations

(a) $\begin{bmatrix} x \\ y \end{bmatrix}' = \begin{bmatrix} -y \\ x^3/2 - x + xy \end{bmatrix}$

(b)° $\begin{bmatrix} x \\ y \end{bmatrix}' = \begin{bmatrix} -y \\ x^4 + 4x^3 - x^2 - 4x + y)/8 \end{bmatrix}$

(i) Find the zeroes of the system, and by linearizing, tell what they are.

(ii) Make a picture of any saddles with their separatrices and proper arrows on each. This can be done in *MacMath* using *DiffEq, Phase Plane*, with no slope marks. When a saddle is located, the separatrices are drawn automatically, the first two pointing out from the saddle and the last two pointing inwards toward the saddle. Print the picture and draw arrows on the separatrices.

(ii*) An alternate way of examining the saddles is to find their eigenvalues and eigenvectors, by hand or with *MacMath's EigenFinder*. If you do not have a computer program that draws separatrices, you may want to locate the other singularities and find their eigenvalues and eigenvectors.

(iii) From this picture of only the saddles and separatrices, with arrows, locate other zeroes and tell what they are. Sketch in some other trajectories, with arrows, in order to fit with the separatrices and their arrows. Show how separatrices determine regions of behavior near the saddle.

(iv) Verify your expectations for the phase portrait with another computer drawing, using slope marks and drawing by computer some trajectories between the separatrices.

8.1#4. For the differential equation of Example 8.1.8,

$$\begin{bmatrix} x \\ y \end{bmatrix}' = \begin{bmatrix} x^2 \\ -y \end{bmatrix}$$

(a) Find the linearization.

(b) Show that an eigenvalue of the linearized equation is zero.

(c) Show that the equation does *not* behave like its linearization.

8.1#5. For the pretty complicated differential equation

$$\begin{bmatrix} x \\ y \end{bmatrix}' = \begin{bmatrix} 2x + \sin y - 4x(x^2 + y^2) \\ \sin y - 3x\sin(x - 2y) - 2y(x^2 + y^2) \end{bmatrix}$$

(i)° Find the linearization near (0,0) to show that the linearization is far simpler than the original equation.

(ii) Compare phase portraits of the nonlinear and linearized systems and estimate how far from the origin the linearization will give a good approximation. (You should not expect the accuracy to be the same in all directions.)

8.1#6. For the following nonlinear systems of differential equations, make sketches by hand to answer the questions:

(a) $\begin{bmatrix} x \\ y \end{bmatrix}' = \begin{bmatrix} y - x \\ xy - 1 \end{bmatrix}$ (d) $\begin{bmatrix} x \\ y \end{bmatrix}' = \begin{bmatrix} \cos(x^2 + y^2) \\ xy \end{bmatrix}$

(b) $\begin{bmatrix} x \\ y \end{bmatrix}' = \begin{bmatrix} x^2 + y^2 - 1 \\ xy \end{bmatrix}$ (e) $\begin{bmatrix} x \\ y \end{bmatrix}' = \begin{bmatrix} \sin(x - y) \\ \cos(x + y) \end{bmatrix}$

(c) $\begin{bmatrix} x \\ y \end{bmatrix}' = \begin{bmatrix} x^2 + y^2 - 2 \\ y^2 - x^4 \end{bmatrix}$ (f) $\begin{bmatrix} x \\ y \end{bmatrix}' = \begin{bmatrix} x(1 - y - x) \\ y(2 - x) \end{bmatrix}$

(i) Sketch in the phase plane the locus where the vector field is vertical and indicate where it is pointing left and where right. Furthermore, sketch the locus where the vector field is horizontal and indicate where it is pointing up and where down. Combine this information in each region bounded by the isoclines to show the resultant direction of the vector field.

(ii) Find all the singularities of the system. Write down the linearization in each case and describe the type of singularity. Find the eigenvector directions for real eigenvalues of the linearization.

(iii) Sketch some trajectories in the phase plane, among them the separatrices, and color the basins of each sink.

8.1#7. Consider the system of differential equations

$$\begin{bmatrix} x \\ y \end{bmatrix}' = \begin{bmatrix} \alpha x - y \\ y - x^2 + x \end{bmatrix}$$

where α is a parameter.

(a) On what curves is the vector field horizontal? Vertical?

(b) In each region bounded by these curves, tell whether the vector field points up or down, left or right. (The description is a bit different when $\alpha > -1$ and $\alpha < -1$, so consider these cases separately).

(c) Find the zeroes of the vector fields and their linearizations. Classify them.

(d) Sketch the trajectories of the solutions, indicating, in particular, the basins of the sinks and cobasins of the sources, for $\alpha = -2, 0, 1$.

8.1#8. Find, by hand or computer, the singularities, linearizations, trajectories near the zeroes, and basins of the sinks for

(a) $\begin{bmatrix} x \\ y \end{bmatrix}' = \begin{bmatrix} \sin y \\ \cos x \end{bmatrix}$

(d) $\begin{bmatrix} x \\ y \end{bmatrix}' = \begin{bmatrix} y \\ (x^2 - 1)y - x \end{bmatrix}$

(b) $\begin{bmatrix} x \\ y \end{bmatrix}' = \begin{bmatrix} x^2 + y^2 - 25 \\ xy - 12 \end{bmatrix}$

(e) $\begin{bmatrix} x \\ y \end{bmatrix}' = \begin{bmatrix} x \sin(x + y) - x \\ y - x^2 \end{bmatrix}$

(c) $\begin{bmatrix} x \\ y \end{bmatrix}' = \begin{bmatrix} 2xy - x \\ -y^2 + 2xy^2 \end{bmatrix}$.

Note: You might get answers in (d) for the type of singularity depending on where you click the mouse. How can you explain this uncertainty?

8.1#9. Consider the nonlinear system of differential equations

$$\begin{bmatrix} x \\ y \end{bmatrix}' = \begin{bmatrix} y \\ -x^2 + y^2 + 1 \end{bmatrix}.$$

(a) Locate the singularities of the system. In each case, write down the linearization and the type of singularity for the linearized system. What can you conclude about the singularities for the nonlinear system from the linearizations?

(b) Draw in the phase plane the curves where the vector field is horizontal and where it is vertical, and indicate where it is pointing up, down, left, and right. Furthermore, indicate the vector field along the line

$$\begin{bmatrix} x \\ y \end{bmatrix} = \begin{bmatrix} -1 \\ 0 \end{bmatrix} + \tau \begin{bmatrix} 1 \\ \sqrt{2} \end{bmatrix}.$$

Why is this line interesting?

(c) Let $(x, y) = (u(t), v(t))$ be a solution to the nonlinear system. Show that $(x, y) = (u(-t), -v(-t))$ is also a solution. What is the geometrical meaning of this? In particular, let the solution be such that $(u(0), v(0))$ is a point in the first quadrant on the curve where the vector field is horizontal. Explain why such a solution has to be periodic.

(d) Finally, sketch the trajectories for the nonlinear system.

8.1#10. Find the linearization of

$$\begin{bmatrix} x \\ y \end{bmatrix}' = \begin{bmatrix} x - xy \\ -y + xy \end{bmatrix}$$

near all its singularities and in each case identify its type. Confirm with a phase portrait, drawn by hand or computer, the behavior of trajectories for the nonlinear system.

8.1#11. For the equations of the pendulum with friction (Example 6.5.4),

$$\begin{bmatrix} \theta \\ y \end{bmatrix}' = \begin{bmatrix} y \\ -\sin \theta - \varepsilon y \end{bmatrix}$$

(a) Use *MacMath's DiffEq Phase Plane* to make computer pictures for different integer values of $\varepsilon > 0$ in order to show at what value(s) of ε the trajectories cease to spiral.

(b) Confirm by algebraic calculation the value of ε at which you expect the spirals to cease.

8.1#12. Suppose that $x(t)$ describes the position of a particle moving in one dimension, under the potential $V(x)$. Show that the vector field in the plane corresponding to this system, as described in Section 6.5, has a zero at (x_0, y_0) if and only if $dV/dx(x_0) = 0$ and $y_0 = 0$. Show that at such a zero, the linearization is a center if $d^2V/dx^2(x_0) > 0$, i.e., if the potential has a nondegenerate minimum. Show that if the potential has a nondegenerate maximum, the linearization has a saddle.

Exercises 8.2–8.3 Saddles, Sinks, Sources

8.2–8.3#1. For

$$\begin{bmatrix} x \\ y \end{bmatrix}' = \begin{bmatrix} x^2 + y^2 - 25 \\ 12 - xy \end{bmatrix},$$

locate any sinks and sources.

8.2–8.3#2°. Consider the system of differential equations

$$\begin{bmatrix} x \\ y \end{bmatrix}' = \begin{bmatrix} -x - xy \\ y - x^2 \end{bmatrix} \qquad (i)$$

and the related single differential equation

$$\frac{dy}{dx} = \frac{y - x^2}{-x - xy}. \qquad (ii)$$

(a) Show that one of three singular points for (i) is a saddle point. Find the eigenvector directions.

(b) We have shown that the shaded region U in Figure 8.3.2,

$$U = \{(x, y) | -1 < x < 0, -x^2 \le y \le x^2\},$$

defines an antifunnel for (ii) satisfying

$$\frac{\partial}{\partial y} \left(\frac{y - x^2}{-x - xy} \right) \ge 0.$$

Show that the symmetric shaded region W,

$$W = \{(x, y) | 0 < x < 1, -x^2 \le y \le x^2\},$$

defines a funnel for (ii) and, when you reverse "time" (x in this case), W defines an antifunnel satisfying the uniqueness criterion.

(c) What does part (b) tell you about the system (i)?

8.2–8.3#3°. Consider an $n \times n$ matrix A with k pairs of conjugate nonreal eigenvalues, so that for any $\varepsilon > 0$, by Theorem L6.18, there exists a basis $w_1, \ldots, w_n$ of $\mathbb{R}^n$ such that in that basis A is k-pseudo-upper-triangular with terms of absolute value $< \varepsilon$ above the pseudo-diagonal. Let $-\Lambda$ equal the largest real part of an eigenvalue. For the proof of Proposition 8.2.3, prove that

$$A\mathbf{x} \cdot \mathbf{x} \le (\Lambda + n\varepsilon) \|\mathbf{x}\|^2.$$

8.2–8.3#4. For the differential equation

$$\begin{bmatrix} x \\ y \end{bmatrix}' = \begin{bmatrix} y \\ -x - y^3 \end{bmatrix}$$

(a) Show that at $(0,0)$ the linearization gives a center, so we need further analysis to determine the nonlinear behavior.

(b) Show that $F(x,y) = x^2 + y^2$ serves as a Liapunov function and use it to analyze the singularity at $(0,0)$ of the differential equation.

(c) Describe all solutions of this equation.

8.2–8.3#5. Consider the differential equation

$$\begin{bmatrix} x \\ y \end{bmatrix}' = \begin{bmatrix} y \\ -e^{-x^2}(x+y) \end{bmatrix}.$$

(a) Show that every solution that enters the region $|y| \leq e^{-x^2/2}$ is attracted to the origin. Hint: use Example 8.2.8 to find an appropriate Liapunov function.

(b) Show that any solution entering the region $x > 0, y > 1/x$ stays in it forever (in particular, is defined for all positive time). Hint: Consider the curve $y = 1/x$ as a fence for the associated equation

$$\frac{dy}{dx} = -e^{-x^2}\frac{x+y}{y}.$$

(c) Use an antifunnel and the symmetries of the equation to describe the basin of the origin with precision.

8.2–8.3#6. Repeat Exercise 8.2–8.3#5 for the equation

$$\begin{bmatrix} x \\ y \end{bmatrix}' = \begin{bmatrix} y \\ -\frac{1}{x^4+1}(x+y) \end{bmatrix},$$

this time finding the appropriate regions yourself.

8.2–8.3#7. The equation

$$\begin{bmatrix} x \\ y \end{bmatrix}' = \begin{bmatrix} y \\ -\frac{1}{x^2+1}(x+y) \end{bmatrix},$$

behaves quite differently from that of Exercise 8.2–8.3#6 (though the picture looks very similar). How and why?

8.2–8.3#8. Give a proof using fences of the following variant of Liapunov's Theorem: If $\mathbf{x}' = \mathbf{f}(\mathbf{x})$ is a differential equation, $\mathbf{f}(x_0) = 0$ and F is a function such that $F(\mathbf{x}_0) = 0$, the point $\mathbf{x}_0$ is an isolated minimum of F, and

$$\nabla F(\mathbf{x}) \cdot \mathbf{f}(\mathbf{x}) \leq -CF(\mathbf{x})$$

for some constant $C > 0$, then $\mathbf{x}_0$ is a sink for the differential equation.

8.2–8.3#9°. Consider the differential equation

$$x' = -ax + P(x, y)$$
$$y' = -ay + Q(x, y),$$

where $a > 0$, and P and Q start with quadratic terms. Show that the solutions attracted to the sink $(0, 0)$ do not spiral around the origin.

Hint: Pass to polar coordinates, and write θ' as in equation (20) in the proof of Theorem 8.2.10. Use the fact that the solutions are attracted exponentially fast to the origin to show that

$$\int_0^\infty \theta' \, dt < \infty.$$

8.2–8.3#10. Consider the differential equation

$$x' = -ax - y + P(x, y)$$
$$y' = -ay + Q(x, y),$$

where $a > 0$, and P and Q start with quadratic terms. Show that the solutions attracted to the sink $(0, 0)$ do not spiral around the origin.

Hint: Consider the associated differential equation

$$\frac{dy}{dx} = \frac{-ay + Q(x, y)}{-ax - y + P(x, y)}$$

and show that it is well defined in the region $-x^{3/2} \leq y \leq x^{3/2}, 0 < x < \varepsilon$ for $\varepsilon > 0$ sufficiently small. Then show that this region is a backward antifunnel.

8.2–8.3#11°. If

$$\begin{bmatrix} x \\ y \end{bmatrix}' = \begin{bmatrix} -\mu x + P(x, y) \\ \nu y + Q(x, y) \end{bmatrix}$$

with P and Q of quadratic or higher order forms, show that the curve $\mu x = P(x, y)$ is twice differentiable and tangent to the y-axis.

8.2–8.3#12. In the proof of Lemma 8.2.3, prove that V is an antifunnel.

8.2–8.3#13. For the Lorenz attractor of Example 6.1.6,

$$\begin{bmatrix} x \\ y \\ z \end{bmatrix}' = \begin{bmatrix} 10(y - x) \\ 28x - y - xz \\ -2.66z + xy \end{bmatrix}$$

find the singularities and draw pictures to show how each works.

8.2–8.3#14. For the following second order equations, express each as a system of first order equations by setting $x' = y$. Then, for $\alpha = -1, 0, 1$, use *MacMath's* program *DiffEq, Phase Plane* to find the singularities and show the behaviors of the trajectories in the xy phase plane. Use arrows to show the directions of the trajectories in forward time and indicate where there seem to be stable solutions. In Exercises 9.1#9 for (a) and 9.5#6 for (b), we shall come back to these particular systems and explore exactly how and when the behaviors change as α changes.

(a) $x'' - \alpha x + x^3 = 0$.

(b) $x'' = -dV/dx$, where $V = \sin x + \alpha x^2$, as in Exercise 6.5#10.

8.2–8.3#15. The system

$$\begin{bmatrix} x \\ y \\ z \end{bmatrix}' = \begin{bmatrix} -\sin y \\ -x \\ -z \end{bmatrix}$$

is easy to understand because the z variable and the variables (x, y) evolve independently of each other.

(a) Locate all the zeroes of the system, and give the linearizations at each.

(b) Show that the origin is a saddle, with one positive and two negative eigenvalues.

(c) Use *MacMath's* program *DiffEq, 3D Views* to draw some sample trajectories, adding arrows, that illustrate the three-dimensional behavior around the saddle.

(d) Describe the solutions that are attracted to the origin when $t \to \pm\infty$.

8.2–8.3#16. For the equation of Exercise 8.1#1(c) and Exercise 8.2–8.3#1,

$$\begin{bmatrix} x \\ y \end{bmatrix}' = \begin{bmatrix} x^2 + y^2 - 25 \\ xy - 12 \end{bmatrix},$$

prove that a separatrix of the upper saddle *must* enter the basin of the sink at lower left. That is, find a region surrounding the sink for which the vector field of the equation everywhere crosses the boundary going into the region; then show that the separatrix of the saddle must enter this region.

Exercises 8.4 Limit Cycles

8.4#1.

(a) Confirm equations (30) for transforming Example 8.4.1

$$\begin{bmatrix} x \\ y \end{bmatrix}' = \begin{bmatrix} y \\ -x \end{bmatrix} + \alpha(1 - x^2 - y^2)\begin{bmatrix} x \\ y \end{bmatrix} \tag{29}$$

into polar coordinates

$$\begin{aligned} r' &= \alpha(1 - r^2)r \\ \theta' &= -1. \end{aligned} \tag{30}$$

(b) Make drawings of the rt-plane to confirm that solutions to equation (30) follow the conclusions of Example 8.4.1: For $\alpha > 0$, all solutions for $r > 0$ are attracted to the constant solution $r = 1$; for $\alpha < 0$, the solution $r = 1$ is repelling, and if $0 \le r(0) < 1$, the solution tends to 0, whereas if $1 < r(0)$, the solution tends to infinity.

8.4#2. Show that

$$\begin{bmatrix} x \\ y \end{bmatrix}' = (x^2 + y^2 - 1)^2 \begin{bmatrix} x \\ y \end{bmatrix}$$

gives an example of an unusual limit cycle that is stable on one side, unstable on the other.

8.4#3°. For the equation of Example 8.4.3,

$$\begin{bmatrix} x \\ y \end{bmatrix}' = \begin{bmatrix} y \\ -x \end{bmatrix} + (x^2 + y^2)^2 \sin \frac{1}{x^2 + y^2} \begin{bmatrix} x \\ y \end{bmatrix},$$

prove the equation has infinitely many periodic solutions, with different trajectories.

Exercises 8.5 Poincaré–Bendixson Theorem

8.5#1°. Let $I = [a, b]$ be a closed and bounded interval. Show that any continuous mapping $f : I \to I$ has a fixed point. Hint: Use the intermediate value theorem, applied to the function $f(x) - x$.

8.5#2. Consider the equation of Example 8.4.3 and Figure 8.4.3,

$$\begin{bmatrix} x \\ y \end{bmatrix}' = \begin{bmatrix} y \\ -x \end{bmatrix} + (x^2 + y^2)^2 \sin \frac{1}{x^2 + y^2} \begin{bmatrix} x \\ y \end{bmatrix}.$$

(a) Draw a phase portrait and make a Poincaré–Bendixson annulus around one of the largest limit cycles.

(b) Find equations for an annulus like you found in part (a).

8.5#3. For the differential equation

$$\begin{bmatrix} x \\ y \end{bmatrix}' = \begin{bmatrix} \sin(x+y) \\ \cos(xy) \end{bmatrix},$$

print a phase portrait and try to draw Poincaré–Bendixson annuli around each limit cycle.

8.5#4°. Consider the Van der Pol equation of Examples 8.4.2 and 8.5.3. Find equations for the annulus as shown in Figure 8.5.5. Hint: It is composed of straight lines plus key portions of the isoclines of zero slope.

8.5#5. Show that the equation

$$\begin{bmatrix} x \\ y \end{bmatrix}' = \begin{bmatrix} y \\ -\sin(x)y - x \end{bmatrix} + (x^2 + y^2)(1 - (x^2 + y^2)^2) \begin{bmatrix} x \\ y \end{bmatrix}$$

has a limit cycle. (Hint: use appropriate solutions of

$$\begin{bmatrix} x \\ y \end{bmatrix}' = \begin{bmatrix} y \\ -\sin(x)y - x \end{bmatrix}$$

as boundaries of an annular region.)

Exercises 8.6 Symmetries; Area-Preserving Equations

8.6#1°. Consider the system of differential equations

$$\begin{bmatrix} x \\ y \end{bmatrix}' = \begin{bmatrix} x - y \\ x^2 - y \end{bmatrix}.$$

(a) Sketch (perhaps using an appropriate computer program) some typical solutions for this equation.

(b) Locate and analyze the singularities.

(c) In part (a), you should have found a region apparently filled by cycles. Show that this is in fact the case.

(d) The saddle apparently has a homoclinic separatrix (an unstable separatrix that curves around and reenters the saddle as a stable separatrix). Show that this is the case, and, for this problem, that you can find an equation for the separatrix!

8.6#2. Consider the system of differential equations

$$\begin{bmatrix} x \\ y \end{bmatrix}' = \begin{bmatrix} y \\ -x^2 + y^2 + 1 \end{bmatrix}.$$

(a) Sketch (perhaps using an appropriate computer program) some typical solutions for this equation.

(b) Locate and analyze the singularities.

(c) In part (a), you should have found a region apparently filled by cycles. Show that this is in fact the case.

The remainder of the exercise attempts to locate this region with greater precision.

(d) Consider the corresponding system of equations for dy/dx. Show that the region $x + 1/2 - 1/x \le y \le x + 1/2$, $t > t_0$, is an antifunnel for an appropriate t_0. Show that this antifunnel satisfies the uniqueness property.

(e) Sketch the solution in this antifunnel.

(f) Show that all solutions in the part of the plane bounded by this solution are cycles.

8.6#3°. If $\mathbf{u}' = A\mathbf{u}$ is the linearization of the differential equation $\mathbf{x}' = \mathbf{f}(\mathbf{x})$ at a zero $\mathbf{x}_0$, show that

$$\operatorname{div} \mathbf{f}(\mathbf{x}_0) = \operatorname{tr} A,$$

in order to support the statement that a volume-preserving vector field cannot have any sources or sinks.

8.6#4. Show that a linear differential equation $\mathbf{x}' = A\mathbf{x}$ is area-preserving if and only if $\operatorname{tr} A = 0$.

8.6#5. Draw a phase portrait for each of the following differential equations. Tell (and prove) whether it exhibits properties of symmetries or an area-preserving map.

(a) $x'' + \sin xx' + x = 0$

(b) $\begin{bmatrix} x \\ y \end{bmatrix}' = \begin{bmatrix} y \\ -x \end{bmatrix} + (x^2 + y^2)^2 \sin \dfrac{1}{x^2 + y^2} \begin{bmatrix} 1 \\ 1 \end{bmatrix}$

(c) $\begin{bmatrix} x \\ y \end{bmatrix}' = \begin{bmatrix} y \\ -x - y \sin x \end{bmatrix}$

8.6#6. Show that if the differential equation

$$\begin{bmatrix} x \\ y \end{bmatrix}' = \begin{bmatrix} f(x, y) \\ g(x, y) \end{bmatrix}$$

is area-preserving (see Exercise 6.6#3), then its zeroes cannot be sinks or sources (and therefore must be centers or saddles).

8.6#7. Let $\mathbf{f}$ be a vector field defined in a region $U \subset \mathbb{R}^2$. Show that the differential equation $\mathbf{x}' = \mathbf{f}(\mathbf{x})$ cannot have two cycles, which together bound a region $V \subset U$ in which div $\mathbf{f} < 0$.

Exercises 8.7 Chaos in Higher Dimensions

8.7#1. Show that the divergence of the vector field

$$\begin{bmatrix} x \\ y \\ z \end{bmatrix}' = \begin{bmatrix} \sigma(y - x) \\ \rho x - y - xz \\ -\beta z + xy \end{bmatrix} \tag{iii}$$

is $-(\sigma + 1 + \beta)$. Why is the Lorenz equation said to be *dissipative*?

8.7#2°. Consider the function $f(x, y, z) = \rho x^2 + \sigma y^2 + \sigma(z - 2\rho)^2$. The object of this exercise is to show that there is a bounded region (in fact, an ellipsoid) which solutions to the Lorenz equations (iii) enter and never leave.

(a) Show that the region $f < a$ is bounded for any a.

(b) Show that the derivative of f along a solution is

$$-2\sigma(\rho x^2 + y^2 + \beta z^2 - 2\rho\beta z).$$

(c) Find a number a such that all solutions to the Lorenz equations (iii) enter the region $f \leq a$ and never leave it.

8.7#3.

(a) Show that the origin is a zero of the differential equation (iii) and that it is a sink when $\rho < 1$, but has two real negative eigenvalues and one positive eigenvalue for $\rho > 1$.

(b) Show that the function $x^2/\sigma + y^2 + z^2$ is a monotone decreasing function of time when $0 < \rho < 1$, so that all solutions to the Lorenz equations (iii) are attracted to 0.

8.7#4.

(a) Show that the origin is the only zero of the vector field for (iii) for $\rho \leq 1$, but that there are two other zeroes for $\rho > 1$, and give their coordinates.

(b) Show that these new zeroes are sinks for

$$\rho \in \left(1, \frac{\sigma(\sigma + \beta + 3)}{\sigma - \beta - 1}\right)$$

and that they have one negative eigenvalue and two complex conjugate eigenvalues with positive real parts for

$$\rho > \frac{\sigma(\sigma + \beta + 3)}{\sigma - \beta - 1}.$$

8.7#5.

(a) Show that the system of differential equations

$$\begin{bmatrix} x \\ y \\ z \end{bmatrix}' = \begin{bmatrix} y + (1 - (x^2 + y^2))x \\ -x + (1 - (x^2 + y^2))y \\ z \end{bmatrix}$$

has the circle $x^2 + y^2 = 1$, $z = 0$ as a cycle.

(b) Compute the section mapping P, using the section $y = 0, x > 0$, and show that at the fixed point of P corresponding to the cycle, the derivative of P has two eigenvalues, one greater than 1 and one less than 1.

(c) Find the separatrices of this fixed point in the section.

(d) Describe the separatrices of the cycle. What behaviors in $\mathbb{R}^2$ do these separatrices separate?

Exercises 8.8. Structural Stability

8.8#1°. Consider the system of differential equations

$$\begin{bmatrix} x \\ y \end{bmatrix}' = \begin{bmatrix} y - x^2 \\ \alpha x - 2 - y \end{bmatrix}.$$

(a) With *MacMath's DiffEq Phase Plane* make and print phase portraits
    for $\alpha = 3.5, 0, -3.5, -6, 10$. To capture all you need to see, set
    bounds as $-20 \le x \le 15, -10 \le y \le 100$.

(b) Describe each phase plane in words, and try to understand what
    happens between one and the next. That is, explain the qualitative
    differences between these pictures.

(c) Experiment by trial and error on the computer to find intermediate
    values of α where the behavior changes.

A generalization of this exercise will be explored a great deal further in
Example 9.6.1.

8*

Structural Stability

In Chapter 8, we mentioned that one of our main goals is to understand how solutions behave "in general." Another goal is to understand when the solutions of two differential equations behave "the same way." Our main result in this direction is in Section 8*.6, but preliminary results occur in Sections 8*.2–8*.5, and they require a bit of terminology. We begin with the groundwork in Section 8*.1, building up to Definition 8*.1.9 for "structural stability".

Section 8*.6 contains two major theorems: the first due to Andronov and Pontryagin, and the second to Peixoto. These results give a real handle on structural stability of differential equations in the plane; the whole chapter builds up to them. Their proofs are an order of magnitude longer and more difficult than anything else in this book so far. We work up to Section 8*.6 with a series of sections, 8*.1–8*.5, each of which can be considered as a Structural Stability Appendix at the end of the corresponding section in the previous Chapter 8. We give preparatory definitions in 8*.1 and prove an essential preliminary result in each of the next four sections. Each of these results, Theorems 8*.2.1, 8*.3.1, 8*.4.1, and 8*.5.6, is important in its own right; Theorem 8*.5.6 is a strong form of the famous Poincaré–Bendixson Theorem 8.5.1.

Finally, Section 8*.7 illustrates why the Poincaré–Bendixson–Pontryagin–Peixoto program fails in dimension three. Some examples will show that any general analysis of a system in $\mathbb{R}^n$ is likely to be extremely complicated, if not impossible. Indeed, as soon as $n \geq 3$, "chaos" has a tendency to set in. There is no universally accepted definition of this term, but some features that should appear in chaotic systems are:

- the presence of infinitely many periodic cycles, which in $\mathbb{R}^3$ will usually be knotted and linked in complicated ways;

- the appearance of fractal attractors and basin boundaries, with complicated structure at all scales;

- sensitive dependence on the initial point: within a bounded attractor, the distance separating two solutions that start nearby grows exponentially with time, until they are uncorrelated.

8*.1. Preliminaries for Structural Stability

Let us begin with some necessary terminology.

Definition 8*.1.1. (Homeomorphisms). Let X and Y be metric spaces. If you do not know what this means, it does not matter: just think that they are two subsets of $\mathbb{R}^n$, that is the only case we will use. A mapping $f : X \to Y$ is a *homeomorphism* if

(1) the mapping f is continuous;

(2) the mapping f is one-to-one and onto, so that f^{-1} exists;

(3) the mapping f^{-1} is continuous.

Saying that two spaces are homeomorphic is saying that they "look alike topologically": to a topologist, they are identical. An old joke says that a topologist is someone who cannot tell the difference between a doughnut and a coffee cup: do you see why? See Figure 8*.1.1.

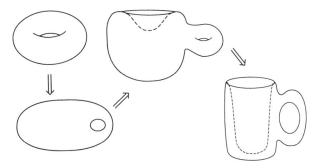

FIGURE 8*.1.1. Topological equivalence of a doughnut and a coffee cup.

Exercise 8*.1#2, with Examples 8*.1.3 and 8*.1.4, give more examples of homeomorphisms.

Definition 8*.1.2 (Topological equivalence). Suppose that $\mathbf{f}_1$ and $\mathbf{f}_2$ are vector fields defined on regions U_1 and U_2 of $\mathbb{R}^n$ for some n. We will say that a homeomorphism $h : U_1 \to U_2$ is a *topological equivalence* of $\mathbf{f}_1$ and $\mathbf{f}_2$ if h sends oriented trajectories of $\mathbf{x}' = \mathbf{f}_1(\mathbf{x})$ to oriented trajectories of $\mathbf{x}' = \mathbf{f}_2(\mathbf{x})$.

Remark. Imagine drawing both vector fields $\mathbf{f}_1$ and $\mathbf{f}_2$ on rubber sheets. Then they are topologically equivalent if one rubber sheet can be stretched and put onto the other so that the trajectories of one equation coincide with the trajectories of the other, as oriented curves.

We are *not* requiring that the parametrization of these curves by time coincide. We are also *not* requiring that the homeomorphism be differentiable, so it can perfectly well send curves that intersect tangentially to curves that intersect transversely, as illustrated by Examples 8*.1.3 and 8*.1.4, illustrated in Figures 8*.1.2 and 8*.1.3.

Example 8*.1.3. Consider the homeomorphism h defined by

$$h : \begin{bmatrix} x \\ y \end{bmatrix} \to \begin{bmatrix} x \\ y^{1/3} \end{bmatrix}.$$

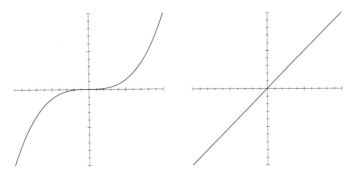

FIGURE 8*.1.2. A homeomorphism h not differentiable at the origin. ▲

Example 8*.1.4. Another homeomorphism h that is not differentiable at the origin is

$$h : \begin{bmatrix} r \\ \theta \end{bmatrix} \to \begin{bmatrix} r \\ \theta + \frac{1}{r} \end{bmatrix}.$$

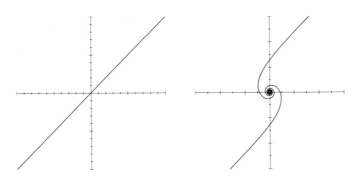

FIGURE 8*.1.3. Another homeomorphism not differentiable at the origin. ▲

Algebraic exploration of these examples are provided in Exercise 8*.1#1; however, we usually do *not* write actual equations for homeomorphisms.

It is not obvious that topological equivalence is a useful equivalence relation. There are at least two others that should be mentioned: *topological conjugacy* and *differentiable conjugacy*.

Definition 8*.1.5 (Topological conjugacy of flows). The homeomorphism h is a *topological conjugacy* of the flows $\phi_{\mathbf{f}_1}(t, \mathbf{x})$ and $\phi_{\mathbf{f}_2}(t, \mathbf{x})$ if

$$h(\phi_{\mathbf{f}_1}(t, \mathbf{x})) = \phi_{\mathbf{f}_2}(t, h(\mathbf{x})). \tag{1}$$

See Figure 8*.1.4.

Flows $\phi_{\mathbf{f}}(t, \mathbf{x})$ are defined and discussed in Section 6.8.

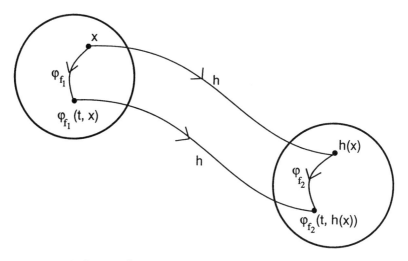

FIGURE 8*.1.4. Topologically conjugating the flows.

Two vector fields have *conjugate flows* if, when drawn on rubber sheets, one sheet can be stretched and superposed on the other so that the trajectories coincide *with their time parametrization*.

Definition 8*.1.6 (Differentiable conjugacy of vector fields). The homeomorphism h is a *differentiable conjugacy* of the vector fields $\mathbf{f}_1(\mathbf{x})$ and $\mathbf{f}_2(\mathbf{x})$ if h and h^{-1} are differentiable, and

$$d_{\mathbf{x}}h(\mathbf{f}_1(\mathbf{x})) = \mathbf{f}_2(h(\mathbf{x})). \tag{2}$$

See Figure 8*.1.5.

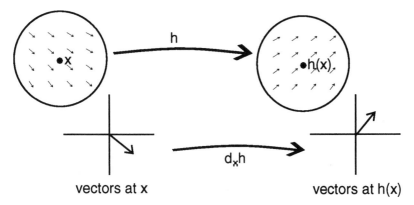

FIGURE 8*.1.5. Differentiable conjugacy of vector fields.

Condition (2) for differentiable conjugacy may look quite complicated, but it is actually exactly what is called "change of variables in differentiable equations." For the authors, it really explains the meaning of such changes of variables. Let us work out explicitly how this works, in the case of Bernoulli equations (Chapter 2 of Part I, Exercises 2.2–2.3#9 and 10a, page 101).

Example 8*.1.7. A Bernoulli equation is one that can be written $x' = -P(t)x + Q(t)x^n$ (where the minus sign is there to be consistent with the notation in Part I). The change of variables suggested in the exercise is $z = x^{1-n}$, so that

$$z' = (1-n)x^{-n}x' = (1-n)x^{-n}(-P(t)x + Q(t)x^n) = (n-1)(P(t)z - Q(t)),$$

which is indeed linear.

To interpret this change of variables in terms of vector fields, let us turn the Bernoulli equation into the system

$$\begin{bmatrix} t \\ x \end{bmatrix}' = \begin{bmatrix} 1 \\ -P(t)x + Q(t)x^n \end{bmatrix}$$

and let

$$h\left(\begin{bmatrix} t \\ x \end{bmatrix} \right) = \begin{bmatrix} t \\ x^{1-n} \end{bmatrix} \overset{\text{def}}{=} \begin{bmatrix} t \\ z \end{bmatrix}.$$

Then equation (2) becomes

$$\begin{bmatrix} 1 & 0 \\ 0 & (1-n)x^{-n} \end{bmatrix} \begin{bmatrix} 1 \\ -P(t)x + Q(t)x^n \end{bmatrix} = \begin{bmatrix} 1 \\ (n-1)(P(t)z - Q(t)) \end{bmatrix}$$

and the differential equation corresponding to the new vector field is

$$\begin{bmatrix} t \\ z \end{bmatrix}' = \begin{bmatrix} 1 \\ (n-1)(P(t)z - Q(t)) \end{bmatrix}.$$

In this example, h is not really a homeomorphism, and this comes back to haunt you when you try to go back from the variable z to the variable x. ▲

You are asked in Exercise 8*.1#3 to check that condition (2) for differentiable conjugacy implies (1) for topological conjugacy, and that (1) implies topological equivalence.

Both conditions (2) and (1) are a bit too fine for our purposes: we want to know that "in general," when two differential equations are "close," they are equivalent. This is not the case for (2), since the eigenvalues at corresponding zeroes would have to coincide (Exercise 8*.1#4a). Condition (1) would require that the periods of corresponding cycles should be equal, which is also unreasonable (Exercise 8*.1#4b).

PERTURBATIONS

This brings us to saying precisely when two differential equations $\mathbf{x}' = \mathbf{f}_1(\mathbf{x})$ and $\mathbf{x}' = \mathbf{f}_2(\mathbf{xff})$ are close or, equivalently, when $\mathbf{g} = \mathbf{f}_1 - \mathbf{f}_2$ is small.

There are serious difficulties if you try to define perturbations on unbounded domains; we will stick to bounded domains. Suppose that $\mathbf{g}$ is defined on the closure $\bar{U}$ of a bounded open subset $U \subset \mathbb{R}^n$. A first guess is simply to require that

$$\sup_{\mathbf{x} \in \bar{U}} \|\mathbf{g}(\mathbf{x})\|$$

be small, but this is not good enough. Indeed, two vector fields that are close in this sense can have zeroes that differ radically.

Example 8*.1.8. Compare the following functions and their derivatives, as shown in Figure 8*.1.6.

$$f_1(x) = x$$
$$f_2(x) = x + \frac{1}{10}e^{-x^2}\sin 5x.$$

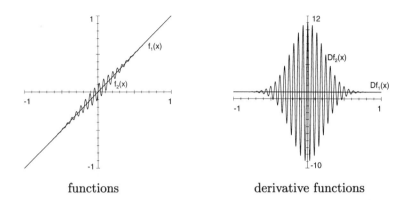

functions   derivative functions

FIGURE 8*.1.6. Functions that are close may have derivatives and fixed points that are not. Note the differences in vertical scale between the two graphs! ▲

The Implicit Function Theorem at the center of multivariable calculus guarantees that if two vector fields $\mathbf{f}_1$ and $\mathbf{f}_2$ are close, *and their derivatives are close*, then the zeroes are as well behaved as can be expected. Let $\mathbf{g}$ be continuous on $\bar{U}$, continuously differentiable on U, with derivative extending continuously to $\bar{U}$. Then $\mathbf{g}$ is small in the C^1 topology if

$$\sup_{\mathbf{x} \in U} \left(\|\mathbf{g}(\mathbf{x})\| + \|d_x \mathbf{g}\| \right),$$

is small.

The Implicit Function Theorem 13.6.4 (Appendix T) says that if the zeroes of $\mathbf{f}_1$ have no zero eigenvalues and if $\mathbf{g}$ is sufficiently small in the C^1 topology, then there is exactly one zero of $\mathbf{f}_2 = \mathbf{f}_1 + \mathbf{g}$ close to every zero of $\mathbf{f}_1$. This is one possible correct notion of perturbation for our problem, and our main theorems are true with this definition.

However, the version of the Implicit Function Theorem we will prove in Chapter 13 (in Part III) is stronger than the standard statement and correspondingly requires a stronger hypothesis. Because the stronger hypothesis simplifies some proofs, we will make the following definition. Let

$$\mathbf{g} = \begin{bmatrix} g_1 \\ \vdots \\ g_n \end{bmatrix}$$

be a twice-differentiable vector field on $\bar{U}$ as above. Define the C^2-norm of this vector field to be

$$\|\mathbf{g}\| = \sup_{\mathbf{x} \in U} \left(\sup_i |\mathbf{g}_i(\mathbf{x})| + \sup_{i,j} \left| \frac{\partial \mathbf{g}_i}{\partial x_j} \right| + \sup_{i,j,k} \left| \frac{\partial^2 \mathbf{g}_i}{\partial x_j \partial x_k} \right| \right).$$

In other words, we will require that "close" means not only that the two functions be close but that their first *and* second derivatives be close.

We can now state the central definition of Chapter 8*.

Definition 8*.1.9 (Structural stability). We will say that a vector field **f** defined on a subset $U \subset \mathbb{R}^n$ is *structurally stable* if there exists $\varepsilon > 0$ such that any perturbation **g** of **f** with $\|\mathbf{g} - \mathbf{f}\| < \varepsilon$ is topologically equivalent to **f**.

Actually, this definition is a bit loose. We will require that U be a closed and bounded subset of $\mathbb{R}^n$, bounded by a smooth $(n-1)$-dimensional surface, and in practice n will be two, so U will be a closed and bounded region in the plane bounded by smooth curves. Our vector fields will be required to have all partial derivatives up to order two continuous on U.

THE DIFFERENTIABILITY OF THE FLOW

Let $\mathbf{f}(\mathbf{x})$ be a smooth vector field on a subset $U \subset \mathbb{R}^n$, and denote by $\phi_{\mathbf{f}}(t, \mathbf{x}) : \mathbb{R} \times U \to \mathbb{R}^n$ the associated flow, as described in Section 6.8. In this section, we will see that the flow is a differentiable mapping, and we will derive some consequences. Just knowing that the flow is differentiable is nice but not very exciting. What makes the result important is the formula for the derivative, which is itself a solution of a *linear* differential equation. This links up the "wild" theory of nonlinear equations with the "tame" theory of linear equations, and the link has proved enormously useful.

In fact, we have already seen at least two instances of the usefulness of this link.

One is the linearization of a differential equation near a zero of the vector field: in that case, the linearization is a linear differential equation with constant coefficients, and linear algebra (eigenvectors and eigenvalues) can be brought to bear; Chapter 8 is filled with the consequences.

The other occurs in Example 5.4.6, where we analyzed the difficulties that "stiffness" introduces to numerical methods. We first studied linear equations, and then saw that by linearizing a nonlinear equation near a given solution, we could transpose the results to an arbitrary equation.

The linearized equation. We first recall the definition of the linearization of $\mathbf{x}' = \mathbf{f}(\mathbf{x})$. Suppose $\mathbf{u}(t)$ is a solution of $\mathbf{x}' = \mathbf{f}(\mathbf{x})$, defined for $0 \le t \le t_0$ and that $\mathbf{v}(t)$ is a small increment to $\mathbf{u}(t)$, so that $\mathbf{x}(t) = \mathbf{u}(t) + \mathbf{v}(t)$ is still a solution. Then

$$\mathbf{u}'(t) + \mathbf{v}'(t) = (\mathbf{u} + \mathbf{v})'(t) = \mathbf{f}(\mathbf{u}(t) + \mathbf{v}(t)) \sim \mathbf{f}(\mathbf{u}(t)) + (d_{\mathbf{u}(t)}\mathbf{f})(\mathbf{v}(t)). \quad (3)$$

Remembering that $\mathbf{u}' = \mathbf{f}(\mathbf{u})$, this says that **v** approximately satisfies the linearized equation

$$\mathbf{w}' = \left(d_{\mathbf{u}(t)}\mathbf{f}\right)\mathbf{w}. \quad (4)$$

You should think of **u** as *known*, so that this equation is of the form **w**$' = A(t)$**w** for some (usually time dependent) matrix $A(t)$.

Example 8*.1.10. Consider the equation $x' = x^2 - 1$. The constant functions $u_1(t) = 1$ and $u_2(t) = -1$ are solutions, where the linearizations are respectively

$$w' = 2w \quad \text{and} \quad w' = -2w,$$

correctly reflecting that solutions pull away from u_1 and tend toward u_2. The function $u_3(t) = -\tanh t$ is also a solution, where the linearization is

$$w' = -2(\tanh t)w.$$

This time the linearization is time dependent, since the solution is nonconstant, and solutions pull away from the zero solution when $t < 0$, i.e., when $u_3(t) \sim u_1(t)$, and are attracted to it when $t > 0$, i.e., when $u_3(t) \sim u_2(t)$.
▲

Example 8*.1.11. We invite the reader to show that the function

$$\mathbf{u}(t) = \begin{bmatrix} \cos t + \cos^2 t \\ \sin t \end{bmatrix}$$

is a solution of the system

$$\begin{bmatrix} x \\ y \end{bmatrix}' = \begin{bmatrix} y - 2xy - y^3 \\ x - 1 + y^2 \end{bmatrix}.$$

The linearized equation along this solution is

$$\mathbf{w}' = \begin{bmatrix} -2\sin t & 1 - 2(\cos t + \cos^2 t) - 3\sin^3 t \\ 1 & 2\sin t \end{bmatrix} \mathbf{w}.$$

▲

Let us interpret **u**, **v**, and **w** above in terms of flows. Suppose that at time 0, we have

$$\mathbf{u}(0) = \mathbf{x}_0, \qquad \mathbf{v}(0) = \mathbf{v}_0 = \mathbf{w}(0);$$

i.e., we start both the equation and the linearized equation with the *same* increment from $\mathbf{u}_0$.

Then equation (3) says that at time t, the increment $\mathbf{v}(t)$ to **u** is approximately $\mathbf{w}(t)$, i.e.,

$$\phi_{\mathbf{f}}(t, \mathbf{x}_0 + \mathbf{v}_0) \sim \phi_{\mathbf{f}}(t, \mathbf{x}_0) + \mathbf{w}(t).$$

This suggests that $\mathbf{w}$ is somehow the derivative of ϕ_f, and this is true. To state it precisely, we want to speak of the flow of the linearized equation. This is not quite well defined, since (4) is time dependent, but let us call $\phi_{\mathbf{u}}(t, \mathbf{v}) : \mathbb{R}^n \to \mathbb{R}^n$ the linear transformation that associates to $\mathbf{v}$ the value at time t of the solution of (4) with initial condition $\mathbf{v}$ at time 0.

Theorem 8*.1.12. *The derivative of the flow of a nonlinear equation is the flow of its associated linear equation. That is,*

$$\left(d_{\mathbf{x}_0}\phi_{\mathbf{f}(t)}\right)(\mathbf{v}) = \phi_{\mathbf{u}_0}(t, \mathbf{v}).$$

We will actually prove a little more: the error in the linear approximation is quadratic. More precisely, there exist constants C and $\delta > 0$ such that

$$\|\phi_{\mathbf{f}}(t, \mathbf{x}_0 + \mathbf{v}) - \phi_{\mathbf{f}}(t, \mathbf{x}_0) - \phi_{\mathbf{u}}(t, \mathbf{v})\| \le C\|\mathbf{v}\|^2$$

when $\|\mathbf{v}\| < \delta$.

Proof. Note that

$$\frac{d}{dt}\left(\phi_{\mathbf{f}}(t, \mathbf{x}_0 + \mathbf{v}) - \phi_{\mathbf{f}}(t, \mathbf{x}_0)\right) = \mathbf{f}(\phi_{\mathbf{f}}(t, \mathbf{x}_0 + \mathbf{v})) - \mathbf{f}(\phi_{\mathbf{f}}(t, \mathbf{x}_0)),$$

whereas

$$\frac{d}{dt}\phi_{\mathbf{u}}(t, \mathbf{v}) = d_{\phi_{\mathbf{f}}(t,\mathbf{x}_0)}\mathbf{f}(\mathbf{v}),$$

and the curves

$$\mathbf{u}_1(t, \mathbf{v}) = \phi_{\mathbf{f}}(t, \mathbf{x}_0 + \mathbf{v}) - \phi_{\mathbf{f}}(t, \mathbf{x}_0)$$

and

$$\mathbf{u}_2(t, \mathbf{v}) = \phi_{\mathbf{u}}(t, \mathbf{v})$$

are solutions of the differential equations

$$\mathbf{v}' = F(t, \mathbf{v}) \quad \text{and} \quad \mathbf{w}' = G(t, \mathbf{w}),$$

respectively, where

$$F(t, \mathbf{v}) = \mathbf{f}(\mathbf{u}(t) + \mathbf{v}) - \mathbf{f}(\mathbf{u}(t)) \quad \text{and} \quad G(t, \mathbf{w}) = \left(d_{\mathbf{u}(t)}\mathbf{f}\right)(\mathbf{w}).$$

The vector fields F and G are close near the origin in $\mathbb{R}^n$, so the task is to say that two solutions of close differential equations are close. This is just what the Fundamental Inequality 4.4.1 (Appendix T) will do for us.

Once we have seen this, the result is straightforward, but you have to be careful to choose the constants in the right order. The first thing is to decide which we will consider as the real solution and which the perturbed equation. It does not make a lot of difference, but let us say that the linearized equation is the "real" equation, so that $\mathbf{u}_2$ is a real solution, and $\mathbf{u}_1(t)$ is an approximate solution of the linearized equation.

Next, choose T, the time at which we will differentiate the flow. Choose R, C, and K so that in $\|\mathbf{x}\| \leq R$, $0 \leq t \leq T$, both equations are Lipschitz with Lipschitz constant K, and so that for $\|\mathbf{x}\| < R$ and $0 \leq t \leq T$, we have

$$\|F(t, \mathbf{x}) - G(t, \mathbf{x})\| = \|\mathbf{f}(\mathbf{u}(t) + \mathbf{x}) - \mathbf{f}(\mathbf{u}(t)) - \left(d_{\mathbf{u}(t)}\mathbf{f}\right)(\mathbf{x})\| \leq C\|\mathbf{x}\|^2.$$

If $\|\mathbf{v}_0\| < Re^{-KT}$, the functions $\mathbf{u}_1(t)$ [and $\mathbf{u}_2(t)$] will satisfy

$$\|\mathbf{u}_1(t)\| \leq e^{Kt}\|\mathbf{v}_0\|$$

for $0 \leq t \leq T$, by the "δ"-term of the Fundamental Inequality bounding how far they stay from the zero solution. Hence, $\mathbf{u}_1$ will satisfy

$$\|\mathbf{u}_1'(t) - G(t, \mathbf{u}_1(t))\| \leq \|\mathbf{v}_0\|^2 e^{2KT}.$$

Now the Fundamental Inequality 4.4.1 says that

$$\|\mathbf{u}_1(T) - \mathbf{u}_2(T)\| < \frac{\|\mathbf{v}_0\|^2 e^{2KT}}{K}\left(e^{KT} - 1\right).$$

$\square$

Two Uses of the Implicit Function Theorem

Later in this chapter, we will require two results, both of which are immediate from the Implicit Function Theorem 13.6.4 (see Appendix T). The first will be used to assert that the zeroes of a vector field depend continuously on the vector field under appropriate circumstances, but the statement does not require you to think of $\mathbf{f}$ as a vector field. Let U be an open subset of $\mathbb{R}^n$, and $\mathbf{f}: U \to \mathbb{R}^n$ a twice continuously differentiable function. Suppose that $\mathbf{f}(\mathbf{x}_0) = 0$ and that $d_{\mathbf{x}_0}\mathbf{f}$ is invertible. Then by Taylor's Theorem, there exists R such that $\|\mathbf{f}(\mathbf{x})\| \neq 0$ is $0 < \|\mathbf{x} - \mathbf{x}_0\| \leq R$ (the linear term does not vanish, and the remainder is too small to cancel it).

Proposition 8*.1.13. *For all $\varepsilon > 0$, there exists $\delta > 0$ such that if $\|\mathbf{g} - \mathbf{f}\| < \delta$, then the equation $\mathbf{g}(\mathbf{x}) = 0$ has a unique solution $\mathbf{x}_0(\mathbf{g})$ satisfying $\|x_0(\mathbf{g}) - \mathbf{x}_0\| < R$.*

Proposition 8*.1.5 is essentially the Inverse Function Theorem; a corollary in Appendix T, of the Implicit Function Theorem 13.6.4.

For our second application of the Implicit Function Theorem, suppose $S \subset \mathbb{R}^n$ is a hypersurface, defined implicitly by the equation $h(\mathbf{x}) = 0$, that $\mathbf{x}' = \mathbf{f}(\mathbf{x})$ is a differential equation with flow $\phi_{\mathbf{f}}(t, \mathbf{x})$, and that $\mathbf{x}_1 = \phi(\tau_0, \mathbf{x}_0) \in S$.

Theorem 8*.1.14. *If $(d_{x_1}h)(\mathbf{f}(x_1)) \neq 0$, then there exists a unique continuously differentiable function $\tau(\mathbf{x})$ defined in a neighborhood of $\mathbf{x}_0$ such that $\phi_{\mathbf{f}}(\tau(\mathbf{x}), \mathbf{x}) \in S$ and $\tau(\mathbf{x}_0) = \tau_0$.*

Figure 8*.1.7 illustrates all of this, in particular, the graph of the function τ.

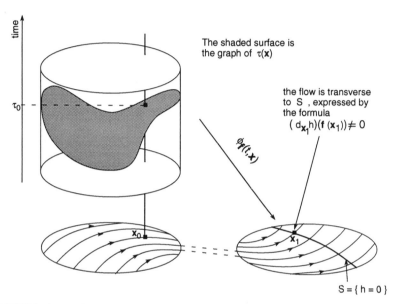

The shaded surface is the graph of $\tau(\mathbf{x})$

the flow is transverse to S , expressed by the formula

$(d_{x_1}h)(\mathbf{f}(x_1)) \neq 0$

$S = \{ h = 0 \}$

FIGURE 8*.1.7. Graph of τ as a function of $(x, y) \in U$, giving the time at which the solution starting at (x, y) hits S.

Proof. The equation $h(\phi_{\mathbf{f}}(t, \mathbf{x})) = 0$, which represents n differentiable equations in the $n + 1$ variables $x_1, \ldots, x_n, t$. Proposition 8*.1.14 asserts that these equations implicitly define t as a function of $x_1, \ldots, x_n$. According to the Implicit Function Theorem, this will be the case near $(\tau_0, \mathbf{x}_0)$ if

$$\frac{\partial}{\partial t} h(\phi_{\mathbf{f}}(t, \mathbf{x})) \mid_{(\tau_0, \mathbf{x}_0)} \neq 0.$$

The chain rule says that

$$\frac{\partial}{\partial t} h(\phi_{\mathbf{f}}(t, \mathbf{x})) \mid_{(\tau_0, \mathbf{x}_0)} = d_{x_1} h \frac{d}{dt} \phi_{\mathbf{f}} \mid_{(\tau_0, \mathbf{x}_0)} = (d_{x_1} h)(\mathbf{f}(x_1)),$$

so the result is true.   □

An exercise in Chapter 13 (Part III) will pin down the use of the Implicit Function Theorem.

What Proposition 8*.1.14 says is that the time it takes for a solution to intersect a hypersurface is a continuously differentiable function of the initial condition if the solution intersects the hypersurface transversely. Note that the condition $\phi_{\mathbf{f}}(\tau(\mathbf{x}), \mathbf{x}) \in S$ can be rewritten $u(\phi_{\mathbf{f}}(\tau(\mathbf{x}), \mathbf{x})) = 0$, which is one equation in the $n+1$ unknowns $t, x_1, \ldots, x_n$; it is not surprising that under appropriate conditions, this defines one of the unknowns (namely, t) implicitly as a function of the others.

8*.2 Structural Stability of Sinks and Sources

This subsection shows the structural stability of sinks and sources: it is one building block for the Structural Stability Theorem of Section 8*.6 (besides being of interest in itself). Theorem 8.2.2 says that if a linear differential equation with a sink is perturbed by higher order terms, the new equation still has a sink. In this subsection, we will pin down more accurately the extent to which the trajectories of a differential equation "do not change" near a sink when the differential equation is perturbed.

Theorem 8*.2.1. (Structural Stability of Sinks). *Let* $\mathbf{f}$ *be a vector field on an open subset* U *of* $\mathbb{R}^n$, *such that* $\mathbf{f}(\mathbf{x}_0) = 0$ *and such that the linearization of* $\mathbf{f}$ *at* $\mathbf{x}_0$ *is a sink. Then there exists a basis of* $\mathbb{R}^n$, $\rho > 0$ *and* $\delta > 0$ *such that if*

(i) V *is the ball around* $\mathbf{x}_0$ *of radius* ρ *(with respect to that basis) and* ∂V *is the boundary of* V,

(ii) $\|\mathbf{g} - \mathbf{f}\| < \delta$ *(think of* $\mathbf{g}$ *as a small perturbation of* $\mathbf{f}$),

(iii) $h : \partial V \to \partial V$ *is any homeomorphism,*

then the homeomorphism h *extends to a homeomorphism* $\bar{h} : V \to V$ *which sends trajectories of* $\mathbf{f}$ *to trajectories of* $\mathbf{g}$.

Remark. We will actually do better: the map $\bar{h}$ will conjugate the flows. However, even if h is very smooth, $\bar{h}$ will generally not be differentiable, and the differential equations will not be differentiably conjugate.

To prove Theorem 8*.2.1, we will use a construction popularized by the physicists under the name *scattering theory*, which is of central importance throughout dynamical systems as the basic way to construct conjugacies.

SCATTERING THEORY

We will denote the flow of the differential equation $\mathbf{x}' = \mathbf{f}(\mathbf{x})$ by $\phi_{\mathbf{f}}(t, \mathbf{x})$. The idea of scattering theory is to choose a small number ρ and to define

$$\bar{h}(\mathbf{x}) = \phi_{\mathbf{g}}(\tau(\mathbf{x}), h(\phi_{\mathbf{f}}(-\tau(\mathbf{x}), \mathbf{x}))), \tag{3}$$

where $\tau(\mathbf{x})$ is the smallest time so that $\|\phi_{\mathbf{f}}(-\tau(\mathbf{x}), \mathbf{x})\| = \rho$.

In other words, as shown in Figure 8*.2.1:

> *Flow out to ∂V, the boundary of V, under the unperturbed equation* $\mathbf{x}' = \mathbf{f}(\mathbf{x})$ *and back in by the perturbed equation.*

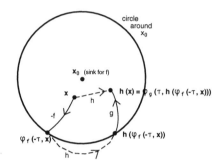

FIGURE 8*.2.1. The idea of scattering theory.

Example 8*.2.2. Let us compute what formula (1) gives when

$$\begin{bmatrix} x \\ y \end{bmatrix}' = \begin{bmatrix} -x \\ -y \end{bmatrix} \quad \text{and} \quad \begin{bmatrix} x \\ y \end{bmatrix}' = \begin{bmatrix} -x \\ -2y \end{bmatrix}$$

are the two differential equations, V is the unit disc, and h is the identity. Flowing out from $\mathbf{x} \in V$ to the unit circle leads to the point $\mathbf{x}/\|\mathbf{x}\|$ at time $-\tau = -\log\|\mathbf{x}\|$, and flowing in from there for time τ under the second differential equation leads to

$$\bar{h}(x, y) = (x, y\sqrt{x^2 + y^2}).$$

This mapping is a homeomorphism of the unit disc to itself, but it is not differentiable at the origin. You are asked in Exercises 8*.2#1–3 to calculate several more examples. ▲

BACK TO THE PROOF OF THE THEOREM

Proof of Theorem 8*.2.1. We have to prove three things to justify the scattering construction suggested by Figure 8*.2.1:

(i) that $\tau(\mathbf{x})$ exists (except when $\mathbf{x} = \mathbf{x}_0$) and is continuously differentiable (this will be shown by Lemma 8*.2.3);

(ii) that the mapping $\bar{h}(\mathbf{x})$, constructed by the scattering argument, extends *continuously* to $\mathbf{x}_0$ (but note: there will be no differentiability here!);

(iii) that the mapping $\bar{h}(\mathbf{x})$ is bijective (one-to-one and onto).

Proof of (i). We begin by normalizing our differential equation as in Theorem 8.2.2; i.e., we will suppose that coordinates on $\mathbb{R}^n$ and R can be chosen so that

(1) $\mathbf{x}_0 = 0$, so that $\mathbf{f}(\mathbf{x}) = A\mathbf{x} + \mathbf{q}(\mathbf{x})$;

(2) $d_{\mathbf{x}_0}\mathbf{f} = A$ satisfies $A\mathbf{x} \cdot \mathbf{x} \le -C_1\|\mathbf{x}\|^2$ for some constant $C_1 > 0$;

(3) $\|\mathbf{q}(\mathbf{x})\| \le C_2\|\mathbf{x}\|^2$ for $\mathbf{x} \le R$.

Let $\tilde{C}_1 = C_1/2$ and $\tilde{C}_2 = 2C_2$; set $\tilde{\rho} = \frac{1}{2}\min\{R, \tilde{C}_1/2\tilde{C}_2\}$ and $\rho = \tilde{\rho}/2$.

In order for formula (3) to make sense, we need to know that the time $\tau(\mathbf{x})$ exists; clearly, it does *not* if $\mathbf{x} = \mathbf{x}_0$, but that is the only exception. We saw in Theorem 8.2.2 that the norm of a nonzero solution decreases faster than an exponential as t increases, or equivalently it increases faster than an exponential as t decreases until it reaches $\tilde{\rho}$, so it must eventually reach ρ.

To show that $\bar{h}$ is continuous, we need to know that $\tau(\mathbf{x})$ is a continuous function of $\mathbf{x}$ for $\mathbf{x} \ne 0$. This follows from Proposition 8*.1.14.

Lemma 8*.2.3. *The function $\tau(\mathbf{x})$ is a continuously differentiable function of $\mathbf{x}$ for $\|\mathbf{x}\| \le \tilde{\rho}$, $\mathbf{x} \ne 0$.*

Proof of Lemma. Let $u(\mathbf{x}) = \mathbf{x} \cdot \mathbf{x}$. According to Proposition 8*.1.14, all we need is $d_{\mathbf{x}}u(\mathbf{f}(\mathbf{x})) \ne 0$ when $\|\mathbf{x}\| = \rho$; this follows from equation (3) from Section 8.2 in the proof of Proposition 8.2.4. □

So for V the ball of radius ρ, formula (3) makes sense and is continuous for $\mathbf{x} \ne 0$. To prove the theorem, we need to show that the mapping $\bar{h}$ extends continuously to $\mathbf{x} = \mathbf{x}_0$ and that the extended $\bar{h}$ is a homeomorphism.

Lemma 8*.2.4. *There exists δ_1 so that the vector field $\mathbf{g}$ has a unique zero $\mathbf{x}_0(\mathbf{g})$ in V, which depends differentiably on $\mathbf{g}$ for $\|\mathbf{f} - \mathbf{g}\| < \delta_1$.*

Proof of Lemma. This follows from Proposition 8*.1.13. We need to check that $d_{\mathbf{x}_0}\mathbf{f}$ is invertible, but our hypothesis is that $\mathbf{x}_0$ is a sink, so all the eigenvalues have negative real part, and none are zero. □

Proof of Theorem 8*.2.1, continued.

Proof of (ii). Set $\mathbf{y} = \mathbf{x} - \mathbf{x}_0(\mathbf{g})$ and write the differential equation $\mathbf{x}' = \mathbf{g}(\mathbf{x})$ in the new coordinates:

$$\mathbf{y}' = A_{\mathbf{g}}\mathbf{y} + \mathbf{q}_{\mathbf{g}}(\mathbf{y}) \tag{4}$$

since the new origin is always a zero of the vector field. Both $A_{\mathbf{g}}$ and $\mathbf{q}_{\mathbf{g}}$ depend continuously on $\mathbf{g}$.

By continuity, there will exist δ_2 satisfying $0 < \delta_2 < \delta_1$ such that

(1) $\|\mathbf{x}_0(\mathbf{g})\| < \rho$,

(2) $A_\mathbf{g}\mathbf{y} \cdot \mathbf{y} \le -\tilde{C}_1\|\mathbf{y}\|^2$

(3) $\|\mathbf{q}_\mathbf{g}(\mathbf{y})\| \le \tilde{C}_2\|\mathbf{y}\|^2$ for $\mathbf{y} \le \tilde{\rho}$

if $\|\mathbf{f} - \mathbf{g}\| < \delta_2$. Note that the third condition above would presumably be false if we had failed to include second derivatives in our norm on perturbations.

Then the proof of Theorem 8.2.2 now carries over to show that if $\|\mathbf{y}(t)\|$ is a solution of (11) with $\|\mathbf{y}(0)\| \le \tilde{\rho}$, then $\mathbf{y}(t) \to 0$ as $t \to \infty$.

The set V is the ball of radius ρ around $\mathbf{x}_0$ and is contained in the balls of radius $\tilde{\rho}$ around both $\mathbf{x}_0$ and $\mathbf{x}_0(\mathbf{g})$. For $\mathbf{x}$ close to $\mathbf{x}_0$, $\tau(\mathbf{x})$ is very large, and formula (1) shows that $\bar{h}(\mathbf{x})$ is obtained by solving $\mathbf{x}' = \mathbf{g}(\mathbf{x})$ for a long time, starting at a point in ∂V, so $\bar{h}(\mathbf{x})$ is very close to $\mathbf{x}_0(\mathbf{g})$. This shows that setting $\bar{h}(\mathbf{x}_0) = \mathbf{x}_0(\mathbf{g})$ provides a continuous extension of $\bar{h} : V \to \mathbb{R}^n$.

Proof of (iii). We still need to show that $\bar{h}$ is a homeomorphism. It is not difficult to see how to do this: set

$$\bar{h}^{-1}(\mathbf{x}) = \phi_\mathbf{f}(\tau_1(\mathbf{x}), h(\phi_\mathbf{g}(-\tau_1(\mathbf{x}), \mathbf{x}))), \tag{5}$$

where $\tau_1(\mathbf{x})$ is the smallest time so that $\|\phi_\mathbf{g}(-\tau_1(\mathbf{x}), \mathbf{x})\| = \rho$. This involves a slightly subtle point: is τ_1 continuous? This requires, according to Proposition 8*.1.9, that ∂V, the boundary of V, be transverse to $\mathbf{g}$. We know that $\mathbf{g}$ is transverse to spheres of radius smaller than $\rho*$ centered at $\mathbf{x}_0(\mathbf{g})$, but ∂V is centered at $\mathbf{x}_0$. But if a vector field is transverse to a closed surface, then so is every sufficiently small perturbation. So we must further restrict $\mathbf{g}$: there exists δ with $0 < \delta \le \delta_2$ such that if $\|\mathbf{g} - \mathbf{f}\| < \delta$, then $\mathbf{g}$ is transverse to ∂V, and $\tau_1(\mathbf{x})$ is a continuous function of $\mathbf{x}$ for $\mathbf{x} \in V$, $\mathbf{x} \ne \mathbf{x}_0(\mathbf{g})$. Then formula (3) extends to $\mathbf{x}_0(\mathbf{g})$ exactly as above, so that $\bar{h}$ is the required homeomorphism. This finishes the proof of Theorem 8*.2.1. $\square$

8*.3   Time to Pass by a Saddle

In this section we prove a more delicate result that will be essential in Section 8*.6. It will be *the* technical result that makes the Pontryagin–Peixoto scheme work. The proof was contributed by A. Douady.

Let $\mathbf{f}(\mathbf{x})$ be a vector field in the plane and $\mathbf{x}_0 \in \mathbb{R}^2$ be a saddle for $\mathbf{f}$. We would like to work in the "first quadrant," so we shall orient the saddle as shown in Figure 8*.3.1, with the stable separatrix on the "horizontal" axis and the unstable separatrix on the "vertical" axis.

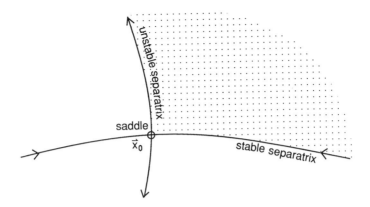

FIGURE 8*.3.1. Labeling saddle for first quadrant.

Choose *sections* (or transversals)

$$\gamma_1 : I_1 \to \mathbb{R}^2 \qquad \text{and} \qquad \gamma_2 : I_2 \to \mathbb{R}^2$$

at $\mathbf{x}_1$ on the right branch of the stable separatrix and at $\mathbf{x}_2$ on the upper branch of the unstable separatrix, respectively. Also choose a *diagonal section* $\gamma : J \to \mathbb{R}^2$ at $\mathbf{x}_0$, i.e., a parametrized curve with nonvanishing derivative, with $\gamma(0) = \mathbf{x}_0$ and $\gamma'(0)$ transverse to both separatrices, as shown in Figure 8*.3.2. We will require that $\gamma'(0)$ point into the first quadrant. We will call r the parameter on J; it will be the main indedendent variable throughout this subsection.

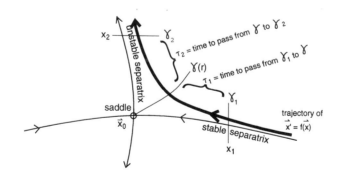

FIGURE 8*.3.2. Transverse sections γ_1 and γ_2; diagonal section γ.

We will start solutions at the point $\gamma(r)$ on the "diagonal section" and study the time these solutions take to cross the sections γ_i, as a function of r. More precisely, there exists $\varepsilon > 0$ such that the *flow times*

$$\tau_1(r) = \inf\{t > 0 \mid \phi_{\mathbf{f}}(-t, \gamma(r)) \in \gamma_1(I_1)\}$$

and
$$\tau_2(r) = \inf\{t > 0 \mid \phi_{\mathbf{f}}(t, \gamma(r)) \in \gamma_2(I_2)\}$$
are well defined and continuously differentiable for $0 < r < \varepsilon$.

Remark. After leaving the neighborhood of $\mathbf{x}_0$ under consideration, a solution might reenter it and intersect the sections I_i again. The infima in the formulas above specify that the intersection we are describing is the first.

As $r \searrow 0$, the trajectory through $\gamma(r)$ passes closer and closer to $\mathbf{x}_0$, so it must go more and more slowly, and we expect $\tau_i(r)$ to tend to infinity as $r \searrow 0$, as shown in Example 8*.3.1 and Figure 8*.3.3.

Example 8*.3.1. Consider $\begin{bmatrix} x \\ y \end{bmatrix}' = \begin{bmatrix} -x \\ 2y \end{bmatrix}$. Its solution is

$$\begin{bmatrix} x \\ y \end{bmatrix}' = \begin{bmatrix} C_1 e^{-t} \\ C_2 e^{2t} \end{bmatrix} = \begin{bmatrix} re^{-t} \\ re^{2t} \end{bmatrix},$$

since at $t = 0$ we want $x = y = r$. For the second section, we want $re^{2\tau_2} = 1$, so $\tau_2 = -(\log r)/2$.

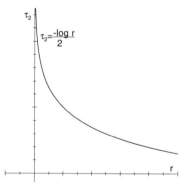

FIGURE 8*.3.3. How τ varies as r. ▲

Theorem 8*.3.2 makes all of this more precise.

Theorem 8*.3.2. *There exist sections $\gamma_i : I_i \to \mathbb{R}^2$, a diagonal section $\gamma : J \to \mathbb{R}^2$, and ε as above such that the functions $\tau_i(r)$ are monotone decreasing on $(0, \varepsilon)$, tending to ∞ as $r \searrow 0$, and their derivatives tend to $-\infty$.*

Proof. We can suppose that the differential equation is the system studied earlier in the proof of Theorem 8.3.2:

$$\begin{aligned} x' &= -\mu x + P(x, y) = F(x, y) \\ y' &= \nu y + Q(x, y) = G(x, y), \end{aligned} \tag{6}$$

and we will take as our sections γ_i segments of the lines $\{x = a\}$ and $\{y = a\}$ with $a > 0$ to be chosen below, and a segment of the diagonal as our "diagonal section," parametrized by $\gamma(r) = (r, r)$. Also, as in Theorem 8.3.2, we will concentrate on the related equation

$$\frac{dy}{dx} = \frac{\nu y + Q(x, y)}{-\mu x + P(x, y)} = H(x, y) \qquad (7)$$

in the region $\{(x, y) \mid |y| < x\}$ in which (7) is defined. This means we will be studying τ_1; the proof for τ_2 is similar.

For $0 < r < a$, let $u(r, x)$ be the solution to (7) passing through the point $\gamma(r) = (r, r)$ on the diagonal. Figure 8*.3.4 shows the picture on which the proof will be based.

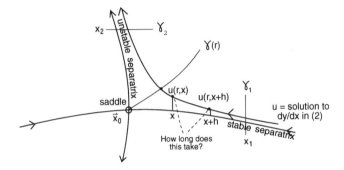

FIGURE 8*.3.4. Central idea for "how long does it take to pass a saddle along a solution?"

The following lemma gives the main tool to prove Theorem 8*.3.2; this is one of Douady's central ideas.

Theorem 8*.3.3. Let $r \leq x_1 \leq x_2 \leq a$. Then the time $T(r, x_1, x_2)$ required to flow from $u(r, x_1)$ to $u(r, x_2)$ is

$$T(r, x_1, x_2) = \int_{x_1}^{x_2} \frac{dx}{|F(x, u(r, x))|}.$$

Remark. The integrand can be thought of as $\frac{dx}{dx/dt} = dt$, so it is not surprising that the integral ends up being in units of time.

Proof of Lemma. Differentiating the formula above with respect to x_2, we need to check that

$$T(r, x, x + h) = \frac{h}{|F(x, u(r, x))|} + o(h).$$

Clearly,

$$\left\| \begin{bmatrix} x \\ u(r, x+h) \end{bmatrix} - \begin{bmatrix} x \\ u(r,x) \end{bmatrix} \right\| = h\sqrt{1 + \left(\frac{\partial u}{\partial x}(r,x)\right)^2} + o(h)$$

$$= h\sqrt{1 + \left(\frac{G(x, u(r,x))}{F(x, u(r,x))}\right)^2} + o(h).$$

On the other hand, the speed of the flow at $(x, u(r,x))$ is

$$\sqrt{F(x, u(r,x))^2 + G(x, u(r,x))^2},$$

and along the graph of u from $u(r,x)$ to $u(r, x+h)$, the speed will vary at most by terms of order h.

Recall from Part I, Appendix A, Asymptotic Development, that little $o(h)$ means smaller than h (for instance, h^2); big $O(h)$ means terms of order at most h.

According to the principle that *time = distance/speed*, this leads to

$$T(r, x, x+h) = \frac{h\sqrt{1 + G^2/F^2} + o(h)}{\sqrt{F^2 + G^2} + O(h)} = \frac{1}{|F|}\left(\frac{h + o(h)}{1 + O(h)}\right) = \frac{h}{|F|} + o(h),$$

which proves the lemma.   □

Proof of Theorem 8*.3.2, continued. Now let us see that $T(r, a)$ tends to ∞ as $r \to 0$ for fixed $x = a$. The function $P(x, y)/x$ is defined and bounded in the region $|y| \le x \le a$, so that we see (Exercise 8*.3#1a) that

$$\int_r^a \frac{dx}{|F(x, u(r,x))|} = \int_r^a \frac{dx}{\mu x - P(x, u(r,x))}$$

$$= -\frac{\log r}{\mu} + \{\text{something bounded as } r \to 0\} \tag{8}$$

and does tend to infinity. The dominant term is determined by r, which measures how close you are passing to the saddle.

To see that the derivative tends to $-\infty$ as $r \to 0$, we differentiate under the integral sign to find

$$\frac{\partial T(r, r, a)}{\partial r} = -\frac{1}{F(r, u(r,r))} - \int_r^a \frac{\partial |F(x, u(r,x))|/\partial r}{|F(x, u(r,x))|^2} dx.$$

The first term is equivalent to $-1/\mu r$; we will show that the absolute value of the second is bounded by $-C \log r$ for an appropriate constant C so is negligible before the first term.

Computing the partial derivative above, still considering x fixed, we find (Exercise 8*.3#1b)

$$\left| \int_r^a \frac{\partial |F|/\partial r}{|F|^2} \right| = \int_r^a \frac{|\partial P/\partial y(x, u(x, r))| \, |\partial u/\partial r(r, x)|}{|\mu x - P(x, u(r, x))|^2} dx. \tag{9}$$

The first factor in the numerator is bounded above by $C_1 x$ in the region $|y| \le x \le a$ for some constant C_1.

Formula (9) from the proof of Lemma 8.3.3 and the computation following it show that the solutions approach each other in the region $\{|y| < x < a\}$ for a sufficiently small; i.e., the function $u(r_1, x) - u(r_2, x)$ is a decreasing function of x for $r_2 < r_1 \le x < a$, so the second factor in the numerator is a decreasing function of x for any fixed r.

Figure 8*.3.5 may help with the final steps of the proof.

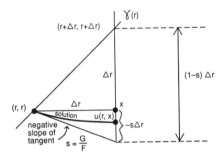

FIGURE 8*.3.5. The final steps of proving Theorem 8*.3.2.

Thus, (Exercise 8*.3#1c) you can see from Figure 8*.3.5 that

$$0 \le \frac{\partial u}{\partial r}(r, x) \le \frac{\partial u}{\partial r}(r, r) = 1 - \frac{G(r, r)}{F(r, r)}, \tag{10}$$

which is a continuous function of r for $0 < r < a$ and tends to the limit $1 + \nu/\mu$ as $r \to 0$; as such, it is bounded, for instance, by $2 + \nu/\mu$ for a sufficiently small. The denominator is bounded below by $\mu^2 x^2/2$ for a sufficiently small, so the whole integral is bounded by

$$\int_r^a \frac{2C_1(2 + \nu/\mu)}{\mu} \frac{dx}{x} \le -C(\log r - \log a) \le -C \log r,$$

setting $C = 2C_1(2 + \nu/\mu)/\mu^2$ and assuming $a < 1$. $\square$

Remark. The proof above makes Theorem 8*.3.2 a rather delicate result, depending on the comparison of the rates at which two functions tend to infinity. The authors do not know of anything simpler.

8*.4 Structural Stability of Limit Cycles

The main business of this section is to prove the Structural Stability Theorem 8*.4.10, which is the analog of Theorem 8*.2.1 for linearly attracting or repelling limit cycles: such cycles are structurally stable. For the purposes of Section 8*.6, we will require a more precise statement. The proofs are very similar in spirit to that for Theorem 8*.2.1; more precisely, the key Lemma 8*.4.2 is by itself an analog of Theorem 8*.2.1 for one-dimensional mappings, as opposed to two-dimensional vector fields.

Before we state and prove the main theorem, however, we must lay some more groundwork.

THE SECTION MAPPING

In Part I, Chapter 5, we encountered the *period mapping* of a periodic differential equation. The *section mapping* introduced in Figure 8.7.4 and described below is closely analogous, and the period mapping is a special case of a section mapping. See Figure 8*.4.1.

Let $\mathbf{f}(\mathbf{x})$ be a vector field on a subset of $\mathbb{R}^n$, with flow $\phi_{\mathbf{f}}(t, \mathbf{x})$.

Let Γ be a cycle of the differential equation $\mathbf{x}' = \mathbf{f}(\mathbf{x})$, $\mathbf{x}_0 \in \Gamma$ a point, and $H \subset \mathbb{R}^n$ an $(n-1)$-dimensional space transverse to Γ through $\mathbf{x}_0$.

In a first reading, the reader may think that $n = 2$, so that Γ is a curve in the plane as in Examples 8.4.1 and 8.4.2, and that H is a line transverse to Γ at $\mathbf{x}_0$. This case does not quite convey the full complexity of the situation, but very little is lost if we imagine $n = 3$, so that Γ is a curve in space and H is a plane transverse to Γ at $\mathbf{x}_0$.

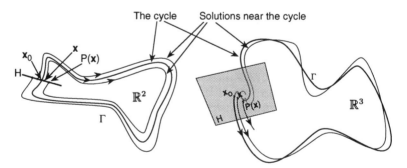

FIGURE 8*.4.1. Section mappings in $\mathbb{R}^2$ and $\mathbb{R}^3$.

For a sufficiently small neighborhood N of $\mathbf{x}_0$ in H, we can define a mapping $P : N \to H$ by starting a solution at $\mathbf{x} \in N$ and setting $P(\mathbf{x})$ to be the next intersection of this solution with H. Figure 8*.4.2 shows this construction. The domain of such a mapping is a bit delicate to define, as the figure suggests. Sometimes solutions close to Γ will return close to $\mathbf{x}_0$ sooner than the "correct" return, and the domain of P must be chosen sufficiently small that the first return is the correct return.

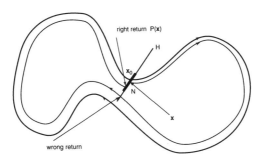

FIGURE 8*.4.2. The inner trajectory crosses H too soon, so the domain for P must be a smaller subset of H, as indicated by the thicker line for N.

Definition 8*.4.1 (Section mapping). $P : N \to H$ is defined as follows. Let T be the period of $\mathbf{x}_0$, i.e., the time it takes for a solution to go around Γ. Choose a neighborhood U of $\mathbf{x}_0$ in H, sufficiently small that the vector field is transverse to H in U and that $U \cap \Gamma = \mathbf{x}_0$. Choose $\varepsilon < T/2$; let $W \subset \mathbb{R}^n$ be the set of $\{\phi_{\mathbf{f}}(t, \mathbf{x})\}$ for $\mathbf{x} \in U$ and $|t| < \varepsilon$. We will assume that ε is chosen sufficiently small so that if $\mathbf{x} \in U$ and $|t_1|, |t_2| < \varepsilon$, then $\phi_{\mathbf{f}}(t_1, \mathbf{x}) \neq \phi_{\mathbf{f}}(t_2, \mathbf{x})$ only when $t_1 = t_2$.

Now consider the set $N \subset U$ defined by

$$N = \{\mathbf{x} \in U \mid \phi_{\mathbf{f}}(T, \mathbf{x}) \in W\}.$$

Since $\phi_{\mathbf{f}}(T, \mathbf{x}_0) = \mathbf{x}_0 \in W$ and $\phi_{\mathbf{f}}(T, \mathbf{x}$ depends continuously on $\mathbf{x}$, we see that N contains a neighborhood of $\mathbf{x}_0$ in H. Define $P(\mathbf{x}) = \phi_{\mathbf{f}}(s(\mathbf{x}) + T, \mathbf{x})$, where $s(\mathbf{x})$ is the unique time satisfying $|s(\mathbf{x})| < \varepsilon$ and $\phi_{\mathbf{f}}(s(\mathbf{x}) + T, \mathbf{x}) \in H$.

In summary, Figure 8*.4.3 shows for $n = 2$ and $n = 3$ the following:

in $\mathbb{R}^{n-1}$:

H = section

U = portion of H where trajectories intersect transversely (are not tangent) to H

N = subset of U in which P is well defined (trajectories always come back)

in $\mathbb{R}^n$:

W = tube of forward and backward trajectories of points in U.

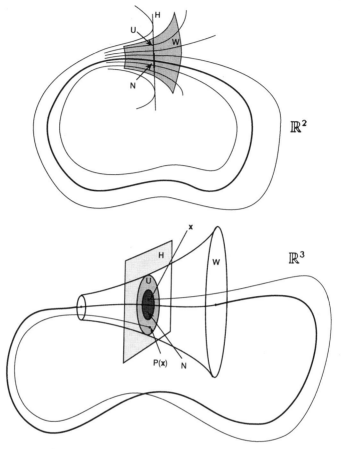

FIGURE 8*.4.3. Regions used in defining section mapping $P : N \to H$ in $\mathbb{R}^2$ (above) and $\mathbb{R}^3$ (below).

Example 8*.4.2. Let us compute explicitly the mapping P for the equation from Example 8.4.1.

$$\begin{bmatrix} x \\ y \end{bmatrix}' = \begin{bmatrix} y \\ -x \end{bmatrix} + \alpha(1 - x^2 - y^2) \begin{bmatrix} x \\ y \end{bmatrix}. \tag{11}$$

In polar coordinates, the radial equation can be rewritten (Exercise 8.4#2a) as

$$\frac{dr}{r(1-r)(1+r)} = \alpha dt, \tag{12}$$

which gives a direction field like Figure 8*.4.4.

Equation (12) can be integrated (Exercise 8*.4#2b) to yield

$$r(t) = \frac{r(0)e^{\alpha t}}{\sqrt{r(0)^2(e^{2\alpha t} - 1) + 1}}. \tag{13}$$

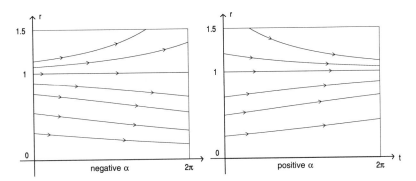

FIGURE 8*.4.4. Solutions for equation (12); $\alpha = -0.1$ (left) and $\alpha = 0.1$ (right).

Take H to be the x-axis, in which case N can be chosen to be the positive x-axis. From the equation $\theta' = 1$, we deduce that the time required to go from the positive x-axis back to itself is 2π, so the mapping P is given by

$$P(x) = \frac{xe^{2\pi\alpha}}{\sqrt{x^2\left(e^{4\pi\alpha} - 1\right) + 1}}. \tag{14}$$

Graphs of such functions are represented in Figure 8*.4.5 for $e^{2\pi\alpha} = 2$ and $e^{2\pi\alpha} = 0.8$. If $\alpha = 0$, the P map is the identity.

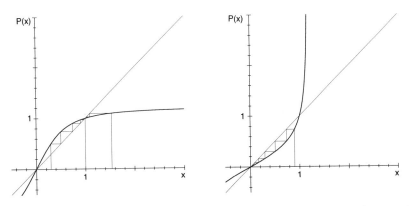

FIGURE 8*.4.5. The section mappings for equation (12) in Example 8*.4.2, with fixed points at $x = 1$: attracting on the left ($\alpha = 0.1$) and repelling on the right ($\alpha = -0.1$). ▲

Proposition 8*.4.3. *The section mapping P is differentiable and one-to-one. If $P(N) = N'$ is the image, then $P^{-1}: N' \to N$ is also differentiable.*

Proof. The fact that P is one-to-one comes from the choice of ε in the construction of P above. More precisely, if you flow a time $-T$ from $P(\mathbf{x})$, you will be in W, and then $\mathbf{x}$ is the unique point of N which you can flow to H in time less than ε. This makes it clear that P is one-to-one and that in fact, P^{-1} is a section mapping for the differential equation $\mathbf{x}' = -\mathbf{f}(\mathbf{x})$.

Theorem 8*.1.14 asserts that there is a differentiable function $\tau(\mathbf{x})$, defined in a neighborhood of $\mathbf{x}_0$, such that $\phi_{\mathbf{f}}(\tau(\mathbf{x}), \mathbf{x}) \in H$ and $\tau(\mathbf{x}_0) = T$. Then P is simply the restriction of

$$\mathbf{x} \mapsto \phi_{\mathbf{f}}(\tau(\mathbf{x}), \mathbf{x})$$

to N, which is a composition of two differentiable maps. □

THE SECTION MAPPING AS A DYNAMICAL SYSTEM

The essential property of P is that *iterating P is equivalent to solving the differential equation.* More precisely, define the return time $T(\mathbf{x}) = s(\mathbf{x}) + T$: this is the function which measures how long it takes for a point of N to return to N. In particular, $T(\mathbf{x}_0) = T$.

Choose $\mathbf{y} \in N$ and set $\mathbf{y}_i = P^{\circ i}(\mathbf{y})$. Of course, this sequence may not be defined for all i because some $\mathbf{y}_i$ may not lie in N, so that $\mathbf{y}_{i+1}$ is not defined.

Proposition 8*.4.4. *Iterates of P and solutions of the differential equation are related by*
$$\mathbf{y}_i = \phi_{\mathbf{f}}(T(\mathbf{y}_0) + \ldots + T(\mathbf{y}_{i-1}), \mathbf{y}_0).$$

Proof. By definition, we have $\phi_{\mathbf{f}}(\mathbf{y}_i, T(\mathbf{y}_i)) = \mathbf{y}_{i+1}$. Now the formula follows from the fundamental property (refer to Section 6.8 on flows):

$$\phi_{\mathbf{f}}(\phi_{\mathbf{f}}(t_2, (t_1, \mathbf{x})) = \phi_{\mathbf{f}}(t_1 + t_2, \mathbf{x}). \quad \square$$

In particular, if $\mathbf{y}_i$ is defined for all $i > 0$, then the times $T(\mathbf{y}_0) + \ldots + T(\mathbf{y}_{i-1})$ tend to infinity, and understanding the orbit of $\mathbf{y}$ under P allows us to say something (in fact, a lot) about solutions for arbitrarily large times.

Proposition 8*.4.4. makes it clear that the mapping P should be viewed as a dynamical system on N, i.e., that it should be iterated. Unfortunately, the mapping P depends not only on $\mathbf{f}$ and Γ, but also on the choice of section. We want to say that it is really independent of the section: more precisely, we will show that different sections lead to conjugate section mappings.

Proposition 8*.4.5. *Let x_0 and x_0' be two points on Γ; choose subspaces H and H' transverse to Γ at x_0 and x_0' and construct section maps $P : N \to H$ and $P' : N' \to H'$ as above. Then there exist subsets N_1 and N_1' and a differentiable mapping $h : N_1 \to N_1'$ with differentiable inverse such that $P' = h^{-1} \circ P \circ h$.*

In other words, the section maps for two different H's are conjugate, as shown in Figure 8*.4.6, which means that P and P' are essentially the same dynamical system.

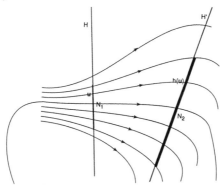

FIGURE 8*.4.6. The mapping h which conjugates P to P'.

Proof. Define h by "flowing" from N to N'; all the extra words in the statement above are just to restrict to a subset on which this is well defined. The details are very similar to Definition 8*.4.1 and will be left to the reader. □

DERIVATIVE OF THE SECTION MAPPING

For the remainder of this section, we will assume that we are in the plane, so that H is a line and N is an interval. The higher-dimensional case is quite a bit more elaborate, and the plane case will be complicated enough.

In the plane, a section mapping P like the one in Example 8*.4.2 is completely typical and just the sort of mapping studied in Chapter 5 (Part I). The mapping $P : N \to H$ is always monotone increasing, and in fact $P' > 0$. The point x_0 is a fixed point of P, which is linearly attracting if $P'(x_0) < 1$ and linearly repelling if $P'(x_0) > 1$.

Note that if x_0 is linearly attracting, then it really does attract a neighborhood of itself: choose a neighborhood I of x_0 in which $P' < 1 - \varepsilon$ for some $\varepsilon > 0$, then by the mean value theorem, $|P(x_0) - P(x)| = P'(y)|x_0 - x|$ for some $y \in [x_0, x] \subset I$, so that $|P(x_0) - P(x)| < (1 - \varepsilon)|x - x_0|$. Thus, applying P to a point in I moves you closer to x_0 by a definite amount.

If $P'(\mathbf{x}_0) < 1$, not only is $\mathbf{x}_0$ an attractive fixed point of P, but Γ attracts a neighborhood of itself. The following result further emphasizes the relation between iterations of P and solutions of the differential equation.

Proposition 8*.4.6. *If $P'(\mathbf{x}_0) < 1$, there is a neighborhood of Γ such that every solution starting in that neighborhood converges to Γ as $t \to \infty$.*

Proof. First assume that the solution starts at $\mathbf{y}_0 \in I$, where I is the neighborhood proved above to be attracted to $\mathbf{x}_0$. Let $\mathbf{y}_i = P^{\circ i}(\mathbf{y}_0)$. Note that since $\mathbf{y}_i$ exists for all i, the solution starting at $\mathbf{y}_0$ is defined for all $t \geq 0$. Moreover, since $\mathbf{y}_i \to \mathbf{x}_0$, the times $T_{\mathbf{y}_i}$ tend to the period T of $\mathbf{x}_0$. Thus, for large t, the solution is always obtained by flowing from a point near $\mathbf{x}_0$ for a time less than $2T$, so it stays close to Γ.

If $\mathbf{y}_0$ is close to Γ but not on I, then the trajectory through $\mathbf{y}_0$ will pass close to $\mathbf{x}_0$ and hence will cross I at some point $\mathbf{y}_1$. You now continue as above. $\quad\square$

Definition 8*.4.7 (Linearly attracting or repelling cycles). If $P'(x_0) < 1$, we will say that the cycle Γ is *linearly attracting*; if $P'(x_0) > 1$, we will say that Γ is *linearly repelling*.

The fact that this depends only on the cycle and not on the section follows from Proposition 8*.4.5. The case where $P'(\mathbf{x}_0) = 1$ will be studied further in Chapter 9.

CHOOSING A NEIGHBORHOOD OF A LIMIT CYCLE

In this subsection, we will construct a nice neighborhood of a limit cycle in the plane which is linearly attracting or repelling. A corresponding neighborhood in the case of a linear sink was found by choosing a basis with respect to which the square length is a Liapunov function, then choosing the region where this Liapunov function is smaller than some small positive constant.

There does not seem to be an analogous construction for cycles, and the result is harder than it appears. We have used a different proof inspired by the Cornell Ph.D. thesis of Salvador Malo, based on the notion of *turning a vector field*. Because this construction is harder to carry over to $\mathbb{R}^n$ for $n > 2$, we still restrict ourselves to the plane.

TURNING A VECTOR FIELD

To "turn" or "rotate" a vector field $\mathbf{f}$ through an angle θ means to create a new vector field

$$\mathbf{f}_\theta = \begin{bmatrix} \cos\theta & -\sin\theta \\ \sin\theta & \cos\theta \end{bmatrix} \mathbf{f}.$$

Examples for positive and negative rotation are shown in Figure 8*.4.7.

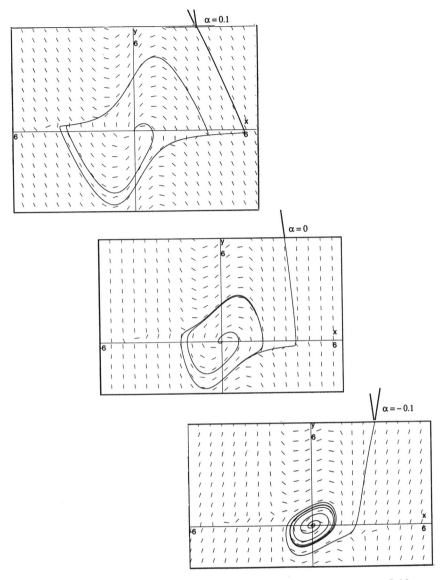

FIGURE 8*.4.7. A vector field is given in the center; turned vector fields are shown for $\alpha = 0.1$ (above), and for $\alpha = -0.1$ (below).   ▲

Example 8*.4.8. Consider the Van der Pol oscillator of Example 8.4.2:

$$x'' + (x^2 - 1)x' + x = 0$$

or

$$x' = y$$
$$y' = (1 - x^2)y - x,$$

as shown in the center of Figure 8*.4.7. The results of rotating this vector field through angles of 0.1 and −0.1, are shown on the left and right, respectively. The angle is easy to observe in the long trajectory coming in at the top of each picture. ▲

Proposition 8*.4.9. Let Γ be a linearly attracting cycle of $x' = f(x)$. Then there exists a neighborhood W of Γ bounded by two smooth simple closed curves, transverse to f, such that every point in W is attracted to Γ.

Proof. Consider the one-parameter family f_θ of vector fields obtained by turning f by an angle θ.

A turned vector field is transverse to f if θ is not a multiple of π. So for small positive θ, it points into or out of the region bounded by Γ; let us suppose it points out (if not, turn the vector field in the other direction). Choose a segment $I = [y_0, z_0]$ transverse to Γ at x_0, attracted to x_0 under $P : I \to I$, as above, where y_0 is the endpoint outside the region bounded by Γ.

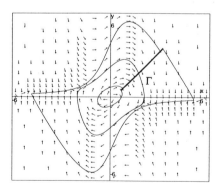

FIGURE 8*.4.8. A suitable neighborhood W of a limit cycle Γ.

The solution to $x' = f(x)$ through y_0 next intersects I at a point y_1 between y_0 and x_0, and for sufficiently small $\theta > 0$, the solution to $x' = f_\theta(x)$ will next intersect I at a point $y_1(\theta)$ between y_0 and y_1. This solution will be one boundary curve of W; we need to round the two corners, as suggested in Figure 8*.4.8, to make it smooth.

The other boundary curve is constructed by turning the vector field in the opposite direction and starting at $\mathbf{z}_0$. $\quad\square$

SRUCTURAL STABILITY OF CYCLES

Let $\mathbf{f}$ be a vector field on an open subset U of $\mathbb{R}^2$, and Γ a linearly attracting cycle of $\mathbf{x}' = \mathbf{f}(\mathbf{x})$. Let W, I, and P be as above:

W is an annular neighborhood of Γ such that every point of W is attracted to Γ, and the vector field $\mathbf{f}$ enters W transversely along its entire boundary; I is a segment transverse to $\mathbf{f}$ joining one boundary curve of W to the other; $P : I \to I$ the corresponding section mapping, with a single linearly attracting fixed point at $\mathbf{x}_0$, which attracts all of I.

Theorem 8*.4.10 (Structural stability of limit cycles). *There exists $\delta > 0$ such that for any homeomorphism $h : \partial W \to \partial W$ which maps each boundary curve to itself, preserving the direction, and any $\mathbf{g}$ vector field on a neighborhood of W with $\|\mathbf{f} - \mathbf{g}\| < \delta$, there exists a homeomorphism $\bar{h} : W \to W$ extending h which maps oriented trajectories of $\mathbf{x}' = \mathbf{f}(\mathbf{x})$ to oriented trajectories of $\mathbf{x}' = \mathbf{g}(\mathbf{x})$.*

Proof. Call $\mathbf{y}_0$ and $\mathbf{z}_0$ the endpoints of I. We ask the reader to believe that there exists a curve $I_1 \subset W$ transverse to $\mathbf{f}$ joining $h(\mathbf{y}_0)$ to $h(\mathbf{z}_0)$. Figure 8*.4.9 suggests how to construct a system of coordinates on W in which I_1 can be taken as a segment of straight line, at least if W is so small that its boundary curves are "close parallels" to Γ and $\mathbf{f}$ has solution curves that are also almost parallel to Γ. Writing the details would distract from the main argument.

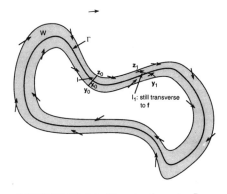

FIGURE 8*.4.9. How to draw in I_1.

Consider the graphs $G_0 = \partial W \cup I$ and $G_1 = \partial W \cup I_1$, as shown in Figure 8*.4.10. We can extend h to W.

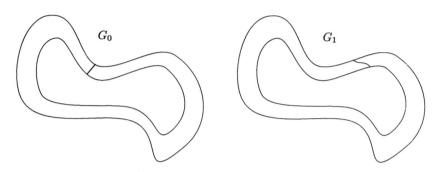

FIGURE 8*.4.10. The graphs G_0 and G_1.

Since every solution in W is attracted to Γ under the flow $\phi_{\mathbf{f}}$ of the differential equation $\mathbf{x}' = \mathbf{f}(\mathbf{x})$, every solution leaving G will intersect G again and, more specifically, I. Let us call this mapping $P_{\mathbf{f}} : G \to I$.

The same is true of the graph G_1, and, more significantly, it is still true for the flow $\phi_{\mathbf{g}}$ if $\|\mathbf{f} - \mathbf{g}\|$ is sufficiently small. Let us call the corresponding mapping $P_{\mathbf{g}} : G_1 \to I_1$.

The following result is the first step toward the construction of $\bar{h}$.

Lemma 8*.4.11. (a) *For any* $\mathbf{g}$ *sufficiently close to* $\mathbf{f}$, $P_{\mathbf{g}}$ *has a unique linearly attractive fixed point on* I_1.

(b) *If* $\mathbf{g}$ *is sufficiently close to* $\mathbf{f}$ *for part* (a) *to hold, then there is a homeomorphism* $\bar{\bar{h}} : G_0 \to G_1$ *extending* h *and conjugating* $P_{\mathbf{f}}$ *to* $P_{\mathbf{g}}$.

Remark. The reader should observe that this statement is not quite obvious even if $\mathbf{f} = \mathbf{g}$.

Proof of Lemma. For part (a), observe that $\mathbf{f}$ induces a section mapping $P_1 : I_1 \to I_1$ which is differentiably conjugate to the section mapping $P_{\mathbf{f}} : I_0 \to I_0$ by Proposition 8*.4.5. Thus, the point $I_1 \cap \Gamma$ is a linearly attractive fixed point of P_1. For $\mathbf{g}$ close to $\mathbf{f}$ (in the technical sense we imposed in Section 8*.1), the mapping $P_{\mathbf{g}}$ will be close to P_1 together with its derivative. In particular, it will have a unique attractive fixed point $\mathbf{x}_1$.

For part (b), the key idea is as follows: We will construct $\bar{\bar{h}}$ by a scattering construction, as usual, but we will need to do it in stages. For any point in I_0, we can solve $\mathbf{x}' = \mathbf{f}(\mathbf{x})$ backward starting at that point. The solution will cross I_0 some number of times (possibly 0) before it escapes W, and we will extend h successively to the subsets where this solution intersects

$0, 1, 2, \ldots$ times. This will still leave the point $\mathbf{x}_0 = \Gamma \cap I_0$, which never escapes and has to be dealt with separately.

For the dedicated reader, we now provide greater detail: On $\partial W \subset G_0$, $\bar{h}$ is already defined (in fact, on the boundary, ∂W, it is h). On $P_{\mathbf{f}}(\partial W)$, define

$$\bar{h}(\mathbf{x}) = P_{\mathbf{g}} h(P_{\mathbf{f}}^{-1}(\mathbf{x})).$$

We invite the reader to think about why this is continuous at the endpoints of I_0. Now extend $\bar{h}$ to $P_{\mathbf{f}}^2(\partial W)$ by

$$\bar{h}(\mathbf{x}) = P_{\mathbf{g}} \bar{h}(P_{\mathbf{f}}^{-1}(\mathbf{x}));$$

this is well defined because $P_{\mathbf{f}}^{-1}(\mathbf{x})$ is in the locus where $\bar{h}$ was defined in the previous step. Proceeding similarly for $P_{\mathbf{f}}^3(\partial W)$, $P_{\mathbf{f}}^4(\partial W), \ldots$, we will define $\bar{h}$ on all of G_0 except $\mathbf{x}_0$.

Set $\bar{h}(\mathbf{x}_0) = \mathbf{x}_1$; we must check that this is continuous. A point in I_0 very close to $\mathbf{x}_0$ will have a backward trajectory under $\mathbf{x}' = \mathbf{f}(\mathbf{x})$ which intersects I_0 a great many times. It will be mapped by $\bar{h}$ to a point whose backward trajectory under $\mathbf{x}' = \mathbf{g}(\mathbf{x})$ intersects I_1 the same number of times before leaving W. Such a point must be close to x_1. □

Proof of Theorem 8*.4.10, continued. Again, we first state the key idea:

Now we must extend $\bar{h}$ to all of W. The idea for this is not difficult, though the formulas are a little awesome. Every point of $\mathbf{p} \in W - G_0$ is on a unique arc of trajectory of $\mathbf{x}' = \mathbf{f}(\mathbf{x})$ going from a point of $\mathbf{q}_1 \in G_0$ to a point $\mathbf{q}_2$. The points $\mathbf{q}_1$ and $\mathbf{q}_2$ are taken by $\bar{h}$ to the points $h(\mathbf{q}_1), h(\mathbf{q}_2) \in G_1$ which are two endpoints of a segment of trajectory under $\mathbf{x}' = \mathbf{g}(\mathbf{x})$. We will map $\mathbf{p}$ to the point on this trajectory that is proportionally as far along as $\mathbf{p}$ was on its segment of trajectory.

To fill out the proof more precisely:

Define the functions $T_0, S_0 : W - G_0 \to \mathbb{R}$, which give the time it takes for a point to reach G_0 under $\mathbf{x}' = \mathbf{f}(\mathbf{x})$ forward and backward, respectively. Similarly, define $T_1, S_1 : W - G_1$ using $\mathbf{x}' = \mathbf{g}(\mathbf{x})$. We extend T_1 to G_1 so that it measures the time it takes to reach the next intersection with G_1 (we could have extended T_0 too, but it is not necessary). Moreover, to lighten the notation (which badly needs it), define the *origin* of $\mathbf{p}$ to be

$$O(\mathbf{p}) = \phi_{\mathbf{f}}(-S_0(\mathbf{p}), \mathbf{p}),$$

i.e., the place where the trajectory through $\mathbf{p}$ last intersected G_0 before reaching **p**. The construction is illustrated in Figure 8*.4.11.

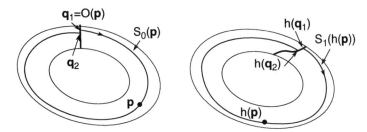

FIGURE 8*.4.11. The key idea of the proof of Theorem 8*.4.10. For **p** one-fifth of the way along the trajectory, $\bar{\bar{h}}(\mathbf{p})$ will also be one-fifth of the way along *its* trajectory.

Now define $\bar{h} : W - G_0 \to W - G_1$

$$\bar{h}(\mathbf{p}) = \phi_{\mathbf{g}} \left(\frac{S_0(\mathbf{p})}{S_0(\mathbf{p}) + T_0(\mathbf{p})} S_1(\bar{\bar{h}}(O(\mathbf{p})), \bar{\bar{h}}(O(\mathbf{p}))) \right).$$

We leave to the reader to check that this ghastly formula really does reflect the geometric construction defined above; once he or she has done this, they will have no trouble convincing themselves that $\bar{h}$ extends $\bar{\bar{h}}$.

That $\bar{h}$ is a homeomorphism follows from the fact that **f** and **g** can be exchanged.   □

Note that in this case the map $\bar{h}$ is not a conjugacy of the flows and that in fact there cannot be a conjugacy if the periods of $\Gamma = \Gamma_{\mathbf{f}}$ and the cycle $\Gamma_{\mathbf{g}}$ for **g**, which we have shown to exist, are different. This is why the equivalence condition requiring topological conjugacy of the flows is too strong.

8*.5 Why Poincaré–Bendixson Rules Out "Chaos" in the Plane

Differential equations in the plane do not exhibit "chaos." You cannot find in $\mathbb{R}^2$ the sort of complication associated with infinitely many periodic orbits, invariant Cantor sets, and other exotic phenomena which are such a common feature of iteration (even in one dimension) and of differential equations in three dimensions or more.

The reason for this simplicity is rooted in the Poincaré–Bendixson Theorem, in its more general form (soon to be stated as Theorem 8*.5.5), which asserts that away from the singularities the most complicated sort of orbit a differential equation in the plane can have is a closed orbit.

To prepare for the classical statement of the Poincaré–Bendixson Theorem 8*.5.5, we must first define the concept of "limit set."

THE LIMIT SETS OF A POINT

A *limit set* of a point $\mathbf{x}_p$ under a differential equation is the set of accumulation points, in the forward or backward direction, of a solution $\mathbf{u}(\mathbf{x}_p)$. More formally, let $\mathbf{x}' = \mathbf{f}(\mathbf{x})$ be a differential equation in $\mathbb{R}^n$, with flow $\phi_{\mathbf{f}}(t, \mathbf{x})$.

Definition 8*.5.1 (Limit set). The ω-*limit* set of a point $\mathbf{x}$ is the set of points to which you come arbitrarily close along the trajectory $\phi_{\mathbf{f}}(t, \mathbf{x})$ as $t \to +\infty$:

$$L_\omega(\mathbf{x}) = \left\{ \lim_{i \to \infty} \phi_{\mathbf{f}}(t_i, \mathbf{x}_p) \mid t_i \to \infty \text{ and the sequence is convergent} \right\}.$$

There is a corresponding definition of the α-*limit* set:

$$L_\alpha(\mathbf{x}) = \left\{ \lim_{i \to \infty} \phi_{\mathbf{f}}(t_i, \mathbf{x}) \mid t_i \to -\infty \text{ and the sequence is convergent} \right\}.$$

Some limit sets are illustrated in Figure 8*.5.1.

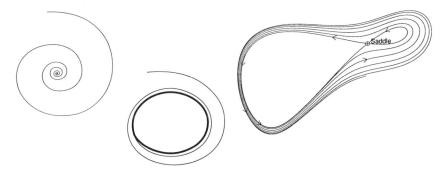

FIGURE 8*.5.1. Some possible limit sets: a point, a simple closed orbit, and a more complicated cycle, formed by two homoclinic orbits of a single saddle.

Alternate definitions of limit sets are explored in Exercise 8*.5#1.

Remark. The names for these limit sets come from the first and last letters of the Greek alphabet: α is the beginning, whence solutions come; ω is the last, whither solutions go.

Clearly, the limit sets of points on the same trajectory coincide, so we can speak of the limit set of a trajectory.

With this definition, the ω-limit set $L_\omega(\mathbf{x})$ is empty if the solution through $\mathbf{x}$ is not defined for all $t \geq 0$ or tends to ∞ with t. Similarly, the α-limit set is empty if the solution is not defined for $t \leq 0$. Furthermore, the limit set has odd properties if the solution does not remain bounded.

Example 8*.5.2. For the differential equation $\mathbf{x}' = \mathbf{x}$ and the solution $\mathbf{u}(t) = e^t$, the ω-limit set is empty and the α-limit set is 0. In fact, all solutions of this equation have the same limit sets, except the constant solution 0 for which both limit sets are the point 0. See Figure 8*.5.2.

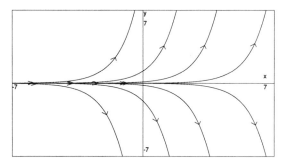

FIGURE 8*.5.2. Solutions for $\mathbf{x}' = \mathbf{x}$. $L_\omega(\mathbf{u}) = \{\emptyset\}$, $L_\alpha(\mathbf{u}) = 0$.   ▲

Example 8*.5.3. For the equation of Figure 8.4.1,

$$\begin{bmatrix} x \\ y \end{bmatrix}' = \begin{bmatrix} y \\ -x \end{bmatrix} + (1 - x^2 - y^2) \begin{bmatrix} x \\ y \end{bmatrix},$$

the ω-limit set of any solution except the constant solution 0 is the unit circle.

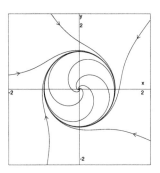

FIGURE 8*.5.3. For any $\mathbf{x} \neq 0$, the $L_\omega(\mathbf{x})$ is the unit circle.   ▲

In proving the strong Poincaré–Bendixson Theorem 8*.5.6, we shall need to know that both limit sets of a point in a limit set are a subset of the limit set. This may not be so surprising for L_ω, but it is certainly not so clear for L_α. Thus, we prove the following proposition:

Proposition 8*.5.4. If $\mathbf{y} \in L_\omega(\mathbf{x})$, then the trajectory through $\mathbf{y}$ is also a subset of $L_\omega(\mathbf{x})$. In particular, $L_\omega(\mathbf{y}) \subset L_\omega(\mathbf{x})$ and $L_\alpha(\mathbf{y}) \subset L_\omega(\mathbf{x})$.

Proof. Let $\mathbf{z} = \phi_{\mathbf{f}}(\tau, \mathbf{y})$. Then for any $\varepsilon > 0$ there exists $\delta > 0$ so that for any $\mathbf{y}_1$ with $\|\mathbf{y}_1 - \mathbf{y}\| < \delta$, we have $\phi_{\mathbf{f}}(\mathbf{y}_1, \tau)$ is defined and $\|\mathbf{z} - \phi_{\mathbf{f}}(\mathbf{y}_1, \tau)\| < \varepsilon$. By (the second) definition of the limit set, there exists a sequence $t_i \in \mathbb{R}$ such that $t_i \to \infty$, and $\phi_{\mathbf{f}}(t_i, \mathbf{x}) \to \mathbf{y}$. For i sufficiently large, $\|\phi_{\mathbf{f}}(t_i, \mathbf{x}_p) - \mathbf{y}\| < \delta$, so we see that

$$\phi_{\mathbf{f}}(t_i + \tau, \mathbf{x}_p) \to \mathbf{z} \quad \text{and} \quad t_i + \tau \to \infty.$$

Thus, points of the forward trajectory through $\mathbf{x}$ will converge to any point of the orbit of $\mathbf{y}$, whether forward or backward. But $L_\omega(\mathbf{x})$ is closed, so it will also contain both limit sets. $\square$

STATEMENT OF THE THEOREMS

The classical statement of the Poincaré–Bendixson Theorem is the following:

Theorem 8*.5.5 (Poincaré–Bendixson). Let $\mathbf{u}(t)$ be a solution to the differential equation $\mathbf{x}' = \mathbf{f}(\mathbf{x})$ in $\mathbb{R}^2$ as above, defined and bounded for $t \geq t_0$. Then if $L_\omega(\mathbf{u})$ contains no zero of $\mathbf{f}$, it is a cycle.

Remark. Our original Poincaré–Bendixson Theorem 8.5.1 concerns a vector field in an annulus, which enters along the entire boundary, and such that the annulus contains no zeroes of $\mathbf{f}$. Theorem 8*.5.5 says that the ω-limit set of every point must be a cycle.

We will prove a stronger result.

Theorem 8*.5.6 Stronger Poincaré–Bendixson). Let $\mathbf{x}$ be a point such that the solution $\phi_{\mathbf{f}}(t, \mathbf{x})$ of the differential equation $\mathbf{x}' = \mathbf{f}(\mathbf{x})$ in $\mathbb{R}^2$ is defined and bounded for $t > 0$. Then either $L_\omega(\mathbf{x})$ is a cycle or, for any $\mathbf{y} \in L_\omega(\mathbf{x})$, $\mathbf{f}$ vanishes identically on $L_\omega(\mathbf{y})$ and $L_\alpha(\mathbf{y})$.

Remark. Of course, one possibility is that $\mathbf{y}$ itself is a zero of $\mathbf{f}$. Reasonable vector fields have isolated zeroes, so that usually $L_\omega(\mathbf{y})$ and $L_\alpha(\mathbf{y})$ will be single points, but it can happen that the limit sets of points $\mathbf{y}$ as above are actually subsets of the plane which are not single points (but on which the vector field vanishes identically, as the theorem says).

LABYRINTHS IN THE PLANE

Theorem 8*.5.6. is hard to understand if you cannot imagine what a counterexample might look like. The reader should run the Lorenz equation, Example 6.1.6, on a computer program such as *DiffEq, 3DViews* in the *MacMath* package, and consider carefully the resulting Lorenz attractor, pictured for instance in Figure 6.1.8.

It might seem "obvious" that nothing like the Lorenz attractor can exist in the plane, but this is just a failure of the imagination. We will show in Example 8*.5.7 that such things, called *labyrinths*, do exist. Nevertheless, the Poincaré–Bendixson Theorem asserts that in $\mathbb{R}^2$ a solution *cannot* wind around such a labyrinth, accumulating on something like the Lorenz attractor.

Example 8*.5.7 (Labyrinth). Consider the region $U \subset \mathbb{R}^2$ bounded by three semi-circles as represented in Figure 8*.5.4. Fill each semi-circle by concentric semi-circles, and imagine the "differential equation" whose flow curves are precisely these arcs. These curves can obviously be continued forever unless they run into one of the centers of the circles or the point θ.

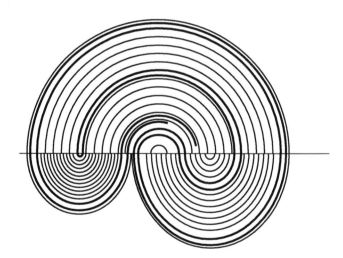

FIGURE 8*.5.4. A labyrinth with an unending orbit, unless you land on a singularity. Part of a typical orbit is drawn as a thicker path, starting and ending near the center. ▲

Example 8*.5.7 becomes especially interesting if θ, the meeting point of the two semi-circles on one side, is irrational. Consider the following two lemmas and their implications.

Lemma 8*.5.8. *If θ is irrational, then for all but countably many $x \in [-1, 1]$, the trajectory through x has limit set equal to U.*

Proof. Consider the two mappings $u_1, u_2 : [-1, 1] \rightarrow [-1, 1]$, which give, for any $x \in [-1, 1]$, the other endpoint of the top or bottom semi-circle, one endpoint of which is x. The bottom map is simply $u_1 : x \mapsto -x$, and the top map is

$$u_2(x) = \begin{cases} -x - 1 + \theta & \text{if } x < \theta \\ -x + 1 + \theta & \text{if } x > \theta \, , \end{cases}$$

so we see that $u_2 \circ u_1(x) = x + 1 + \theta \bmod 2$. Exercise 8*.5.2 asks you to prove the result from here. □

Remark. You can use the *MacMath* program *Analyzer* to iterate this function, even though it is discontinuous. Enter the function

$$0.5 * ((\text{sgn}(x - \theta) + 1) * (-x + \theta + 1) + (\text{sgn}(\theta - x) + 1) * (-x + \theta - 1)),$$

say with $\theta = \pi/11$, and then with $\theta = 3/11$, and see the difference, as shown in Figure 8*.5.5.

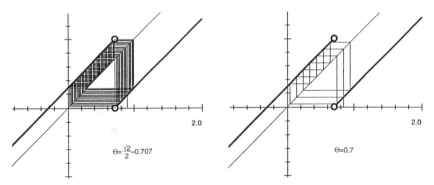

FIGURE 8*.5.5. Different iteration behaviors, to illustrate Lemma 8*.5.8; θ is *irrational* on the left, *rational* on the right.

The labyrinth of Example 8*.5.7 is not quite a counterexample to the Poincaré–Bendixson Theorem, because U contains singularities. In order to remove the singularities, try Example 8*.5.9.

Example 8*.5.9. Start with the idea of Example 8*.5.7, but open up the exceptional trajectories to a union of tadpole-shaped regions V, with narrowing infinitely long tails, as indicated in Figure 8*.5.6, and fill in the opened region, also as indicated. Now a trajectory in $U - V$ has limit set

$\overline{U-V}$ and, in particular, contains no singularities. See Exercise 8*.5.3 for further exploration of this example.   ▲

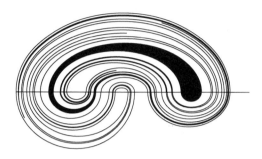

FIGURE 8*.5.6. A labyrinth without singularities, to illustrate Example 8*.5.9.

Why is the labyrinth of Example 8*.5.9 still not a counterexample to the Poincaré–Bendixson Theorem 8*.5.6? Because there is *no direction* to these flow curves, you cannot put arrows on the curves in a coherent way! So the Poincaré–Bendixson Theorem has met the labyrinth challenge in $\mathbb{R}^2$. Another labyrinth example is given in Exercise 8*.5#4 which especially shows how wrong the Poincaré–Bendixson Theorem must be for differential equations in $\mathbb{R}^3$.

Remark. There is a great similarity between this labyrinth "counterexample" and the cardboard model for the Lorenz equations of Example 8.7.1. Nothing like Lemma 8*.5.10, which excludes the labyrinth, holds in $\mathbb{R}^3$, and "chaos" can therefore rear its ugly head in the Lorenz attractor.

PROOF OF THE POINCARÉ–BENDIXSON THEOREM

The first step in proving the Poincaré–Bendixson Theorem 8*.5.6 is Lemma 8*.5.10, the key result that avoids chaos in the plane.

Lemma 8*.5.10 (Monotonicity). *If I is a segment in $\mathbb{R}^2$ transverse to $\mathbf{f}$, which a solution $\mathbf{u}$ of $\mathbf{x}' = \mathbf{f}(\mathbf{x})$ crosses at three points $A_i = \mathbf{u}(t_i)$ with $t_1 < t_2 < t_3$, then A_2 is between A_1 and A_3 on I.*

Proof of Lemma. The construction is shown in Figure 8*.5.7. The set $\mathbf{u}[t_1, t_2] \cup [A_1, A_2]$ forms a simple closed curve, which bounds a region U. The vector field either enters or leaves U. Changing the sign of the vector field if necessary, we can suppose it enters.

Then $\mathbf{u}(t_2, \infty)$ is contained in the interior of U, but the part of $I - A_2$ containing A_1 is in the boundary of U or outside. Hence, A_3 is in the component of $I - A_2 \in \mathbf{u}(t_2, \infty)$ belonging to the component of $I - A_2$ not containing A_1.   □

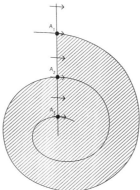

FIGURE 8*.5.7. Section crossed by $\mathbf{u}$ at three points.

Remark. As said at the beginning of Chapter 8, we are assuming Lipschitz vector fields. So even if the boundary of a region is a solution rather than being crossed by the vector field, it nevertheless prevents any solutions from escaping.

Proof of Theorem 8*.5.6. Let $\mathbf{x}$ be a point with bounded forward trajectory, $\mathbf{y} \in L_\omega(\mathbf{x})$, and $\mathbf{z} \in L_\omega(\mathbf{y})$ or $\mathbf{z} \in L_\alpha(\mathbf{y})$. We shall illustrate the steps of the proof with Figure 8*.5.8.

If $\mathbf{z}$ is not a zero of $\mathbf{f}$, there exists a section $I_\mathbf{z}$ through $\mathbf{z}$. The essential thing to see is that $\phi_\mathbf{f}([0, \infty), \mathbf{y}) \cap I_\mathbf{z} = \{\mathbf{z}\}$, i.e., the trajectory through $\mathbf{y}$ cannot intersect $I_\mathbf{z}$ in more than one point.

Indeed, suppose $\phi_\mathbf{f}([0, \infty), \mathbf{y}) \cap I_\mathbf{z}$ has two points $\mathbf{y}_1$ and $\mathbf{y}_2$. By Proposition 8*.5.4, both $\mathbf{y}_1$ and $\mathbf{y}_2$ belong to $L_\omega(x)$, so there exist sequences $t_i \to \infty$ and $s_i \to \infty$ such that $\phi_\mathbf{f}(t_i, \mathbf{x})) \to \mathbf{y}_1$ and $\phi_\mathbf{f}(s_i, \mathbf{x}) \to \mathbf{y}_2$. There is a neighborhood V of $I_\mathbf{z}$ such that any point in it can be moved onto $I_\mathbf{z}$ by flowing a time $< \varepsilon$; so there exist sequences $t_i' \to \infty$ and $s_i' \to \infty$ with $\phi_\mathbf{f}(t_i', \mathbf{x}))$, $\phi_\mathbf{f}(s_i', \mathbf{x}) \in I$ and still $\mathbf{u}_x(t_i') \to \mathbf{y}_1$ and $\mathbf{u}_x(s_i') \to \mathbf{y}_2$. If $\mathbf{y}_1$ and $\mathbf{y}_2$ are distinct, this is incompatible with the Monotonicity Lemma 8*.5.10.

Now there exist t_1 and t_2 such that $\phi_\mathbf{f}(t_1, \mathbf{y})$ and $\phi_\mathbf{f}(t_2, \mathbf{y})$ are both in V and, in fact, on the arc of trajectory through $\mathbf{z}$ in V; moreover, we may assume $t_2 - t_1 > 2\varepsilon$.

Then by modifying t_1 and t_2 by less than ε, you can find t_1' and t_2' such that $\phi_\mathbf{f}(t_1', \mathbf{y}) = \phi_\mathbf{f}(t_2', \mathbf{y}) = z$, and $t_1' \neq t_2'$. This shows that $\phi_\mathbf{f}(t_2' - t_1', \mathbf{y}) = \mathbf{y}$, which implies that the trajectory through $\mathbf{y}$ is a cycle.

We still need to show that if the trajectory through $\mathbf{y}$ is a cycle, then this cycle is the whole ω-limit set of $\mathbf{x}$. Choose a section I at $\mathbf{y}$ and let t_i be the times for which $\phi_\mathbf{f}(t_i, \mathbf{x}) \in I$; the $\phi_\mathbf{f}(t_i, \mathbf{x})$ form a sequence on I

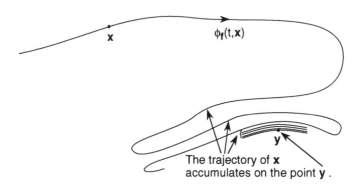

The trajectory of **x**
accumulates on the point **y** .

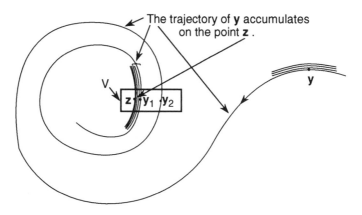

The trajectory of **y** accumulates
on the point **z** .

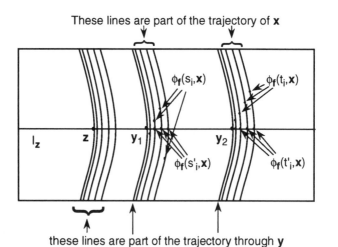

FIGURE 8*.5.8. Illustration of the proof of Theorem 8*.5.6.

converging monotonically to $\mathbf{y}$. Since the solution through $\mathbf{y}$ is a cycle of some period $\tau > 0$, we have $\phi_\mathbf{f}(\tau, \mathbf{y}) = \mathbf{y}$, so that $\phi_\mathbf{f}(t_i + \tau, \mathbf{x})$ is close to $\mathbf{y}$ for large i and, in particular, t_{i+1} is close to $t_i + \tau$. But this means that the trajectory through $\mathbf{x}$ is close to the trajectory through $\mathbf{y}$ for all sufficiently large times. $\square$

8*.6 Structurally Stable Equations in the Plane

MORSE–SMALE VECTOR FIELDS IN THE PLANE

In this section we will attempt to pin down the fact that vector fields *in the plane* do not exhibit "chaos." More specifically, we will define the set of "nice" (Morse–Smale) vector fields by a few negative properties. Then we will show that these vector fields are *structurally stable* and *dense*.

The first of these properties means that

> *If you perturb a nice vector field a little bit, for instance by adding on a small vector field, then the essential structure of the phase plane does not change: the perturbed differential equation is topologically equivalent to the original one.*

The second property means that

> *Any vector field, however nasty, for instance the zero vector field, can be perturbed an arbitrarily small amount and become nice.*

Definition 8*.6.1 (Morse–Smale vector field). A vector field $\mathbf{f}$ on a region $U \subset \mathbb{R}^2$ is *Morse–Smale* (abbreviated M-S) if:

(1) the linearizations of $\mathbf{f}$ at the zeroes are sinks, saddles, or sources;

(2) the limit cycles are all linearly attracting or repelling;

(3) there are no saddle connections.

Historical Remark. In the West, after the momumental work of Poincaré at the turn of the century, the study of differential equations more or less dried up (with the notable exception of George Birkhoff, and some very deep work in celestial mechanics by Carl Ludwig Siegel). In the Soviet Union, work on differential equations never slowed, spearheaded by such mathematicians as A. A. Andronov and L. Pontryagin.

To the authors of this book, these names are historical. Not so the names of Marston Morse, Stephen Smale, Raoul Bott, and John Milnor, who have been at the heart of current developments. Morse's main contribution is not in differential equations, but in a topic called *the calculus of variations in the large*, now usually called Morse theory. In the 1930s, he proposed to take something like a surface with a function f and try to understand the topology of the surface by flowing down the gradient curves of the function, eventually deforming the surface into the unstable separatrices of the gradient flow. Even if one is not centrally interested in differential equations, thinking about surfaces forces one to think about gradient flows.

This program was spectacularly applied by Bott in the early 1950s to understand the topology of the space of loops on groups like $SO(3)$. This brought the subject to the forefront of mathematical research. In particular, Milnor wrote a book called *Morse theory*, which was extremely influential. Smale, who had been a student of Bott in the 1950s, used Morse theory to prove the Poincaré conjecture in dimensions at least five. This earned him a Fields medal, the mathematical equivalent of the Nobel prize. After this, his work centered on dynamical systems, and he started a whole school, of which the authors are in some sense a part. He is still very active in the subject, and it is clear that he was much influenced by gradient flows in his more general investigations.

Such gradient flows have very nice properties, and Morse–Smale vector fields are those that share these properties.

For differential equations in the plane, the long-term behavior of trajectories of $\mathbf{x}' = \mathbf{f}(\mathbf{x})$ is quite easy to understand. We have suggested that to study a differential equation in the plane, the main objective is to locate the basins of the sinks and attracting limit cycles (usually bounded by stable separatrices and repelling limit cycles), and the "antibasins" of the sources and repelling limit cycles (usually bounded by unstable separatrices and attracting limit cycles). Theorem 8*.6.2 justifies this philosophy when the vector field is M-S.

Theorem 8*.6.2. *Let U be a bounded region of the plane, bounded by smooth curves, and $\mathbf{f}$ be a M-S vector field in U transverse to its boundary, ∂U. Then as $t \to \infty$, every solution of $\mathbf{x}' = \mathbf{f}(\mathbf{x})$ either*

(1) *leave U, or*

(2) *tends to a sink or an attracting limit cycle, or*

(3) *is a stable separatrix of a saddle or a repelling limit cycle.*

Similarly, as $t \to -\infty$, every solution either

(1) *leaves U, or*

(2) *emanates from a source or a repelling limit cycle, or*

(3) *is an unstable separatrix of a saddle or an attracting limit cycle.*

Proof. This is a corollary of the Poincaré–Bendixson theorem (in its strong form Theorem 8*.5.6). Suppose the trajectory $\phi_{\mathbf{f}}(t, \mathbf{x})$ does not leave U as $t \to \infty$, that $L_\omega(\mathbf{x})$ is not a cycle, and let $\mathbf{y} \in L_\omega(\mathbf{x})$. We wish to show that $\mathbf{y}$ is a zero of $\mathbf{f}$ (clearly either a sink or a saddle unless $\mathbf{x}$ was itself a source). Suppose that $\mathbf{f}(\mathbf{y}) \neq 0$. The vector field $\mathbf{f}$ vanishes identically on $L_\alpha(\mathbf{y})$ and on $L_\omega(\mathbf{y})$ by Theorem 8*.5.6. But both $L_\alpha(\mathbf{y})$ and $L_\omega(\mathbf{y})$ must be saddles: clearly, $L_\omega(\mathbf{y})$ cannot be a source; if it is a sink, then that sink is also $L_\omega(\mathbf{x})$. Similarly, $L_\alpha(\mathbf{y})$ cannot be a source or a sink. So the trajectory through $\mathbf{y}$ is a saddle connection. A similar argument can be made for $L_\alpha(\mathbf{x})$. $\square$

Remark. The notion of a Morse–Smale vector field is not restricted to the plane. However, Theorem 8*.6.2 does not follow from the definition we have given of Morse–Smale in higher dimensions; and the appropriate modification of the conclusion of Theorem 8*.6.2 simply becomes part of the definition of M-S. With this modified definition, structural stability is still true, and the proof we give almost carries over. On the other hand, the density result fails drastically, so the notion of M-S is much less useful: a "general" vector field in $\mathbb{R}^n$ is not M-S when $n > 2$.

THE STRUCTURAL STABILITY THEOREM

Theorem 8*.6.3 is due to Andronov and Pontryagin. It had a deep influence on the whole theory of dynamical systems, and for quite a while, the hope was that something similar would be true in higher dimensions: that structurally stable vector fields would be open and dense in general. Although this hope was dashed [Smale, 1966,Amer. J. Math], the underlying philosophy is still a central part of the way people think about dynamical systems.

Theorem 8*.6.3 (Structural Stability Theorem). *Let U be a bounded region of the plane, bounded by smooth curves, and $\mathbf{f}$ be a Morse–Smale vector field on U transverse to the boundary curves. Then $\mathbf{f}$ is structurally stable.*

Proof. Without being enormously difficult, the proof is quite long and an order of magnitude more elaborate than anything encountered in this book so far. There are four preliminary lemmas, so hang on for a long ride!

For each sink or source $\mathbf{x}_i$, choose a small neighborhood $W_{\mathbf{x}_i}$ and $\delta_{\mathbf{x}_i} > 0$ as given by Theorem 8*.2.1. For each limit cycle Γ_j, choose a neighborhood W_{Γ_j} and δ_{Γ_j} as given by Theorem 8*.4.10. We will take all these neighborhoods disjoint. Let V be U with the interiors of these neighborhoods deleted. Then the boundary ∂V of V is a finite union of simple closed curves, and naturally breaks up into $\partial V = \partial V' \cup \partial V''$, where $\partial V'$ is the part of the boundary where solutions enter and $\partial V''$ is the part where they leave.

Lemma 8*.6.4. *Every trajectory enters and leaves V, except the constant solutions at the saddles and their separatrices. The stable separatrices enter V transversely to $\partial V'$ and the unstable separatrices leave V transversely to $\partial V''$.*

Proof of Lemma. By Theorem 8*.6.2, any solution that does not leave is the stable separatrix of a saddle, and any solution that does not enter is the unstable separatrix of a saddle; all other behaviors were excluded by removing neighborhoods of the sinks, sources, and cycles. Moreover, no trajectory can be both a stable and an unstable separatrix, since there are no saddle connections. $\square$

So we see that we can define the *arrival time*, *departure time*, and the *transit time* respectively as Arr, Dep, Trans $: V \to \mathbb{R} \cup \infty$ by setting

$$\mathrm{Arr}(\mathbf{x}) = \sup\{t \mid \phi_{\mathbf{f}}([-t, 0], \mathbf{x}) \subset V\},$$
$$\mathrm{Dep}(\mathbf{x}) = \sup\{t \mid \phi_{\mathbf{f}}([0, t], \mathbf{x}) \subset V\},$$
$$\mathrm{Trans}(\mathbf{x}) = \mathrm{Arr}(\mathbf{x}) + \mathrm{Dep}(\mathbf{x}).$$

Let a stable separatrix of a saddle $\mathbf{x}_0$ intersect $\partial V'$ at $\mathbf{x}$; let I be a neighborhood of $\mathbf{x}$ in ∂V homeomorphic to an interval; $I - \{\mathbf{x}\}$ will consist of two intervals I^+ and I^-.

Lemma 8*.6.5. *If I is sufficiently short, the function $\mathrm{Trans}(\mathbf{y})$ is a monotone function on each of I^+ and I^-, tending to ∞ as $\mathbf{y} \to \mathbf{x}$.*

Proof of Lemma. Choose a neighborhood of $\mathbf{x}_0$, two sections $\gamma_1 : I_1 \to V$ and $\gamma_2 : I_2 \to V$ at points of the stable and unstable separatrices, respectively, and a diagonal segment $\gamma_J : J \to V$ at $\mathbf{x}_0$, all sufficiently small so that Proposition 8*.3.2 holds. The construction is shown in Figure 8*.6.1. We will call s_1 and s_2 the variables of I_1 and I_2, and s the variable of J, so that the s_i live in a neighborhood of 0, whereas $0 < s < \varepsilon$. Again, as in Proposition 8*.3.2, the key to success is to write everything in terms of s.

If necessary, making the intervals I_i short, and the J shorter yet, the flow $\phi_{\mathbf{f}}$ defines injective maps $\alpha_i : J \to I_i$ and positive functions $\tau_i : J \to \mathbb{R}$ and $\sigma_i : I_i \to \mathbb{R}$ such that

$$\phi_{\mathbf{f}}\left((-1)^i \tau_i(s), \gamma_J(s)\right) = \gamma_i(\alpha_i(s))$$

and

$$\phi_{\mathbf{f}}\left((-1)^i \sigma_i(s_i), \gamma_i(s_i)\right) \in \partial V.$$

In words, τ_i is the time it takes to flow from the diagonal to the transverse sections $\gamma_i(I_i)$, and σ_i is the time it takes to flow from these to the entering and exiting boundary.

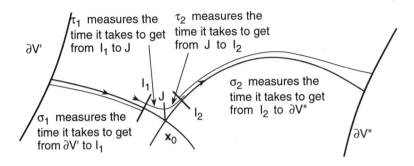

FIGURE 8*.6.1.

With this description, clearly

$$\text{Trans}(\gamma_J(s)) = \sum_{i=1,2} \tau_i(s) + \sigma_i(\alpha_i(s)).$$

Proposition 8*.3.2 tells us that $\tau_i(s)$ tends to infinity as $s \searrow 0$, with derivative tending to $-\infty$, and that α_i has a bounded derivative. On the other hand, σ_i is defined and differentiable in a neighborhood of 0, so it certainly has a bounded derivative in a neighborhood of 0. By the chain rule,

$$\frac{d}{ds}\text{Trans}(\gamma_J(s)) = \sum_{i=1,2} \frac{d\tau_i}{ds}(s) + \frac{d\sigma_i}{ds_i}(\alpha_i(s))\frac{d\alpha_i}{ds}(s)$$

is the sum of a function tending to $-\infty$ and of a bounded function, and hence tends to $-\infty$. Thus, $\text{Trans}(\gamma_J(s))$ is a monotone function tending to infinity on J if J is chosen sufficiently short.

This proves the lemma, since the flow defines a homeomorphism from J to a one-sided neighborhood of $\mathbf{x}$ in ∂V. $\square$

Choose, for each intersection $\mathbf{x}$ of a separatrix of a saddle and ∂V, a neighborhood $I_{\mathbf{x}} \subset \partial V$ of $\mathbf{x}$ homeomorphic to an interval, and such that these neighborhoods are disjoint.

We can now state precisely what perturbations $\mathbf{g} = \mathbf{f} + \mathbf{k}$ of $\mathbf{f}$ we will allow. They will be required to satisfy five conditions:

(1) $\mathbf{g}$ is transverse to ∂V, entering on $\partial V'$ and leaving on $\partial V''$;

(2) for each sink or source $\mathbf{x}_i$, the perturbation is smaller on the neighborhood $W_{\mathbf{x}_i}$ than $\delta_{\mathbf{x}_i} > 0$, so that Theorem 8*.2.1 applies to the restriction of $\mathbf{g}$ to $W_{\mathbf{x}_i}$;

(3) for each limit cycle Γ_j, the perturbation is smaller on the neighborhood W_{Γ_j} than δ_{Γ_j}, so that Theorem 8*.4.1 applies to the restriction of $\mathbf{g}$ to W_{Γ_j};

(4) for each saddle $\mathbf{x}_0$ of $\mathbf{f}$, there exists a saddle $\mathbf{x}^*$ of $\mathbf{g}$, with separatrices intersecting ∂V in exactly the same four segments $I_{\mathbf{x}}$ as the separatrices of $\mathbf{x}_0$;

(5) there are no other zeroes of $\mathbf{g}$ in V.

Let $\mathbf{g}$ be a perturbation of $\mathbf{f}$ satisfying the conditions above. We will denote by $\mathrm{Trans}_{\mathbf{f}}$, etc., the constructions above as applied to $\mathbf{f}$.

Lemma 8*.6.6. *There exists a homeomorphism $h : \partial V' \to \partial V'$ which is the identity on $\partial V' - \cup I_{\mathbf{x}}$ and such that, for each intersection $\mathbf{x}$ of a separatrix with ∂V, there exist subintervals $K_{\mathbf{x}}, K_{\mathbf{x}}^* \subset I_{\mathbf{x}}$ on which transit times are preserved, i.e.,*

$$\mathrm{Trans}_{\mathbf{f}}(\mathbf{y}) = \mathrm{Trans}_{\mathbf{g}}(h(\mathbf{y})) \tag{15}$$

for all $\mathbf{y} \in K_{\mathbf{x}}$.

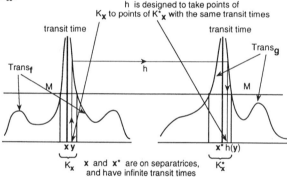

FIGURE 8*.6.2. Choosing h so that corresponding points have equal transit times.

Proof of Lemma. Lemma 8*.6.5 also applies to $\mathbf{x}^*$. So for M sufficiently large, there exist $K_{\mathbf{x}}$ and $K_{\mathbf{x}^*}$ neighborhoods of $\mathbf{x}$ and $\mathbf{x}^*$ respectively, such that $\mathrm{Trans}_{\mathbf{f}}$ maps each component of $K_{\mathbf{x}} - \{\mathbf{x}\}$ bijectively to (M, ∞), and similarly for $K_{\mathbf{x}^*}$ and $\mathbf{g}$. Taking M larger if necessary, we may assume $K_{\mathbf{x}}, K_{\mathbf{x}^*} \subset I_{\mathbf{x}}$. Equation (15) and the fact that h should preserve the orientation of ∂V now impose $h : K_{\mathbf{x}} \to K_{\mathbf{x}^*}$. It can now easily be extended to $I_{\mathbf{x}}$, for instance as suggested by Figure 8*.6.2. $\square$

We can now extend $h : \partial V' \to \partial V'$ to a homeomorphism $\bar{h} : V \to V$. To make the formula more readable, we define

$$\mathrm{Entr}(\mathbf{x}) = \phi_{\mathbf{x}}(-\mathrm{Arr}(\mathbf{x}), \mathbf{x}) \quad \text{and} \quad \mathrm{Exit}(\mathbf{x}) = \phi_{\mathbf{f}}(\mathrm{Dep}(\mathbf{x}), \mathbf{x}),$$

so that $\text{Entr}(\mathbf{x})$ is the point of ∂V where the trajectory through $\mathbf{x}$ enters V and $\text{Exit}(\mathbf{x})$ is the point where it leaves. The point $\text{Entr}(\mathbf{x})$ is defined except at the saddles and their unstable separatrices, and the point $\text{Exit}(\mathbf{x})$ is defined except at the saddles and their stable separatrices.

Now our extension is given by the following formula:

$$\bar{h}(\mathbf{x}) = \phi_{\mathbf{g}}\left(\frac{\text{Arr}_{\mathbf{f}}(\mathbf{x})}{\text{Trans}_{\mathbf{f}}(\mathbf{x})}\text{Trans}_{\mathbf{g}}(h(\text{Entr}(\mathbf{x}))), h\left(\text{Entr}(\mathbf{x})\right)\right). \qquad (16)$$

This formula is *not well defined on the separatrices*, since the times $\text{Arr}(\mathbf{x})$ and $\text{Trans}(\mathbf{x})$ are not well defined on the separatrices. Let us denote by $S(\mathbf{f})$ the union of the saddles of $\mathbf{f}$ and of their separatrices, and similarly for $S(\mathbf{g})$.

Lemma 8*.6.7. *Formula (16) defines a homeomorphism* $\bar{h} : V - S(\mathbf{f}) \to V - S(\mathbf{g})$, *which extends continuously to a homeomorphism* $V \to V$.

Proof of Lemma. Clearly, $\bar{h}$ is defined and continuous on $V - S(\mathbf{f})$ and maps $V - S(\mathbf{f})$ to $V - S(\mathbf{g})$. The first part of the statement simply comes from observing that the roles of $\mathbf{f}$ and $\mathbf{g}$ can be interchanged, leading to an inverse of $\bar{h}$.

For the second part, first choose $\mathbf{p} \in S_s(\mathbf{x}_0)$, a point on a stable separatrix entering V at $\mathbf{x}$. Then for $\mathbf{y}$ in a sufficiently small neighborhood of $\mathbf{x}_0$, the point $\text{Entr}(\mathbf{y})$ will be in $K_{\mathbf{x}}$, so that $\text{Trans}_{\mathbf{f}}(\text{Entr}(\mathbf{y}) = \text{Trans}_{\mathbf{g}}(h(\text{Entr}(\mathbf{y}))$. In this neighborhood, formula (17) becomes

$$\bar{h}(\mathbf{y}) = \phi_{\mathbf{g}}(\text{Arr}_{\mathbf{f}}(\mathbf{y}), h\left(\text{Entr}(\mathbf{y})\right)),$$

which is a continuous function of $\mathbf{y}$.

To see the result for points on the unstable separatrices, we must write formula (17) in terms of the leaving point $\text{Exit}(\mathbf{x})$ rather than the entering point. First observe that the homeomorphism $h : \partial V' \to \partial V'$ induces a homeomorphism $k : \partial V'' \to \partial V''$ of the leaving boundary by the formula

$$k(\mathbf{y}) = \text{Exit}(h(\text{Entr}(\mathbf{y}))).$$

This is obviously well defined and continuous except at the intersections of the unstable separatrices with the boundary. There is a unique such point in each interval $I_{\mathbf{x}} \subset \partial V''$ for both $\mathbf{f}$ and $\mathbf{g}$; the first must be sent to the second.

Now we leave to the reader the verification that

$$\bar{h}(\mathbf{x}) = \phi_{\mathbf{g}}\left(\frac{\text{Dep}_{\mathbf{f}}(\mathbf{x})}{\text{Trans}_{\mathbf{f}}(\mathbf{x})}\text{Trans}_{\mathbf{g}}(k(\text{Exit}(\mathbf{x}))), k\left(\text{Exit}(\mathbf{x})\right)\right);$$

with this formula, the proof works as above for points on the unstable separatrices.

This leaves the saddles. If a sequence of points $\mathbf{x}_i$ approaches a saddle $\mathbf{x}_0$, and these points are not on the separatrices, then the $\mathbf{x}_i$ have entering and leaving points that must approach the intersections of the separatrices with ∂V; moreover, their arrival and departure times tend to infinity. The sequence $\bar{h}(x_i)$ must then have the same property, and this shows that it also converges to the saddle $\mathbf{x}^*$ of $\mathbf{g}$ corresponding to $\mathbf{x}_0$. If the sequence $\mathbf{x}_i$ is contained in the separatrices of the saddle $\mathbf{x}_0$ and converges to $\mathbf{x}_0$, then the image sequence is contained in the separatrices of $\mathbf{x}^*$, and since the times tend to infinity, such a sequence must must also converge to $\mathbf{x}^*$. Any sequence converging to $\mathbf{x}_0$ is formed of two subsequences, one of each of the types above, and since the images of both converge to $\mathbf{x}^*$, the image sequence converges. $\quad\square$

Finally, the homeomorphisms h and k can be extended to the interiors of the V_i and W_j by Theorems 8*.2.1 and 8*.4.1. Denote still by $\bar{h}$ the homeomorphism $U \to U$ obtained in this way; it is a homeomorphism sending oriented trajectories of $\mathbf{f}$ to trajectories of $\mathbf{g}$. This ends the proof of the Structural Stability Theorem 8*.6.3. $\quad\square$

The Density Theorem

We next want to claim that any vector field can be approximated by a structurally stable one. The next theorem is attributed to M. M. Peixoto, a contemporary Brazilian mathematician.

Theorem 8*.6.8 (Peixoto's Density Theorem). *For any twice-differentiable vector field* $\mathbf{f}$ *in a region* U *of the plane, there exists a M-S vector field* $\mathbf{g}$ *for* $\|\mathbf{f} - \mathbf{g}\|$ *arbitrarily small.*

Proof. We will proceed in several steps:

(i) First perturb the vector field so that the zeroes are isolated and are sinks, sources, and saddles.

(ii) Then destroy the saddle connections.

(iii) Finally, adjust the cycles so that they are linearly attracting or repelling.

Each of these steps is rather delicate, because we must be sure that we have not undone the steps already accomplished.

Proof of (i). Approximate $\mathbf{f}$ by a vector field with isolated zeroes. We will outline in Exercise 8*.6.#2 a proof that any function defined and k times continuously differentiable on a closed subset of $\mathbb{R}^n$ can be approximated on that subset together with its first k derivatives by a polynomial function.

For our purposes, approximate both components f_1 and f_2 of the vector field together with their first two derivatives by polynomials p_1 and p_2.

Remark. It is not really any easier, but it is more standard to show that functions can be approximated by trigonometric polynomials. For our purposes, this would do just as well. Exercise 8*.6#3 outlines this proof.

Polynomials vanish on algebraic curves, and algebraic curves, if they do not have whole components in common, intersect at finitely many points. If p_1 and p_2 have an algebraic curve Z of zeroes in common, take any polynomial p that does not vanish identically on Z; then p_1 and $p_2 + \varepsilon p$ will have finitely many common zeroes for arbitrarily small ε.

Essentially the same argument shows that the zeroes can be taken to be sinks, saddles, or sources. We will take them one at a time, each time taking a much smaller perturbation than before, so as not to disturb the work already done. So suppose $\mathbf{x}_0$ is a common zero of p_1 and p_2; by translation, we can assume $\mathbf{x}_0 = 0$. If we modify p_1 and p_2 by adding on an arbitrarily small linear vector field, for instance choose $p_1^*(x,y) = p_1(x,y) + \varepsilon_1 x$ and $p_2^*(x,y) = p_2(x,y) + \varepsilon_2 y$, then ε_1 and ε_2 can be chosen arbitrarily small so that neither the trace nor the determinant of the linearization vanish.

Proof of (ii). We next deal with the saddle connections, which are discussed at length in Section 9.3. Some pictures of saddle connections are shown in Figures 9.3.1 and 9.3.2 in Figure 8*.6.3; note that a saddle connection may connect a saddle to another saddle (*heteroclinic*) or to itself (*homoclinic*).

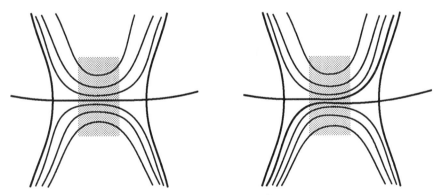

FIGURE 8*.6.3. A heteroclinic saddle connection (on the left), slightly perturbed in the shaded region (as shown on the right).

Note in Figure 8*.6.3 that this break creates a *global* change in the behavior of solutions that were near the saddle connection.

There are only finitely many saddle connections; choose a point $\mathbf{x}_0$ on one of them, and a neighborhood V of $\mathbf{x}_0$, obtained by choosing a section

$\gamma : I_1 \to \mathbb{R}_2$ and considering the image of the map $I_1 \times I_2$ given by $(s, t) \mapsto \phi_{\mathbf{f}}(t, \gamma(s))$. We will make infinitely many modifications to $\mathbf{f}$ but will keep the same $V's$ throughout.

Lemma 8*.6.9. *There exists* $\mathbf{g}$ *with* $\|\mathbf{g} - \mathbf{f}\|$ *arbitrarily small such that the vector field* $\mathbf{g}$ *has no saddle connection intersecting* V.

It seems obvious that you can break saddle connections by locally modifying the vector field near a saddle connection as suggested by Figure 8*.6.4.

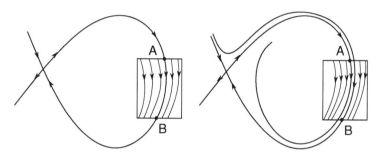

FIGURE 8*.6.4. A small perturbation of a vector field breaks a homoclinic saddle connection. Note that a limit cycle is created in the homoclinic case.

Example 8*.6.10. The vector field of Figure 8*.6.5 illustrates the difficulties that may arise in when carrying out this construction: if a separatrix of a saddle has a saddle connection as ω-limit set, as in Figure 8*.6.5, then an arbitrarily small perturbation of the vector field in V may cut the original saddle connection, but create a new one at the same time.

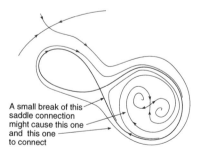

A small break of this
saddle connection
might cause this one
and this one
to connect

FIGURE 8*.6.5. A bad saddle connection where a break creates another saddle connection. ▲

Proof of Lemma. The following argument, suggested by John Gucken-heimer, will show two statements:

(a) For any integer k and any perturbation $\mathbf{f}_k$ of $\mathbf{f}$ that coincides with $\mathbf{f}$ outside V and such that every saddle connection crossing V crosses V at least k times, there exists an arbitrarily small perturbation of $\mathbf{f}_{k+1}$ such that every saddle connection for $\mathbf{f}_{k+1}$ crossing V crosses V at least $k+1$ times.

(b) For any k and any $\mathbf{f}_k$ of $\mathbf{f}$ that coincides with $\mathbf{f}$ outside V and such that every saddle connection crossing V crosses V at least k times, there exists $\varepsilon > 0$ such that any perturbation $\mathbf{g}$ of $\mathbf{f}_k$ with $\|\mathbf{g} - \mathbf{f}_k\| < \varepsilon$, any saddle connection for $\mathbf{g}$ crossing V crosses at least k times.

Using these two statements, we prove Lemma 8*.6.10 as follows:

Given $\delta > 0$, find a sequence of perturbations $\mathbf{f}_2$ of $\mathbf{f}_1 = \mathbf{f}$, then $\mathbf{f}_2$ of $\mathbf{f}_1$, then $\mathbf{f}_3$ of $\mathbf{f}_2$, and so on, and a sequence of numbers $\delta_1 = \delta, \delta_2, \delta_3, \ldots$, such that

(1) $\|\mathbf{f}_k - \mathbf{f}_{k-1}\| < \delta_{k-1}/2, k = 2, 3, \ldots,$

(2) every saddle connection for $\mathbf{f}_k$ crosses V at least k times,

(3) for any $\mathbf{g}$ with $\|\mathbf{f}_k - \mathbf{g}\| < \delta_k$, all saddle connections for $\mathbf{g}$ crossing V cross V at least k times,

(4) $\delta_k < \delta_{k-1}/2.$

You can choose $\mathbf{f}_k$ satisfying (1) and (2) by property (a), and you can choose δ_k satisfying (3) and (4) by property (b).

Properties (1) and (4) guarantee that the $\mathbf{f}_k$ converge to a vector field $\mathbf{g}$ with $\|\mathbf{g} - \mathbf{f}\| < \delta$, and properties (3) and (4) guarantee that $\mathbf{g}$ has no saddle connections crossing V. This reduces the proof of Lemma 8*.6.10 to proving (a) and (b).

Remark. In this proof, we made separatrices cross V more and more times. In the limit, they must cross V infinitely many times. This implies that their ω-limit sets are cycles.

The proof of (a) goes as follows: Mark on the boundary of V the points where separatrices enter and leave V on their first k crossings (if they occur): this is a finite set of points $Z = Z' \sqcup Z''$, where Z' is the subset of Z on saddle connections crossing V in exactly k crosses and $Z'' = Z - Z'$. Next, choose a neighborhood $W \subset V$ of all k crossings of all the saddle connections crossing V in exactly k segments and intersecting no trajectory segment through a point of Z''.

Now perturb the vector field in W so little that the trajectories first entering V at points of Z' cross V k times in W, but after k crosses they do not connect points of Z' to points of Z'.

This perturbation satisfies our requirement: any saddle connection crossing V must enter at some point of Z; if it entered at a point of Z'', its trajectory has not been modified for the first k crossings of V (if they occur), and after k such crossings, it does not connect with another point Z. If it enters at a point of Z', then after k crossings it does not connect with a point of Z, so again it must take more than k crossings to connect to a saddle. This proves (a).

Part (b) is much easier. Mark on the boundary of V the points where stable separatrices depart to go to saddles without reentering V, and the first k exit points of unstable separatrices. Our hypothesis says that these two finite sets of points are disjoint, but both move continuously with the vector field, so they will still be disjoint after a sufficiently small perturbation. □

Using Lemma 8*.6.10, it is easy to make an arbitrarily small perturbation of the vector field so that it will have no saddle connections. By the Poincaré–Bendixson Theorem (in its strong form, Theorem 8*.5.6), we see that all separatrices must now either go to sinks or sources, accumulate on cycles, or leave the region U.

Now we need to make the cycles linearly attracting or repelling. There may well be infinitely many limit cycles, but they can be classified into finitely many classes as follows:

Declare two cycles to be equivalent ("homotopic in the complement of the zeroes of $\mathbf{f}$" is the standard expression) if they separate the same zeroes of $\mathbf{f}$. Then there are only finitely many classes, since the (finite) set of zeroes of $\mathbf{f}$ can be partitioned in only finitely many ways. Moreover, for each equivalence class Γ, we can consider V_Γ the union of the annular regions bounded by cycles $\gamma \in \Gamma$.

Lemma 8*.6.11. *Either the class Γ contains a single cycle, or V_Γ is an annular region bounded by two elements of Γ (the innermost and the outermost cycles in Γ).*

Remark. For this to be true, it is essential that we first eliminate the saddle connections, for otherwise the boundaries of such regions V_Γ might well be made up of saddle connections on which cycles in Γ accumulate. That would make the next argument much more delicate.

Proof of Lemma. Suppose $\mathbf{x}_i \in \gamma_i$ are points of $\gamma_i \in \Gamma$ converging to $\mathbf{x} \in \partial V_\Gamma$. Now apply the Poincaré–Bendixson theorem to $\mathbf{x}$: both the $L_\alpha(\mathbf{x})$ and $L_\omega(\mathbf{x})$ must either be cycles or contain a zero of $\mathbf{f}$.

Suppose first that $L_\omega(\mathbf{x})$ is a cycle δ, so that there exists a sequence $t_i \to \infty$ such that $\phi_{\mathbf{f}}(t_i, \mathbf{x})$ converges to a point $\mathbf{y} \in \delta$, so $\phi_{\mathbf{f}}(t_i, \mathbf{x}_i)$ also converges to $\mathbf{y}$. Call $\mathbf{y}_i = \phi_{\mathbf{f}}(t_i, \mathbf{x}_i)$.

Next observe that the periods T_i of γ_i converge to the the period T_δ of δ (and not to ∞, as one might fear). Indeed, take a transversal J to δ, which we may assume will contain the y_i. Note that by monotonicity, a cycle can intersect J in at most one point. Then $\phi_{\mathbf{f}}(T_\delta, \mathbf{y}) = \mathbf{y}$, so $\phi_{\mathbf{f}}(T_\delta, \mathbf{y}_i)$ is close to $\mathbf{y}$ and will intersect J again within a time which goes to 0 as $i \to \infty$; since the next intersection is also $\mathbf{y}_i$, this shows that the periods converge.

We may then assume that the t_i are bounded and in fact a convergent sequence, converging to some t_0. Then

$$\mathbf{x} = \lim_{i \to \infty} \phi_{\mathbf{f}}(-t_i, \mathbf{y}_i) = \phi_{\mathbf{f}}(-t_0, \mathbf{y}) \in \delta,$$

and this shows that if $L_\omega(\mathbf{x})$ is a cycle, then every point of ∂V_Γ is on a cycle. The same argument applies if $L_\alpha(\mathbf{x})$ is a cycle.

If neither is a cycle, then both limit sets must contain zeroes of $\mathbf{f}$. In fact, both limit sets must be just one point, which is a saddle of $\mathbf{f}$. Indeed, by Theorem 8*.5.6, $\mathbf{f}$ will vanish on both limit sets of any nontrivial trajectory contained in $L_\omega(\mathbf{x})$, and that would make such a trajectory a saddle connection. But this means that the trajectory through $\mathbf{x}$ is itself a saddle connection unless $\mathbf{x}$ is itself a saddle of $\mathbf{f}$.

If this is the case, then γ_i must accumulate not only on $\mathbf{x}$, but also on at least two of the separatrices emanating from $\mathbf{x}$; we just saw that that means that these are saddle connections, which contradicts our hypothesis. $\square$

Proof of (iii). Now we will modify the flow by a small perturbation near each V_Γ. Choose a section $\gamma : I \to \mathbb{R}^2$ to the flow which goes from inside the inner boundary to outside the outer boundary of V_Γ. There is a subinterval $I_1 \subset I$ also crossing V_Γ such that the Poincaré section mapping $P : I_1 \to I$ is well defined. Note that the cycles of Γ, and in particular the boundary of V_Γ, intersect I_1 in fixed points of P. Let us approximate P by a C^∞ mapping $P^* : I_1 \to I$ which agrees with P outside a small neighborhood of $V_\Gamma \cap I_1$, with all fixed points transverse and having at least one fixed point.

Remark. This last condition is to avoid creating new saddle connections when we eliminate cycles; in Section 9.3 we will see that this is a real possibility.

Lemma 8*.6.12. *There exists a vector field which agrees with $\mathbf{f}$ outside a small neighborhood of V, which is close to $\mathbf{f}$ when P^* is close to P, and with section mapping P^*.*

Proof of Lemma. Embed the rectangle $I_1 \times [0, a]$ into V by $(s, t) \mapsto \phi_{\mathbf{f}}(t, \gamma(s))$; we will modify the vector field only in this rectangle. There are two flows from the right side to the left side of this rectangle: backward through the inside of the rectangle, and forward through the outside. We will use the outside identification to identify the right edge with a subinterval of $J \subset I$, so that the flow through the rectangle connects $s \in I$ to $P(s) \in J$.

We may think of the trajectory through $(0, s)$ as the graph of a function $\alpha_s(t)$. We will replace this trajectory by the graph of the function $(1 - \eta(t))\alpha_s(t) + \eta(t)P^*(s)$, where $\eta(t) \geq 0$ is a C^∞ function defined on $[0, a]$ which is identically 0 near 0 and identically 1 near a. We leave it to the reader that these graphs are indeed trajectories of a vector field $\mathbf{f}^*$, which is an arbitrarily small perturbation of $\mathbf{f}$ when P^* is sufficiently close to P. Moreover, the Poincaré map for this new vector field is P^*, so all cycles of $\mathbf{f}^*$ in V_Γ are linearly attracting or repelling. $\quad\square$

No new saddle connections were created by this construction. The vector field was not modified outside a small neighborhood of the V_Γ, so there can be no saddle connection there. Moreover, each boundary of V_Γ is attracting or repelling, according to whether the graph of P is above or beneath the diagonal on $I_1 - V_\Gamma$. We did not modify P near the edges of I_1, so even after modification, a separatrix that enters V_H on one side can never leave it on the same side. And it can not leave V_H on the other side, since there is a cycle in V_Γ separating the two boundary components for the modified equation. This proves the Density Theorem 8*.6.8. $\quad\square$

Chapter 8* Exercises

Exercises 8*.1 Preliminaries of Structural Stability

8*.1#1.

(a) Find formulas for the inverse of the mappings h_A and h_B in Examples 8*.1.3 and 8*.1.4.

(b) Verify that h_A does indeed map the curve of equation $y = x^3$ to the diagonal. Verify that h_B does map a half-line $\theta = \theta_0$ to a curve which spirals infinitely many times around the origin.

8*.1#2. Write a formula for a homeomorphism $f: \mathbb{R}^2 \to \mathbb{R}^2$ which will map

(a) every ring to a spiral of finite length.

(b) every ring to a spiral of infinite length.

(c) circles centered at the origin to squares.

8*.1#3. With reference to Definitions 8*.1.2, 5 and 6,

(a) Show that differentiable conjugacy of vector fields implies topological conjugacy of the flows.

(b) Show that topological conjugacy of flow implies topological equivalence.

8*.1#4.

(a) Show that if h is a differentiable conjugacy between the vector fields f and g, with $f(x_0) = 0$, then

   (i) $g(h(x_0)) = 0$, and

   (ii) the eigenvalues of the linearizations of f at x_0 and of g at $h(x_0)$ coincide. Hint: Show that

$$d_{x_0} h d_{x_0} f = d_{h(x_0)} g d_{x_0} h$$

by differentiating equation (2) in Definition 8.1.6.

(b) Show that if h is a homeomorphism conjugating the flows of the vector fields f and g, and if γ is a cycle for f of period T, then $h(\gamma)$ is a cycle for g of the same period T.

Exercises 8*.2 Structural Stability of Sinks and Sources

8*.2#1.

(a) Find a homeomorphism $h_\alpha : \mathbb{R} \to \mathbb{R}$ that conjugates the flows of $x' = -x$ and $x' = -\alpha x$ for $\alpha > 0$. That is, flow out the first equation to some point like $x = 1$, then flow back in by the second equation.

(b) When is h_α differentiable? When is h_α^{-1} differentiable? When are both differentiable? How do these results relate to Exercise 8*.1#4?

8*.2#2. Find a homeomorphism $h : \mathbb{R} \to \mathbb{R}$ which conjugates the flows of $x' = -x$ and $x' = -x^3$ for $\alpha > 0$. Show that this function is even less differentiable than the ones found in Exercise 8*.2#1, in the sense that $h(x)/x$ tends to infinity faster.

8*.2#3.

(a) Find a homeomorphism $h_\alpha : I \to \mathbb{R}$, where I is an appropriate neighborhood of the origin, which conjugates the flows of $x' = -ax$ and $x' = -\sin x$ for $a > 0$. You might try flowing out the first equation to $x = \pi/3$, then flowing back in by the second equation. If you do this, the formula you obtain will contain an arccos, and you have to be careful about which branch of the arccos you are using.

(b) Draw graphs of h_α for $\alpha = 1/2, 1, 2$. What is the largest interval I on which h_α is defined? What is the image of h_α? Why can h_α not be extended to a larger interval?

(c) Are h_α and h_α^{-1} differentiable? How do these results relate to Exercise 8*.1#4?

If we set $f_\alpha(x) = -x$ and $g(x) = -\sin x$ (thought of as vector fields on the line), we might also try to find such a homeomorphism by computing

$$\lim_{t \to \infty} \phi_{f_\alpha}(-t, \phi_g(t, x)).$$

(d) Say carefully what the difference is between this formula and the hint in part (a).

(e) Show that the limit above exists only for one value of α. Compute it in that case, and relate your formula to the one found in a.

8*.2#4. In this exercise, we will explore the difference between differentiable conjugacy of vector fields and topological conjugacy of the flows. Let the reader be warned that this exercise is quite difficult, and that both the computations and the concepts involved are pretty deep.

(a) Write explicitly the flows $\phi_\mathbf{f}$, $\phi_\mathbf{g}$, $\phi_\mathbf{k}$, of the differential equations

$$\begin{bmatrix} x \\ y \end{bmatrix}' = \begin{bmatrix} -x \\ -2y \end{bmatrix}, \quad \begin{bmatrix} x \\ y \end{bmatrix}' = \begin{bmatrix} -x \\ -2y + x^3 \end{bmatrix},$$

and

$$\begin{bmatrix} x \\ y \end{bmatrix}' = \begin{bmatrix} -x \\ -2y + x^2 \end{bmatrix}.$$

Note the difference that arises from the fact that undetermined coefficients do not work for the last equation.

(b) Show that the first two vector fields are differentiably conjugate, by computing

$$h(\mathbf{x}) = \lim_{t \to \infty} \phi_\mathbf{f}(-t, \phi_\mathbf{g}(\mathbf{x}, t)).$$

Verify that $h : \mathbb{R}^2 \to \mathbb{R}^2$ is a differentiable homeomorphism with differentiable inverse, and provides a change of variables which turns the second equation into the first.

(c) Try to do the same for the first and the third equation, and show that the corresponding limit does not exist. Try to explain this as follows: the trajectories of the first equation are the curves $y = Cx^2$. What is the equation for the trajectories of the third equation? Can a differentiable homeomorphism with differentiable inverse map the first family of curves to the second?

(d) Show that the flows of the first and third equation are none the less conjugate by a homeomorphism, by

    i. showing that each trajectory crosses the curve C of equation $x^4 + y^2$ in a single point;

    ii. finding the time $T(\mathbf{x})$ such that $\phi_{\mathbf{f}}(T(\mathbf{x}), \mathbf{x}) \in C$;

    iii. writing the formula for $\phi_{\mathbf{k}}(-T(\mathbf{x}), \phi_{\mathbf{f}}(T(\mathbf{x}), \mathbf{x}))$.

    iv. showing that this formula is a homeomorphism $h : \mathbb{R}^2 \to \mathbb{R}^2$.

(e) How differentiable is h at the origin?

Exercises 8*.3 Time to Pass by a Saddle

8*.3#1. In the proof of Theorem 8*.3.2:

8*.3#2. Let D be the unit disc, and $\mathbf{f}$ be the linear differential equation

$$x' = ax$$
$$y' = -by$$

for $a, b > 0$. Find the entering and leaving parts of the boundary, and compute the arrival, departure and transit time.

8*.3#3. Repeat Exercise 8*.3#2 for the differential equation

$$x' = ax$$
$$y' = -by + x^2$$

for $a, b > 0$. Find the entering and leaving parts of the boundary, and compute the arrival, departure and transit time.

Exercises 8*.4 Structural Stability of Limit Cycles

8*.4#1. Consider the system of differential equations

$$x' = y + \frac{\sin(x^2 + y^2)}{x^2 + y^2}x$$

$$y' = -x + \frac{\sin(x^2 + y^2)}{x^2 + y^2}y$$

(a) Show that the positive x-axis is transverse to the flow, and compute the section mapping $P(x)$. Hint: pass to polar coordinates.

(b) What are the fixed points of P, and what are the derivatives of P at these fixed points?

(c) Relate the results of (b) to the limit cycles of the differential equation.

8*.4#2. Confirm the calculations for Example 8*.4.2, the radial equation and its solution.

(a) First confirm the radial equation (12). Hint: to translate into polar coordinates, see equation (18) in Example 8.4.1.

(b) Second, confirm the solution (13). Hint: use partial fractions.

Exercises 8*.5 Poincaré–Bendixson Rules Out Chaos in the Plane

8*.5#1. There is another way of defining $L_\omega(\mathbf{x}_p)$:

$$L_\omega(\mathbf{x}_p) = \cap_{s\to\infty} \overline{\phi_{\mathbf{f}}([s, \infty), \mathbf{x}_p)},$$

and similarly

$$L_\alpha(\mathbf{u}) = \cap_{s\to-\infty} \overline{\phi_{\mathbf{f}}((-\infty, s], \mathbf{x}_p)}.$$

8*.5#2. Complete the proof of Lemma 8*.5.8.

8*.5#3. For Example 8*.5.9, show that the white regions crossing the axis form a Cantor set.

8*.5#4. For a labyrinth in $\mathbb{R}^3$, make a can X out of lined paper, so that the lines are vertical on the sides and let $2\pi\alpha$ be the angle between the lines on the top and the bottom. The space X is topologically a sphere, and there are trajectories on it gotten by following the lines. Which four of these trajectories are exceptional?

Exercises 8*.6 Structural Stability in the Plane

8*.6#1. The object of this exercise is to prove a strong form of the Weierstrass approximation theorem, which is an important and rather difficult result.

Theorem (Strong Weierstrass). *If f is a k-times continuously differentiable function on $\mathbb{R}^n$, then given any bound $R > 0$, the function f can be approximated on $\|\mathbf{x}\| \leq R$ by a polynomial, and further, the derivatives of this polynomial up to order k approximate the derivatives of f.*

(a) Show that given such a function f and $R > 0$, there exists a k-times continuously differentiable function f_1 such that

$$f_1(\mathbf{x}) = \begin{cases} f(\mathbf{x}) & \text{if } \|\mathbf{x}\| \leq R/2 \\ 0 & \text{if } \|\mathbf{x}\| \geq R. \end{cases}$$

Try this first in dimension one, and then think of polar coordinates in dimension two.

We will assume from here on that $f = f_1$, so that $f(\mathbf{x}) = 0$ when $\|\mathbf{x}\| \geq R$.

(b) Let f be a continuous function of one variable and $p(x)$ a polynomial. Show that the function

$$g(x) = \int_{-\infty}^{\infty} p(x-y)f(y)dy$$

is a polynomial.

(c) Do the same if f is a function of $\mathbf{x} = (x_1, \ldots, x_n)$ vanishing when $\|\mathbf{x}\| \geq R$ and p is a polynomial in $(x_1, \ldots, x_n)$.

Thus the idea is to find polynomials $p_\epsilon(\mathbf{x})$ such that

$$g_\epsilon(\mathbf{x}) = \int_{-\infty}^{\infty} \cdots \int_{-\infty}^{\infty} p_\epsilon(\mathbf{x} - \mathbf{y})f(\mathbf{y})dy_1 \ldots dy_n$$

satisfies $|g_\epsilon(\mathbf{x}) - f(\mathbf{x})| \leq \epsilon$ for any $\mathbf{x}$ with $\|\mathbf{x}\| \leq R$. One way to do this is to try to make a polynomial bump function, with graph something like a tower in the middle of a plain. With polynomials this is impossible, since polynomials always tend to ∞ at ∞; but if our tower is surrounded by a big enough plain, then the surrounding mountains will not influence $g(\mathbf{x})$ when $\mathbf{x} \leq R$.

(d) Let $p(x) = 1 - x^2$. Show that for any $\varepsilon > 0$, there exist N, ρ and C such that the function

$$p_N(x) = C(p(\rho x))^N$$

satisfies

$$\int_{-2R}^{2R} p_N(x)dx = 1, \qquad |p_N(x)| < \varepsilon \text{ when } \varepsilon < |x| < 2R.$$

(e) Show the n-dimensional analog of (d) for the polynomial

$$q_N(x_1, \ldots x_n) = p_N(x_1) \cdot \ldots \cdot p_N(x_n).$$

(f) Show that

$$g_\varepsilon(\mathbf{x}) = \int_{-\infty}^{\infty} \cdots \int_{-\infty}^{\infty} q_N(\mathbf{x} - \mathbf{y}) f(\mathbf{y}) dy_1 \ldots dy_n$$

is a polynomial which approximates f for $\|\mathbf{x}\| \le R$.

(g) Show that the derivatives of g_N of order at most K approximate the derivatives of f for $\|\mathbf{x}\| \le R$. by differentiating under the integral sign.

8*.6#2. The object of this exercise is to outline a proof of the following theorem.

Theorem. *If f is a k times continuously differentiable function on $\mathbb{R}^n$, then for any $R > 0$ there exists a trigonometric polynomial which approximates f on $\|f\| \le R$ together with its first k derivatives.*

This should remind you of Dirichlet's theorem giving the convergence of Fourier series, and the proof is closely related. However, it is a bit fussy to go directly from Dirichlet's theorem to this result, and we will sketch a different method, very similar to the proof above. The first step (a), to replace f by a function which vanishes when $\|\mathbf{x}\| > R$, is identical.

(b) Show that for any $0 < r < R$ and $\varepsilon > 0$, there exists N and C such that the function

$$p_N(\mathbf{x}) = C \left(\cos\left(\frac{\pi}{2R} \|x\|\right) + 1 \right)^N$$

satisfies

$$\int_{B_R} p_{N,\rho}(\mathbf{x}) dx_1 \ldots dx_n = 1,$$

where $B_R = \{\mathbf{x} \in \mathbb{R}^n \mid \|\mathbf{x}\| < R\}$, and $p_{N,\rho}(\mathbf{x}) < \varepsilon$ when $r \le \|\mathbf{x}\| < R$. Can you compute C exactly when $n = 1$?

(c) Show that the function g_N defined by

$$g_N(x) = \int_{\mathbb{R}^n} f(\mathbf{x}) p_N(\mathbf{y} - \mathbf{x}) dx_1 \ldots dx_n$$

converges to f as $N \to \infty$, together with its first k derivatives.

(d) Show that $g_N(x)$ is a trigonometric polynomial.

9

Bifurcations

Consider an autonomous differential equation in $\mathbb{R}^2$, with a parameter α:

$$\frac{dx}{dt} = f(x, y, \alpha)$$
$$\frac{dy}{dt} = g(x, y, \alpha),$$

or, more generally, several parameters:

$$\frac{dx}{dt} = f(x, y, \alpha_1, \alpha_2, \ldots)$$
$$\frac{dy}{dt} = g(x, y, \alpha_1, \alpha_2, \ldots).$$

In general, the phase portrait changes gradually as the parameters vary. But there will usually be values of the parameter for which the phase portrait changes drastically, "revolutionary" values where the entire social order is changed in an instant. For instance, if a sink becomes a source, the solutions starting at initial values in the basin of the sink will have to change their minds and decide to do something else. These abrupt changes in the phase portrait due to a changing parameter are called *bifurcations*.

Before going further, we must insist on the distinction between the *dynamical variables* x and y, and the *parameters* $\alpha_1, \alpha_2, \ldots$. The number of dynamic variables is here restricted to *two*. It makes perfectly good sense to try to understand bifurcations for differential equations in any number of dimensions, and we have seen it in one dimension as in the hunting Example 2.5.2 in Part I. However, we do not understand bifurcations in more than two dimensions: there is no classification of the bifurcations in even three dimensions, and no clear indication that there ever will be such a classification. The number of parameters is much less critical; essentially for reasons of convenience, we will consider only one- and two-parameter families, but similar techniques would go through (with greater effort) for families depending on more parameters.

In order for a parameter value to be revolutionary, at least one of the following four exceptional behaviors must occur:

(1) a zero has linearization with a zero eigenvalue;

(2) a zero has linearization with a pair of purely imaginary eigenvalues;

(3) there is a saddle connection;

(4) there is a cycle that is not linearly attracting or repelling.

In Chapter 8.8 we explained why, when none of these occur, flow of the vector field is stable, and we proved the results in Chapter 8*.6.

We will see that corresponding to each of the four events listed, there is a bifurcation that can occur "in general" in one-parameter families. They are, respectively,

    (1) saddle-node bifurcations: Section 9.1

    (2) Andronov-Hopf bifurcations: Section 9.2

    (3) saddle connection: Section 9.3

    (4) semi-stable limit cycles: Section 9.4

The first two *local bifurcations* were already mentioned as exceptions to Principle 8.1.6 and studied in Examples 8.1.7 and 8.1.8. Saddle connections and degenerate limit cycles are harder to study, because they are *global bifurcations* of the equation; they concern the global behavior of the equation, not just the behavior near a zero of the equation.

Figures 9.0.1 and 9.0.2 illustrate how the presence of a saddle connection or a degenerate limit cycle can cause revolutionary change in the phase plane portrait of a differential equation.

A stable separatrix spirals backward to a source,

then saddle connects (solutions accumulate on the saddle connection),

then an unstable separatrix accumulates on a newly created attracting limit cycle

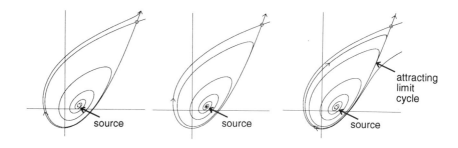

$\longrightarrow$ varying parameter $\longrightarrow$

FIGURE 9.0.1. A saddle connection before, during, and after bifurcation. The "revolutionary" value of the parameter that creates the center picture separates different behaviors "before" and "after" the bifurcation.

Two limit cycles   that collide   and disappear.

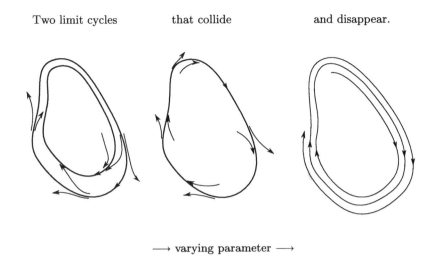

$\longrightarrow$ varying parameter $\longrightarrow$

FIGURE 9.0.2. Two limit cycles that collide and disappear, causing a degenerate limit cycle at the "revolutionary" value of the parameter, shown in the center picture.

You will see that limit cycles often occur in bifurcations; therefore, wherever you find a limit cycle is a good place to look for bifurcation behavior.

BIFURCATION DIAGRAMS

The main object of this chapter is not so much to study the individual bifurcations for their own sake but to *understand families of differential equations that depend on parameters*. In practice, this means drawing the *bifurcation diagram* for such a parametrized family of differential equations: finding the locus of "revolutionary values."

We have already studied one bifurcation diagram: the one corresponding to linear differential equations

$$\begin{bmatrix} x \\ y \end{bmatrix}' = A \begin{bmatrix} x \\ y \end{bmatrix} = \begin{bmatrix} a & b \\ c & d \end{bmatrix} \begin{bmatrix} x \\ y \end{bmatrix}$$

which was analyzed in detail in Section 7.5.

This really forms a four-parameter family (the parameters are a, b, c, and d), and even here, trying to imagine the locus in $\mathbb{R}^4$ where the determinant vanishes (i.e., given in $\mathbb{R}^4$ by the equation $ad - bc = 0$) is quite a mind-stretcher, so we will study a two-parameter family instead, which displays most of the features of the general case [since the behavior of the linear equation is almost always controlled by the trace (α) and the determinant (β) of A, as you can recall from Section 7.5 and Figure 7.5.7, the bifurcation diagram for two-dimensional linear systems].

Example 9.0.1. Consider the two-parameter family of differential equations

$$\begin{bmatrix} x \\ y \end{bmatrix}' = \begin{bmatrix} \tau & \delta \\ -1 & 0 \end{bmatrix} \begin{bmatrix} x \\ y \end{bmatrix}, \tag{1}$$

cooked up so that the trace is τ and the determinant is δ. Many different phase portraits occur for this family, depending on the value of the parameters τ and δ; a few are shown in Figure 9.0.3. The picture in the (τ, δ) parameter plane comes from Figure 7.5.7 and is shown in Figure 9.0.4.

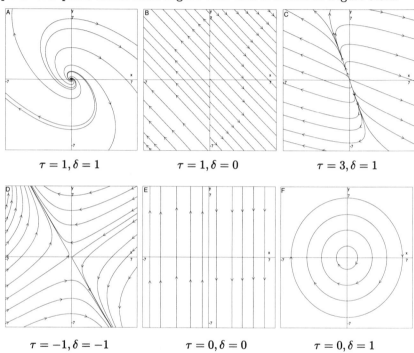

$\tau = 1, \delta = 1$	$\tau = 1, \delta = 0$	$\tau = 3, \delta = 1$
$\tau = -1, \delta = -1$	$\tau = 0, \delta = 0$	$\tau = 0, \delta = 1$

FIGURE 9.0.3. Different phase plane behaviors for $x' = \tau x + \delta y$, $y' = x$, for different values of the parameters τ and δ.

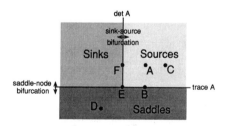

FIGURE 9.0.4. The bifurcation diagram for equation (1). The six labeled points give values of the parameters corresponding to the phase portraits in Figure 9.0.3.

▲

For nonlinear systems,

$$\frac{dx}{dt} = f(x, y, \alpha_1, \alpha_2, \ldots, \alpha_n)$$

$$\frac{dy}{dt} = g(x, y, \alpha_1, \alpha_2, \ldots, \alpha_n),$$

the bifurcations will usually occur on hyperspaces in the n-dimensional parameter space (points if $n = 1$, curves if $n = 2$, surfaces if $n = 3$, etc.). Suppose $n = 2$; then we will be interested in finding, in the $\alpha_1\alpha_2$-parameter plane, the bifurcation curves along which the differential equation changes abruptly. These curves define regions in the parameter space in each of which all the corresponding differential equations have roughly the same behavior (shown earlier, for example, in Figure 8.8.1, where we discussed *structural stability*). That is:

In each region defined by bifurcation curves in parameter space, the corresponding differential equations have the same numbers of sinks, sources, saddles, limit cycles, and basins, related to each other in the same ways.

OVERVIEW

We will devote each of Sections 9.1–9.4 to one of the four bifurcations. Then we summarize and put them all together in Section 9.5, where we will study several equations depending on a single parameter, and show how and where the phase diagrams bifurcate. In Section 9.6, we will introduce *two*-parameter families, which have more degenerate bifurcations, and construct global *bifurcation diagrams* for these situations. We close with a grand example in Section 9.7.

Our goal with Chapter 9 is to show that for a general two-dimensional system of differential equations depending on parameters, there is in the parameter space a bifurcation drawing, somewhat similar to Figure 9.0.4, when there are *two* parameters. Finding this bifurcation diagram for a given family is tantamount to "understanding the family of differential equations."

PREVIEW SUMMARY

For easy reference we have created here a summary chart of the bifurcations that will be discussed in each of the next four sections. Do not expect to understand this summary all at once! But as you read each of the sections that are devoted to explaining one of the bifurcations, you will be able to use this chart to get some perspective and keep track of what is going on.

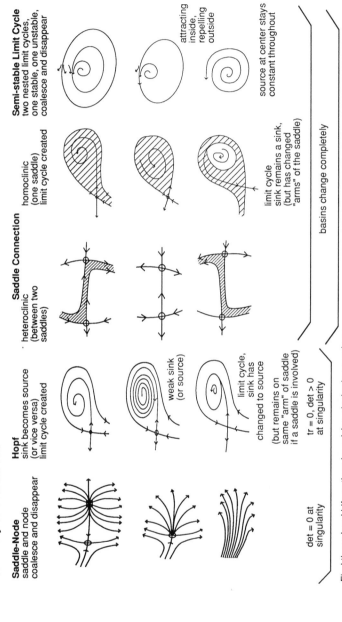

FIGURE 9.0.5.

9.1 Saddle-Node Bifurcation

The saddle-node bifurcation is the most important of all: it is the bifurcation in which zeroes of vector fields are created and destroyed.

Definition 9.1.1 (Saddle-node bifurcation). A *saddle-node bifurcation* is what occurs, in general, when a zero of a vector field has a linearization with *zero determinant*.

A zero determinant implies that at least one of the eigenvalues is zero. Later in this section, the definition's phrase "in general" will be amplified by specific Nondegeneracy Conditions 9.1.6 and 9.1.7. Meanwhile, we shall explain the generic saddle-node, where we can assume one eigenvalue is zero and the other is not:

Suppose a vector field depends on some parameter α, and $(x(\alpha), y(\alpha))$ is a zero of the vector field. Then there is a corresponding linearization with coefficient matrix $A(\alpha)$, which in turn defines a point

$$(\operatorname{tr} A(\alpha), \det A(\alpha))$$

which moves along a curve in the (trace, determinant)-plane as α varies. You might think from the bifurcation diagram for linear differential equations in $\mathbb{R}^2$ (as in Example 9.0.1) that if this curve hits the line $\det A = 0$ for some value α_0 of the parameter, the zero of the vector field just changes from being a saddle to being a node, but this is infinitely exceptional. Usually, the zero under consideration collides with some other zero of the vector field, and the two destroy each other like a particle and an antiparticle, disappearing into thin air.

More precisely, in the most usual case:

(i) For α to one side of α_0, there are two zeroes of the vector field; a node and a saddle both exist.

(ii) At $\alpha = \alpha_0$ there is only one zero; the node and saddle coalesce.

(iii) For α to the other side of α_0, these zeroes of the vector field no longer exist; the node and saddle simply disappear.

Example 9.1.2. Consider the differential equation

$$\frac{dx}{dt} = y$$
$$\frac{dy}{dt} = x^2 - y + \alpha,$$

which undergoes bifurcation for $\alpha_0 = 0$, as shown in Figure 9.1.1.

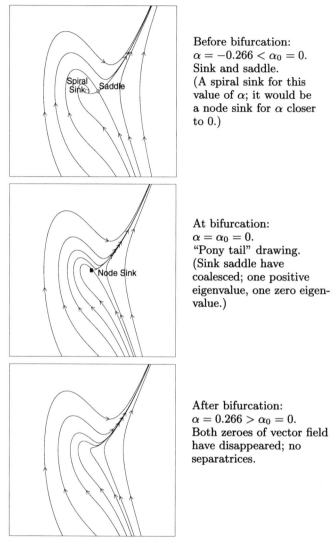

Before bifurcation:
$\alpha = -0.266 < \alpha_0 = 0$.
Sink and saddle.
(A spiral sink for this
value of α; it would be
a node sink for α closer
to 0.)

At bifurcation:
$\alpha = \alpha_0 = 0$.
"Pony tail" drawing.
(Sink saddle have
coalesced; one positive
eigenvalue, one zero eigenvalue.)

After bifurcation:
$\alpha = 0.266 > \alpha_0 = 0$.
Both zeroes of vector field
have disappeared; no
separatrices.

FIGURE 9.1.1. Saddle-node bifurcation for $x' = y$, $y' = x^2 - y + \alpha$.

You are asked in Exercise 9.1.#3 to check that the zeroes of the vector
field are of the type described and to explore the "pony tail" behavior,
which you may recall from Part I, Section 2.7.   ▲

Figure 9.1.2 reproduces the middle picture of Figure 9.1.1. for closer
examination of the actual saddle-node: The original stable separatrix from
the saddle remains as a stable separatrix of the saddle-node, tangent to

the eigendirection for the nonzero eigenvalue. But the original unstable separatrix from the saddle exists only on one side, and we call that the exceptional saddle node trajectory. On the other side is the so-called pony tail of undistinguished trajectories.

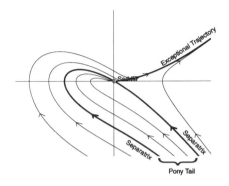

FIGURE 9.1.2. Anatomy of a saddle-node.

All solutions in the pony tail are tangent to the eigendirection with eigenvalue zero, as is the exceptional trajectory.

In Figure 9.1.2, this tangency to the horizontal axis may be hard to see; Exercise 9.1.#3c asks you to blow up further the equations of the example. Look ahead to Figure 9.1.8 for the generic saddle-node picture; Figure 9.1.2 shows how it can be twisted.

In this case, there is no *distinguished* solution in the ponytail:

No *solution in the pony tail naturally continues the exceptional solution,*

in the sense that it is given by the same convergent power series. [For this example, this was proved in Chapter 2 (in Part I), as you can show in Exercise 9.1.#d]. Such behavior is usually but not always the case. For instance, Exercise 9.1#5 gives an example of a saddle-node in which one solution in the pony tail and the exceptional solution are given by a single formula.

Note that there are two kinds of saddle-nodes. Figure 9.1.2 is an example where the unstable separatrix of the saddle (corresponding to a positive nonzero eigenvalue) becomes the exceptional trajectory. Reversing all the arrows will give the opposite type, where it is the stable separatrix of the saddle (corresponding to a negative nonzero eigenvalue) that becomes the exceptional trajectory.

FINDING SADDLE-NODES

It is quite easy to understand how two singularities can coalesce and disappear if you consider the isoclines along which the vector field is horizontal and vertical. The zeroes of a vector field are the points where these isoclines intersect; possibilities are shown in Figure 9.1.3. If the isoclines for two different slopes (e.g., horizontal and vertical) are smooth curves depending on a parameter α, the number of intersections remains constant unless the parameter passes through a value α_0 where the curves are tangent. At such a point of tangency, in general, two intersection points coalesce and disappear, as in the last picture of Figure 9.1.1.

FIGURE 9.1.3. Possible interactions of isoclines of horizontal and vertical slopes: intersection, tangency, nonintersection.

In Figure 9.1.3 the phase plane for the left-hand possibility has five different incline regions (see Section 6.1); then as you move through the middle possibility, which defines four regions, to the right-hand possibility, which defines only three regions, the phase plane pictures become simpler, until there is no singularity at all.

Example 9.1.3. For the differential equation of Example 9.1.2, with values of α before, at, and after saddle-node bifurcation, the isoclines of horizontal and vertical slopes in the phase plane look like Figure 9.1.4.

> *you can expect a saddle-node bifurcation any time the isoclines for horizontal and vertical slopes have an ordinary tangency.*

More elaborate tangencies, for instance tangencies where the curves cross, lead to different bifurcations, which do not, in general, appear in one-parameter families. They will occur, however, if the equation has symmetries or satisfies other restrictions. See the subsection below on degenerate saddle nodes, and Section 9.6, with Exercises 9.1#8–10, 9.1#12, and 9.6#6.

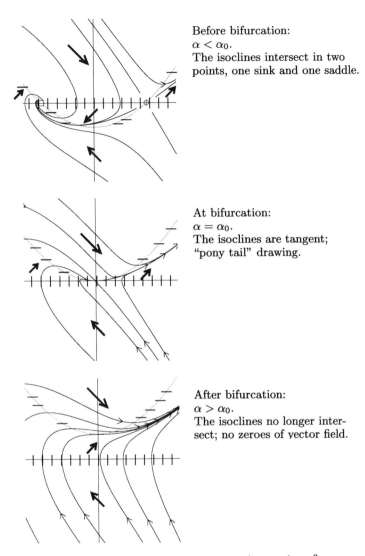

Before bifurcation:
$\alpha < \alpha_0$.
The isoclines intersect in two
points, one sink and one saddle.

At bifurcation:
$\alpha = \alpha_0$.
The isoclines are tangent;
"pony tail" drawing.

After bifurcation:
$\alpha > \alpha_0$.
The isoclines no longer intersect; no zeroes of vector field.

FIGURE 9.1.4. How saddle-node bifurcation occurs for $x' = y$, $y' = x^2 - y + \alpha$.

SADDLE-NODES IN ONE DIMENSION

Saddle-nodes are not restricted to dimension two; they occur in all dimensions, and the simplest is dimension one. We have not emphasized sinks and sources in dimension one, largely because they are so simple; they are covered as a special case in Section 8.3. A zero of $x' = f(x)$ is, of course, a value x_0 of x such that $f(x_0) = 0$; it will be a sink if $f'(x_0) < 0$ and a source if $f'(x_0) > 0$. In the tx-plane, for any $\varepsilon > 0$, the lines $x = x_0 + \varepsilon$ and $x = x_0 - \varepsilon$ will then delimit a funnel (in the case of a sink) or an antifunnel (in the case of a source), as shown in Figure 9.1.5. Furthermore, the antifunnel will have the uniqueness property of Theorem 1.4.5.

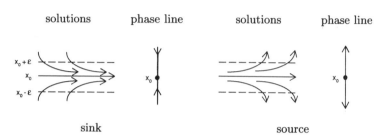

FIGURE 9.1.5. One-dimensional sinks and sources, with phase lines (on the right in each case).

To a one-dimensional autonomous equation $x' = f(x)$ there corresponds a (pretty dull) drawing in the phase line, with sinks and sources appearing as in the vertical lines at the right of each solutions drawing in Figure 9.1.5. (In dimension one, there is no phase plane, just a phase *line*, parametrized by the dependent variable x.)

In one dimension, it would be more reasonable to talk of sink-source bifurcation, instead of a saddle-node: In one dimension, there are no saddles, and the bifurcation occurs when a sink and a source collide and annihilate each other.

Example 9.1.4. The simplest family of equations for which saddle-node bifurcation occurs is

$$x' = \alpha - x^2.$$

Clearly, this equation has no equilibria if $\alpha < 0$, and two if $\alpha > 0$, at $\pm\sqrt{\alpha}$; $+\sqrt{\alpha}$ is a sink and $-\sqrt{\alpha}$ is a source. Computer drawings in the tx-plane for $\alpha = -0.5, 0, 0.5$ look like Figure 9.1.6, with phase lines shown to the right of each solutions picture.

It is clear from these pictures, and not much harder to verify by explicitly calculating solutions, that when α decreases to 0, the source and the sink coalesce, forming an equilibrium at $x = 0$ which is semi-stable, in the sense that solutions that start at a positive initial value are attracted to it and those that start at a negative initial value are repelled.

solutions drawing   phase line

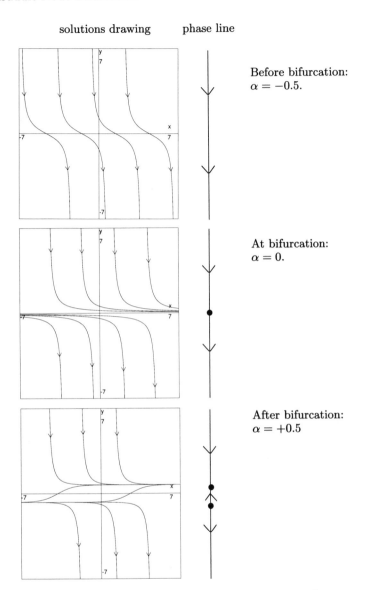

Before bifurcation:
$\alpha = -0.5$.

At bifurcation:
$\alpha = 0$.

After bifurcation:
$\alpha = +0.5$

FIGURE 9.1.6. Sink-source bifurcation, for $x' = \alpha - x^2$. ▲

GENERIC VERSUS DEGENERATE SADDLE-NODES

In this section, we will discuss what "in general" means. We will not give a precise definition of the term, but it should be fairly easy to get an intuitive grasp of the notion, and the discussion below should help.

Genericity is easiest in one dimension. If we have a differential equation

$$x' = f_\alpha(x)$$

depending on the parameter α, then the equilibria or zeroes for any value of α are the solutions of $f_\alpha(x) = 0$.

It should be clear to the reader that, in general, these zeroes will be nondegenerate; i.e., when $f_\alpha(x_0) = 0$, then $f'_\alpha(x_0) \neq 0$. But sometimes, as α varies, the graph of f_α will become tangent to the x-axis and $f'_\alpha(x_0) = 0$; this is unavoidable in one-parameter families, where a graph cannot deform or move from a position where it intersects the x-axis to a position where it does not without going through a position where they are tangent, as shown in Figure 9.1.7.

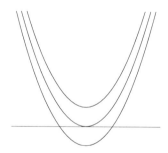

FIGURE 9.1.7. As $f_\alpha(x)$ moves from crossing the x-axis twice to not crossing it at all, it must go through a position where it is tangent to the x-axis.

However, when $f_\alpha(x_0) = f'_\alpha(x_0) = 0$ occurs at x_0 for some parameter value α_0, we would expect $f''_{\alpha_0}(x_0) \neq 0$ in general. When this is the case, the differential equation $x' = f_\alpha(x)$ undergoes a saddle-node bifurcation, exactly as in Example 9.1.4.

Of course, in an exceptional family, at some zero the second derivative might vanish when the first derivative vanishes; for instance, $f_\alpha(x) = x^3 - \alpha x$ has, when $\alpha = 0$, $f_0(0) = f'_0(0) = f''_0(0) = 0$. The geometric significance is that three zeroes for $\alpha > 0$ coalesce at 0 when $\alpha = 0$, leaving just one root for $\alpha < 0$. Thus, the differential equation

$$x' = x^3 - \alpha x$$

has a more complicated bifurcation than a saddle-node: *three* zeroes coalesce rather than two. This is a *degenerate* bifurcation (in fact, a *pitchfork* bifurcation, such as will be shown in Exercises 9.1#8–10). But such degenerate or unusual behavior will not happen in a *generic* family depending on one parameter.

In two dimensions, genericity is a bit more complicated. It is possible to define degeneracy in terms of appropriate first and second derivatives, but this will be easier if we first put the equation into a standard form. We may assume, by a translation and linear *change of variables*, that

- the equilibrium point is at 0;

- the x-axis is the eigendirection for the eigenvalue 0;

- the y-axis is the eigendirection for the eigenvalue λ.

Then the system of differential equations can be written

$$x' = 0 + P(x, y)$$
$$y' = \lambda y + Q(x, y), \tag{2}$$

where $P(x, y)$ and $Q(x, y)$ both start with at least quadratic terms. In particular,

$$P(x, y) = p_{2,0}x^2 + p_{1,1}xy + p_{0,2}y^2 + \text{(higher order terms)}.$$

Generic will mean that two numbers, λ and $p_{2,0}$, do not vanish.

9.1.5 First Nondegeneracy Condition: *The value 0 is a simple eigenvalue of the linearization, the other eigenvalue λ is nonzero.*

9.1.6 Second Nondegeneracy Condition: *The coefficient $p_{2,0}$ of x^2 in the Taylor expansion of x' is nonzero.*

There are two basic configurations for a saddle-node bifurcation, depending on the sign of λ, as you can show in Exercise 9.1#6. Reflection about the y-axis (i.e., changing to the variables $x_1 = -x$, $y_1 = y$) will change the sign of $p_{2,0}$, and reversing the direction of time changes the sign of λ. For convenience, we shall assume $\lambda < 0$ and $p_{2,0} > 0$, as pictured in Figure 9.1.8. System (2) is based on this configuration, but you can adapt to one of the others by reversing t or reversing x.

Removing the nondegeneracy conditions opens a Pandora's box of possibilities. Practically anything might happen, but some behaviors are more common than others; these are discussed in Section 9.6 on two-parameter families. Here we will just give a brief comment about each.

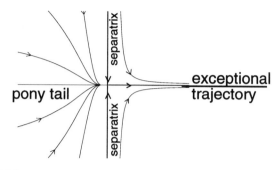

FIGURE 9.1.8. Generic saddle-node for $\lambda < 0$, $p_{2,0} > 0$.

If the First Nondegeneracy Condition 9.1.5 fails, both eigenvalues are zero. The linearization is then usually not diagonalizable, so the form (2) is not a reasonable normal form. The example

$$x' = y$$
$$y' = x^2 + xy$$

is quite typical, and the corresponding phase plane is represented in the first drawing in Figure 9.1.9.

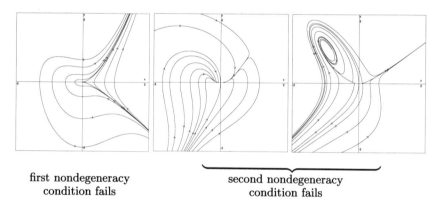

first nondegeneracy
condition fails

second nondegeneracy
condition fails

FIGURE 9.1.9. Degenerate saddle-nodes: some examples.

If the Second Nondegeneracy Condition 9.1.6 fails, then the locus where the vector field is vertical will have equation

$$p_{1,1}xy + p_{0,2}y^2 + \ldots = 0,$$

which will, when $p_{1,1} \neq 0$, consist of two curves C_1^V and C_2^V, through the origin, with C_1^V having slope 0 at the origin and the other not. The

locus where the vector field is horizontal is a smooth curve C^H through
the origin when $\lambda \neq 0$, usually having an ordinary tangency with C_1^V. See
Figure 9.1.10 and Exercise 9.1#7.

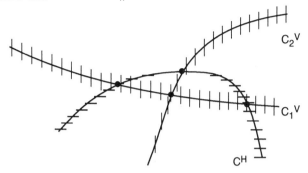

FIGURE 9.1.10. Isocline of horizontal slope C^H intersecting isoclines of vertical
slope C_1^V nd C_2^V.

Then it is clear that a small perturbation of the vector field, for instance
adding a small constant to y', will create *three* zeroes, so that this cannot be
a standard saddle-node. The second and third drawings of Figure 9.1.9 give
typical pictures: they look like a node sink or a saddle, with greatly flattened
tangencies. Exercise 9.1.#12 shows that which picture to expect depends
on the sign of $q_{2,0}p_{1,1} + p_{3,0}$. Warning: Exercise 9.1#12 is designed to check
your understanding of the serious explanation of saddle node behavior that
follows, going through similar computations in a more difficult setting.
Exercise 9.6#6 gives a different proof of the same result.

SERIOUS EXPLANATION OF SADDLE-NODE BEHAVIOR

Theorem 9.1.7. *Under the two nondegeneracy assumptions 9.1.5 and
9.1.6, with $\lambda < 0$ and $p_{2,0} > 0$, there exist unique trajectories which tend to
0 tangentially to the y-axis from both sides.*

(i) *These trajectories, together with the origin, form a smooth curve
called the separatrix of the saddle node.*

(ii) *All trajectories to the left of this separatrix tend to 0 tangentially to
the x-axis, forming the pony tail.*

(iii) *There is a unique exceptional trajectory to the right which emanates
from the origin, also tangential to the x-axis.*

In other words, Theorem 9.1.7 gives the generic saddle-node of Figure
9.1.8.

Proof. The proof of Theorem 9.1.7 is essentially a replay of that for Theorem 8.3.2 about saddles; we repeatedly use funnels and antifunnels. Rather

than prove everything, we will sketch the main points; the details are left as exercises, corresponding to the similar proof of Theorem 8.3.2.

First, we turn the system of equations (2) into two first order equations, the first for y as a function of x, and the second for x as a function of y.

$$\frac{dy}{dx} = \frac{\lambda y + Q(x,y)}{P(x,y)}, \tag{3}$$

$$\frac{dx}{dy} = \frac{P(x,y)}{\lambda y + Q(x,y)}. \tag{4}$$

Of course, we need to see where these equations are defined, since the denominators can vanish. For any $\eta, \beta > 0$, let S be the square $|x|, |y| \leq \eta$, and let the curves $y = \pm\beta x^2$ subdivide S into regions U_1, U_2, U_3, and U_4 as indicated in Figure 9.1.11.

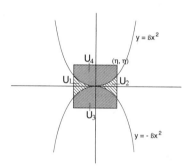

FIGURE 9.1.11. Regions defined by $y = \pm\beta x^2$.

We proceed to prove the theorem in stages, with three lemmas. Lemma 9.1.8 tells where everything is defined.

Lemma 9.1.8. *There exists $\beta_0 > 0$ such that if $\beta > \beta_0$, then there exists $\eta_0(\beta)$ such that if $\eta < \eta_0(\beta)$, then*

(i) *the function $P(x,y)$ does not vanish in U_1 or in U_2 except at the origin;*

(ii) *the function $\lambda y + Q(x,y)$ does not vanish in U_3 or U_4 except at the origin.*

Proof of Lemma. By Taylor's Theorem we can write

$$P(x,y) = p_{2,0}x^2 + p_{1,1}xy + p_{0,2}y^2 + h(x,y),$$

where $h(x,y)$ satisfies the following property:

For all $\varepsilon > 0$, there exists $\eta_1 > 0$ such that if $|x|, |y| < \eta_1$, then

$$|h(x,y)| \leq \varepsilon(x^2 + y^2).$$

For any ε and β, you can choose $\eta_1(\varepsilon, \beta)$ such that for the corresponding U_1 and U_2, we have $|h(x,y)| \leq \varepsilon(1+\beta)x^2$ on $U_1 \cup U_2$. Clearly, there exists $\eta_2(\varepsilon, \beta)$ such that $|p_{1,1}xy + p_{0,2}y^2| \leq \varepsilon x^2$ in $U_1 \cup U_2$. Thus,

$$P(x,y) = p_{2,0}x^2 + p_{1,1}xy + p_{0,2}y^2 + h(x,y) \geq (p_{2,0} - \varepsilon(2+\beta))x^2 \geq 0, \quad (5)$$

which equals zero only when $x = 0$.

Similarly, in U_3 and U_4 we have that $|x| \leq \sqrt{|y/\beta|}$. Write

$$Q(x,y) = q_{2,0}x^2 + q_{1,1}xy + q_{0,2}y^2 + \text{ higher order terms.}$$

The term $q_{2,0}x^2$ contributes to the sum something of order at most y, and this term together with the term λy comprises the dominant terms of $\lambda y + Q(x,y)$. So we have

$$|\lambda y + q_{2,0}x^2| \geq (|\lambda y| - \frac{|q_{2,0}y|}{\beta^2}.$$

Choose β_0 sufficiently large that $(|\lambda| - |q_{2,0}|/\beta_0^2) > 0$; for any $\beta > \beta_0$, there exists $\varepsilon(\beta) > 0$ such that

$$p_{2,0} - \varepsilon(\beta)(2+\beta) > 0 \quad \text{and} \quad |\lambda| - |q_{2,0}/\beta| - \varepsilon(\beta) > 0.$$

By Taylor's Theorem, there exists $\eta_3(\beta)$ such that for the corresponding U_3 and U_4,

$$|\lambda y + Q(x,y)| \geq (|\lambda| - |q_{2,0}/\beta| - \varepsilon(\beta))|y|. \quad (6)$$

Finally, set $\eta_0(\beta)$ to be the smaller of $\eta_1(\varepsilon, \beta)$, $\eta_2(\varepsilon, \beta)$, and $\eta_3(\beta)$. Thus, inequality (5) proves part (i); inequality (6) proves part (ii). $\quad \square$

Thus, Lemma 9.1.8 has shown that equation (3) is well defined in U_1 and U_2, and equation (4) is well defined in U_3 and U_4.

Next we need to show, by Lemma 9.1.9, that each of these regions is an appropriate funnel or antifunnel, forward or backward, as shown in Figure 9.1.12.

Lemma 9.1.9. *For $\beta > 0$ sufficiently large and $\eta < \eta_0(\beta)$ sufficiently small, we have*

(i) *the region U_1 is a funnel for (3), and all trajectories entering U_1 approach 0 tangentially;*

(ii) *the region U_2 is a backward antifunnel for (3) satisfying the (first) uniqueness condition of Theorem 4.7.4. Hence, there is exactly one trajectory leaving the origin in U_2, and it leaves tangent to the x-axis at the origin;*

(iii) *the regions U_3 and U_4 are an antifunnel and a backward antifunnel, respectively, satisfying the (second) uniqueness property of Theorem 4.7.5.*

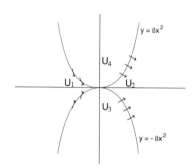

FIGURE 9.1.12. Funnels and antifunnels for saddle-node bifurcation.

Proof of Lemma. Parts (i) and (ii) are relatively easy (Exercise 9.1#11), but the uniqueness claim in (iii) requires a bit of work. The cases of U_3 and U_4 are similar; we will deal with U_3. A careful look at Theorem 4.7.5 will show that we need to find a function $W(y)$ defined for $-\eta < y < 0$, such that

$$\frac{\partial}{\partial x}\left(\frac{P(x,y)}{\lambda y + Q(x,y)}\right) > W(y)$$

in U_3 and

$$\int_{-\eta}^{0} W(y)dy > -\infty.$$

The partial derivative yields

$$\frac{\partial}{\partial x}\left(\frac{P(x,y)}{\lambda y + Q(x,y)}\right) = \frac{(\lambda y + Q(x,y))\,\partial P/\partial x - P\,\partial Q/\partial x}{(\lambda y + Q(x,y))^2}.$$

In U_3, the denominator is bounded below by $C_1 y^2$ for some $C_1 > 0$, as shown in inequality (6), and the leading terms of the numerator are of order $|y|^{3/2}$, satisfying an inequality

$$\left|(\lambda y + Q(x,y))\frac{\partial P}{\partial x} - P\frac{\partial Q}{\partial x}\right| < C_2\,|y|^{3/2}$$

for some $C_2 > 0$, since $|x| \leq (y/\beta)^{1/2}$. Thus, the partial derivative is bounded below by $-C_2/(C_1\sqrt{|y|})$, and this function has a finite integral over $(-\eta, 0)$. $\square$

Remark. We could not use the first uniqueness theorem for antifunnels, since the leading terms above do not have signs that we can control.

Lemma 9.1.9 proves all the *uniqueness* required by the theorem; that is, everything except that the solutions leading to the origin in U_3 and U_4 do so tangentially to the y-axis. To see this, in Lemma 9.1.10 we essentially turn Figure 9.1.12 by 90° and consider the regions V_3 and V_4, defined by $|x| \leq \gamma y^2$ with $|x|, |y| \leq \eta$, as shown in Figure 9.1.13.

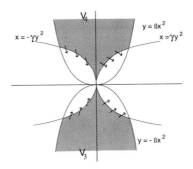

FIGURE 9.1.13. More funnels and antifunnels for saddle-node bifurcation.

Lemma 9.1.10. *For γ sufficiently large, regions V_3 and V_4 are a narrowing antifunnel and a narrowing backward antifunnel, respectively, for equation* (4).

Proof of Lemma. This is again a little delicate. The leading term of

$$\frac{P(\pm\gamma y^2, y)}{\lambda y + Q(\pm\gamma y^2, y)}$$

is $(p_{0,2}/\lambda)y$ (or higher degree terms in y if $p_{0,2} = 0$), and we know nothing about its sign. But it does not matter: the slope of the bounding curve is $\pm 2\gamma y$, and if we choose γ so that $2\gamma > |p_{0,2}/\lambda|$, the required inequalities will be satisfied for η sufficiently small.

Thus, the unique solution in U_3 and U_4 must in fact be in V_3 and V_4, respectively. This forces them to be tangent to the y-axis. $\square$

Proof of Theorem 9.1.7, continued. The proof is completed by Lemmas 9.1.8, 9.1.9, and 9.1.10, giving where things are defined, where there is uniqueness, and where there are tangencies, respectively. □

See Exercises 9.1.#12 and 9.6#6 to work through what happens to this proof when degeneracies occur.

HOMOCLINIC SADDLE-NODES

When Nondegeneracy Conditions 9.1.5 and 9.1.6 are satisfied, saddle-nodes will have, in the eigendirection of eigenvalue zero, a pony tail of solutions tending to the singularity on one side and a single trajectory leaving on the other side. Something peculiar happens when that exceptional trajectory leads back into the pony tail: for nearby values of the parameter for which there are no zeroes of the vector field, there will be *limit cycles*.

Example 9.1.11. Consider the system of differential equations

$$x' = \alpha + x^2 - y^3$$
$$y' = -y + x^3.$$

For $\alpha = 0$, a computer drawing of the phase plane looks like Figure 9.1.14.

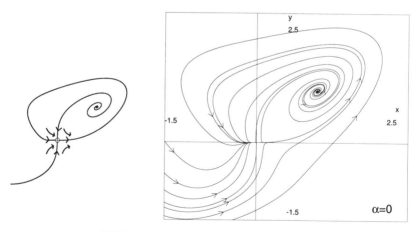

FIGURE 9.1.14. Homoclinic saddle node.

The origin is a saddle-node, satisfying both nondegeneracy hypotheses, and the drawing shows that the trajectory leaving the saddle-node goes back into the pony tail.

If (x_0, y_0) is a zero of the vector field, then x_0 is a solution of the equation

$$\alpha + x_0^2 - x_0^9 = 0,$$

for which there are three solutions when α is small and negative, and only one when α is positive, as shown in Figure 9.1.15.

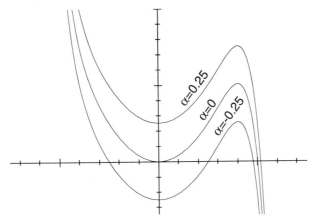

FIGURE 9.1.15. The three possibilities for the roots of $\alpha + x_0^2 - x_0^9 = 0$.

So the zero of the vector field has bifurcated out of existence for $\alpha > 0$, in particular when $\alpha = 0.2$, and the phase plane now looks like Figure 9.1.16. A *limit cycle* has appeared.

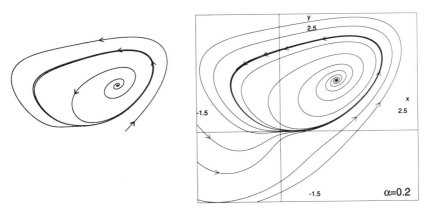

FIGURE 9.1.16. Limit cycle created by homoclinic saddle-node bifurcation. Note the *source* in the center of both phase portraits. ▲

It is quite easy to see why the limit cycle of Example 9.1.11 must be there, as will be shown with Figure 9.1.17. Choose coordinates near the saddle-node so that the differential equation is written in "generic form" as system (2).

Take a small rectangle around the saddle-node, so that the vector field points into it on all sides except the right side, and the exceptional trajectory enters it on the left.

If the points A and B are chosen along the top and bottom of the rectangle, sufficiently close to the trajectory approaching the origin tangent to the eigenvector with eigenvalue $\lambda < 0$, then the trajectories starting at A and B will also reenter the rectangle on the left, as shown.

If we now perturb the equation slightly to make the saddle-node zero of the vector field disappear, the shaded region, bounded by the rectangle and the new solutions through A and B, satisfies the conditions of the Poincaré–Bendixson Theorem 8.5.1; thus, a limit cycle within that shaded region is guaranteed.

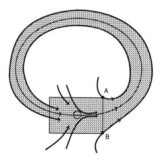

FIGURE 9.1.17. Poincaré–Bendixson region just "after" saddle-node bifurcation, when saddle and node singularities have disappeared.

In conclusion of our discussion of saddle-node behavior, we add yet another possibility of degenerate behavior. Under *symmetry*, saddle nodes may also exhibit degeneracies. An example is the *pitchfork* bifurcation, which is explored in Exercises 9.1#8–10.

9.2   Andronov–Hopf Bifurcations

The Hopf bifurcation is a bit more exotic than the saddle-node bifurcation: it is really a phenomenon that requires at least two dimensions and has no analog in dimension one. Its discovery and analysis are usually attributed in the West to Eberhardt Hopf, a Dutch mathematician. We will follow the customary terminology, although apparently A.A. Andronov, a Russian mathematician, has a better claim to its discovery in the 1930s, when Hopf and Andronov were both active. In any case, Poincaré apparently knew all about it 40 years earlier.

Definition 9.2.1. (Andronov–Hopf bifurcation) An *Andronov–Hopf bifurcation* occurs in general when a zero of a vector field has a linearization with *zero trace* and *positive determinant*.

Thus, Hopf bifurcation occurs in general if an equilibrium is a "center." Of course, a linear center is surrounded by periodic solutions. In Section 8.7, we saw that if a vector field is area-preserving, the zeroes will often be surrounded by periodic solutions, but this is not what happens in general. In general, a zero of a vector field whose linearization is a center will be a "weak" source or sink, in the sense that the nonlinear center attracts or repels nearby points, but at a slower rate than the exponential rate associated with linear sinks and sources.

Associated with Hopf bifurcations are limit cycles. It is easy to see why. Suppose that a nonlinear center is a weak sink: nearby solutions are attracted to it; the solutions with the same initial condition will still come back inside itself after a sufficiently small perturbation. If this perturbation makes the center a source, then such a solution is trapped: it must head toward the zero of the vector field, but it cannot get there. This is just the kind of situation to which the Poincaré–Bendixson Theorem 8.5.1 applies, and implies the existence of a limit cycle, as shown in Figure 9.2.1.

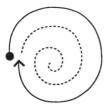
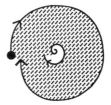

before perturbation   after perturbation

FIGURE 9.2.1. Existence of the limit cycle created by Hopf bifurcation, as indicated by the Poincaré–Bendixson Theorem.

The typical scenario is that as a parameter α approaches a bifurcation value α_0, a sink (or source) becomes a weak sink (or source) and then, after bifurcation, changes to a source (or sink). At the same time, a limit cycle grows out of the equilibrium point, as the following example shows.

Example 9.2.2. Consider the differential equation

$$x' = y$$
$$y' = (\alpha - x^2)y - x.$$

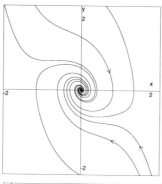

Before bifurcation:
$\alpha = -0.5 < \alpha_0 = 0$.
The origin is a linear sink to
which everything is attracted.
(Several trajectories.)

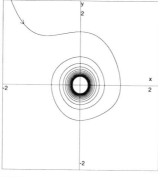

At bifurcation:
$\alpha = \alpha_0 = 0$.
The linearization at the origin
is a center, but the origin is a
weak sink, very slowly attracting all solutions.
[All this is one trajectory, still
heading toward $(0,0)$.]

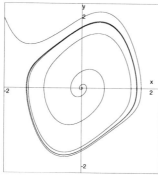

After bifurcation:
$\alpha = 0.5 > \alpha_0 = 0$.
The origin is now a source.
However, the small modification
of the equation has not much
modified its behavior far from
the origin. Solutions from afar
still head inward and are attracted to a limit cycle.
(Two trajectories shown—one
inside, one outside limit cycle.)

FIGURE 9.2.2. Hopf bifurcation for Example 9.2.2.

For $\alpha = -1$, this is the Van der Pol equation studied in Example 8.5.2.
There is clearly a unique zero of the vector field, at $(0,0)$, where the linearization has coefficient matrix

$$A = \begin{bmatrix} 0 & 1 \\ -1 & \alpha \end{bmatrix}.$$

Since $\det A = 1$ and $\operatorname{tr} A = \alpha$, the equation has a Hopf bifurcation at $\alpha = 0$.

The computer gives for $\alpha = -0.5$, 0, 0.5 the phase plane pictures shown
in Figure 9.2.2.

The following computation shows that for $\alpha = 0$, the zero is a weak sink: If $(x(t), y(t))$ is a solution, then

$$\frac{d}{dt}(x(t)^2 + y(t)^2) = 2xx' + 2yy' = -2x^2y^2 \leq 0.$$

Thus, $x(t)^2 + y(t)^2$ is a decreasing positive function of t, which must have a limit as $t \to \infty$. Theorem 9.2.3 will give a way to improve on this function by adding cubic and fourth degree terms to avoid unpleasant places where this derivative vanishes.

Thus, all solutions are attracted to 0 for $\alpha \leq 0$. ▲

SERIOUS EXPLANATION OF HOPF BIFURCATION BEHAVIOR

Justifying our description of the Hopf bifurcation is really quite complicated. The main task is to show that when the linearization of a vector field at a zero is a center, then the zero in question is usually a weak source or a weak sink. As we saw in Example 8.7.1, an equilibrium point where the linearization is a center may be far more complicated than that; for instance it may have infinitely many limit cycles in every neighborhood of the equilibrium. But we shall see that this is exceptional rather than ordinary.

Suppose $\mathbf{x}' = \mathbf{f}_\alpha(\mathbf{x})$ is a one-parameter family of differential equations in the plane, that $\mathbf{x}_0$ is a zero of $\mathbf{f}_\alpha$, and that the linearization of $\mathbf{x}' = \mathbf{f}_{\alpha_0}(\mathbf{x})$ at $\mathbf{x}_0$ is a *center*. By a translation, we may assume that $\mathbf{x}_0 = (0, 0)$, and by a linear change of variables we may assume that the linearization has coefficient matrix

$$A = \begin{bmatrix} 0 & 1 \\ -1 & 0 \end{bmatrix}.$$

In Exercise 9.2.#6, you will see how to make a linear change of variables in x and y, and also in t, to bring the equation to this form; Exercise 9.2.#7 works out an example (which is a harder computation than one might expect). Thus, we assume that at $\alpha = \alpha_0$ our differential equation can be written as a perturbation of the equation that gives centers,

$$\begin{aligned} x' &= F(x, y) = y + F_2(x, y) + F_3(x, y) + \cdots \\ y' &= G(x, y) = -x + G_2(x, y) + G_3(x, y) + \cdots, \end{aligned} \tag{7}$$

where F_2, F_3, G_2, and G_3 are the second and third degree terms of F and G, and so on. In Theorem 9.2.3, we will use the coefficients of the F_i and G_i and will write

$$F_i(x, y) = \sum_{j=0}^{i} \mu_{j,(i-j)} x^j y^{i-j}$$

$$G_i(x, y) = \sum_{j=0}^{i} \nu_{j,(i-j)} x^j y^{i-j}.$$

Theorem 9.2.3. (Liapunov test for weak sink or source in Hopf bifurcation) *If* $\mathbf{x}' = \mathbf{f}_\alpha(\mathbf{x})$ *is a one-parameter family of differential equations in the plane such that for* $\alpha = \alpha_0$ *the equation can be written*

$$x' = y + \mu_{2,0}x^2 + \mu_{1,1}xy + \mu_{0,2}y^2 + \mu_{3,0}x^3 + \mu_{2,1}x^2y + \mu_{1,2}xy^2$$
$$+ \mu_{0,3}y^3 + \cdots,$$
$$y' = -x + \nu_{2,0}x^2 + \nu_{1,1}xy + \nu_{0,2}y^2 + \nu_{3,0}x^3 + \nu_{2,1}x^2y + \nu_{1,2}xy^2$$
$$+ \nu_{0,3}y^3 + \cdots,$$

as described directly above, then define the Liapunov coefficient

$$L \equiv 3\mu_{3,0} + \mu_{1,2} + \nu_{2,1} + 3\nu_{0,3} - \mu_{2,0}\mu_{1,1} + \nu_{1,1}\nu_{0,2} \tag{8}$$
$$- 2\mu_{0,2}\nu_{0,2} - \mu_{0,2}\mu_{1,1} + 2\mu_{2,0}\nu_{2,0} + \nu_{1,1}\nu_{2,0}.$$

(i) *If* L *is positive, the origin is a weak source for* $\alpha = \alpha_0$, *and if* L *is negative, then the origin is a weak sink; if* L *is zero, you cannot tell.*

(ii) *If* $L > 0$, *the differential equation will have an unstable limit cycle for all values of* α *near* α_0 *for which the zero is a sink, and if* $L < 0$, *it will have a stable limit cycle for all values of* α *near* α_0 *for which the zero is a source.*

Note: Formula (8) for L is pretty awful, but it may help to think of it as (a weighted average of coefficients of cubic terms) minus (a weighted average of coefficients of quadratic terms).

It is important to realize that when a real system undergoes a Hopf bifurcation, then weak sinks or weak sources correspond to very different behaviors. Suppose a "real" system has a "control" parameter a (a knob), and for some value a_0 undergoes a Hopf bifurcation; we will imagine that there is a sink for $a < a_0$, and that the parameter is increased. Then if there is a weak sink for $a = a_0$, there will be a small limit cycle for a slightly larger than a_0, and the system simply changes from rest at the sink to small oscillations.

But if there is a weak source for $a = a_0$, the behavior is quite different: There will be nothing near the source to "catch" the system, which will presumably crash or blow up as soon as a hits a_0.

In the literature, Hopf bifurcations are often called by the ghastly names "subcritical" in the case of a weak source and "supercritical" in the case of a weak sink: we will not use this exceptionally obscure and unsuggestive terminology.

Example 9.2.4. In Example 9.2.2, at $\alpha = 0$, all the $\mu_{i,j}$ and $\nu_{i,j}$ are 0 except $\nu_{2,1} = -1$; hence, the Liapunov coefficient $L = -1$ and the zero at $(0,0)$ is a weak sink, as it should be. ▲

Proof of Theorem 9.2.3. (i) Since we are perturbing a center, we will seek a *Liapunov function* (Definition 8.2.6) of a form to give a perturbation of circles,

$$f(x,y) = (x^2 + y^2)/2 + f_3(x,y) + f_4(x,y), \qquad (9)$$

with f_3 and f_4 homogeneous of degrees three and four, respectively; i.e., we will try to choose a function that is strictly increasing or decreasing along trajectories near $(0,0)$. Note that we have chosen these quadratic terms because the linear terms of the differential equation were selected to make the solutions turn on round circles. Since any such function f has an isolated minimum at $(0,0)$, if it is decreasing along trajectories, this will imply that the origin will be a sink, as indicated in Figure 9.2.3.

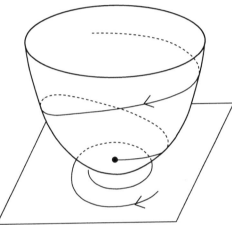

FIGURE 9.2.3. The surface is the graph of $f(x,y)$ in $\mathbb{R}^3$; the curve on the surface is the unique projection from the plane up onto the surface, where each point on the curve has the form $(x(t),\ y(t),\ f(x(t),\ y(t)))$. This figure illustrates a weak sink; if the singularity were a weak source, the arrows would be reversed.

Incidentally, there is no good a priori reason to limit oneself to homogeneous polynomials of degree four; it just seems to be what works, and the proof naturally leads to consideration of such a function.

By the chain rule, if $(x(t), y(t))$ is a solution of the differential equation, we see that along this path, on the surface $f(x,y)$ described by equation (9),

$$\frac{d}{dt}(f(x(t), y(t))) = \left(\frac{\partial f}{\partial x}\right) F + \left(\frac{\partial f}{\partial y}\right) G$$

$$= \left(x + \frac{\partial f_3}{\partial x} + \frac{\partial f_4}{\partial x}\right)(y + F_2(x, y) + F_3(x, y))$$

$$+ \left(y + \frac{\partial f_3}{\partial y} + \frac{\partial f_4}{\partial y}\right)(-x + G_2(x, y) + G_3(x, y))$$

$$+ \text{higher order terms}$$

$$= [xy - xy] \qquad \text{(2nd order terms)}$$

$$+ \left[xF_2 + yG_2 + y\frac{\partial f_3}{\partial x} - x\frac{\partial f_3}{\partial y}\right] \qquad \text{(3rd order terms)}$$

$$+ \left[xF_3 + yG_3 + F_2\frac{\partial f_3}{\partial x} + G_2\frac{\partial f_3}{\partial y} + y\frac{\partial f_4}{\partial x} - x\frac{\partial f_4}{\partial y}\right]$$

$$\text{(4th order terms)}$$

$$+ \text{higher order terms.} \qquad (10)$$

In (10), the terms of degree two cancel; this was the point of the original changes of variables (7). If we want df/dt to be of constant sign, we must make the terms of degree three disappear; we will show that f_3 can be chosen uniquely so that the terms of degree three vanish. However, f_4 cannot be chosen so as to make the terms of degree four vanish in general.

This is really a problem in linear algebra: solving linear equations. Let P_k be the space of homogenous polynomials of degree k in two variables; it is a vector space of dimension $k + 1$ and can be identified with $\mathbb{R}^{k+1}$ by identifying

$$\begin{bmatrix} a_0 \\ \vdots \\ a_k \end{bmatrix} \quad \text{to} \quad a_0 x^k + a_1 x^{k-1} y + \ldots + a_k y^k.$$

Now consider the linear transformation $T_k \colon P_k \to P_k$ given by

$$T_k(p) = y\frac{\partial p}{\partial x} - x\frac{\partial p}{\partial y},$$

where the polynomials $T_k(p)$, $\partial p/\partial x$, and $\partial p/\partial y$ are all evaluated at (x, y).

Lemma 9.2.5. *If k is odd, T_k is an isomorphism. If k is even, T_k has kernel of dimension one, with basis $(x^2 + y^2)^{k/2}$, and its image has dimension k.*

Proof of Lemma. It is quite possible to give a conceptual proof of this lemma; such a proof is sketched in Exercise 9.2.#8. We will only need the result for $k = 3$ and 4, and it is easier to simply write down the matrix of T_3 and T_4 and check.

With the identification of P_k with $\mathbb{R}^{k+1}$ given above, the matrices are the following:

$$T_3 \text{ has matrix } \begin{bmatrix} 0 & -1 & 0 & 0 \\ 3 & 0 & -2 & 0 \\ 0 & 2 & 0 & -3 \\ 0 & 0 & 1 & 0 \end{bmatrix},$$

$$T_4 \text{ has matrix } \begin{bmatrix} 0 & -1 & 0 & 0 & 0 \\ 4 & 0 & -2 & 0 & 0 \\ 0 & 3 & 0 & -3 & 0 \\ 0 & 0 & 2 & 0 & -4 \\ 0 & 0 & 0 & 1 & 0 \end{bmatrix}.$$

The first is invertible: it has nonzero determinant. The second is not invertible: it has determinant zero. But it has rank 4, since the 4×4 submatrix in the upper right corner has determinant $\neq 0$, and the vector

$$\begin{bmatrix} 1 \\ 0 \\ 2 \\ 0 \\ 1 \end{bmatrix},$$

corresponding to the polynomial $(x^2 + y^2)^2$, is in the kernel.

Remark. The function $3a_0 + a_2 + 3a_4$ vanishes on each column vector, hence on the image. $\square$

Proof of Theorem 9.2.3, continued. Using Lemma 9.2.5, we see that we can choose f_3 uniquely so as to make the cubic term of equation (10) vanish. Further, we can choose f_4 so that if we write

$$xF_3 + yG_3 + F_2\frac{\partial f_3}{\partial x} + G_2\frac{\partial f_3}{\partial y} = a_0 x^4 + a_1 x^3 y + a_2 x^2 y^2 + a_3 xy^3 + a_4 y^4$$

and set $L = 3a_0 + a_2 + 3a_4$, then

$$xF_3 + yG_3 + F_2\frac{\partial f_3}{\partial x} + G_2\frac{\partial f_3}{\partial y} + y\frac{\partial f_4}{\partial x} - x\frac{\partial f_4}{\partial y} = \frac{L}{8}(x^2 + y^2)^2,$$

since

$$xF_3 + yG_3 + F_2\frac{\partial f_3}{\partial x} + G_2\frac{\partial f_3}{\partial y} - \frac{L}{8}(x^2 + y^2)^2$$

is in the image of T_4.

Clearly, with this choice of function f, the quantity

$$\frac{d}{dt}(f(x(t), y(t))) = \frac{L}{8}(x(t)^2 + y(t)^2)^2 + \text{higher order terms}$$

has the same sign as L for $(x(t), y(t))$ sufficiently small. This proves the first part of the theorem.

(ii) This second part is almost an immediate consequence of the Poincaré–Bendixson Theorem 8.5.1. Suppose the linearization of $\mathbf{x}' = \mathbf{f}_{\alpha_0}(\mathbf{x})$ at x_0 is a center; we will give the proof when the number L above is negative. A similar argument takes care of the case when L is positive. Construct the function $f(x, y)$ as in part (a) and choose a level curve of f sufficiently close to the origin so that the vector field points strictly inside it.

> *The vector field will still point in along that curve for nearby values of α: a small perturbation cannot change the direction in which a vector field crosses a curve if it cuts the curve transversely.*

Now consider a value α_1 very close to α_0 for which x_0 is a source. There is then a much smaller curve around x_0 such that the vector field f_{α_1} points out along that curve, as a result of Theorem 8.2.2. The two curves above bound an annular region to which the Poincaré–Bendixson Theorem 8.5.1 can be applied and that will contain a limit cycle. This completes the proof for the second part of the theorem. □

Theorem 9.2.3 has shown that if a vector field undergoes a Hopf bifurcation that is a weak sink, then the differential equation will have a limit cycle for nearby values of the parameter for which the zero of the vector field has become a source. If it is a weak source, then the limit cycles will exist for parameter values for which the zero is a sink.

The following examples demonstrate some of the possibilities for nonlinear systems where linearization at a zero produces a center.

Example 9.2.6. Consider the differential equation

$$x' = y + x^2$$
$$y' = -x.$$

Then the Liapunov number L of Theorem 9.2.4 is zero. The origin is a zero at which the linearization is a center. In fact, the computer quite likely will tell you that the origin is also a center for the nonlinear system; i.e., that all solutions close to the origin are periodic. Exercise 9.2#5 asks you to prove this.

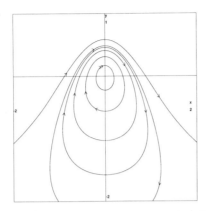

FIGURE 9.2.4. Phase plane for Example 9.2.6. ▲

Example 9.2.7. Consider the family of differential equations

$$x' = y + \alpha x + x^2 + x^3$$
$$y' = -x.$$

For $\alpha = 0$, the origin is a zero at which the linearization is a center. The number L is 3 in this case, so the origin is a weak source.

If $\alpha < 0$, the origin is a sink, so there should be limit cycles for α very small and negative.

Again you are asked to verify these facts by computer in Exercise 9.2.#3, showing that Hopf bifurcation occurs in this family.

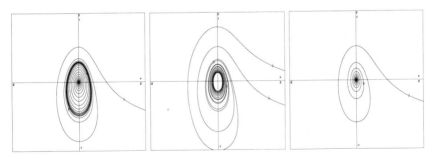

FIGURE 9.2.5. Phase plane for Example 9.2.7 for $\alpha = -0.1$, 0, and 0.1 from left to right. ▲

Example 9.2.8. Consider the system

$$x' = y + xy^2$$
$$y' = -x + xy - y^2.$$

In this case again, $L = 0$, so you cannot tell, but Figure 9.2.6 shows that the origin is a very weak source. Proving this will wait until later; see Exercise 9.6#3.

Exercise 9.2#4 asks you to show that a small perturbation of this equation can be made to have *two* limit cycles, thus showing that something else is going on, which we will meet in Section 9.4.

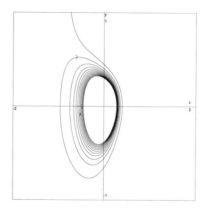

FIGURE 9.2.6. Phase plane for Example 9.2.8.   ▲

9.3   Saddle Connections

We now turn our attention to *saddle connection* bifurcations. Unlike saddle-node or Hopf bifurcations, these are *global bifurcations*, and they cannot be detected by studying the zeroes of the vector field.

Saddle connections are harder to locate than saddle-nodes or Hopf bifurcations (Sections 9.1 and 9.2, respectively), for which you must solve some algebraic (nondifferential) equations. You must actually solve the *differential* equation in order to know that a saddle connection occurs. That is, to draw the pictures and see what is happening, you (or the computer) must be able to draw the separatrices of the saddles.

In general, the stable separatrix of a saddle is not the unstable separatrix of the same or any other saddle. However, this does usually occur for some parameter value in a one-parameter family, and when it does, the phase portrait tends to undergo big changes: we will give some examples below, and there are many more in the exercises. Figure 9.3.1 shows a *heteroclinic* saddle connection between *two* saddles.

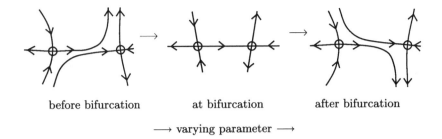

before bifurcation   at bifurcation   after bifurcation

$\longrightarrow$ varying parameter $\longrightarrow$

FIGURE 9.3.1. Heteroclinic saddle connection at bifurcation.

When a stable and an unstable separatrix of *one* saddle coincide, the saddle is called *homoclinic*, as shown in Figure 9.3.2.

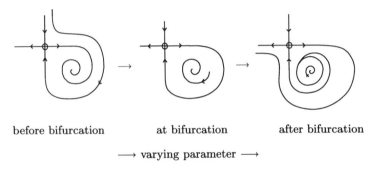

before bifurcation   at bifurcation   after bifurcation

$\longrightarrow$ varying parameter $\longrightarrow$

FIGURE 9.3.2. Homoclinic saddle connection at bifurcation. Note that the sink remains a sink throughout, so a limit cycle is created.

Quite often, the homoclinic loop of a homoclinic saddle connection encloses a unique zero of the vector field. This zero will be a source or a sink, unless a Hopf or a saddle-node bifurcation occurs for the same value of the parameter. Suppose this zero is a source; it will then still be a source for nearby values of any parameters. But the overall spiral behavior "changes arms"—if a stable separatrix goes "in" before bifurcation, then an unstable one does after bifurcation. The stable separatrix will usually emanate from the source, but the unstable one cannot go to a source and must do something else. Most often it will be attracted to a limit cycle, which is created in the homoclinic saddle connection.

In this case of a homoclinic saddle connection with a limit cycle, there often is a Hopf bifurcation nearby in the parameter space, where the limit cycle dies; this comment will be examined in Section 9.5.

An important aspect of both heteroclinic and homoclinic saddle connections is that this bifurcation drastically changes the *basins* as shown in Figures 9.3.3 and 9.3.4, elaborations of Figures 9.3.1 and 9.3.2.

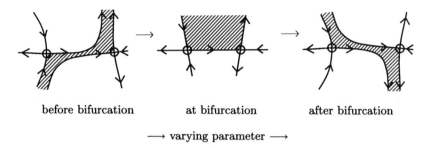

before bifurcation   at bifurcation   after bifurcation

$\longrightarrow$ varying parameter $\longrightarrow$

FIGURE 9.3.3. Heteroclinic saddle connection, showing change of basins.

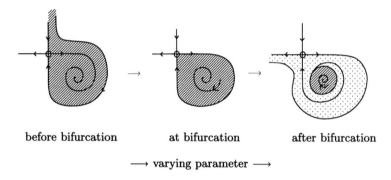

before bifurcation   at bifurcation   after bifurcation

$\longrightarrow$ varying parameter $\longrightarrow$

FIGURE 9.3.4. Homoclinic saddle connection, showing change of basins.

Example 9.3.1. Consider the family of differential equations

$$x' = y$$
$$y' = x^3 - x + \alpha x^2 - 0.1y$$

depending on the parameter α. The computer shows in Figure 9.3.5 the phase planes for $\alpha = 0.1$ and $\alpha = 0.2$, on either side of a saddle connection.

We shall concentrate henceforth on these "before" and "after" pictures, since that is what you look for when you are trying to locate a saddle connection.

As Exercise 9.3.#1, you should experiment with a computer to find to two or three significant figures the value of α_0, for $0.1 < \alpha_0 < 0.2$, where this saddle connection occurs.

In Figure 9.3.5, note that the region A from the first picture, which is in the basin of the sink at the origin, has no corresponding region in the second picture.

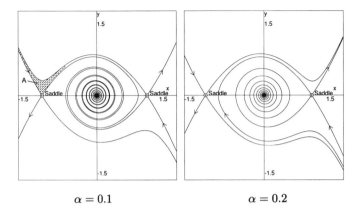

$$\alpha = 0.1 \qquad\qquad\qquad\qquad \alpha = 0.2$$

FIGURE 9.3.5. Before and after a *heteroclinic* saddle connection for Example 9.3.1, showing the change of basins of the spiral sink.

The system of differential equations of this example is fairly easy to understand from mechanics, as discussed in Section 6.5: it corresponds to a particle moving in the potential

$$V_\alpha(x) = \frac{x^2}{2} - \alpha\frac{x^3}{3} - \frac{x^4}{4},$$

with a little bit of friction. The graph of such a function looks like Figure 9.3.6.

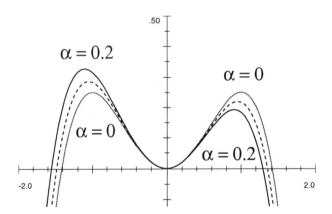

FIGURE 9.3.6. Potential $V_\alpha(x)$ for Example 9.3.1. The parameter α controls the difference in height of the two peaks and the x-coordinate of the two corresponding saddles. The solid curve is for $\alpha = 0.2$, dashed for $\alpha = 0.1$, dotted for $\alpha = 0$.

The potential energy $V_\alpha(x)$ has two maxima, with the left-hand one higher than the right-hand one for $\alpha > 0$. As α increases, the bump on the left gets higher and the one on the right gets lower. Before the saddle connection, a particle falling from rest at the top of the left bump loses too much energy due to friction to make it over the right bump and gets caught by the sink. After the saddle connection, the same initial position leads to a motion which goes over the right-hand bump and escapes. Now the solutions starting in A, correspond to initial position far to the left and initial velocity positive, slightly larger than what is required to make it over the left-hand maximum. For $\alpha = 0$, such a solution is captured by the sink; for $\alpha = 0.3$, it bounces over the right-hand maximum and escapes. ▲

Example 9.3.2. Consider the system of equations

$$x' = y$$
$$y' = x^3 - x + 0.2x^2 + (\beta x^2 - 0.2)y.$$

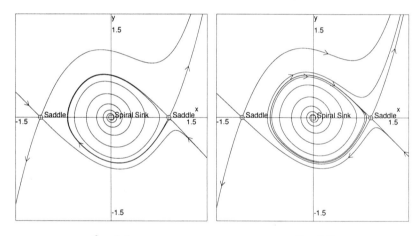

$\beta = 1.4$	$\beta = 1.5$
The unstable separatrix from the right-hand saddle near $(1, 0)$ is attracted to the sink at $(0, 0)$.	The basin of the sink is now bounded by a repelling limit cycle.

FIGURE 9.3.7. Before and after a *homoclinic* saddle connection (on the right-hand saddle) for Example 9.3.2. ▲

As Exercise 9.3#2a, you should experiment with a computer to find to two or three significant figures the value of β_0, for $1.4 < \beta_0 < 1.5$, where this homoclinic saddle connection occurs, and to confirm that it is indeed a *sink* at the origin for *all* values of β in this interval. As Exercise 9.3#2b, you can experiment further with the computer to make the left-hand saddle of Figure 9.3.7 connect, either to itself or to the right-hand saddle.

SERIOUS MATHEMATICAL EXPLANATION OF A
HOMOCLINIC SADDLE CONNECTION

We will now show that in a homoclinic saddle connection bifurcation, the
appearance of limit cycles is unavoidable. This requires a bit of terminology,
illustrated by Figures 9.3.8 and 9.3.9. A homoclinic saddle connection is
a loop bounding a region (shaded) of the plane. A *perturbation* f_α of the
equation $x' = f_\alpha(x)$ at bifurcation (that is, for α near α_0) will be called
inward if the unstable separatrix which at bifurcation led back to the zero
now goes into the loop region, *outward* otherwise, as shown in Figure 9.3.8.

The saddle connection will be called *attracting* if solutions just inside it
spiral toward the homoclinic loop, *repelling* otherwise, as shown in Figure
9.3.9 on the next page.

Theorem 9.3.3. *Let* $x' = f_\alpha(x)$ *be a family of differential equations depending on a parameter* α, *and suppose that for some value* α_0 *of the parameter, the vector field has a zero* x_0 *which is a saddle with a homoclinic
saddle connection. Let* $u' = Au$ *be the linearization of* $x' = f_{\alpha_0}(x)$ *at* x_0.

(i) *If* tr $A > 0$, *the saddle connection is repelling.*
   If tr $A < 0$, *the saddle connection is attracting.*
   If tr $A = 0$, *you cannot tell.*

(ii) *If the saddle connection is repelling, and* α *is close to* α_0 *with the
    perturbation outward, there will be for that value of* α *a repelling
    limit cycle.*

(iii) *If the saddle connection is attracting, and* α *is close to* α_0 *with the
    perturbation inward, there will be for that value of* α *an attracting
    limit cycle.*

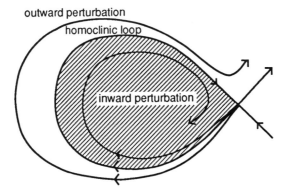

FIGURE 9.3.8. Inward and outward perturbations of a homoclinic orbit, as shown
on the right saddle of each picture in Figure 9.3.2.

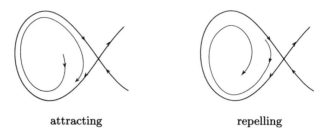

attracting   repelling

FIGURE 9.3.9. Attracting and repelling homoclinic loops.

Example 9.3.4. Go back to the equations of Example 9.3.2,

$$x' = y$$
$$y' = x^3 - x + 0.2x^2 + (\beta x^2 - 0.2)y.$$

For $\beta \approx 1.4$, for which there is a homoclinic saddle connection, we find that
the saddle point to the right is the point $(x_0, 0)$, where x_0 is the positive
solution of the equation $x^2 + 0.2x - 1 = 0$, i.e., $x_0 = -0.1 + \sqrt{1.01} \approx 0.9$.
The linearization matrix

$$A = \begin{bmatrix} 0 & 1 \\ 3x_0^2 - 0.4x_0 - 1 + 2\beta x_0 y & \beta x_0^2 - 0.2 \end{bmatrix}$$

has $\text{tr}(A) = \beta x_0^2 - 0.2 \approx (1.4)(0.9)^2 - 0.2 > 0$. Thus, the saddle connection
is *repelling*, and *repelling* limit cycles appear for the *outward* perturbations,
which occur for $\beta > \beta_0$, as was shown on the right of Figure 9.3.7.   ▲

Proof of Theorem 9.3.3. The hard part is part (i); parts (ii) and (iii)
then follow by an easy application of the Poincaré–Bendixson Theorem
8.5.1.

We will deal with the case $\text{tr}(A) < 0$; the proof for $\text{tr}(A) > 0$ is similar.

Consider a solution inside the attracting saddle connection and break it
up into the part near the saddle and the remainder, as indicated in Figure
9.3.10.

We claim that part I of the solution, the part near the saddle, is important, and controlled by $\text{tr}(A)$, and part II, away from the saddle, is
benign.

To be more precise, we need to introduce some notation. First, let $\lambda_1 > 0$,
$\lambda_2 < 0$ be the two eigenvalues at the saddle, so that the trace being negative
gives $\lambda \equiv |\lambda_2/\lambda_1| > 1$. Choose transversals T_S and T_U to the homoclinic
loop near the saddle, as indicated in Figure 9.3.11, parametrized by u and
v, respectively, so that $u = 0$ and $v = 0$ are the intersections with the
separatrices.

Suppose a solution starts on T_S at $u = \delta_0$, then intersects T_U at $v = \delta_1$,
and then intersects T_S at $u = \delta_2$.

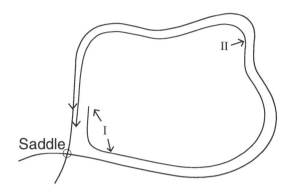

FIGURE 9.3.10. Homoclinic saddle connection; inner solution near (I) and away (II) from saddle.

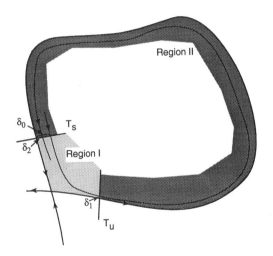

FIGURE 9.3.11. Defining transversals and deltas for a typical attracting homoclinic saddle connection.

Then part I is essential in the sense that

$$\delta_1 \sim C_1 \delta_0^\lambda \quad \text{for some } C_1 \neq 0,$$

and part II is benign in the sense that

$$\delta_2 \sim C_2 \delta_1 \quad \text{for some } C_2 \neq 0.$$

Raising to the power λ dominates multiplication by a constant,

$$\delta_2 \sim C_2 \delta_1 \sim C_1 C_2 \delta_0^\lambda < \delta_0,$$

when $\delta_0 > 0$ is sufficiently small, so the loop is attracting.

Armed with this notation, we return to proving the theorem, which consists of three parts—looking at region I first with the linear approximation to the saddle, then as a nonlinear saddle, and then meandering through region II. We will actually prove inequalities rather than equivalences, which is enough for our purposes.

Swinging by the Saddle, Part I: the Linear Approximation. To see why the passage near the saddle has the dominant effect, let us study how the linear equation behaves; i.e., let us consider

$$u' = \lambda_1 u$$
$$v' = \lambda_2 v,$$

in some region $|u|, |v| \leq \eta$, centered at $\mathbf{x}_0$. The solutions when $\lambda_1 = 1$ and $\lambda_2 = -2$ appear in Figure 9.3.12.

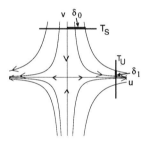

FIGURE 9.3.12. A linear saddle when $\lambda_1 = 1$, $\lambda_2 = -2$. Note that $\delta_1 \sim \delta_0^2$.

The solution $(u(t), v(t))$, with $(u(0), v(0)) = (\delta, \eta)$ is given explicitly by

$$u(t) = \delta\, e^{\lambda_1 t}$$
$$y(t) = \eta\, e^{\lambda_2 t}.$$

This solution crosses the line $u = \eta$ when $t = 1/\lambda_1 \log(\eta/\delta)$; the corresponding value of v is

$$\delta_1 = \eta\, e^{\lambda_2/\lambda_1 \log(\eta/\delta)} = \eta^{1-\lambda}\delta^{\lambda}.$$

So we see that passage near a linear saddle does indeed raise the distance to the homoclinic loop by a power > 1.

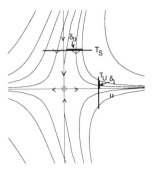

FIGURE 9.3.13. A nonlinear saddle when $\lambda_1 = 1$, $\lambda_2 = -2$. Note that $\delta_1 \sim \delta_0^2$.

Swinging by the Saddle, Part II: the Nonlinear Case. Now we must see that passage near a nonlinear saddle leads to the same behavior, as shown in Figure 9.3.13. This requires a lemma.

Lemma 9.3.5. *There exist variables (u, v) centered at $\mathbf{x}_0$ such that in these variables the equation becomes*

$$u' = \lambda_1 u(1 + g_1(u, v))$$
$$v' = \lambda_2 v(1 + g_2(u, v)),$$

where $g_1(0, 0) = g_2(0, 0) = 0$, so both begin with at least linear terms.

Remark. In these coordinates, the stable separatrix is the v-axis and the unstable separatrix is the u-axis.

Proof of Lemma. By a linear change of variables, it is easy to find co-ordinates (u_1, v_1) centered at x_0 in which the differential equation has the form

$$u_1' = \lambda_1 u_1 + h_1(u_1, v_1)$$
$$v_1' = \lambda_2 v_1 + h_2(u_1, v_1),$$

where both h_1 and h_2 start with *quadratic* terms. The unstable separatrix is then a curve tangent to the u_1-axis, and the stable separatrix is a curve tangent to the v_1-axis, by Theorem 8.3.2. The first is then the graph of a function $v_1 = \alpha(u_1)$ and the second is the graph of a function $u_1 = \beta(v_1)$, where both α and β start with terms that are at least quadratic.

Now set $u = u_1 - \beta(v_1)$ and $v = v_1 - \alpha(u_1)$. It requires the Inverse Function Theorem to see that these form coordinates near $\mathbf{x}_0$: the Jacobian matrix at (u_1, v_1) is

$$\begin{bmatrix} 1 & -\partial\beta/\partial v_1 \\ -\partial\alpha/\partial v_1 & 1 \end{bmatrix},$$

which is the identity, hence, the matrix is invertible at $(u_1, v_1) = (0, 0)$.

It should be clear that in these coordinates, the stable and unstable manifolds become exactly the v- and u-axes, respectively, and this implies that the differential equation has the required form.   □

In these coordinates, we can compare the nonlinear equation to a linear system by a fence argument rather similar to several we have seen in Chapter 8 and in Section 9.2. Choose μ_1 and μ_2 so that

$$\lambda_2 < \mu_2 < 0 < \lambda_1 < \mu_1,$$

and $\mu_1 + \mu_2 < 0$, so that $\mu = |\mu_2/\mu_1| > 1$. Next, switch to the equation for v as a function of u:

$$\frac{dv}{du} = \frac{\lambda_2(v(1 + g_2(u, v)))}{\lambda_1 u(1 + g_1(u, v))} \tag{11}$$

and compare it to the equation coming from the linear system

$$v' = \frac{\mu_2 v}{\mu_1 u} = -\mu \frac{v}{u}. \tag{12}$$

The first (and, of course, the second) is well defined near the origin if $u > 0$; i.e., $\lambda_1 u(1 + g_1(u, v)) \neq 0$ if $u \neq 0$ and both u and v are sufficiently small, since g_1 vanishes at the origin.

Next, we claim that the solutions of (12) are upper fences for (11) sufficiently close to the origin; i.e.,

$$\left| \frac{\lambda_2 v(1 + g_2(u, v))}{\lambda_1 u(1 + g_1(u, v))} \right| \geq \left| \frac{\mu_2 v}{\mu_1 u} \right|, \tag{13}$$

again because g_1 and g_2 vanish at the origin.

We can now complete our analysis of passing near a nonlinear sink. Choose a neighborhood U of $\mathbf{x}_0$ in which (u, v) form coordinates, and a number $\eta > 0$ so small that the fence property (13) holds in the region $2|u|, 2|v| \leq \eta$. Let $\mathbf{x}_{\delta_0}(t)$ be the solution of the differential equation with initial condition $\mathbf{x}_{\delta_0}(0)$ as the point with coordinates $u = \eta$, $v = \delta_0$, and suppose this solution crosses the line $u = \eta$ at (η, δ_1). Since the solution to equation (12) with the same initial condition is

$$v = \eta \delta_0^\mu u^{-\mu}$$

with value $\eta^{1-\mu} \delta_0^\mu$ at $u = \eta$, we see that

$$\delta_1 \leq \eta^{1-\mu} \delta_0^\mu. \tag{14}$$

Meandering away from the saddle. Going back to the notation at the beginning of this proof, let T_S and T_U be the transversals $v = \eta$ and $u = \eta$, respectively [in the coordinates (u, v) above]. Note that the coordinates u and v do parametrize these transversals, as required.

The homoclinic loop itself is a perfectly nice solution leading from T_U to T_S. The same proof as in Section 8*.4.3 says that the mapping $T_U \to T_S$ defined near the point $v = 0$ of T_U, which consists of starting a solution at a point of T_U and following it until it intersects T_S, is a differentiable mapping with differentiable inverse.

Lemma 9.3.6. *There exists a constant C such that the solution starting at the point $(u, v) = (\eta, \delta_1)$ next intersects the line T_S at the point (δ_2, η), where $\delta_2 \le C\delta_1$.*

Formula (14) and Lemma 9.3.6 together prove part (i) of Theorem 9.3.3. The main message is that passing by the saddle of an attracting homoclinic saddle connection squeezes the solution *much* closer to the unstable separatrix.

For parts (ii) and (iii), find a solution inside the saddle connection for which the solution returns closer to the saddle connection than it started. Then this will still be true for a small perturbation; if the saddle connection is inward, a segment of the unstable separatrix and of the solution above bound an annular region such that the vector field points into it everywhere and which must contain a limit cycle. $\square$

9.4 Semi-Stable Limit Cycles

Another bifurcation that cannot be understood locally, by studying the zeroes of the vector field, is the *coalescence of limit cycles*. Again, you must solve the differential equation in order to know that this bifurcation occurs.

Example 9.4.1. A simple but rather artificial example of such a bifurcation is given by the family of equations

$$x' = -y + \left(\frac{(x^2 + y^2)^2}{4} - \frac{x^2 + y^2}{2} + \alpha\right) x$$
$$y' = x + \left(\frac{(x^2 + y^2)^2}{4} - \frac{x^2 + y^2}{2} + \alpha\right) y$$

The computer pictures for this family are shown in Figure 9.4.1, on the next page.

This equation is actually rather easy to understand: in polar coordinates the equation becomes

$$r' = (r^4/4 - r^2/2 + \alpha)r$$
$$\theta' = 1.$$

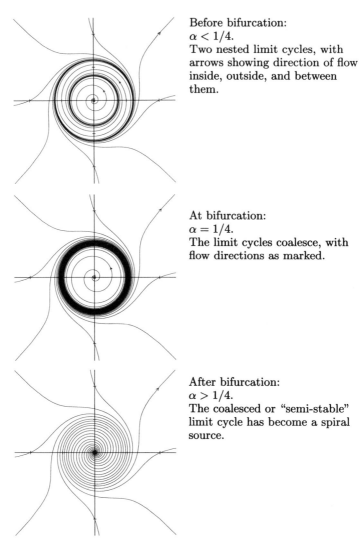

Before bifurcation:
$\alpha < 1/4$.
Two nested limit cycles, with arrows showing direction of flow inside, outside, and between them.

At bifurcation:
$\alpha = 1/4$.
The limit cycles coalesce, with flow directions as marked.

After bifurcation:
$\alpha > 1/4$.
The coalesced or "semi-stable" limit cycle has become a spiral source.

FIGURE 9.4.1. Semi-stable limit cycle bifurcation for Example 9.4.1.

Since the zeroes of the polynomial $r^4/4 - r^2/2 + \alpha$ occur at $r^2 = 1 \pm \sqrt{1 - 4\alpha}$, we see that there are two zeroes of r' for $0 < \alpha < 1/4$, one degenerate zero for $\alpha = 1/4$, and none for $\alpha > 1/4$. Thus, there are two cycles when $0 < \alpha < 1/4$, one when $\alpha = 1/4$, and none when $\alpha > 1/4$.

For $0 < \alpha < 1/4$, r' is negative between the zeroes and positive elsewhere, so solutions spiral toward the inner limit cycle from both sides and away from the outer one from both sides. For $\alpha = 1/4$, $r' \geq 0$ everywhere, vanishing only when $r = 1$; consequently, the unit circle $r = 1$ is a semi-stable limit cycle, attracting solutions inside it and repelling solutions outside it.

For $\alpha > 1/4$, $r' > 0$ whenever $r > 0$, so all solutions, except the equilibrium at the origin, spiral to infinity.

This analysis completely explains the observed behavior.   ▲

It is not so very easy to come up with examples of semi-stable limit cycle bifurcations: even though they occur in many families, they tend to occur for rather restricted values of the parameter. We will show in Exercises 9.4#1 and 9.4#2 that near the indeterminate cases of both Theorem 9.2.3, where the Liapunov coefficient $L = 0$, and Theorem 9.3.3, where a homoclinic saddle connection occurs and the trace of the linearization at the saddle vanishes, this bifurcation usually can be found.

It is also possible to perturb area-preserving vector fields so as to create limit cycles, and examples are given in Exercise 9.4.#2. The next example is of this form, since it can be thought of as a perturbation of

$$x' = y$$
$$y' = -x.$$

Since many differential equations of practical importance can be thought of as conservative mechanical systems, slightly perturbed by friction and slightly perturbed by a driving force, such examples show up quite frequently in applications.

Example 9.4.2. Consider the family of differential equations

$$x' = y$$
$$y' = -x + (\alpha + \beta \cos x)y.$$

Figure 9.4.2 is the computer result for $\alpha = 0$ and $\beta = 0.5$.

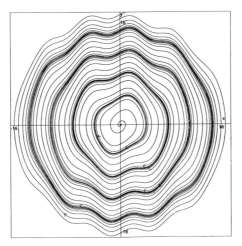

FIGURE 9.4.2. Four, of the infinitely many, limit cycles for Example 9.4.2.

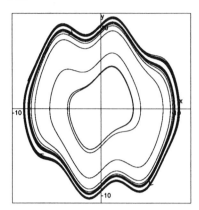

Before bifurcation:
$\alpha = 0.063$.
Three limit cycles: one irrelevant inner one, and two very
close outside.
Two trajectories shown.

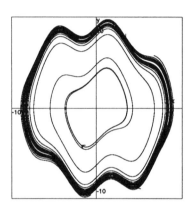

At bifurcation:
$\alpha = 0.065$.
Too outer cycles appear to
have coalesced; too close to call.
Drawn as two trajectories, one
on the outside, one on the
inside.

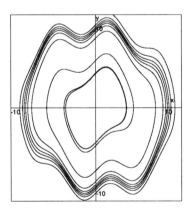

After birfucation:
$\alpha = 0.07$.
Although solutions take a while
getting through the region where
the limit cycles were, there certainly are not any there anymore; this is all a single trajectory.

FIGURE 9.4.3. One semi-stable limit cycle bifurcation for Example 9.4.2.

It seems that there are infinitely many cycles as you zoom out. This
was proved by Zhang Zhi-fen in 1980; the reader can find a proof and a
discussion of the literature in *The Theory of Limit Cycles* (Ye Yan-Qian,
AMS, 1984, pp. 166, 172).

More precisely, Zhang proves that there are precisely n limit cycles in the vertical strip $|x| < (n+1)\pi$. Note that Figure 9.4.2 corresponds to the region $|x| < 15.71 \sim 5\pi$, and that there are four limit cycles, as there should be. Observe that the limit cycles are alternately attracting and repelling, with the innermost one attracting.

Suppose we keep $\beta = 0.5$ and increase α. Exercise 9.4#5a asks you to show that the origin is a source for all $\alpha > 0$ and that $r' > 0$ when $\alpha > \beta$, so that there can then be no limit cycles. What happens to the cycles?

Figure 9.4.3 shows what happens: the attracting limit cycles move out and the repelling limit cycles move in, until they merge and disappear in pairs.

Exercise 9.4#5e asks you to prove that this description is correct. It seems likely that the outer ones disappear first and that the innermost pair disappears last, though the authors have not proved this.   ▲

Another natural way to find coalescing limit cycles is to consider periodic differential equations in one variable. In fact, we have already seen in Chapter 2 (Part I) an example of such a thing.

Example 9.4.3. Example 2.5.2 was the population problem that incorporated both competition and hunting:

$$x' = (2 + \cos t)x - 0.5x^2 - \alpha.$$

When we changed the values of the "hunting" parameter α, our pictures looked like Figure 9.4.4.

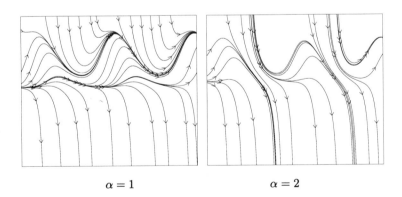

$$\alpha = 1 \qquad\qquad\qquad \alpha = 2$$

FIGURE 9.4.4. Results of changing hunting parameter in Example 9.4.3. ▲

If we now wrap these pictures around a cylinder of circumference equal to one period, we realize that they are the first and third stages of a semi-stable limit cycle bifurcation, as shown in Figure 9.4.5.

Before bifurcation: Two limit cycles, with arrows showing direction of flow inside, outside, and between them.

At bifurcation: The limit cycles coalesce, with flow directions as marked.

After bifurcation: *All* solutions simply go downward.

FIGURE 9.4.5. Illustration of Example 9.4.3 as semi-stable limit cycle bifurcation.

9.5 Bifurcation in One-Parameter Families

In preparation for this section, you might review the summary presented in Figure 9.0.5, of the four types of bifurcations described in the four earlier sections. We now proceed to give some examples of how you find the various possible bifurcations in a family of differential equations with one parameter. In the process, we will explain what understanding a differential equation that depends on a parameter means.

Example 9.5.1. Consider the system of differential equations

$$x' = x^2 - y^2 + 1$$
$$y' = y - x^2 - \alpha.$$

In this case, it is quite easy to understand the isocline picture. The curve where the vector field is vertical is the hyperbola of equation $x^2 - y^2 = -1$, and the curve where the vector field is horizontal is the parabola of equation $y = x^2 + \alpha$, which moves up and down with α.

It is clear that these two curves are tangent exactly when $\alpha = \pm 1$, at the points $(0, 1)$ and $(0, -1)$, respectively. It should also be clear that the vector field has no zeroes for $\alpha > 1$, two zeroes for $-1 < \alpha < 1$, and four if $\alpha < -1$.

At a zero (x_0, y_0) of the vector field, the linearization matrix A is

$$A = \begin{bmatrix} 2x_0 & -2y_0 \\ -2x_0 & 1 \end{bmatrix}.$$

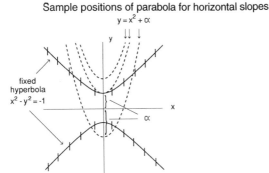

Sample positions of parabola for horizontal slopes
$y = x^2 + \alpha$

FIGURE 9.5.1. Isoclines of horizontal and vertical slope for $x' = x^2 - y^2 + 1$, $y' = y - x^2 - \alpha$.

From Figure 9.5.1 it is clear that saddle nodes occur only when the (moving) parabola is tangent to the (fixed) hyperbola at the points $(0, -1)$ and $(0, 1)$; you can check this by noting that $\det A$ vanishes only when $x_0 = 0$ and when $y_0 = 1/2$; this latter line does not intersect the hyperbola $x^2 - y^2 + 1 = 0$, hence it contributes no tangencies. Computer drawings confirming saddle nodes at $\alpha = -1$ and $\alpha = +1$ are in Figure 9.5.2.

Similarly, we can look for Hopf bifurcations by looking at $\text{tr}\, A = 1 + 2x_0$, which vanishes when $x_0 = -1/2$. There are two points on the hyperbola with x-coordinate $-1/2$; they are $(-1/2, \pm\sqrt{5}/2)$. If $(-1/2, +\sqrt{5}/2)$ is a zero of the vector field for some α, then $\det A = \sqrt{5} - 1 > 0$, so at such a value of α, a Hopf bifurcation will indeed occur. It is easy to show that this happens for the unique value $\alpha_0 = \sqrt{5}/2 - 1/4 \approx 0.868$. With a computer, you can confirm that this Hopf bifurcation lies just to the right of $\alpha = 0.868$, for which the phase plane is shown in Figure 9.5.2.

A similar analysis shows that if $(-1/2, -\sqrt{5}/2)$ is a zero of the vector field, then it will be a saddle, so it contributes no bifurcation.

There are two more bifurcations, which you cannot find by analyzing the isocline picture or the linearization matrix, but which can be found by computer exploration: a *homoclinic saddle connection* for some value $\alpha_1 \approx 0.722$, and a *heteroclinic saddle connection* for $\alpha_2 \approx -1.513$, as shown in Figure 9.5.2. Both the Hopf bifurcation and the homoclinic saddle connection are associated with *limit cycles*; indeed, this equation has such a cycle for $\alpha = 0.868$, as shown in Figure 9.5.2. Apparently, limit cycles exist for no values of α outside the small interval $(0.722, 0.86803\ldots)$ (though we do not quite know how to prove that). A clue to how and where to search for global bifurcations will be provided by closer examination of the numbers and types of singularities and limit cycles in each region of the parameter space defined by the bifurcation diagram.

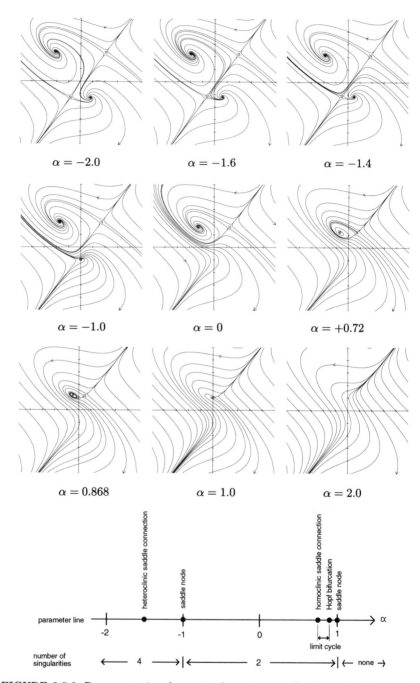

FIGURE 9.5.2. Representative dynamic plane pictures for Example 9.5.1, with parameter line bifurcation diagram. The equations are $x' = x^2 - y^2 + 1, y' = y - x^2 - \alpha$.

Pictures of the phase plane for this equation, for some representative values of the parameter, are shown (arranged in order of increasing α) in Figure 9.5.2. The accompanying bifurcation diagram on the parameter line is shown at the bottom of the same page. ▲

CAPSULES FOR NUMERICAL INVARIANTS

It is helpful to label the parameter line in a more informative way. The parameter line of Example 9.5.1 is broken up by the five bifurcation points into six regions, and within each of these regions, the numbers of sinks, sources, saddles, and attracting and repelling limit cycles does not change, as we saw in Section 8.8 and more formally in the Structural Stability Theorem 8*.6.3. We call these numbers the numerical invariants of the region, and we will write them as the following "capsules":

$$\begin{bmatrix} \# \text{ sinks} & & \# \text{ sources} \\ & \# \text{ saddles} & \\ \# \text{ attr lim cycles} & & \# \text{ repel lim cycles} \end{bmatrix}.$$

The numerical symbols for adjacent regions are related to each other in quite specific ways, depending on the bifurcation that separates them. In Exercise 9.5#7, you are asked to summarize Sections 9.1–9.4 by making a complete list of these possible changes.

Thus, the parameter line in Figure 9.5.2 should be labeled as in Figure 9.5.3; the reader should check that the capsules of adjacent regions are related as Exercise 9.5#7 prescribes. For example, with blowups, you can see that the limit cycle between the homoclinic saddle connection and Hopf bifurcation is repelling, or you can deduce that from the fact that inside the limit cycle the zero is a sink, as identified from the trace of the determinant of the linearization.

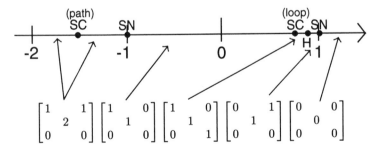

FIGURE 9.5.3. Repeat of the bifurcation diagram for Figure 9.5.2, with capsules describing numerical invariants in each of the six intervals defined by the bifurcation values of α.

Note that a heteroclinic saddle connection does not change the numbers in a capsule.

Example 9.5.1 brings out a very common phenomenon: saddle connections are often related to Hopf bifurcations in the sense that a limit cycle is born in one bifurcation and dies in the other; moreover, the relevant values of the parameter are often very close, and you often find parameters for which limit cycles exist only after trying to come up with a consistent capsule labeling of the parameter space.

Example 9.5.2. Consider the differential equation

$$x' = y - x + (x - \alpha)^2$$
$$y' = y - x^3 + 3x.$$

For $\alpha = 2$, the point $(2, 2)$ is an equilibrium point whose linearization is a center. However, the nonlinear behavior of this equilibrium makes it a weak source; you will note in Figure 9.5.4 when $\alpha = 2$ how slowly the solution converges backward to the equilibrium point. For $\alpha > 2$, the equilibrium point is an ordinary *sink*; after Hopf bifurcation, the equilibrium point is an ordinary *source*, with solutions spiralling out from it toward a *limit cycle*. The limit cycle disappears with a homoclinic saddle connection at $\alpha \approx 1.77$. ▲

As explained in Section 8.7, symmetries in differential equations can lead to other, nongeneric, behaviors that will be discussed further in Section 9.6. An example is provided in Exercise 9.5#2, where a double saddle connection appears.

9.6 Bifurcation in Two-Parameter Families

In this section, we will present examples of bifurcation behavior for two-parameter systems of the form

$$x' = f(x, y, \alpha, \beta)$$
$$y' = g(x, y, \alpha, \beta).$$

That is, we will break up the $\alpha\beta$ *parameter plane* according to the dynamical behavior of the corresponding differential equation. A detailed treatment of this topic is really beyond the scope of this book; there is a large variety of possible behaviors that can occur, and we will not provide an exhaustive list. But we can provide a good introduction, and you can go to the References for more detail.

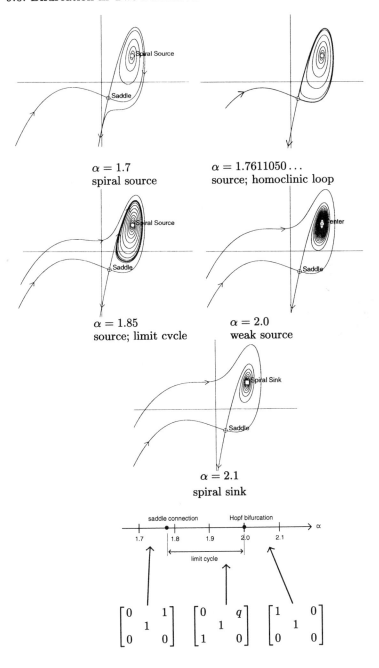

FIGURE 9.5.4. Representative dynamic plane pictures for Example 9.5.2. At the bottom is the parameter line bifurcation diagram, with capsules describing numerical invariants. The equations are $x' = y - x + (x - \alpha)^2$, $y' = y - x^3 + 3x$.

The main thing to realize is that each of the four bifurcations presented in Sections 9.1–9.4 will now occur along a *curve* in the $\alpha\beta$-*parameter plane*. These curves cut up the parameter plane into regions in which all phase planes "look the same," in particular, have the same "capsule," as defined in Section 9.5.

These bifurcation curves can intersect and meld in various ways; the object of this section is to describe some of the ways in which this can happen. One very innocent way in which they can intersect occurs simply when two bifurcations occur independently in different parts of the *xy phase plane*, even though they collide in the parameter plane. Things are much more complicated when two bifurcations collide in the phase plane, as well as in the parameter plane. These are called *degenerate bifurcations*; we will be concerned with bifurcations of codimension two, meaning that they will inevitably occur in systems with two parameters and cannot be eliminated or simplified by small modifications of the equations. Our examples in this section will explore some of the possibilities.

What would we expect of degenerate bifurcations? One answer lies in the fact that each of the codimension one bifurcations—saddle-nodes, Hopf bifurcations and homoclinic saddle connections, and semi-stable limit cycles—had a nondegeneracy hypothesis: some number had to be nonzero in order for the description to be correct. Along the curve corresponding to that bifurcation, the number in question might change sign, going from positive to negative. You must expect something strange to happen at such values.

This leads to the following degenerate bifurcations (Figures 9.6.1–9.6.5):

Case (a): A saddle-node with *both eigenvalues zero* at the zero of the vector field, creating failure of the First Nondegeneracy Condition 9.1.6 and simultaneous Hopf bifurcation (Example 9.6.1).

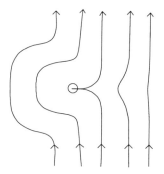

FIGURE 9.6.1.

Case (b): A saddle-node with *one eigenvalue zero* at the zero of the vector field, *the other nonzero*, and Failure of the Second Nondegeneracy Condition 9.1.7, resulting in a double pony tail (Example 9.6.5).

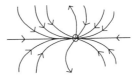

FIGURE 9.6.2.

Case (c): A Hopf bifurcation for which *the first Liapunov coefficient L* (Theorem 9.2.3) *vanishes*, resulting in a weak weak sink or source; we will see that there are arbitrarily small perturbations with two limit cycles near the zero of the vector field (Example 9.6.2 and Exercise 9.2#4).

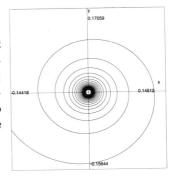

FIGURE 9.6.3.

Case (d): A homoclinic saddle connection for *a saddle with linearization having trace* 0 at the zero of the vector field; we will see that when this happens there are arbitrarily small perturbations with two nested limit cycles near the homoclinic saddle connection (Example 9.6.2 and Exercise 9.4#1).

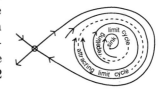

FIGURE 9.6.4.

Case (e): *Three limit cycles coalescing into one.*
(Example 9.6.6, which has a great deal of discussion, and Exercise 9.6#13).

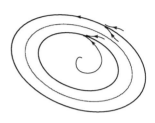

FIGURE 9.6.5.

A bit of thought will show that there are many other degenerate possibilities (Figures 9.6.6–9.6.8):

Case (f): A sort of *double saddle connection*
in which a stable separatrix leading from one
saddle to another may occur at the same time
as a stable separatrix from the second leads
back to the first (Exercise 9.5#2).

FIGURE 9.6.6.

Case (g): Another sort of *double saddle connection* involving a single saddle with two homoclinic loops (Exercise 9.6#1).

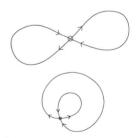

FIGURE 9.6.7.

Case (h): An *exceptional homoclinic saddle-node*
where the exceptional solution comes back not
in the pony tail but as one of the separatrices
of the saddle (Exercises 9.6#6 and 9.6#12).

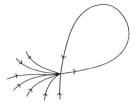

FIGURE 9.6.8.

This is not the end of the list of possible degenerate bifurcations: Exercise
9.6#15 suggests several more. Space does not allow us to look at all of these
carefully, but we will examine a few.

Degenerate case (a) is perhaps the most entertaining. Note that it is a
degenerate case of both a Hopf bifurcation and a saddle-node bifurcation,
so we would expect it to occur where the curves of saddle-node bifurcations
and Hopf bifurcations meet. It turns out that a curve of saddle connections
comes to the same point of the parameter plane! The following example
represents what happens in general.

Example 9.6.1.
$$x' = y - x^2$$
$$y' = \alpha x + \beta - y. \tag{15}$$

The bifurcation diagram is represented with relevant phase plane figures
in Figure 9.6.9. This is one rather exceptional case in which most of it can
be calculated by hand.

The first thing to do is to find the zeroes and to see when their linearizations have a zero or purely imaginary eigenvalue. This is quite easy: the
zeroes of the vector field are the intersections of the parabola $y = x^2$ and
the line $y = \alpha x + \beta$, which have coordinates

$$x_0 = \frac{\alpha \pm \sqrt{\alpha^2 + 4\beta}}{2}, \qquad y_0 = x_0^2. \tag{16}$$

The linearization at (x_0, y_0) is given by

$$\begin{bmatrix} -2x_0 & 1 \\ \alpha & -1 \end{bmatrix}.$$

(i) The saddle-node locus

Saddle-nodes occur when one eigenvalue of the linearization is zero or,
alternately, when the determinant of the linearization is zero. In our case,
this gives

$$2x_0 - \alpha = (\alpha \pm \sqrt{\alpha^2 + 4\beta}) - \alpha = 0,$$

i.e., along the curve in the parameter plane of equation $\alpha^2 + 4\beta = 0$. The reader should see that

$$\alpha^2 + 4\beta = 0 \qquad (17)$$

precisely when the line of equation $y = \alpha x + \beta$ is tangent to the parabola $y = x^2$. So we see that at least for this example, along the saddle-node locus, zeroes of the vector field are created or annihilated.

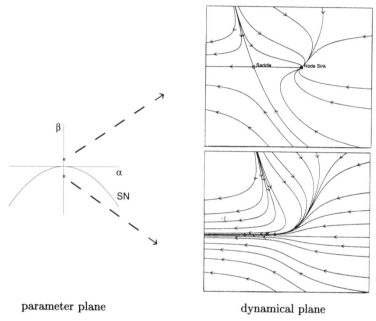

parameter plane   dynamical plane

FIGURE 9.6.9. Saddle-node bifurcation locus (in parameter plane).

In summary, as shown in Figure 9.6.9,

The saddle-node locus is the parabola of equation $\alpha^2 + 4\beta = 0$ in the parameter plane; inside the parabola, the vector field has no zeroes; outside the parabola, the vector field has two zeroes.

We will now analyze the types of these zeroes.

(ii) The Hopf bifurcation locus

Hopf bifurcation occurs if there is a zero of the vector field whose linearization (2) has a pair of purely imaginary eigenvalues or, equivalently, for a 2×2 matrix, if the determinant is positive and the trace is zero.

In our case, these conditions read

$$2x_0 = -1, \qquad 2x_0 - \alpha > 0. \qquad (18)$$

The first equation of (18) is satisfied if $\alpha^2 + 4\beta = (\alpha+1)^2$, which reduces to the equation of the line

$$\beta = \frac{\alpha}{2} + \frac{1}{4}. \tag{19}$$

Substituting the second equation of (18) into the first gives $\alpha < -1$, which describes part of the line (19).

At the point $(\alpha, \beta) = (-1, -1/4)$, the Hopf bifurcation curve meets tangentially with the saddle-node curve, as shown in Figure 9.6.10. The corresponding zero of the vector field is at $(x_0, y_0) = (-1/2, 1/4)$.

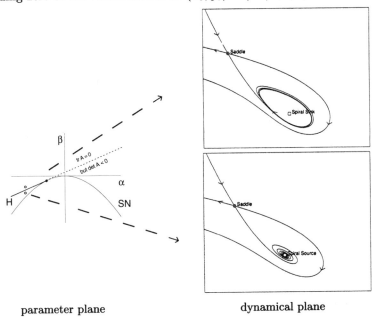

parameter plane   dynamical plane

FIGURE 9.6.10. Hopf bifurcation locus added to parameter plane.

In Exercise 9.6#2b you are asked to show that in this example the Hopf bifurcations are always weak sources, so that case (c) bifurcations do not occur.

(iii) Saddle connections

Figure 9.6.10 is not the complete description of the parameter plane: if we attempt to put capsules in the various regions, we will find

$$\begin{bmatrix} 1 & 0 \\ & 1 \\ 0 & 1 \end{bmatrix} \quad \text{and} \quad \begin{bmatrix} 1 & 0 \\ & 1 \\ 0 & 0 \end{bmatrix},$$

just above the line of Hopf bifurcations and just to the right of the parabola, respectively. Something must happen to make the limit cycle disappear, and it is quite easy to see what.

If we try the values $\alpha = -2$, $\beta = -0.7$ and $\alpha = -2$, $\beta = -0.5$, we find the pictures of Figure 9.6.11 in the dynamical plane. A careful look at the handedness of the separatrices shows that a homoclinic saddle connection must occur between these two values.

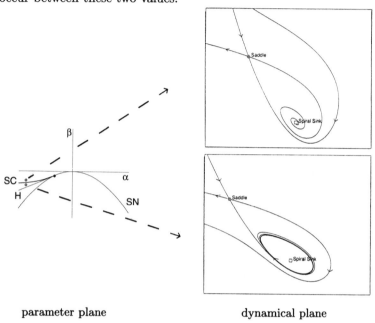

<table>
<tr><td>parameter plane</td><td>dynamical plane</td></tr>
</table>

FIGURE 9.6.11. Saddle connection bifurcation locus added to parameter plane.

The curves of Hopf bifurcations and saddle-nodes could be computed explicitly in this case. It is much harder to understand the curve of saddle connections, since this requires solving the differential equation. But this bifurcation locus can be found approximately with a bit of computer experimentation, and the program *Planar Systems* will draw it. It turns out to be another curve tangent to the previous two at their point of contact $(\alpha, \beta) = (-1, -1/4)$, as is also shown in Figure 9.6.11.

Exercise 9.6#2a asks you to extend the locus for saddle connections by computer experiment, such as by fixing α at -3, then -4, then -5, and so on, in each case varying β to locate where the saddle connection occurs for that value of α. The results are shown, with capsules, in Figure 9.6.12.

We claim to have now found the complete bifurcation diagram. We have not actually proved it, but the labeling by capsules is now coherent.

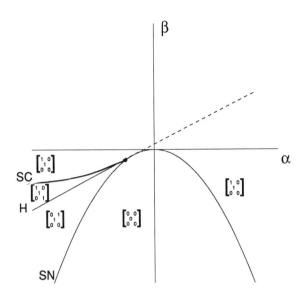

FIGURE 9.6.12. Parameter plane for Example 9.6.1, with "matrix" notation added for numbers of sources, sinks, saddles, and attracting and repelling limit cycles. ▲

Example 9.6.1 has demonstrated what usually happens under degenerate possibility (a):

> *If a vector field depends on two parameters and for some value*
> (α_0, β_0) *of these parameters a zero has linearization with both*
> *eigenvalues zero, then a smooth curve of saddle nodes goes through*
> (α_0, β_0); *furthermore, both a curve of Hopf bifurcations and a*
> *curve of saddle connections touch this saddle-node curve tangen-*
> *tially at the same point and die there. There are limit cycles for*
> *parameter values between these two curves.*

When for a parametrized family of differential equations we have produced a picture like Figure 9.6.12, we think we can fairly say that we understand that family. It may still be quite a task to prove the figure complete, but in our experience, the compatibility test is a good indication that it is. In any case, that is our objective in this chapter.

You know that in passing from one structurally stable region to another, these numerical invariants can vary only in quite restricted ways (Exercise 9.5#7), and it often occurs that a first attempt at labeling does not satisfy these rules, indicating that some bifurcation curves have not been found.

The next example explores degenerate possibilities (c) and (e):

Example 9.6.2. Consider the system

$$x' = \beta x + y + \alpha(x^2 + xy) + x^3$$
$$y' = -x + y^2. \tag{20}$$

In Exercise 9.7#11, you are asked to explore the bifurcation diagram in the large for this system of equations, but one feature is clear: the origin is a zero of the vector field for all values of (α, β), and the line $\beta = 0$ is a line of Hopf bifurcations. Moreover, the first Liapunov coefficient (defined in Theorem 9.2.3) is $L = 3 - \alpha^2$, which vanishes at $\alpha = \pm\sqrt{3}$; we will examine the bifurcation diagram near $\alpha = -\sqrt{3}$.

An easy computer experiment, or, more rigorously, a difficult computation (better done using a computer algebra system such as *Maple* or *Mathematica*) will show that this point is a weak weak sink. This means that for $(\alpha, \beta) = (-\sqrt{3}, 0)$ you can find a unique function

$$f(x, y) = (x^2 + y^2)/2 + f_3 + f_4 + f_5 + f_6, \tag{21}$$

with f_i homogeneous of degree i, and a unique number B such that

$$\frac{d}{dt}(f(x(t), y(t))) = B(x^2 + y^2)^3 + \text{higher order terms}.$$

The number B is called the *second Liapunov coefficient*; it is only defined if the first one vanishes. Clearly, if $B \neq 0$, the function f is a Liapunov function, making the origin a weak weak sink or source depending on whether B is negative or positive.

This statement about the second Liapunov coefficient is fairly easy to prove using Lemma 9.2.5. Finding such a function f really comes down to making the cubic, quartic, and quintic terms of

$$\frac{d}{dt}f(x(t), y(t))$$

vanish, which involves solving three systems of linear equations for the coefficients of f_3, f_4, and f_5. The system of equations to find f_4 is singular but has a solution anyway: that is precisely what the vanishing of the first Liapunov exponent means. Then there is an "obstruction" to making the terms of degree six vanish, and this obstruction is the number B.

Using *Maple* for equation (21) on system (20) (with many thanks to Allen Back), we find

$$f(x, y) = \frac{x^2}{2} + \frac{y^2}{2} + \left(-\frac{2}{3} + \frac{\sqrt{3}}{3}\right) x^3 - \sqrt{3}x^2 y - xy^2 - \frac{2\sqrt{3}}{3}y^3$$

$$+ \left(2\sqrt{3} - 2\right) x^3 y + \left(\frac{3}{2} + \sqrt{3}\right) x^2 y^2 + 2\sqrt{3}xy^3 + \left(\frac{1}{4} + \frac{3\sqrt{3}}{4}\right) y^4$$

$$+ \left(\frac{64}{15} - \frac{34\sqrt{3}}{15} \right) x^5 + \left(\sqrt{3} - 2 \right) x^4 y + \left(\frac{5}{3} - \frac{11\sqrt{3}}{3} \right) x^3 y^2$$

$$+ \left(-\frac{35}{3} + 3\sqrt{3} \right) x^2 y^3 + \left(-1 - 3\sqrt{3} \right) x y^4 + \left(-\frac{88}{15} + \frac{12\sqrt{3}}{5} \right) y^5$$

$$+ \left(\frac{407}{12} - \frac{167\sqrt{3}}{8} \right) x^5 y + \left(8 - \frac{11\sqrt{3}}{3} \right) x^4 y^2 + \left(\frac{574}{9} - \frac{97\sqrt{3}}{3} \right) x^3 y^3$$

$$+ \left(\frac{151}{12} - \frac{5\sqrt{3}}{12} \right) x^2 y^4 + \left(\frac{353}{12} - \frac{299\sqrt{3}}{24} \right) x y^5 + \left(\frac{181}{36} - \frac{71\sqrt{3}}{36} \right) y^6$$

and

$$B = \frac{1}{12} - \frac{11}{24}\sqrt{3}.$$

Now we will see that near the point $(-\sqrt{3}, 0)$ in the (α, β)-plane, there must be parameter values for which the system has *two* limit cycles. We have seen examples of this sort of thing in Exercise 9.4#2. Indeed, since the origin is a weak weak sink, we can find a solution which comes back inside itself; then we perturb the parameters along the Hopf bifurcation curve, very slightly so that the origin is now a weak source, but the solution we were considering still comes back inside itself. Now since the origin is a weak *source*, we can find a solution closer to the origin which comes back outside itself, and these two solutions bound a region to which the Poincaré–Bendixson Theorem 8.5.1 can be applied, guaranteeing the existence of one limit cycle.

Now we further perturb the parameters, off the Hopf bifurcation curve, so the origin becomes a linear sink, but we perturb it sufficiently little that the two solutions considered above still come back inside or outside themselves as before. We can now find another solution, much closer to the origin yet, which comes back inside itself, and this third solution, together with the second, bounds a new region to which the Poincaré–Bendixson Theorem 8.5.1 can also be applied. Hence, the existence of a second limit cycle is also guaranteed.

Figure 9.6.13 illustrates the construction described in this example; note, however, that the phenomena described are often rather hard to observe, because a weak weak sink is so nearly a center that it is rather hard to see the solution coming back clearly inside itself, as opposed to on itself. The effect being observed is so small that the skeptic would be well justified in wondering whether he is really observing a phenomenon connected with the weak weak sink or whether the numerical method introduces errors larger than the effect under study.

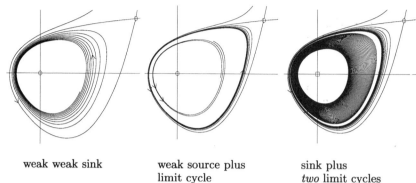

weak weak sink   weak source plus   sink plus
                        limit cycle   *two* limit cycles

FIGURE 9.6.13. Three steps for going from a weak weak sink to two limit cycles.
▲

Note also the word "small" as applied to the perturbations of Example
9.6.2. These pairs of limit cycles only exist in a very small region of the
parameter space, and if the trace at the linear sink is further decreased,
the limit cycles collide and disappear. Thus, as shown in Figure 9.6.14:

> *Near a degenerate Hopf bifurcation, you can expect to see a smooth
> curve of Hopf bifurcations and a curve of semi-stable limit cycles,
> which touches the curve of Hopf bifurcations tangentially and dies.*

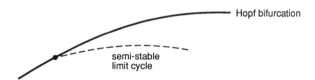

FIGURE 9.6.14. Locus of semi-stable limit cycle bifurcation ending tangentially
at Hopf bifurcation locus.

Exactly the same argument as the last can be applied to a homoclinic
saddle connection if the trace of the linearization at the saddle vanishes.
The saddle loop will then usually be weakly attracting or repelling; two
perturbations—the first keeping the saddle connection but making it of
the opposite type, as it was for the degenerate value of the parameter,
will create one limit cycle, and the second to vary it in or out will create
another. Thus, as shown in Figure 9.6.15:

> *near a degenerate saddle connection, you also expect a smooth
> curve of saddle connections and a curve of semi-stable limit cycles
> touching the saddle connection curve tangentially and dying there.*

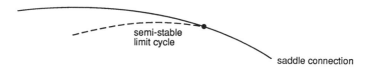

FIGURE 9.6.15. Locus of semi-stable limit cycle bifurcation ending tangentially at saddle connection bifurcation locus.

The two events above often occur together. We have seen that a curve of Hopf bifurcations often has a companion curve of saddle connections, together bounding a region of the parameter plane in which the differential equation has a limit cycle. If the Hopf bifurcation becomes degenerate, its companion curve must move to the other side but will not, in general, cross the curve of Hopf bifurcations exactly at the degenerate Hopf bifurcation, but somewhere nearby. Until you understand the discussion above, this would appear to lead to inconsistencies as to the sides of the curves of Hopf bifurcations and saddle connections on which limiit cycles appear. These inconsistencies are resolved by Figure 9.6.16.

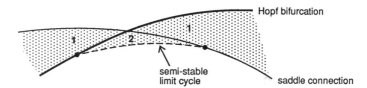

limit cycles occur in shaded areas, in number indicated

FIGURE 9.6.16. Typical fitting together of Hopf, saddle connection, and semi-stable limit cycle bifurcation loci.

All three bifurcation curves together as described is precisely what happens in our last example; let us return to it.

Example 9.6.3. For the equation of Example 9.6.2,

$$x' = \beta x + y + \alpha(x^2 + xy) + x^3$$
$$y' = -x + y^2,$$

Figure 9.6.17 represents the parameter space near $(\alpha, \beta) = (-\sqrt{3}, 0)$.

The α-axis of Figure 9.6.17 is a curve of Hopf bifurcations, degenerate when $\alpha = -\sqrt{3}$; the computer also calculated a curve of saddle connections, which becomes degenerate near $(-1.265, -0.0465)$. The computer, furthermore, calculated a curve of semi-stable limit cycles, which connects the two degenerate bifurcations. This curve, together with the curves of

Hopf bifurcations and saddle connections, delimits a tiny region in the parameter plane in which the equation has two limit cycles. This sort of figure in parameter space is quite common, but a look at its size will tell you why you seldom see it.

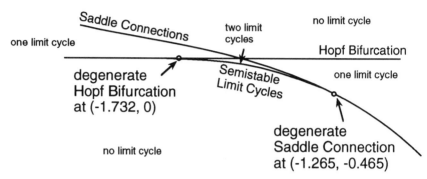

FIGURE 9.6.17. A tiny piece of parameter space, near two degenerate bifurcations, for $x' = \beta x + y + \alpha(x^2 + xy) + x^3$, $y' = -x + y^2$.

Figure 9.6.17 is a blowup of just a small part of the parameter space for Example 9.6.3. Exercise 9.6#12 gives you the larger picture and asks you to determine the capsule labels for each region. This is an excellent way to test your understanding of bifurcations.   ▲

In order to examine cases (b) and (d), we shall need the following lemma, which is proved in Exercise 9.6#14.

Lemma 9.6.4. *The discriminant of the polynomial $P(x) = x^3 + \alpha x + \beta$ is $\Delta = 4\alpha^3 + 27\beta^2$. Then P has*

> *three real roots if $\Delta < 0$,*
>
> *one real root if $\Delta > 0$, and*
>
> *two real roots, of which one is double, if $\Delta = 0$.*

The curve $\Delta = 0$ in the parameter plane is represented in Figure 9.6.18; note that it has a *cusp* at the origin.

The following example shows one of the two possible generic possibilities when degenerate possibility (b) occurs, the other is explored in Exercise 9.6#10.

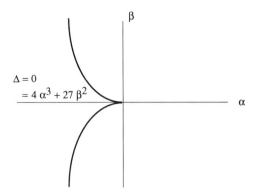

FIGURE 9.6.18. Parameter plane for Examples 9.6.5 and 9.6.6.

Example 9.6.5. Consider the system of differential equations

$$x' = y - x^3$$
$$y' = y - \alpha x + \beta.$$

At $(\alpha, \beta) = (0,0)$, there are two pony tails tangent to the x-axis, and, in addition, there are unique trajectories approaching the origin tangent to the other eigendirection; in the example, these are the upper and lower parts of the y-axis.

Nearby differential equations depend on the value of Δ; inside the cusp of Figure 9.6.18, there are three zeroes, and the differential equation looks like Figure 9.6.19. Outside the cusp, there is a single zero, which is near the origin a node sink. Along the curve $\Delta = 0$, there are saddle-nodes.

In the parameter space, we see that *degenerate possibility* (b) simply means that *the curve of saddle nodes has a cusp.*

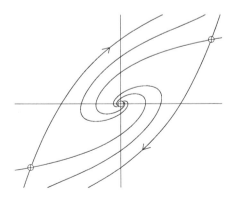

FIGURE 9.6.19. On the way to a double pony tail; phase plane for $(\alpha, \beta) = (-1, 0)$, in the Δ cusp of Example 9.6.5. ▲

Degenerate possibility (d), where three limit cycles coalesce, is rather similar to degenerate possibility (b). Consider the following example, which also looks rather artificial.

Example 9.6.6. Let $r^2 = x^2 + y^2$ and consider the system of differential equations

$$\begin{bmatrix} x \\ y \end{bmatrix}' = \begin{bmatrix} y \\ -x \end{bmatrix} + ((r^2 - 1)^3 + \alpha(r^2 - 1) + \beta) \begin{bmatrix} x \\ y \end{bmatrix}.$$

This equation has limit cycles $r^2 = C$ at those numbers C such that

$$(C - 1)^3 + \alpha(C - 1) + \beta = 0.$$

By Lemma 9.6.4 we know $C-1$ has three real roots if $\Delta = 4\alpha^3 + 27\beta^2 < 0$.

Figure 9.6.18 serves as parameter plane for this example also. We find that inside the cusp there are three limit cycles, outside there is just one, and on the curve $\Delta = 0$, there is a semi-stable limit cycle plus one other limit cycle.

Phase plane pictures are shown in Figure 9.6.20. The calculations comprise Exercise 9.6#8.

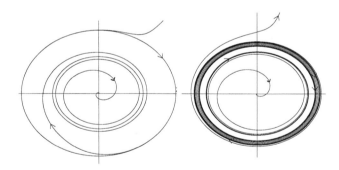

FIGURE 9.6.20. Creation of three limit cycles for Example 9.6.6.   ▲

Thus, we see in the parameter space that *the curve of semi-stable limit cycles may also have cusps*, under degenerate possibility (d).

9.7   Grand Example

If you wish to understand a system of differential equations depending on two (or for that matter any number of) parameters, the basic idea is to

> locate in parameter space the curves (hypersurfaces in parameter space of higher dimensions) along which bifurcations occur,

more particularly, the places where these curves meet. Then attempt to

> label the components of your drawing by the numbers of sources, sinks, saddles, and attracting or repelling limit cycles.

We shall use the capsules described in Section 9.5, telling how many of each type of feature a differential equation has for parameter values located within a given region of the parameter plane:

$$\begin{bmatrix} \text{sinks} & & \text{sources} \\ & \text{saddles} & \\ \text{attracting cycles} & & \text{repelling cycles} \end{bmatrix}.$$

This section is devoted to the following example, which shows some of the complications even simple systems can display. The authors learned a lot from its study, and we have tried to recapture the spirit of exploration in the text.

Example 9.7.1. Consider the system of differential equations

$$x' = (y - \alpha)\cos\theta - (y - x^3 + 3x)\sin\theta$$
$$y' = (y - \alpha)\sin\theta + (y - x^3 + 3x)\cos\theta,$$

for which a few phase portraits are shown in Figure 9.7.1—enough to show that we will definitely find some interesting bifurcations.

$$\begin{bmatrix} 1 & 0 \\ & 0 & \\ 1 & 1 \end{bmatrix} \qquad \begin{bmatrix} 2 & 0 \\ & 1 & \\ 1 & 2 \end{bmatrix} \qquad \begin{bmatrix} 0 & 1 \\ & 0 & \\ 0 & 0 \end{bmatrix}$$

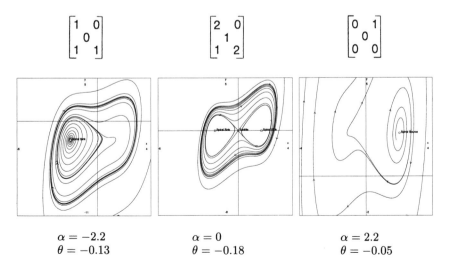

$\alpha = -2.2$	$\alpha = 0$	$\alpha = 2.2$
$\theta = -0.13$	$\theta = -0.18$	$\theta = -0.05$

FIGURE 9.7.1. Very different phase planes for three different (α, θ) values.

TURNING A VECTOR FIELD

Example 9.7.1 is an instance of an interesting general construction: *turning* or *rotating* a vector field, already encountered in Section 8*.4. Various properties of turned vector fields are explored in Exercise 9.7#1. If

$$x' = f(x, y), \qquad y' = g(x, y)$$

is any vector field, the *turned vector field* is the one-parameter family

$$x' = f(x, y) \cos \theta - g(x, y) \sin \theta$$
$$y' = f(x, y) \sin \theta + g(x, y) \cos \theta.$$

It can easily be imagined by simply rotating, at every point (x, y), the vector at that point by θ. A nice illustration was given by Figure 8*.4.7.

In particular, the zeroes of a rotated vector field do not move, and if $u' = Au$ is the linearization of the old vector field at such a zero, then the turned vector field has linearization (Exercise 9.7.#1b)

$$\mathbf{u}' = \begin{bmatrix} \cos \theta & -\sin \theta \\ \sin \theta & \cos \theta \end{bmatrix} A\mathbf{u}.$$

Since the rotation matrix has determinant one, we see that the determinant is not changed, so that: saddles stay saddles, saddle-nodes stay saddle-nodes, sources can become sinks, and vice versa.

More specifically, you are asked in Exercise 9.7#1e to show that for any zero of the original vector field that is a sink, source, or center, there are precisely two values of $\theta \in [-\pi, \pi)$ for which a Hopf bifurcation occurs and that these differ by π.

In Exercise 9.7#1f, you are asked to show that if the zero of the vector field is a nondegenerate saddle-node, then there are precisely two values of θ in $[-\pi, \pi)$ for which both eigenvalues of the linearization vanish and that these differ by π.

LOCAL BIFURCATIONS OF EXAMPLE 9.7.1

It is quite easy to see (in Exercise 9.7#5), when $\theta = 0$ or $2n\pi$, what the zeroes of

$$x' = y - \alpha, \qquad y' = y - x^3 + 3x \qquad (22)$$

look like: there are three zeroes for $-2 < \alpha < 2$, two of which collide at $\alpha = \pm 2$ in a saddle-node, leaving only one for $|\alpha| > 2$. Figure 9.7.2

illustrates this basically geometric argument. Thus, the two lines $\alpha = \pm 2$ are the saddle-node locus in parameter space.

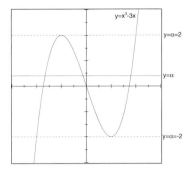

FIGURE 9.7.2. Isoclines of horizontal and vertical slope for $x' = y - \alpha$, $y' = y - x^3 + 3x$.

A little computation or experimentation will show that when there is only one zero, it is a sink or a source, and when there are three (necessarily aligned on the line $y = \alpha$), the middle one is a saddle and the two others are sinks or sources.

To find the parameter space locus for Hopf bifurcation, we must find zeros of the vector field where *the trace of the linearization is zero*. This means we must solve the system (22) in conjunction with the equation for zero trace:
$$3x^2 \sin \theta = 2 \sin \theta - \cos \theta. \tag{23}$$
It is easy to eliminate x and y from equations (22) and (23) to get
$$\alpha = \pm \sqrt{\frac{2 - \cot \theta}{3}} \left(\frac{7 + \cot \theta}{3} \right).$$

In Exercises 9.7#6 and 9.7#7, you are asked to fill in the details. The graph of this curve for $\alpha = f(\theta)$ looks like Figure 9.7.3.

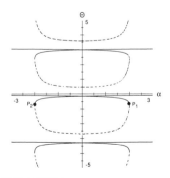

FIGURE 9.7.3. $\alpha = \pm \sqrt{2 - \cot \theta}/3 (7 + \cot \theta)/3$. The solid parts of these curves give the locus for Hopf bifurcation.

The points marked P_1 and P_2 in Figure 9.7.3 have coordinates $(\alpha, \theta) = (\pm 2, -\pi/4)$: they are two of the points where the saddle-node becomes degenerate with a double eigenvalue zero [the others are $(\pm 2, 3\pi/4)$]. (Recall Exercise 9.7#1f.) The part of the curve above P_1 and P_2 (i.e., $-\pi/4 < \theta < 0$) corresponds to a node having trace zero; in other words, they are Hopf bifurcations; the remainder of the curve (dotted) concerns saddles and hence is not a bifurcation curve.

All this information about local bifurcations is displayed in Figure 9.7.4.

THE GLOBAL BIFURCATIONS OF EXAMPLE 9.7.1

So far, we know where the saddle nodes and Hopf bifurcations occur; it remains to find the saddle connections and semi-stable limit cycles. These can no longer be computed by hand; finding the global bifurcations requires an entirely different approach.

However, even without further computer work, note that we do know something: the points of tangency of the curve of saddle nodes and Hopf bifurcations, at $(\pm 2, -\pi/4)$, are places where degenerate bifurcation possibility (a), the double zero eigenvalue, occurs (Section 9.6). You can confirm this in Exercises 9.7#8,9.

In particular, there should be a curve of saddle connections emanating from these points, and there should be a region in the parameter plane between these curves and the Hopf bifurcation curves for which the corresponding dynamical plane pictures show limit cycles surrounding the appropriate nodes.

If you have a program like *Planar Systems* from *Extensions of MacMath* draw these curves, you will see that they cross on the line $\alpha = 0$, at a point Q where θ is very close to -0.19, and where there are two homoclinic saddle connections simultaneously, creating degenerate bifurcation possibility (f). It is a good bit harder to figure out what they do after that, and the authors only found out after considerable experimentation. It turns out that the curve of saddle connections starting at $(2, -\pi/4)$ approaches the line $\alpha = -2$, at a critical homoclinic saddle-node where $\theta \in [-0.139, -0.1385]$, as shown in Figure 9.7.5.

The typical behavior near such a parameter value was studied in Exercise 9.6#6. Such a point must be at one end of a segment in the curve of saddle-nodes in which the saddle-node is homoclinic. By computer, we can attempt to find the other end of the segment, and find that there is another critical homoclinic saddle node with $\theta \in [-0.1405, -0.14]$, and that between these two values the saddle-node is homoclinic. Notice that this range (less than 0.002, tiny compared to the natural θ range of 2π) is so small that we never would have found it if we had not been looking for a place for the curve of saddle connections to go.

But this leads to another question: what happens at the other end of this interval? A bit of thought and experimentation will show that there is

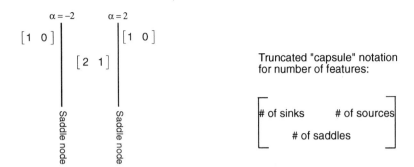

Truncated "capsule" notation
for number of features:

$$\begin{bmatrix} \text{\# of sinks} \quad\quad \text{\# of sources} \\ \text{\# of saddles} \end{bmatrix}$$

First, showing loci for saddle node
bifurcations at $\alpha = \pm 2$ showing total
numbers for sinks and sources, ignoring
numbers of saddles or limit cycles

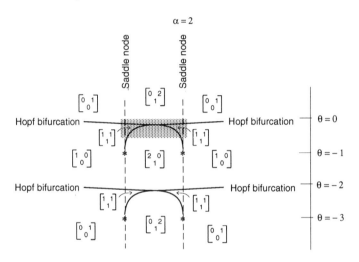

Second, adding loci for Hopf bifurcation showing number of sinks, sources,
and saddles, ignoring numbers of limit cycles

All the "action" for other bifurcations occurs in a region like that shaded,
which is blown up in the remaining figures of Section 9.7.

FIGURE 9.7.4. Summarizing the bifurcation loci for local bifurcations.

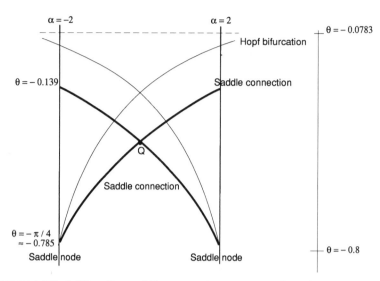

FIGURE 9.7.5. Adding first saddle connection curves to the bifurcation diagram for saddle-node and Hopf bifurcation. The region illustrated is a blowup shown by the shaded region of Figure 9.7.3.

yet another curve of saddle connections emanating from this point! Again, this possibility only occurred to the authors as they tried to understand the question above. Now we must follow this new curve of saddle connections: where does it go? The answer is that it goes right back to the point Q, where the original saddle connection curves met, as shown in Figure 9.7.6.

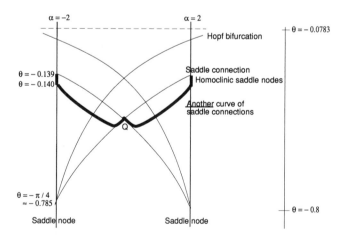

FIGURE 9.7.6. Adding yet another curve for saddle connections to the bifurcation diagram.

In fact, if two for some value of the parameters a saddle has two homoclinic loops [degenerate possibility (g)], there will usually be in the parameter space *two* curves of homoclinic saddle-nodes which intersect transversely, and two half-curves corresponding to "funny" saddle connections. Typical phase plane pictures are shown in Figure 9.7.7.

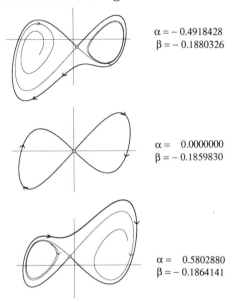

$\alpha = -0.4918428$
$\beta = -0.1880326$

$\alpha = 0.0000000$
$\beta = -0.1859830$

$\alpha = 0.5802880$
$\beta = -0.1864141$

FIGURE 9.7.7. Phase planes for degenerate and "funny" saddle connections in Example 9.7.1.

SEMI-STABLE LIMIT CYCLES IN EXAMPLE 9.7.1

When attempting to label components of the drawing so far by the number of limit cycles, we discover inconsistencies. For instance, if you draw the phase diagram for $\alpha = 2.1$, $\theta = -0.13$, you will find two limit cycles, both surrounding the unique sink that exists for this value. On the other hand, if you try $\alpha = 2.1$, $\theta = -0.2$, you will find no limit cycles. Something must have happened, and a bit of experimentation will show that there is a semi-stable limit cycle for some $\theta \in [-0.155, -0.15]$. Of course, this is just one point of a curve, and we need to find out where the curve goes. Such a curve could meld with a curve of saddle connections, but only at a point where the trace of the corresponding saddle is zero, and we have seen that there are none. It could also meld with a curve of Hopf bifurcations at a point where the first Liapunov coefficient vanishes, and as far as we know, this may happen for some large values of α; in principle, the Liapunov coefficient can be computed, but the algebra is daunting. The program *Planar Systems* in *Extensions of MacMath* will draw this curve of semi-stable limit cycles, if properly coached, as shown in Figure 9.7.8. It turns out to be a nearly horizontal curve that crosses the entire picture without interacting with any other bifurcation curve.

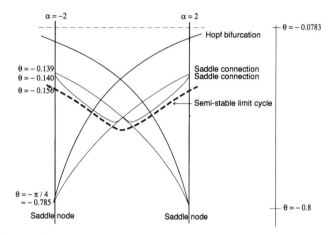

FIGURE 9.7.8. Adding the curves for semi-stable limit cycles to the bifurcation diagram.

Figure 9.7.9 collects all the bifurcation curves we have found in this grand example, complete with capsules showing the number of singularities and limit cycles in each region found by the bifurcation curves.

Cycles at Infinity in Example 9.7.1

There is one more curve of "bifurcations" that must be understood in order to have a coherent bifurcation diagram, and as far as we know, this one cannot be understood in terms of the theory developed so far. There are values of (θ, α) such that when you approach them, a limit cycle surrounding everything appears to grow without bounds. We have not succeeded in analyzing this phenomenon theoretically, but the computer indicates that there is a curve close to the line $\theta = 0$ along which this occurs.

THE FINAL BIFURCATION DIAGRAM FOR EXAMPLE 9.7.1

Adding the global bifurcation curves to a blow-up of the shaded region in Figure 9.7.4, stretched vertically:

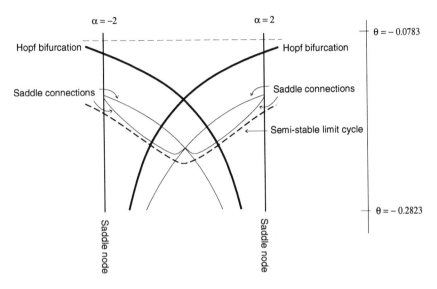

And finally, adding the "capsules" showing the number of features:

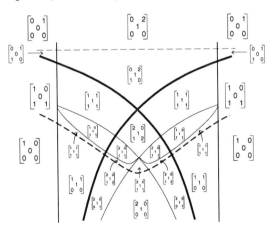

FIGURE 9.7.9. The complete bifurcation diagram for Example 9.7.1.

Chapter 9   Exercises

Exercises 9.1   Saddle-Node Bifurcation

9.1#1. Each curve in the following pairs represents an isocline of horizontal or vertical slope and its associated phase plane influence. Put each pair together in the three possible ways (intersecting twice, tangent, or nonintersecting) and sketch the resulting phase plane trajectories (as in Example 6.1.5) to illustrate the saddle-node bifurcation.

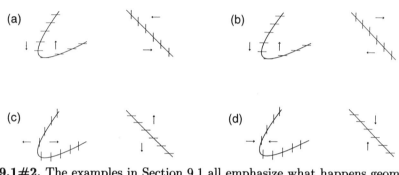

9.1#2. The examples in Section 9.1 all emphasize what happens geometrically for saddle-node bifurcation to occur; here we focus on the algebraic approach, which may be useful when it gives solvable equations. For each of the following systems of differential equations, find algebraically the values of α for which the system undergoes a saddle-node bifurcation. For each segment of the α-axis in the complement of these points, say how many saddles and how many sinks or sources there are.

(a) $\begin{aligned} x' &= y \\ y' &= x^2 - y + \alpha \end{aligned}$ (Example 9.1.2)

(b) $\begin{aligned} x' &= y - x^3 \\ y' &= x - y + \alpha \end{aligned}$

(c)° $\begin{aligned} x' &= y - x^3 + xy \\ y' &= 4x - y - \alpha \end{aligned}$

(d) $\begin{aligned} x' &= x^2 + \alpha - y^3 \\ y' &= -y + x^3 \end{aligned}$ (Example 9.1.12)

Hint: you will need to write three equations in x, y, and α, stating the locations of the singularities as well as the condition for the bifurcation. It is then quite easy to eliminate x and y in parts (a) and (b), but for parts (c) and (d), you will probably want a computer to solve a cubic equation (unless you happen to know Cardano's formulas).

9.1#3. For the system of Example 9.1.2,

$$x' = y$$
$$y' = x^2 - y + \alpha.$$

(a) Confirm algebraically that for $\alpha = -0.266, 0$, and $+0.266$, the zeroes are of the type described in Figure 9.1.1.

(b) Find, by computer experiment and algebraic calculation, a negative value of α closer to zero that gives a node sink rather than a spiral sink.

(c) Blow up a computer picture for $\alpha = 0$, further than in Figure 9.1.2, to show the tangencies of solutions near the origin. Identify the separatrix of the saddle-node, the exceptional solution, and the pony tail.

(d) Show that when $\alpha = 0$ there is no curve $y = f(x)$ given by a power series near zero, with $f(0) = f'(0) = 0$, that contains the exceptional solution in its graph. Hint: Look at Part I, Example 2.7.4.

9.1#4. For each of the following systems, find a value of α for which the system undergoes saddle-node bifurcation. Explain what happens for values of α close to α_0 (bifurcation value). Verify your answers with the computer. Make a set of "before, at, after" pictures to use in the next two exercises.

(a) $\begin{aligned} x' &= \alpha - x^2 \\ y' &= -y \end{aligned}$

(b) $\begin{aligned} x' &= \cos x - y + \alpha \\ y' &= x^4 - y \end{aligned}$

9.1#5°. In Exercise 9.1#4a, the differential equation can be solved explicitly. Do so in order to provide equations for the various curves occurring in your pictures. Show how (in this case as opposed to Example 9.1.2) the exceptional solution has a special hair in the pony tail, because one analytic formula holds all across the domain.

9.1#6. Show that in the system of equations (2)

$$x' = 0 + P(x, y)$$
$$y' = \lambda y + Q(x, y).$$

if $\lambda > 0$, solutions on the exceptional trajectory leave the origin, and if $\lambda < 0$, solutions on the exceptional trajectory approach the origin. Hint: Read Theorem 9.1.7.

9.1#7. For each of the four systems described by

$$x' = xy \pm y^2$$
$$y' = -y \pm x^2,$$

draw the loci where the vector field is horizontal and where it is vertical. In all cases find arbitrarily small perturbations where the differential equation has three zeroes near the origin.

9.1#8. Suppose $V(x)$ is an even potential, so $V(x) = V(-x)$, and $dV/dx(0) = 0$. Consider the differential equation

$$x'' + \alpha x' + \frac{dV}{dx} = 0.$$

(a) For $\alpha > 0$, show that the origin is a sink if $\frac{d^2V}{dx^2} > 0$, and a saddle if $\frac{d^2V}{dx^2} < 0$.

The question we will explore is what happens when $\frac{d^2V}{dx^2} = 0$. Note that $\frac{d^3V}{dx^3} = 0$ (why?). We shall suppose that $\frac{d^4V}{dx^4} > 0$.

(b) Make a phase portrait for the equation if $V(x) = x^4$.

(c) Show that for $V_\beta(x) = V(x) + \beta x^2$, when $\beta > 0$, the differential equation has a single zero in the neighborhood of the origin, and when $\beta < 0$, it has three zeroes.

(d) Make phase portraits to show the effects of (c).

(e) Draw a graph of $\{(x, \beta) \mid V_\beta(x) = 0\}$. This should explain why this bifurcation is called a "pitchfork."

9.1#9. Return to the second order equation of Exercise 8.2–8.3#14a,

$$x'' - \alpha x + x^3 = 0,$$

and treat it as a system of first order equations with $x' = y$. At what values of α does the behavior of the trajectories change? Explain these changes.

9.1#10. A pendulum in $\mathbb{R}^3$, rotating around a vertical axis at a constant speed ω with friction coefficient $a > 0$, satisfies the equation

$$\theta'' + a\theta' + [1 - \omega^2 \cos\theta] \sin\theta = 0.$$

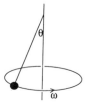

(a) Write this equation as a system of two first order equations.

(b) Show that this is analogous to the scenario of Exercise 9.1#8 for a potential $V_\beta(\theta)$. Find $V_\beta(\theta)$.

(c) Find the value ω_0 of ω for which this system has a pitchfork bifurcation. Explain physically the meaning of the two new zeroes that arise after bifurcation.

(d) Make phase plane drawings for $\omega < \omega_0$ and for $\omega > \omega_0$. Explain the difference in the motions.

9.1#11. Verify parts (i) and (ii) of Lemma 9.1.10.

9.1#12°. (A serious exercise!) In order to see what happens when the Second Nondegeneracy Condition 9.1.6 fails, consider the differential equation

$$x' = f(x,y) = p_{1,1}xy + p_{0,2}y^2 + p_{3,0}x^3 + \text{higher order terms},$$
$$y' = g(x,y) = -y + q_{2,0}x^2 + q_{1,1}xy + \text{higher order terms}.$$

Suppose that $q_{2,0}p_{1,1} + p_{3,0} \neq 0$.
 Consider also the associated differential equation

$$\frac{dy}{dx} = \frac{g(x,y)}{f(x,y)} = \frac{-y + q_{2,0}x^2 + q_{1,1}xy + \cdots}{p_{1,1}xy + p_{0,2}y^2 + p_{3,0}x^3 + \cdots} = G(x,y).$$

(a) Show that for any $A_1 < A_2$, there exists an $\varepsilon > 0$ such that in the set $U_{A_1,A_2} = \{(x,y)|\, q_{2,0}x^2 + A_1x^3 < y < q_{2,0}x^2 + A_2x^3,\quad 0 < x < \varepsilon\}$, $G(x,y)$ is defined [i.e., $f(x,y) \neq 0$].

(b) Show that such numbers A_1, A_2 can be chosen so that U_{A_1,A_2} is a backward antifunnel if $C = q_{2,0}p_{1,1} + p_{3,0} > 0$, and a backward funnel if $C < 0$.

(c) Show that if $C < 0$, there is a unique solution in the antifunnel.

(d) Formulate and prove analogies of (a), (b), and (c) for $-\varepsilon < x < 0$.

(e) Show that if $C > 0$, the origin is a sink, and that all trajectories approach the origin in the funnels of parts (b) and (d), except for two trajectories that approach the origin tangent to the y-axis, from above and below, respectively.

(f) Show that if $C < 0$, there are two exceptional trajectories that emanate from the origin tangent to the y-axis and two trajectories that approach the origin tangent to the y-axis; all others leave a neighborhood of the origin both backward and forward.

Another approach to this problem is given in Exercise 9.6#6.

Exercises 9.2 Hopf Bifurcation

9.2#1. Often a Hopf bifurcation will occur near a saddle. Start with this phase plane picture, assuming that the spiral singularity is a sink.

(a) Draw appropriate arrows on all the separatrices of the saddle and add a few trajectories, with arrows.

(b) Make two more drawings, with arrows, showing how this sink will Hopf bifurcate into a source, showing both what happens at the bifurcation value of parameter and what will happen on the other side of the bifurcation value.

9.2#2°. Consider the equation of Exercise 9.1#2c,

$$x' = y - x^3 + xy$$
$$y' = 4x - y - \alpha.$$

(a) Using the algebraic approach, find where Hopf bifurcation occurs.

(b) Explain how the saddle-node and Hopf bifurcations fit together as α varies. Use the computer to make a picture of each of the possible types of phase planes that occur.

9.2#3. Verify the facts of Example 9.2.7 for

$$x' = y + \alpha x + x^2 + x^3$$
$$y' = -x.$$

By computer experiment, verify that Hopf bifurcation occurs for α small and negative. Find a sufficiently small perturbation of α to produce a single limit cycle.

9.2#4. Consider the equations of Example 9.2.9,

$$x' = y + xy^2$$
$$y' = -x + xy - y^2.$$

Find a computer perturbation such that the phase plane shows *two* limit cycles (a phenomenon that will be further explored in Exercise 9.4#2). There is a strategy to this exercise, and an interactive graphics program is most advisable.

First, check that the first Liapunov coefficient L indeed vanishes, and use the computer to see that the origin is a weak weak source.

Next you will have to introduce new terms, with a parameter, for each of the following two steps. There are many possible ways to do this, but your choices should accommodate the following stipulations:

(a) *First, a limit cycle can be made to occur by creating a Hopf bifurcation.* Add a nonlinear term so that the origin remains a center, but so that the first Liapunov coefficient L changes. (For example, add to the x' equation the term αx^3.) Vary the parameter α as you look at the phase portraits until the origin becomes a weak sink. If α is sufficiently small, you will find a limit cycle.

(b) *Then you want to change the linearization to make the origin become a source again.* That is, add a linear term, such as βx to x'. Show that if the origin becomes a source, and if β is sufficiently small, there will be a second limit cycle.

Save your results, which you will use in Exercise 9.4#2.

9.2#5. For the differential equation of Example 9.2.6,

$$x' = y + x^2$$
$$y' = -x,$$

(a) Show that the Liapunov coefficient at the origin is zero, by using Theorem 9.2.4.

(b) Show that all solutions starting near the origin are periodic. Hint: Consider the symmetries of the differential equation.

(c) What is the relation between parts (a) and (b)?

Analysis of a similar problem will be extended in Exercise 9.6#4.

9.2#6. Suppose that a differential equation

$$x' = f(x, y)$$
$$y' = g(x, y)$$

has a zero at the origin and that the linearization there is a center. Show that by a linear change in x, y, and t, the differential equation can be brought to the form

$$x' = y + P(x, y)$$
$$y' = -x + Q(x, y),$$

where P and Q start with quadratic terms. Hint: In Appendix L6, you will find how to bring the linear part to the form

$$A = \begin{bmatrix} 0 & \beta \\ -\beta & 0 \end{bmatrix}$$

for an appropriate β, by making a linear change of coordinates in x, y. Then you should rescale time.

9.2#7°. As illustration of Theorem 9.2.3, consider the system of differential equations

$$x' = (1 + \alpha)x - 2y + x^2$$
$$y' = x - y.$$

(a) Verify that for $\alpha = 0$, the linearization at the origin is a center.

(b) Compute the Liapunov coefficient at the origin, in order to know whether the nonlinear singularity is a weak sink or a weak source. Hint: to use Theorem 9.2.3, you will need to make a linear change of variables to bring the linear terms to

$$\begin{bmatrix} u \\ v \end{bmatrix}' = \begin{bmatrix} 0 & 1 \\ -1 & 0 \end{bmatrix} \begin{bmatrix} u \\ v \end{bmatrix}.$$

The trick to do this is described in Exercise 9.2#6.

(c) Would you expect to find limit cycles for small positive or negative values of α? Check your prediction on a computer.

9.2#8.

(a) Show that if f is a differentiable function on $\mathbb{R}^2$ satisfying

$$y\frac{\partial f}{\partial x} - x\frac{\partial f}{\partial y} = 0,$$

then f is constant on all circles $x^2 + y^2 = r^2$, so f depends only on the radius.

(b) Show that if p is a homogeneous polynomial of odd degree satisfying $y\frac{\partial p}{\partial x} - x\frac{\partial p}{\partial y} = 0$, then p vanishes identically.

(c) Show that if p is a homogeneous polynomial of even degree $2n$, satisfying $y\frac{\partial p}{\partial x} - x\frac{\partial p}{\partial y} = 0$, then

$$p(x,y) = A(x^2 + y^2)^{2n} \quad \text{for some } A.$$

(d) Prove Lemma 9.2.5 for all k.

Exercises 9.3 Saddle Connections

9.3#1. For the family of differential equations in Example 9.3.1,

$$x' = y$$
$$y' = x^3 - x + \alpha x^2 - 0.1y,$$

experiment with a computer to find to two or three significant figures the value of α_0, for $0.1 < \alpha_0 < 0.2$ where a (heteroclinic) saddle connection occurs.

9.3#2. For the family of differential equations in Example 9.3.1,

$$x' = y$$
$$y' = x^3 - x + 0.2x^2 + (\beta x^2 - 0.2)y$$

(a) Experiment with a computer to find to two or three significant figures the value of β_0, for $1.4 < \beta_0 < 1.5$ where a homoclinic saddle connection occurs.

(b) Experiment further with the computer to make the left-hand saddle connect, either to itself or to the right-hand saddle. This is harder than Exercise 9.2#4, where you had conditions on the linearization matrix to guide you. But we will tell you that at the very least you can find a value of β in the current equation that will indeed cause a saddle connection.

9.3#3. Consider the system of equations

$$x' = x^2 - xy + x - 0.4y + \beta$$
$$y' = y^2 + xy - y + \alpha.$$

Set $\alpha = -5$.

(a) For sequence of $\beta = -0.1, 0, +0.1, +0.2, +0.3, \ldots$ make enough computer drawings to locate the β interval in which a saddle connection occurs.

(b) Show on your drawings the difference this bifurcation makes in global behavior of solutions. Also sketch by hand the saddles and separatrices before and after bifurcation, exaggerating, if you like, to show how the changes in global behavior have occurred.

(c) Using blow-ups, try to locate the actual saddle connection to at least one more decimal place.

9.3#4°. Consider the system of equations

$$x' = y - x + \alpha xy + \beta$$
$$y' = 0.8y - x^3 + 3x.$$

Set $\beta = -.0.4$.

(a) For $\alpha = -0.24, -0.26, \ldots, -0.34$ make enough computer drawings to locate the α interval in which a homoclinic saddle connection occurs. You should blow up the area of interest in the first picture and keep that new window for this exploration.

(b) Explain with hand drawings the global behavior of solutions before, at, and after bifurcation.

9.3#5. For the differential equation

$$x'' = (1 - x^2)x' - x + \alpha x^3$$

(a) Turn the second order equation into a system of first order equations.

(b) Find to at least two decimal places the value of α between 0 and 1 where a heteroclinic saddle connection occurs.

(c) Print out phase portraits for a set of "before," "at," and "after" values of α and color in the cobasins of the source in each of the three views to demonstrate how the global behavior changes. Show also any other places that the behavior changes.

(d) Explain the various possible behaviors as a Van der Pol oscillator (Example 8.4.2) to which a term $-\alpha x^4/4$ has been added to the potential.

9.3#6. With an asymmetric potential equation, you can get a system of equations

$$x' = y + \alpha x^2$$
$$y' = (1 - x^2)y - x$$

similar to those of Exercise 9.3#5.

(a) Show that a bifurcation again occurs for a value of α between 0 and 1, but that this one gives a homoclinic saddle connection.

(b) Print out phase portraits for a set of "before," "at," and "after" values of α and color in the cobasins of the source in each of the three views to demonstrate how the global behavior changes. Show also any other places that the behavior changes.

Exercises 9.4 Semi-stable Limit Cycles

9.4#1°. Consider the system of equations

$$x' = y + \beta(x^2 + y^2)$$
$$y' = x + y^2 + y^3.$$

(a) Show that the origin is a critical saddle for all values of the parameter β.

(b) Show that the separatrices for this saddle have a homoclinic saddle connection for a value β_0 satisfying $1.2 < \beta_0 < 1.21$. Find another decimal of β_0. Show (using a computer) that this critical saddle connection is weakly attracting.

Now introduce another parameter α: consider the system of differential equations

$$x' = \alpha x + y + \beta(x^2 + y^2)$$
$$y' = x + y^2 + y^3.$$

(c) For values of (α, β) for which there is still a saddle connection, show that the homoclinic loop is attracting or repelling, depending on the sign of α. For which sign would one expect limit cycles?

(d) Experiment on the computer with $|\alpha| = 0.3$ and of the appropriate sign until you find a good approximation to a saddle connection. Observe the homoclinic loop. Is it attracting or repelling?

(e) Now perturb α and break the saddle connection so as to create two limit cycles. Is the new limit cycle attracting or repelling?

(f) Vary the parameter β so as to make the cycles merge and disappear.

9.4#2°. Return to the equations of Exercise 9.2#4 and Example 9.2.9,

$$x' = y + xy^2$$
$$y' = -x + xy - y^2.$$

There you found with a computer some perturbation of the equations such that the phase plane showed *two* limit cycles. Vary your equations further and use the computer to find a semi-stable limit cycle.

9.4#3. Consider the following differential equation (which is a *turned* vector field, as discussed in Example 8*.4.8 and at the beginning of Section 9.7):

$$x' = y \cos \beta - (-x + y \cos x) \sin \beta$$
$$y' = y \sin \beta + (-x + y \cos x) \cos \beta.$$

(a) Make a phase portrait for $\beta = 0.01$, with bounds $-20 \leq x \leq 20, -20 \leq y \leq 20$, to show that there exist four limit cycles.

(b) Make another phase portrait for $\beta = 0.02$ to show now that there exist only two limit cycles.

(c) Find to two significant figures where the semi-stable limit cycle bifurcation occurs.

(d) Find to two significant figures where another semi-stable limit cycle bifurcation will occur, eliminating the last two limit cycles.

(e) Estimate, and support with computer experimentation, a value of β above which the phase portrait will show *six* limit cycles.

9.4#4. For the equations of Exercise 9.4#3, conjecture and prove how many limit cycles can exist for this equation. Hint: You might try something to make the original solutions become inward and outward fences. For example, you might turn the vector field by an angle that is alternately positive and negative.

9.4#5. Consider the family of equations

$$x' = y$$
$$y' = -x + (\alpha + \beta \cos x)y.$$

(a) Show that the origin is a source if $\beta > 0$ and a sink if $\beta < 0$.

(b) Show that if $\alpha < \beta < 0$, then r^2 is a Liapunov function on all of $\mathbb{R}^2$. Deduce that there can be no cycles when $|\alpha| > |\beta|$.

(c) Suppose that γ is a cycle for some value (α, β) of the parameter. Show that if $\alpha_1 > \alpha$, then solutions for the parameter value (α_1, β) cross γ from the inside to the outside.

(d) Suppose now that γ_1 and γ_2 are two distinct cycles for some value $(\alpha > 0, \beta > 0)$ of the parameter. Suppose γ_1 attracting, γ_2 repelling, and γ_1 inside γ_2. Show that if $\alpha_1 > \alpha$ and $\alpha_1 - \alpha$ is sufficiently small, then the equation for (α_1, β) has at least two limit cycles in the region bounded by γ_1 and γ_2.

(e) Fix $\beta = 0.5$ (for instance) and let γ_α be a cycle for the parameter value (β, α), defined and linearly attracting for $0 \le \alpha < \alpha_0$, and that α_0 is maximal for this property. Show that $\lim_{\alpha \uparrow \alpha_0} \gamma_\alpha$ exists and is not linearly attracting or repelling.

Exercises 9.5 Bifurcation in One-Parameter Families

9.5#1. Consider the system of differential equations

$$x' = y + \alpha x^2$$
$$y' = (1 - x^2)y - x,$$

which we will study for $\alpha > 0$.

(a) Analyze the zero at $(0, 0)$ for all α.

(b) Show that the differential equation has exactly one other zero for all $\alpha > 0$.

(c) Find and analyze the extra zero for $\alpha = 1/24$ and $\alpha = 1/6$.

(d) You should find in (c) that it is a saddle for both values. Show that this is true for all $\alpha > 0$. (Hint: this can be done either by computing the determinant of the linearization or by an antifunnel argument).

(e) Use the computer to sketch the separatrices of this saddle for $\alpha = 1/6$ and $\alpha = 1/24$, and deduce that there is a number α_0 with $1/24 < \alpha_0 < 1/6$ for which there is a saddle connection.

(f) By successive attempts, find α_0 to three significant digits.

(g) Show that the change of variables $x_1 = -x$, $y_1 = -y$ changes the differential equation above into itself, except that α becomes $-\alpha$.

(h) For what values of α, $-\infty < \alpha < \infty$, does the equation admit a limit cycle?

9.5#2. Consider the system of differential equations

$$x' = y$$
$$y' = (1 - x^2)y - x + \alpha x^3,$$

which we will study for $\alpha > 0$.

(a) Analyze the zero at $(0, 0)$ for all α.

(b) Show that the system has exactly two other zeroes for all $\alpha > 0$.

(c) Show that these two extra zeroes are saddles for all $\alpha > 0$.

(d) You should find in (c) that the two extra zeroes have precisely the same behavior. This can be better expressed by saying that the change of variables $x_1 = -x$, $y_1 = -y$ preserves the differential equation. What does this say about the solutions?

(e) Use the computer to sketch the separatrices of these saddles for $\alpha = 0.1$ and $\alpha = 0.3$ and deduce from this and the symmetry that there is a number α_0 with $0.1 < \alpha_0 < 0.3$ for which there are two saddle connections.

(f) By successive attempts, find α_0 to three significant digits.

9.5#3°. For each of the following differential equations,

(a) $\begin{aligned} x' &= y - x^3 + 3x \\ y' &= x + \alpha - y^2 \end{aligned}$

(b) $\begin{aligned} x' &= x + \alpha - y^2 \\ y' &= y - x^3 + 3x \end{aligned}$

(c) $\begin{aligned} x' &= (x - \alpha)^2 + y^2 - 4 \\ y' &= y - x^3 + 4x \end{aligned}$

(i) Sketch the phase plane, showing the isoclines of horizontal and vertical slope. One of these isoclines is fixed in each case; the other depends on α. Show the possible different arrangements of the two isoclines and how many zeroes are produced by each.

(ii) Find any saddle-node bifurcations. From the pictures in (i), you can tell where on the α parameter line to look. Calculate (from the determinant of the linearization, from the isocline intersections, or from computer experimentation) the values of α at which saddle-node bifurcations occur, and use the program *DiffEq* to make computer pictures to verify each result.

(iii) Find any saddle connections. You can use the computer pictures from (ii) to locate and explore possibilities for saddle connections; in two of the three cases, you should find some. Hint: The easiest way to look closely for saddle connections is to clear slope marks and draw saddles with separatrices.

(iv) Find any Hopf bifurcations. You can use the linearization matrix at the singularities, the pictures, or any information from the previous parts.

Note: Parts (a) and (b) exhibit a common phenomenon: switching the functions for x' and y' leads to a picture of entirely different character.

9.5#4. Investigate the following system:

$$\begin{bmatrix} x \\ y \end{bmatrix}' = \begin{bmatrix} y \\ -4x^3 + 2x \end{bmatrix} + (x^4 - x^2 - \alpha) \begin{bmatrix} 4x^3 - 2x \\ y \end{bmatrix}.$$

Draw the phase plane for $\alpha = -1/2, -1/4, -1/8, 0, 1/2$. Can you explain why the solutions look the way they do?

9.5#5. Consider the nonlinear system of differential equations

$$x' = y$$
$$y' = -\sin x - \varepsilon y,$$

from Example 6.5.4, describing a pendulum with constant $\varepsilon \geq 0$ depending on friction.

(a) For a given ε, locate the singularities of the system. Show that there are two types of singularities for each ε.

(b) For a given ε, find the linearization of the system at the singular point $(x_0, y_0) = (0, 0)$. Call the relevant matrix A_ε and draw in the (tr A, det A)-plane the curve made up of the points (tr A_ε, det A_ε) for varying $\varepsilon \geq 0$.

(c) Use the principle that near $(x_0, y_0) = (0, 0)$ the solutions to the nonlinear system look like those of

$$\begin{bmatrix} x \\ y \end{bmatrix}' = A_\varepsilon \begin{bmatrix} x \\ y \end{bmatrix}''$$

to describe near $(0, 0)$ the different qualitative behaviors of the pendulum due to different friction.

(d) For a given ε, find the linearization of the system at the singular point $(x_0, y_0) = (\pi, 0)$. Call the relevant matrix B_ε and draw in the (tr A, det A)-plane the curve made up of the points (tr B_ε, det B_ε) for varying $\varepsilon \geq 0$.

(e) What can you say about different qualitative behaviors near $(\pi, 0)$?

9.5#6. Return to the second order equation of Exercises 6.5#10 and 8.2–8.3#14b,

$$x'' = -\frac{dV}{dx}, \quad \text{where} \ V = \sin x + \alpha x^2,$$

and treat it as a system of first order equations with $x' = y$. You can predict at which values of α the trajectories will change behavior using the following geometric method.

(a) Show that the zeroes of the system occur when the graph of $\cos x$ intersects the graph of $2\alpha x$ and sketch the different possibilities within $-10 < x < 10$.

(b) With the help of *MacMath's Analyzer*, or another graphing program, find, using the computer, the exact values α^* of α where the number of intersections changes. Then, for each α^*, use *DiffEq* to make a series of careful drawings before, at, and after bifurcation to confirm these results. Mark all separatrices with arrows showing the direction of the trajectories in forward time. Explain exactly which type of bifurcation occurs each time and discuss the stability of the equilibria.

9.5#7. Consider a one-parameter family of differential equations

$$x' = f_\alpha(x, y)$$
$$y' = g_\alpha(x, y).$$

(a) Suppose that the system undergoes a saddle-node bifurcation at α_0. If the capsule of the equation is

$$\begin{bmatrix} a & & b \\ & c & \\ d & & e \end{bmatrix}$$

on the side of α_0 with the fewer zeroes, show that on the other side it must be one of

$$\begin{bmatrix} a+1 & & b \\ & c+1 & \\ d & & e \end{bmatrix}, \quad \begin{bmatrix} a & & b+1 \\ & c+1 & \\ d & & e \end{bmatrix},$$

$$\begin{bmatrix} a+1 & & b \\ & c+1 & \\ d-1 & & e \end{bmatrix}, \quad \begin{bmatrix} a+1 & & b \\ & c+1 & \\ d & & e-1 \end{bmatrix},$$

where the latter two correspond to homoclinic saddle nodes.

(b) Now suppose that at α_0 the system undergoes a Hopf bifurcation. If the capsule of the equation is

$$\begin{bmatrix} a & & b \\ & c & \\ d & & e \end{bmatrix}$$

on the side of α_0 where the bifurcating zero is a sink, show that on the other side the capsule must be either

$$\begin{bmatrix} a-1 & & b+1 \\ & c & \\ d+1 & & e \end{bmatrix} \quad \text{or} \quad \begin{bmatrix} a-1 & & b+1 \\ & c & \\ d & & e-1 \end{bmatrix},$$

depending on whether the zero is for $\alpha = \alpha_0$ a weak sink or a weak source.

(c) Now suppose that at α_0 the system undergoes a homoclinic saddle connection. If the capsule of the equation is

$$\begin{bmatrix} a & & b \\ & c & \\ d & & e \end{bmatrix}$$

on the side of α_0 where there is no limit cycle near the homoclinic loop, then show that on the other side of α_0, the capsule is either

$$\begin{bmatrix} a & & b \\ & c & \\ d+1 & & e \end{bmatrix} \quad \text{or} \quad \begin{bmatrix} a & & b \\ & c & \\ d & & e+1 \end{bmatrix},$$

depending on whether the trace of the linearization of the relevant saddle is negative or positive.

(d) Show that heteroclinic saddle connections do not change the capsule.

(e) Finally, suppose that at α_0 the system undergoes a semi-stable limit cycle. If the capsule of the equation is

$$\begin{bmatrix} a & & b \\ & c & \\ d & & e \end{bmatrix}$$

on the side of α_0 where there are no limit cycles near the semi-stable limit cycle, show that the capsule must be

$$\begin{bmatrix} a & & b \\ & c & \\ d+1 & & e+1 \end{bmatrix}$$

on the other side of α_0.

Exercises 9.6 Bifurcation in Two-Parameter Families

9.6#1. Consider the two-parameter family of differential equations

$$x' = y + \beta x^3$$
$$y' = x - x^3 + \alpha y.$$

(a) Set $\alpha = -0.1$ and make phase portraits for $\beta = 0.03, 0.05$, showing the behavior at all the singularities.

(b) Find by computer experiment a value of β that gives a double saddle connection. Hint: Think ahead about what information about the singularities will be your sign that such a connection has occurred.

(c) Continue the computer experiment to find some of the bifurcation locus for saddle connections, e.g., by fixing β above the bifurcation value found in (b) and varying α; repeat for β below the bifurcation value.

9.6#2. For Example 9.6.1,

$$x' = y - x^2$$
$$y' = \alpha x + \beta - y,$$

(a) Extend the locus for saddle connections by computer experiment, such as by fixing α at -3, then -4, then -5, and so on, in each case varying β to locate where the saddle connection occurs for that value of α.

(b) Show that the Hopf bifurcations are always weak sources and that therefore degenerate bifurcations type (c) can never happen.

9.6#3. Consider again the system of equations studied in Exercise 9.2#5,

$$x' = x^2 - \alpha - \beta y^3 + 0.015x^4$$
$$y' = -y + \beta x^3.$$

What does part (b) of that exercise say about the second Liapunov coefficient? Compute this coefficient directly and confirm.

9.6#4. For the system of equations

$$x' = y + xy^2$$
$$y' = -x + xy - y^2,$$

show that the first Liapunov coefficient vanishes but the second does not. Predict whether the origin is a very weak source or very weak sink. Confirm with a computer picture.

9.6#5. Carry out the calculations to confirm the statements in Example 9.6.7.

9.6#6. Consider the system of differential equations

$$x' = x^2 - \alpha - \beta y^3 + 0.015x^4$$
$$y' = -y + \beta x^3.$$

(a) Show that for $\alpha = 0$ this equation has a saddle-node at the origin, which is nondegenerate for all values of β. Further, show that if $\alpha > 0$, there are no zeroes of the vector field near the origin, and if $\alpha < 0$, there are two.

(b) Draw the dynamical plane by computer for $\beta = -2.5$ and $\beta = -1.5$, when $\alpha = 0$. For which of these is the saddle-node homoclinic? For which is there a limit cycle?

(c) Find, still by computer, a value β_0 of β that separates homoclinic saddle-nodes from nonhomoclinic ones. State exactly what happens to the exceptional trajectory and the separatrices of the saddle-node for that value.

(d) Show that for β_1 slightly smaller than β_0 (for instance, $\beta_1 = -2.3$), there is a value of α negative but close to zero for which there is a saddle connection. Follow the curve of saddle connections and "show" that it approaches the line $\alpha = 0$ tangentially at $\beta = \beta_0$.

9.6#7. For the following two-parameter families of equations, investigate where bifurcations can occur and confirm them with computer drawings.

(a) $\begin{aligned} x' &= \alpha x - \beta y^2 \\ y' &= \alpha(x - y) - y^2 \end{aligned}$

(b) $\begin{aligned} x' &= x^2 - xy + x + \alpha \\ y' &= y^2 + xy - y + \beta \end{aligned}$

9.6#8°. For the differential equation of Example 9.6.6,

$$x' = y + ((r^2 - 1)^3 + \alpha(r^2 - 1) + \beta)x$$
$$y' = -x + ((r^2 - 1)^3 + \alpha(r^2 - 1) + \beta)y,$$

(a) Show that the equation decouples in polar coordinates.

(b) Show that the limit cycles correspond to the circles of radius r, where

$$(r^2 - 1)^3 + \alpha(r^2 - 1) + \beta = 0.$$

(c) Find the locus of Hopf bifurcation.

(d) Compute the first Liapunov number along this locus. Where does it vanish?

(e) Find the semi-stable limit cycle locus, using parts (a) and (b), and Lemma 9.6.4. Show that it is tangent, at the point found in part (d), to the locus found in part (c). What exactly happens at that point?

9.6#9. Consider the system of differential equations

$$x' = x - \alpha y - x(x^2 + y^2)$$
$$y' = \alpha x + y - y(x^2 + y^2) - \beta,$$

the subject of pp. 70-73 in J. Guckenheimer and P. Holmes, *Nonlinear Oscillations, Dynamical Systems, and Bifurcations of Vector Fields* (Springer-Verlag, 1983). Use the computer program *Planar Systems* (or *MacMath's DiffEq, Phase Plane* to draw actual pictures for trajectories with (α, β) chosen from various different regions in the $\alpha\beta$-parameter plane.

9.6#10. Explore the family of equations

$$x' = x^3 + \alpha x + \beta$$
$$y' = y,$$

which is very similar to Example 9.6.5 but shows the other possibility for degenerate possibility (b).

9.6#11. This problem is devoted to another way to understand the behavior of saddle nodes when the Second Nondegeneracy Hypothesis (9.1.6) fails. This question was first considered as Exercise 9.1#12.

Consider the system of equations

$$x' = p_{1,1}xy + p_{0,2}y^2 + p_{3,0}x^3 + \cdots$$
$$y' = -y + q_{2,0}x^2 + \cdots.$$

We will show that if the number $p_{3,0} + p_{1,1}q_{2,0}$, is positive, the trajectories behave like the trajectories of a saddle, whereas if $p_{3,0} + p_{1,1}q_{2,0} < 0$, the trajectories behave like those of a node sink.

(a) Show that if you make the change of variable $u = x$, $v = y + q_{2,0}x^2$, the differential equation becomes

$$u' = p_{1,1}uv + p_{0,2}v^2 + (p_{3,0} + p_{1,1}q_{2,0})x^3 + \cdots = P_1(u, v)$$
$$v' = -v + q_{1,1}uv + q_{0,2}v^2 + \cdots = -v + Q_1(u, v),$$

where $Q_1(u, v)$ has no term in u^2.

(b) Show that for sufficiently small ε and δ, $(-v + Q_1(u, v))/P_1(u, v)$ is well defined in the region $\{|x|, |y| \le \varepsilon; |y| \le \delta x^2\}$ and $P_1(u, v)/(-v + Q_1(u, v))$ is well defined in the region $\{|x|, |y| \le \varepsilon; |y| \ge \delta x^2\}$. Why is the preliminary change of variables (a) necessary?

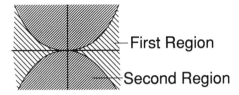

First Region

Second Region

(c) If $p_{3,0} + p_{1,1}q_{2,0} < 0$, show that the first region of (b) is the union of a funnel and a backward funnel for the differential equation

$$\frac{dv}{du} = \frac{-v + Q_1(u,v)}{P_1(u,v)}$$

and that the second region of (b) is the union of an antifunnel and a backward antifunnel for the differential equation

$$\frac{du}{dv} = \frac{P_1(u,v)}{-v + Q_1(u,v)},$$

to which the uniqueness criteria of Section 4.7 of Part I apply.

(d) If $p_{3,0} + p_{1,1}q_{2,0} > 0$, show in a manner like part (c) that each of the regions of (b) are formed by a forward and backward antifunnel for an appropriate differential equation, to which a uniqueness criterion applies.

(e) Use the computer to draw the trajectories for a differential equation

$$x' = 2xy + x^3$$
$$y' = -y \pm x^2,$$

and relate the results to the analysis above.

9.6#12°.

(a) Given the full bifurcation diagram for Example 9.6.3, add labels for the capsules showing the numbers of sinks, sources, saddles, and cycles in each region defined by the bifurcation curves.

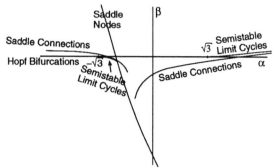

(b) Analyze the points where the curve of saddle-nodes meets the curves of saddle connections; show that these are examples of degenerate possibility (h).

9.6#13. Find the saddle-node locus for the family of equations

$$x' = y - x^3$$
$$y' = y - ax - b,$$

and show that it has a cusp at the origin. Show what this implies for the phase plane behavior.

9.6#14. In this exercise, we will prove Lemma 9.6.4. Consider the function $f(x) = x^3 + \alpha x + \beta$.

(a) Show that the equation $f(x) = 0$ will have three real roots if and only if f has a positive maximum and a negative minimum.

(b) Show that there are no maxima or minima if $\alpha > 0$, and that if $\alpha \le 0$, there is a maximum at $-\sqrt{-\alpha/3}$ and a minimum at $\sqrt{-\alpha/3}$.

(c) Show that the maximum is positive and the minimum is negative if

$$\beta > \frac{2\alpha}{3}\sqrt{-\frac{\alpha}{3}} \quad \text{and} \quad \beta < -\frac{2\alpha}{3}\sqrt{-\frac{\alpha}{3}}.$$

(d) Show that these conditions together are equivalent to $4\alpha^3 + 27\beta^2 < 0$.

(e) The *trigonometric solution of the cubic* gives an entertaining way of computing the three real roots when they exist. Make the change of variables $x = cy$, where $c = 2\sqrt{-\alpha/3}$, and set $A = 3/(c\alpha)$ to find

$$A(x^3 + \alpha x + \beta) = A((cy)^3 + \alpha cy + \beta) = 4y^3 - 3y - \frac{3\beta}{2\alpha}\sqrt{-\frac{3}{\alpha}} = 0.$$

Show that

$$\left| -\frac{3\beta}{2\alpha}\sqrt{-\frac{3}{\alpha}} \right| < 1$$

when the discriminant is negative, so that there exists an angle θ with

$$\cos\theta = \frac{3\beta}{2\alpha}\sqrt{-\frac{3}{\alpha}}.$$

Use the formula $\cos 3\phi = 4\cos^3\phi - 3\cos\phi$ to find

$$y_1 = \cos\frac{\theta}{3}, \quad y_2 = \cos\frac{\theta + 2\pi}{3}, \quad y_3 = \cos\frac{\theta + 4\pi}{3}.$$

(f) When the discriminant is positive, it is easy to find a formula for the unique positive root. Make the substitution $x = y - \alpha/(3y)$ to obtain a quadratic equation for y^3, with real solutions when the discriminant is positive. Use the quadratic formula to find y, then x.

9.6#15. There are many ways in which the exceptional curve and the separatrices of a saddle-node can interact with the separatrices of a saddle or lead into the pony tail of another saddle-node.

(a) Make a list of such possibilities, which occur in codimension two. The authors found six distinct ones, and are not sure that the list is complete.

(b) For each, find a two-parameter family of differential equations in which it occurs generically. What curves of codimension one bifurcations lead to this parameter value?

Exercises 9.7 Grand Example

9.7#1. Let

$$\mathbf{f} = \begin{bmatrix} f_1 \\ f_2 \end{bmatrix}$$

be a vector field, and set $\mathbf{f}_\theta$ to be $\mathbf{f}$ turned by θ, i.e.,

$$\mathbf{f}_\theta = \begin{bmatrix} \cos\theta & -\sin\theta \\ \sin\theta & \cos\theta \end{bmatrix} \begin{bmatrix} f_1 \\ f_2 \end{bmatrix}.$$

(a) Show that if $\mathbf{x}_0$ is a zero of $\mathbf{f}$, then it is a zero of $\mathbf{f}_\theta$ for all θ. (This is obvious; it is included for completeness.)

(b) Show that if $\mathbf{u}' = A\mathbf{u}$ is the linearization of $\mathbf{x}' = \mathbf{f}(\mathbf{x})$ at a zero $\mathbf{x}_0$, then the vector field turned by θ has linearization

$$\mathbf{u}' = \begin{bmatrix} \cos\theta & -\sin\theta \\ \sin\theta & \cos\theta \end{bmatrix} A\mathbf{u};$$

i.e., the linearization of the turned vector field is the turn of the linearization.

(c) If $A = \begin{bmatrix} a & b \\ c & d \end{bmatrix}$, then find a formula for the trace of the linearization of $\mathbf{f}_\theta$ at $\mathbf{x}_0$.

(d) Show that if $\mathbf{x}_0$ is a zero of a vector field that is a saddle-node, then $\mathbf{x}_0$ is still a saddle node for the vector field turned by any angle.

(e) Show that if $\mathbf{x}_0$ is a linear sink or a linear source, then as you turn the vector field, there will be two opposite angles for which the linearization at $\mathbf{x}_0$ is a center (usually these values are Hopf bifurcations).

(f) Show that if $\mathbf{x}_0$ is a zero of a vector field that is a saddle-node, then as you turn the vector field, there will be two opposite angles for which the linearization at $\mathbf{x}_0$ has a double eigenvalue 0.

9.7#2. Let

$$x' = x^2 - y^2$$
$$y' = x + 2y + x^2.$$

(a) Imagine turning this vector field through an angle θ. Find the trace of the linearization of $\mathbf{f}_\theta$ at the origin.

(b) For what values of θ is the origin a sink, a source, or a center?

(c) For the values of θ for which the origin is a center for the linearization, is it a weak sink or a weak source in the original nonlinear equation?

(d) On which side of these values does the equation have a limit cycle?

9.7#3. Let $\mathbf{f}$ be a vector field with a linearly attracting limit cycle γ.

(a) Show that the vector field $\mathbf{f}_\theta$ that has been turned through an angle θ still has a limit cycle γ_θ for $|\theta| < \varepsilon$, where ε is sufficiently small.

(b) Show that γ does not intersect γ_θ if $\theta \neq 0$.

(c) Show that γ_{θ_1} and γ_{θ_2} are on opposite sides of γ if

$$-\varepsilon < \theta_1 < 0 < \theta_2 < \varepsilon.$$

Hint: Think of the Poincaré–Bendixson Theorem 8.5.1.

9.7#4. For Example 9.7.1,

$$x' = (y - \alpha)\cos\theta - (y - x^3 + 3x)\sin\theta$$
$$y' = (y - \alpha)\sin\theta + (y - x^3 + 3x)\cos\theta,$$

show that there will always be values of θ for which the zero becomes a spiral source or a spiral sink. Will there always be values for which it becomes a node source or node sink?

9.7#5. For Example 9.7.1, locate and demonstrate the saddle-node bifurcation for any θ.

9.7#6. For $\alpha = 0$ in Example 9.7.1, show that the zero at $(\sqrt{3}, 0)$ undergoes Hopf bifurcation at angles θ where $\tan\theta = -1/7$. This locates precisely those points where the Hopf bifurcation locus (Exercise 9.7#7) crosses itself, as shown in Figure 9.7.3.

9.7#7.

(a) Show that Example 9.7.1 has a zero at which the trace of the linearization vanishes precisely if

$$\alpha = \pm\sqrt{\frac{2 - \cotan\theta}{3} \frac{7 + \cotan\theta}{3}}.$$

(b) Draw a graph of the curve above, which gives the locus of Hopf bifurcation for Example 9.7.1.

(c) Identify the part of this curve where the zero in question is a saddle.

9.7#8. For Example 9.7.1, show that if the zero of the vector field is a nondegenerate saddle-node, then there are precisely two values of θ in $[-\pi, \pi)$ for which both eigenvalues of the linearization vanish [degenerate behavior (a)]. Show also that these differ by π.

9.7#9. Show that for $\alpha = 2$ in Example 9.7.1, the double zero eigenvalue occurs for angles θ such that $\tan\theta = -1$. This shows exactly where you can begin to look for the locus of saddle connections.

9.7#10. Consider the two-parameter family of vector fields obtained by *turning* the system of equations

$$x' = y - x^2, \qquad y' = y - \alpha.$$

(a) Find the locus where the equation has a saddle node.

(b) When is the saddle node degenerate?

(c) Find the locus where the equation has a Hopf bifurcation.

(d) Find a value of the parameters where the equation ha a limit cycle.

(e) Find an approximate value where the equation has a saddle connection.

9.7#11. Explore the system

$$x' = \beta x + y + \alpha(x^2 + xy) + x^3$$
$$y' = -x + y^2,$$

as discussed in Example 9.6.2 as a general problem, using the sort of analysis illustrated in the Grand Example of Section 9.7.

Appendix L

Linear Algebra

Linear algebra is a language and a collection of results used throughout mathematics and in a great many applications. Elementary linear algebra is not very difficult; in fact, mathematicians tend to consider a problem solved when they can say: "and now it's just linear algebra." For all that, the field involves a number of ideas that are typical of modern mathematics and rather foreign to students whose background is strictly calculus.

Furthermore, although basic linear algebra is rather simple, the problem for the beginner is that it takes so many words to describe the simple procedures and results that the subject may seem tedious and deadly. So let us try to put it in a lighter perspective to keep you going while we lay out the essential tools.

Linear algebra is a bit like a classical symphony, with three main movements and a minuet thrown in for diversion. The main movements are:

   I. the theory of linear equations (lento ponderoso);

  II. the geometry of inner product spaces (andante grazioso);

 III. the theory of eigenvalues and eigenvectors (allegro).

The minuet is the theory of determinants (allegretto) and is played between the second and third movements.

Only the third part is really a branch of the theory of dynamical systems, which includes differential equations, but the others are necessary there as well as in practically every other aspect of mathematics.

Each of the parts above has its *key words* and *themes*: for the first,

> *vector space, linear independence, span, dimension, basis*;

for the second,

> *orthonormal basis, quadratic form*;

finally, for the third,

> *eigenvalue, eigenvector, diagonalizability, invariant subspace*.

The key words above all relate to the theoretical aspect of linear algebra, but there is a different way of looking at the field. Especially in these days of computers, linear algebra is an essential computational tool, and one can also organize linear algebra around the *algorithms used in computations*. The main algorithm of the first part (and perhaps of all of linear mathematics) is *row reduction.*

The key algorithm for the second part is the

Gram–Schmidt orthogonalization process.

Finally, the third part depends on a variety of more recent algorithms. The decision as to which ones are key is perhaps not yet final; however, the

QR algorithm

is emerging as central for arbitrary matrices, and

Jacobi's method

appears to be an excellent way to approach symmetric matrices.

The remainder of this appendix on linear algebra is arranged in the following eight sections:

L1   The theory of linear equations: in practice   first movement
L2   The theory of linear equations: in theory

L3   Vector spaces with inner product spaces   second movement
L4   Linear transformations and inner products

L5   The theory of determinants   minuet

L6   The theory of eigenvalues and eigenvectors   third movement
L7   Finding eigenvalues: the QR method
L8   Finding eigenvalues: Jacobi's method

The first movement is so "ponderous" as to require splitting into two sections. Nevertheless, they share the common theme of "existence and uniqueness of solutions to systems of linear equations." The second movement and the algorithms of the third movement are also lengthy enough that they are better as separate sections.

After L8 we return to the overall Appendix L to give a very brief summary of all eight subsections and some exercises. You can use the summaries for a quick reference or review, then refer to the specified sections for details.

L1 Theory of Linear Equations: In Practice

All readers of this book will have solved systems of simultaneous linear (nondifferential) equations. Such problems keep arising all over mathematics and its applications, so a thorough understanding of the problem is essential. What most people encounter in high school is systems of n equations in n unknowns, where n might be general or restricted to $n \leq 3$. Such a system usually has a unique solution, but sometimes something goes wrong and some equations are "consequences of others" or "incompatible with others," and in these cases there will be infinitely many solutions or no solutions, respectively. This section is largely concerned with making these notions systematic.

A language has evolved to deal with these concepts, using the words "linear transformation," "linear combination," "linear independence," "kernel," "span," "basis," and "dimension." These words may sound unfriendly, but they correspond to notions that are unavoidable and actually quite transparent if thought of in terms of linear equations. They are needed to answer questions like: "How many equations are consequences of the others?"

The relationship of all these words with linear equations goes further. Throughout this section and the next, there is just one method of proof:

> *Reduce the statement to a statement about linear equations, row-reduce the resulting matrix, and see whether the statement becomes obvious.*

If so, the statement is true, and otherwise it is likely false.

This is not the only way to do these proofs; some people might prefer abstract induction proofs. But we use this method in order to hook into what most students more readily understand.

L1.1. INTRODUCING THE ACTORS: VECTORS AND MATRICES

Much of linear algebra takes place within $\mathbb{R}^n$ or $\mathbb{C}^n$. These are the spaces of ordered n-tuples of numbers, real for $\mathbb{R}^n$ and complex for $\mathbb{C}^n$.

Such ordered n-tuples, or *vectors*, occur everywhere, from grades on a transcript to prices on the stock exchange. But the most important examples of *vector spaces* are also the most familiar, namely $\mathbb{R}^2$ and $\mathbb{R}^3$, which everyone has encountered when studying analytic geometry.

These spaces are important because they have a geometric interpretation, as the plane and space, for which almost everyone has some intuitive feel. Much of linear algebra consists of trying to extend this geometric feel to higher-dimensional spaces; the results and the language of linear algebra are largely extrapolated from these cases. We will often speak of things like "three-dimensional subspaces of $\mathbb{R}^5$." In case these words sound scary, you should realize that even the experts understand such objects only by "educated analogy" to objects in $\mathbb{R}^2$ or $\mathbb{R}^3$; the authors cannot actually "visualize $\mathbb{R}^4$" and they believe that no one really can.

We will write our n-tuples, the vectors, as *columns*. Some people (especially typists) are bothered by writing vectors as columns rather than rows. But there really are reasons for doing it as columns, such as in expressing systems of linear equations (this section) and in interpreting determinants

(Section L4). So if you stick to columns religiously you will avoid endless confusion later.

As you know from basic calculus, an important operation unique to vectors is the scalar or *dot product*, formally defined as follows:

$$\mathbf{x} \cdot \mathbf{y} \equiv \sum_{k=1}^{n} x_k y_k, \quad \text{or,} \quad \begin{bmatrix} x_1 \\ x_2 \\ \vdots \\ x_n \end{bmatrix} \cdot \begin{bmatrix} y_1 \\ y_2 \\ \vdots \\ y_n \end{bmatrix} = x_1 y_1 + x_2 y_2 + \cdots + x_n y_n.$$

The dot product is a special case of the inner product, to be studied in Section L3.

The other concrete objects that occur in linear algebra are *matrices*. An $m \times n$ matrix is simply a rectangular array, m high and n wide. A vector $\mathbf{v} \in \mathbb{R}^m$ is also an $m \times 1$ matrix. Usually our matrices will be arrays of numbers, real or complex, but sometimes one might wish to consider matrices of polynomials or of more general functions. Clearly, the space of $m \times n$ real or complex matrices is just $\mathbb{R}^{mn}$ (or $\mathbb{C}^{mn}$). Putting the entries into a rectangular array allows another operation to be performed with matrices: *matrix multiplication*.

You cannot multiply just any two matrices A and B; for the product AB to be defined, the length of A must be equal to the height of B, and then the resulting matrix has height A and length B.

The formal definition is as follows: if $A = (a_{i,j})$ is an $m \times n$ matrix, and $B = (b_{i,j})$ is an $n \times p$ matrix, then $C = AB$ is the $m \times p$ matrix with entries

$$c_{i,j} = \sum_{k=1}^{n} a_{i,k} b_{k,j}.$$

There is a nice way of remembering this formula, which the authors recommend using whenever multiplying matrices: write B above and to the right of A, then C will fit in the space to the right of A and below B, and each entry is the dot product of the row of A and the column of B that it is on.

Example L1.1. To compute $AB = C$, write

$$\begin{array}{cc} & \begin{bmatrix} B \end{bmatrix} \\ \begin{bmatrix} A \end{bmatrix} & \begin{bmatrix} C \end{bmatrix} \end{array} = \begin{array}{cc} & \begin{bmatrix} 1 & 4 & -2 \\ 3 & 0 & 2 \end{bmatrix} \\ \begin{bmatrix} 2 & -1 \\ 3 & 2 \end{bmatrix} & \begin{bmatrix} -1 & 8 & -6 \\ 9 & 12 & -2 \end{bmatrix} \end{array}. \quad \blacktriangle$$

Observe that this method applies equally well to repeated multiplications. An "explanation" of what makes matrix multiplication a "natural" operation to perform appears in Theorem L2.20. Exercise L1#1 provides practice on matrix multiplication if you need it.

Once you have matrix multiplication, you can write a system of linear equations much more succinctly, as follows:

$$
\begin{array}{ccccc}
a_{1,1}x_1 & +\cdots+ & a_{1,n}x_n & = b_1 \\
\vdots & \cdots & \vdots & \vdots \\
\vdots & \cdots & \vdots & \vdots \\
a_{m,1}x_1 & +\cdots+ & a_{m,n}x_n & = b_m
\end{array}
$$

is the same as

$$
\begin{bmatrix}
a_{1,1} & \cdots & a_{1,n} \\
\vdots & & \vdots \\
\vdots & & \vdots \\
a_{m,1} & \cdots & a_{m,n}
\end{bmatrix}
\begin{bmatrix}
x_1 \\
\vdots \\
\vdots \\
x_n
\end{bmatrix}
=
\begin{bmatrix}
b_1 \\
\vdots \\
\vdots \\
b_m
\end{bmatrix},
$$

which can be written far more simply as

$$
A\mathbf{x} = \mathbf{b}.
$$

The $m \times n$ matrix A is comprised of the coefficients on the left of the equations; the vector $\mathbf{x}$ in $\mathbb{R}^n$ represents the unknowns, the operation between A and $\mathbf{x}$ is matrix multiplication, and the vector $\mathbf{b}$ in $\mathbb{R}^m$ represents the constants on the right of the equations.

L1.2. THE MAIN ALGORITHMS: ROW REDUCTION

Given a matrix A, a *row operation* on A is one of the following three operations:

(1) *multiplying a row by a nonzero number;*

(2) *adding a multiple of a row onto another row;*

(3) *exchanging two rows.*

There are two good reasons why these operations are so important. The first is that they only involve *arithmetic*, i.e., addition, subtraction, multiplication, and division. That is just what computers do well; in some sense, it is all they can do. And they spend a lot of their time doing it: row operations are fundamental to most other mathematical algorithms.

The other reason is that row operations are closely connected to linear equations. Suppose $A\mathbf{x} = \mathbf{b}$ represents a system of m linear equations in n

unknowns. Then the $m \times (n+1)$ matrix $[A, \mathbf{b}]$ composed of the coefficients on the left and the constants on the right,

$$
\begin{bmatrix}
a_{1,1} & \vdots & a_{1,n} & b_1 \\
\vdots & \cdots & \vdots & \vdots \\
a_{m,1} & \cdots & a_{m,n} & b_m
\end{bmatrix} = [A, \mathbf{b}],
$$

sums up all the crucial numerical information in each equation.

Now the key property is the following:

Theorem L1.2. *If the matrix $[A', \mathbf{b}']$ is obtained from $[A, \mathbf{b}]$ by a sequence of row operations, then the set of solutions of $A\mathbf{x} = \mathbf{b}$ and of $A'\mathbf{x} = \mathbf{b}'$ coincide.*

Proof. This fact is not hard to see: the row operations correspond to multiplying one equation through by a nonzero number, adding a multiple of one equation onto another, and exchanging two equations. Thus, any solution of $A\mathbf{x} = \mathbf{b}$ is also a solution of $A'\mathbf{x} = \mathbf{b}'$. On the other hand, any row operation can be undone by another row operation (Exercise L1#2), so any solution $A'\mathbf{x} = \mathbf{b}'$ is also a solution of $A\mathbf{x} = \mathbf{b}$.   □

Theorem L1.2 suggests trying to solve $A\mathbf{x} = \mathbf{b}$ by using row operations on $[A, \mathbf{b}]$ to bring it to the most convenient form. The following example shows this idea in action.

Example L1.3. Let us solve (we shall explain how over the next two pages)

$$
\begin{aligned}
2x + y + 3z &= 1 \\
x - y + z &= 1 \\
x + y + 2z &= 1.
\end{aligned}
$$

By row operations, the matrix

$$
\begin{bmatrix}
2 & 1 & 3 & 1 \\
1 & -1 & 1 & 1 \\
1 & 1 & 2 & 1
\end{bmatrix}
$$

can be brought to

$$
\begin{bmatrix}
\underline{1} & 0 & 0 & -2 \\
0 & \underline{1} & 0 & -1 \\
0 & 0 & \underline{1} & 2
\end{bmatrix}, \tag{1}
$$

so $\begin{bmatrix} -2 \\ -1 \\ 2 \end{bmatrix}$ is the unique solution of the equation, because the matrix (1) is equivalent to the system

$$x = -2$$
$$y = -1$$
$$z = 2. \quad \blacktriangle$$

Most people agree that the "echelon" form (1) at the center of Example L1.3 is best for solving systems of linear equations (though there are many variants, and the echelon form is not actually best for all purposes).

A matrix is in *echelon form* if

(a) for every *row*, the first nonzero entry is 1, called a *leading* 1;

(b) the leading 1 of a lower row is always further to the right then the leading 1 of a higher row;

(c) for every *column* containing a leading 1, all other entries are 0.

Examples L1.4 and L1.5 show matrices that respectively are and are not in echelon form.

Example L1.4. The following matrices are in echelon form; the leading 1's are underlined:

$$\begin{bmatrix} \underline{1} & 0 & 0 & 2 \\ 0 & \underline{1} & 0 & -1 \\ 0 & 0 & \underline{1} & 1 \end{bmatrix}, \quad \begin{bmatrix} \underline{1} & 1 & 0 & 0 \\ 0 & 0 & \underline{1} & 0 \\ 0 & 0 & 0 & \underline{1} \end{bmatrix}, \quad \begin{bmatrix} 0 & \underline{1} & 2 & 0 & 0 & 3 & 0 & -3 \\ 0 & 0 & 0 & \underline{1} & -1 & 1 & 0 & 1 \\ 0 & 0 & 0 & 0 & 0 & 0 & \underline{1} & 2 \end{bmatrix}. \quad \blacktriangle$$

Examples L1.5. The following matrices are *not* in echelon form:

$$\begin{bmatrix} 1 & 0 & 0 & 2 \\ 0 & 0 & 1 & -1 \\ 0 & 1 & 0 & 1 \end{bmatrix}, \quad \begin{bmatrix} 1 & 1 & 0 & 1 \\ 0 & 0 & 2 & 0 \\ 0 & 0 & 0 & 1 \end{bmatrix}, \quad \begin{bmatrix} 0 & 1 & 2 & 0 & 0 & 3 & 0 & -3 \\ 0 & 0 & 0 & 1 & -1 & 1 & 1 & 1 \\ 0 & 0 & 0 & 0 & 0 & 0 & 1 & 2 \end{bmatrix}.$$

The first matrix violates rule (b); the second violates rules (a) and (c); the third violates rule (c). Exercise L1#3 asks you to find row operations that will bring them to echelon form. ▲

The following result is absolutely fundamental:

Theorem L1.6. *Given any matrix A, there exists a unique matrix $\tilde{A}$ in echelon form which can be obtained from A by row operations.*

The proof of Theorem L1.6 is more important than the result: it is an explicit algorithm to compute $\tilde{A}$. This algorithm, called *row-reduction*, or *Gaussian elimination* (or several other names), is the main tool in linear equations.

Row-reduction algorithm. To bring a matrix to echelon form:

(1) Look down the first column until you find a nonzero entry (called a *pivot*). If you do not find one, then look in the second column, etc.

(2) Move the row containing the pivot to the first row position, and then divide that row by the pivot to make that entry a *leading* 1, as defined above.

(3) Add appropriate multiples of this row onto the other rows to cancel the entries in the first column of each of the other rows.

Now look down the next column over (and then the next column if necessary, etc.), starting beneath the row you just worked with, and look for a nonzero entry (the next pivot). As above, exchange its row with the second row, divide through, etc.

This proves existence of a matrix in echelon form which can be obtained from a given matrix. Uniqueness is more subtle (and less important) and will have to wait. $\square$

Example L1.7. The following row operations row-reduce the original matrix:

$$
\begin{bmatrix} 1 & 2 & 3 & 1 \\ -1 & 1 & 0 & 2 \\ 1 & 0 & 1 & 2 \end{bmatrix}
\begin{array}{c} \rightarrow R_1 + R_2 \\ R_3 - R_1 \end{array}
\begin{bmatrix} \underline{1} & 2 & 3 & 1 \\ 0 & 3 & 3 & 3 \\ 0 & -2 & -2 & 1 \end{bmatrix}
\rightarrow R_2/3
\begin{bmatrix} \underline{1} & 2 & 3 & 1 \\ 0 & \underline{1} & 1 & 1 \\ 0 & -2 & -2 & 1 \end{bmatrix}
$$

$$
\begin{array}{c} R_1 - 2R_2 \\ \rightarrow \\ R_3 + 2R_2 \end{array}
\begin{bmatrix} \underline{1} & 0 & 1 & -1 \\ 0 & \underline{1} & 1 & 1 \\ 0 & 0 & 0 & 3 \end{bmatrix}
\begin{array}{c} \rightarrow \\ R_3/3 \end{array}
\begin{bmatrix} \underline{1} & 0 & 1 & -1 \\ 0 & \underline{1} & 1 & 1 \\ 0 & 0 & 0 & \underline{1} \end{bmatrix}.
$$

The row operations are labelled with R_i's, which refer in each case only to the rows of the immediately preceding matrix.

Note that after the fourth matrix, we were unable to find a nonzero entry in the third column and below the second, so we had to look in the next column over, where there is a 3. ▲

A word about practical implementation of the row-reduction algorithm: real matrices as generated by computer operations often have very small entries, which are really zero entries but round-off error has made them nonzero (perhaps 10^{-50}). Such an entry would be a poor choice for a pivot, because you will need to divide its row through by it, and the row will then contain very large entries. When you then add multiples of that row onto another row, you will be commiting the basic sin of computation: adding numbers of very different sizes, which leads to loss of precision. So, what do you do? You skip over that amost-zero entry and choose another pivot (that is, you set a tolerance below which the computer will treat the entry as an actual zero).

Example L1.8. If you are computing to 10 significant digits, then $1 + 10^{-10} = 1.0000000001 = 1$. So consider the system of equations

$$10^{-10}x + 2y = 1$$
$$x + y = 1,$$

the solution of which is $x = 1/(2 - 10^{-10})$, $y = (1 - 10^{-10})/(2 - 10^{-10})$. If you are computing to 10 significant digits, this is $x = y = 0.5$. If you actually use 10^{-10} as a pivot, the row reduction, to 10 significant digits, goes as follows:

$$\begin{bmatrix} 10^{-10} & 2 & 1 \\ 1 & 1 & 1 \end{bmatrix} \rightarrow \begin{bmatrix} \underline{1} & 2\cdot10^{10} & 10^{10} \\ 1 & 1 & 1 \end{bmatrix} \rightarrow \begin{bmatrix} \underline{1} & 2\cdot10^{10} & 10^{10} \\ 0 & -2\cdot10^{10} & -10^{10} \end{bmatrix}$$
$$\rightarrow \begin{bmatrix} \underline{1} & 0 & 0 \\ 0 & \underline{1} & .5 \end{bmatrix}.$$

The "solution" shown by the last matrix reads $x = 0$, which is very wrong: x is supposed to be .5. If instead we treat 10^{-10} as zero and use the second entry of the first column as a first pivot, we find

$$\begin{bmatrix} 10^{-10} & 2 & 1 \\ 1 & 1 & 1 \end{bmatrix} \rightarrow \begin{bmatrix} \underline{1} & 1 & 1 \\ 0 & 2 & 1 \end{bmatrix} \rightarrow \begin{bmatrix} \underline{1} & 0 & .5 \\ 0 & \underline{1} & .5 \end{bmatrix},$$

which is right. In Exercise L1#5, you are asked to analyze precisely where the troublesome errors occurred. All computations have been carried out to 10 significant digits only. ▲

There is no reason to choose the first nonzero entry in a given column; in practice, one always chooses the largest.

L1.3. SOLVING EQUATIONS USING ROW REDUCTION

If you want to solve the system of linear equations $Ax = b$, form the matrix $[A, b]$ and row-reduce it to echelon form, giving $[\tilde{A}, \tilde{b}]$. The solutions can then be read right off, as already shown in Example L1.3.

To be more precise, we can state the following, showing what Theorem L1.6 does for us:

Theorem L1.9. *Consider the system of linear equations* $Ax = b$, *where A is an* $m \times n$ *matrix,* x *is a vector in* $\mathbb{R}^n$, b *is a vector in* $\mathbb{R}^m$, *and the matrix* $[A, b]$ *row-reduces to echelon form* $[\tilde{A}, \tilde{b}]$. *Then one of the following must occur:*

(a) *If* $\tilde{b}$ *contains a leading 1, then there are no solutions.*

(b) *If* $\tilde{b}$ *does not contain a leading 1, and if every column of* $\tilde{A}$ *contains a leading 1, then there is exactly one solution.*

(c) *If* $\tilde{\mathbf{b}}$ *does not contain a leading* 1, *and if some column of* $\tilde{A}$ *contains no leading* 1, *then there are infinitely many solutions. They form a family that depends on as many parameters as the number of columns of* $\tilde{A}$ *not containing leading* 1*'s.*

Before discussing details of this theorem, let us consider the instance where the results are most intuitive, where $n = m$. Case (b) of Theorem L1.9 has been illustrated by Example L1.3; Examples L1.10 and L1.11 illustrate cases (a) and (c), respectively.

Example L1.10. Let us solve

$$2x + y + 3z = 1$$
$$x - y = 1$$
$$x + y + 2z = 1.$$

The matrix

$$\begin{bmatrix} 2 & 1 & 3 & 1 \\ 1 & -1 & 0 & 1 \\ 1 & 1 & 2 & 1 \end{bmatrix}$$

row-reduces to

$$\begin{bmatrix} \underline{1} & 0 & 1 & 0 \\ 0 & \underline{1} & 1 & 0 \\ 0 & 0 & 0 & \underline{1} \end{bmatrix},$$

so the equations are incompatible and there are no solutions. ▲

Example L1.11. Let us solve

$$2x + y + 3z = 1$$
$$x - y = 1$$
$$x + y + 2z = 1/3.$$

The matrix

$$\begin{bmatrix} 2 & 1 & 3 & 1 \\ 1 & -1 & 0 & 1 \\ 1 & 1 & 2 & 1/3 \end{bmatrix}$$

row-reduces to

$$\begin{bmatrix} \underline{1} & 0 & 1 & 2/3 \\ 0 & \underline{1} & 1 & -1/3 \\ 0 & 0 & 0 & 0 \end{bmatrix}$$

and there are infinitely many solutions. You can choose z arbitrarily, and then the vector

$$\begin{bmatrix} 2/3 - z \\ -1/3 - z \\ z \end{bmatrix}$$

is a solution, the only one with that value of z.   ▲

For instances where $n \neq m$, examples for Theorem L1.9 are provided in Exercises L1#4 and L1#6e.

These examples have a geometric interpretation. As the reader surely knows, and can verify in Exercise L1#7, two equations in two unknowns

$$a_1 x + b_1 y = c_1$$
$$a_2 x + b_2 y = c_2$$

are incompatible if and only if the lines ℓ_1 and ℓ_2 in $\mathbb{R}^2$ with equations $a_1 x + b_1 y = c_1$ and $a_2 x + b_2 y = c_2$ are parallel. The equations have infinitely many equations if and only if $\ell_1 = \ell_2$.

The case of three equations in three unknowns is a bit more complicated. The three equations each describe a plane in $\mathbb{R}^3$.

There are two ways for the equations in $\mathbb{R}^3$ to be incompatible, which means that the planes never meet in a single point. One way is that two of the planes are parallel, but this is not the only, or even the usual way: they will also be incompatible if no two are parallel, but the line of intersection of any two is parallel to the third, as in Figure L1.1A. This latter possibility occurs in Example L1.10.

There are also two ways for equations in $\mathbb{R}^3$ to have infinitely many solutions. The three planes may coincide, but again this is not necessary or usual. The equations will also have infinitely many solutions if the planes intersect in a common line, as shown in Figure L1.1B. This second possibility occurs in Example L1.11.

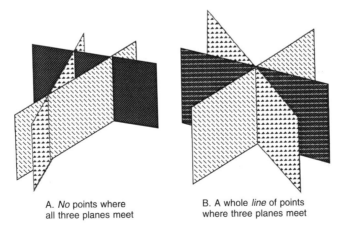

A. *No* points where   B. A whole *line* of points
all three planes meet   where three planes meet

FIGURE L1.1.

The phrase in part (c) of Theorem L1.9 concerning the number of parameters is actually a statement about the "dimension of the solution space," a concept that will be explained in Subsections L2.1 and L2.2.

Theorem L1.9 has additional spin-offs, such as the following:

Remark: If you want to solve several systems of n linear equations in n unknowns with the same matrix, e.g., $A\mathbf{x}_1 = \mathbf{b}_1, \ldots, A\mathbf{x}_k = \mathbf{b}_k$, you can deal with them all at once using row reduction. Form the matrix

$$[A, \mathbf{b}_1, \ldots, \mathbf{b}_k]$$

and row-reduce it to get

$$[\tilde{A}, \tilde{\mathbf{b}}_1, \ldots, \tilde{\mathbf{b}}_k].$$

Various cases can occur:

(a) If $\tilde{A}$ is the identity, then $\tilde{\mathbf{b}}_i$ is the solution to the ith system $A\mathbf{x}_i = \mathbf{b}_i$.

(b) If $\tilde{\mathbf{b}}_i$ has a nonzero entry in a row and all the entries of $\tilde{A}$ in that row are zero, then the ith equation has no solutions.

(c) If $\tilde{\mathbf{b}}_i$ has only zeroes in the rows in which $\tilde{A}$ has only zeroes, then the ith equation has infinitely many solutions, depending on as many parameters as the number of columns of $\tilde{A}$ not containing leading 1's.

This remark will be very useful when we come to computing *inverses* of matrices, in the next subsection.

L1.4. INVERSE AND TRANSPOSE OF A MATRIX

The *identity matrix* I_n is the $n \times n$-matrix with 1's along the diagonal and 0's elsewhere:

$$I_2 = \begin{bmatrix} 1 & 0 \\ 0 & 1 \end{bmatrix} \quad \text{and} \quad I_3 = \begin{bmatrix} 1 & 0 & 0 \\ 0 & 1 & 0 \\ 0 & 0 & 1 \end{bmatrix}.$$

It is called an identity matrix because multiplication by it does not change the matrix being multiplied: If A is an $n \times m$ matrix, then

$$I_n A = A I_m = A.$$

The columns $\mathbf{e}_1, \ldots, \mathbf{e}_n$ of I_n, i.e., the vectors with just one 1 and all other entries 0, will play an important role, particularly in Appendix L2.

For any matrix A, its *matrix inverse* is another matrix A^{-1} such that

$$AA^{-1} = A^{-1}A = I, \quad \text{the identity matrix.}$$

The existence of a matrix inverse gives another possible tack to solving equations. That is because $A^{-1}A\mathbf{x} = \mathbf{x}$, so the solution of $A\mathbf{x} = \mathbf{b}$ is then

x = A^{-1}**b**. In practice, computing matrix inverses is not often a good way of solving linear equations, but it is nevertheless a very important construction.

Only square matrices can have inverses: Exercise L1#8 asks you

(a) to derive this from Theorem L1.9, and

(b) to show an example where $AB = I$, but $BA \neq I$. Such a B would be only a "one-sided inverse" for A; a "one-sided inverse" can give uniqueness *or* existence of solutions to A**x** = **b**, but not both.

Example L1.12. The inverse of a 2×2 matrix

$$A = \begin{bmatrix} a & b \\ c & d \end{bmatrix} \quad \text{is} \quad A^{-1} = \frac{1}{ad - bc} \begin{bmatrix} d & -b \\ -c & a \end{bmatrix},$$

as Exercise L1#9 asks you to confirm by matrix multiplication of AA^{-1} and $A^{-1}A$.   ▲

Notice that this 2×2 matrix A has an inverse if and only if $ad - bc \neq 0$; we will see much more about this sort of thing in the section about determinants. There are analogous formulas for larger matrices, but they rapidly get out of hand.

The effective way of computing matrix inverses for larger matrices is by *row reduction*:

Theorem L1.13. *If A is a square $n \times n$ matrix, and if you construct the $n \times 2n$ matrix $[A \mid I]$ and row-reduce it, then there are two possibilities:*

(1) *the first n columns row-reduce to the identity; in that case the last n columns are the inverse of A;*

(2) *the first n columns do not row-reduce to the identity, in which case A does not have an inverse.*

Proof. (1) Suppose $[A \mid I]$ row-reduces to $[I \mid B]$. Then, since the columns $(n + 1, \ldots, 2n)$ behave independently, the ith column of B is the solution $\mathbf{x}_i$ to the equation $A\mathbf{x}_i = \mathbf{e}_i$. Putting the columns together, this shows $A[\mathbf{x}_1, \mathbf{x}_2, \ldots, \mathbf{x}_n] = AB = I$.

In order to show $BA = I$, start with the matrix $[B \mid I]$. Undoing the row operations that led from $[A \mid I]$ to $[I \mid B]$ will lead from $[B \mid I]$ to $[I \mid A]$, so $BA = I$.

(2) If row-reducing $[A \mid I]$ does not reduce to $[I \mid B]$ but to $[A' \mid A'']$, with the bottom row of A' all zeroes, then there are two possibilities. If the bottom row of A'' is also all zeroes, there are infinitely many solutions; if there is a nonzero element in the bottom row of A'', then there is no solution. In either case, A is noninvertible.   □

Remark. The careful reader will observe that we have shown that if $[A \mid I]$ row-reduces to $[I \mid B]$, then $AB = I$. We have not shown that $BA = I$, although this also is true.

Example L1.14.

$$A = \begin{bmatrix} 2 & 1 & 3 \\ 1 & -1 & 1 \\ 1 & 1 & 2 \end{bmatrix} \quad \text{has inverse} \quad A^{-1} = \begin{bmatrix} 3 & -1 & -4 \\ 1 & -1 & -1 \\ -2 & 1 & 3 \end{bmatrix}$$

because

$$\begin{bmatrix} 2 & 1 & 3 \cdot 1 & 0 & 0 \\ 1 & -1 & 1 \cdot 0 & 1 & 0 \\ 1 & 1 & 2 \cdot 0 & 0 & 1 \end{bmatrix}$$

row-reduces to

$$\begin{bmatrix} \underline{1} & 0 & 0 \cdot & 3 & -1 & -4 \\ 0 & \underline{1} & 0 \cdot & 1 & -1 & -1 \\ 0 & 0 & \underline{1} \cdot & -2 & 1 & 3 \end{bmatrix}.$$

You can confirm in Exercise L1#9 that $AA^{-1} = A^{-1}A = I$ and that you can use this inverse matrix to solve the system of Example L1.3. ▲

Example L1.15.

$$A = \begin{bmatrix} 2 & 1 & 3 \\ 1 & -1 & 0 \\ 1 & 1 & 2 \end{bmatrix}$$

has *no* inverse A^{-1} because

$$\begin{bmatrix} 2 & 1 & 3 \cdot 1 & 0 & 0 \\ 1 & -1 & 0 \cdot 0 & 1 & 0 \\ 1 & 1 & 2 \cdot 0 & 0 & 1 \end{bmatrix}$$

row-reduces to

$$\begin{bmatrix} \underline{1} & 0 & 1 \cdot & 1 & 0 & -1 \\ 0 & \underline{1} & 1 \cdot & -1 & 0 & 2 \\ 0 & 0 & 0 \cdot & -2 & 1 & 3 \end{bmatrix}.$$

This is the matrix of Examples L1.10 and L1.11, for two systems of linear equations, neither of which has a unique solution. ▲

Note: Examples L1.14 and L1.15 are unusually "simple" in the sense that the row reduction involved no fractions; this is the exception rather than the rule (as you might guess from Example L1.12). So do not be alarmed when your calculations look a lot messier.

Finally, to complete our list of matrix definitions, we note that one more matrix related to a matrix A is its *transpose* $A^{\top}$, the matrix that interchanges all the rows and columns of A.

Example L1.16. If $B = \begin{bmatrix} 1 & 4 & -2 \\ 3 & 0 & 2 \end{bmatrix}$, then $B^\top = \begin{bmatrix} 1 & 3 \\ 4 & 0 \\ -2 & 2 \end{bmatrix}$. ▲

A frequently used result involving the transpose is the following:

Theorem L1.17. *The transpose of a product is the product of the transposes in reverse order:*
$$(AB)^\top = B^\top A^\top.$$

The proof of Theorem L1.17 is straightforward and is left as Exercise L1#11.

The importance of the transpose will be discussed in Section L3.3.

L2 Theory of Linear Equations: Vocabulary

We now come to the most unpleasant chore of linear algebra: defining a fairly large number of essential concepts. These concepts have been isolated by generations of mathematicians as the right way to speak about the phenomena involved in linear equations. For all that, their usefulness, or for that matter meaningfulness, may not be apparent at first.

The most unpleasant of all is the notion of vector space. This is the arena in which "linear phenomena" occur; that is, the structure imposed by a vector space is the bare minimum needed for such phenomena to occur.

L2.1. VECTOR SPACES

A *vector space* is a set V, the elements of which can be added and multiplied by numbers, and satisfying all the rules which you probably consider "obvious." Being specific about what this means will probably seem both pedantic and mysterious, and it is. In practice, one rapidly gets an intuitive feeling for what a vector space is, and never verifies the axioms explicitly.

A vector space is a set endowed with two operations, *addition* and *multiplication by scalars*. A *scalar* is a number, and in this book the scalars will always be either the real numbers or the complex numbers.

For a vector space V, these two operations must satisfy the following ten rules:

(1) *Closure under addition.*
    For all $\mathbf{v}, \mathbf{w} \in V$, we have $\mathbf{v} + \mathbf{w} \in V$.

(2) *Closure under multiplication by scalars.*
    For any $\mathbf{v} \in V$ and any scalar α, we have $\alpha\mathbf{v} \in V$.

(3) *Additive identity.*
    There exists a vector $\mathbf{0} \in V$ such that for any $\mathbf{v} \in V$, $\mathbf{0} + \mathbf{v} = \mathbf{v}$.

(4) *Additive inverse.*
   For any $\mathbf{v} \in V$, there exists a vector $-\mathbf{v} \in V$ such that $\mathbf{v} + (-\mathbf{v}) = \mathbf{0}$.

(5) *Commutative law for addition.*
   For all $\mathbf{v}, \mathbf{w} \in V$, we have $\mathbf{v} + \mathbf{w} = \mathbf{w} + \mathbf{v}$.

(6) *Associative law for addition.*
   For all $\mathbf{v}_1, \mathbf{v}_2, \mathbf{v}_3 \in V$, we have $\mathbf{v}_1 + (\mathbf{v}_2 + \mathbf{v}_3) = (\mathbf{v}_1 + \mathbf{v}_2) + \mathbf{v}_3$.

(7) *Multiplicative identity.*
   For all $\mathbf{v} \in V$, we have $1\mathbf{v} = \mathbf{v}$.

(8) *Associative law for multiplication.*
   For all scalars α, β and all $\mathbf{v} \in V$, we have $\alpha(\beta\mathbf{v}) = (\alpha\beta)\mathbf{v}$.

(9) *Distributive law for scalar addition.*
   For all scalars α, β and all $\mathbf{v} \in V$, we have $(\alpha + \beta)\mathbf{v} = \alpha\mathbf{v} + \beta\mathbf{v}$.

(10) *Distributive law for vector addition.*
   For all scalars α and $\mathbf{v}, \mathbf{w} \in V$, we have $\alpha(\mathbf{v} + \mathbf{w}) = \alpha\mathbf{v} + \alpha\mathbf{w}$.

We shall now give four examples of vector spaces, which with their variants will be the main examples used in this book.

Example L2.1. The basic example of a vector space is $\mathbb{R}^n$, with the obvious componentwise addition and multiplication by scalars. Actually, you should think of the ten rules as some sort of "essence of $\mathbb{R}^n$," abstracting from $\mathbb{R}^n$ all its most important properties. ▲

Example L2.2. The more restricted set of vectors

$$\begin{bmatrix} x \\ y \\ z \end{bmatrix} \in \mathbb{R}^3 \quad \text{such that} \quad 2x - 3y + z = 0$$

(or any similar homogeneous linear relation on the variables of $\mathbb{R}^n$) forms another vector space, because all ten rules are satisfied by such x, y, z triples (Exercise L2#1). These are important examples also, since they are subspaces of the vector space $\mathbb{R}^n$. ▲

A *subspace* of a vector space is a subset that is also a vector space under the same operations as the original vector space. Exercise L2#2 asks you to show that all requirements for a vector space are automatically satisfied if just the following two statements are true for any elements $\mathbf{w}_i, \mathbf{w}_j$ of a subset W of V and any scalar α:

i) $\mathbf{w}_i + \mathbf{w}_j \in W$;   ii) $\alpha\mathbf{w}_i \in W$;

The "vectors" in our third example will probably seem more exotic.

Example L2.3. Consider the space $\mathcal{C}([0,1])$, which denotes the space of continuous real-valued functions $f(x)$ defined for $0 \le x \le 1$, with addition and multiplication by scalars the usual operations. That is, the "vectors" are now functions $f(x)$, so addition means $f(x) + g(x)$ and multiplication means $\alpha\, f(x)$. Although these functions are not geometric vectors in the sense of our previous example, this space also satisfies all ten requirements for a vector space. ▲

Example L2.4. Consider the space of twice-differentiable functions $f(x)$ defined for all $x \in \mathbb{R}$ such that $d^2 f/dx^2 = 0$. This is a subspace of the vector space of the preceding example and is a vector space itself. But since a function has a vanishing second derivative if and only if it is a polynomial of degree 1, we see that this space is the set of functions

$$f_{a,b}(x) = a + bx.$$

In some sense, this space "is" $\mathbb{R}^2$, by identifying $f_{a,b}$ with $\begin{bmatrix} a \\ b \end{bmatrix} \in \mathbb{R}^2$; this was not obvious from the definition. ▲

L2.2. Linear Combinations, Linear Independence and Span

If $\mathbf{v}_1, \ldots, \mathbf{v}_k$ is a collection of vectors in some vector space V, then a *linear combination* of the $\mathbf{v}_i$ is a vector $\mathbf{v}$ of the form

$$\mathbf{v} = \sum_{i=1}^{k} a_i \mathbf{v}_i.$$

The *span* of $\mathbf{v}_1, \ldots, \mathbf{v}_k$ is the set of linear combinations of the $\mathbf{v}_i$.

Examples L2.5. The standard unit vectors $\mathbf{i}$ and $\mathbf{j}$ span the plane, because any vector in the plane is a linear combination

$$a_1\, \mathbf{i} + a_2\, \mathbf{j}.$$

The vectors $\mathbf{u}$ and $\mathbf{v}$ *also* span the plane, as illustrated in Figure L2.1.

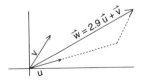

FIGURE L2.1. Any vector can be expressed as the sum of components in the directions $\mathbf{u}$ and $\mathbf{v}$.

A set of vectors $\mathbf{v}_1, \ldots, \mathbf{v}_k$ *spans* a vector space V if every element of V is in the span. For instance, $\mathbf{i}$ and $\mathbf{j}$ span $\mathbb{R}^2$ but not $\mathbb{R}^3$.

The vectors $\mathbf{v}_1, \ldots, \mathbf{v}_k$ are *linearly independent* if there is only one way of writing a given linear combination, i.e., if

$$\sum_{i=1}^{k} a_i \mathbf{v}_i = \sum_{i=1}^{k} b_i \mathbf{v}_i \text{ implies } a_1 = b_1, \ a_2 = b_2, \ldots, a_k = b_k.$$

Many books use the following as the definition of linear independence, then prove the equivalence of our definition: A set of vectors $\mathbf{v}_1, \ldots, \mathbf{v}_k$ is linearly independent if and only if the only solution to

$$a_1 \mathbf{v}_1 + a_2 \mathbf{v}_2 + \cdots + a_{n+1} \mathbf{v}_{n+1} = \mathbf{0} \text{ is } a_1 = a_2 = \cdots + a_{n+1} = 0.$$

Another equivalent statement is to say that none of the $\mathbf{v}_i$ is a linear combination of the others (Exercise L2#3). If one of the $\mathbf{v}_i$ is a linear combination of the others, they are called *linearly dependent*.

Remark. Geometrically, *linear independence* means the following:

(1) One vector is linearly independent if it is not the zero vector.

(2) Two vectors are linearly independent if they do not lie on a line, i.e., if they are *not collinear*.

(3) Three vectors are linearly independent if they do not lie in a plane, i.e., if they are *not coplanar*.

The following theorem is basic to the entire theory.

Theorem L2.6. *In $\mathbb{R}^n$, $n + 1$ vectors are never linearly independent, and $n - 1$ vectors never span.*

Proof. The idea is to reduce both statements in the theorem to linear equations.

For the first part, let the $n + 1$ vectors in $\mathbb{R}^n$ be

$$\mathbf{v}_1 = \begin{bmatrix} v_{1,1} \\ \vdots \\ v_{n,1} \end{bmatrix}, \ \mathbf{v}_2 = \begin{bmatrix} v_{1,2} \\ \vdots \\ v_{n,2} \end{bmatrix}, \ldots, \mathbf{v}_{n+1} = \begin{bmatrix} v_{1,n+1} \\ \vdots \\ v_{n,n+1} \end{bmatrix}.$$

Linear independence for $n + 1$ vectors can be written as the vector equation $a_1 \mathbf{v}_1 + a_2 \mathbf{v}_2 + \cdots + a_{n+1} \mathbf{v}_{n+1} = \mathbf{0}$ or as a system of linear equations

$$a_1 v_{1,1} + \cdots + a_{n+1} v_{1,n+1} = 0$$
$$\vdots \quad \cdots \quad \vdots \quad = \vdots$$
$$a_1 v_{n,1} + \cdots + a_{n+1} v_{n,n+1} = 0,$$

which is n equations in $n+1$ unknowns $a_1, \ldots, a_{n+1}$, with $v_{1,1}, \ldots, v_{n,n+1}$ as coefficients (exactly the reverse of the usual interpretation of such a system). These equations are certainly compatible, since $a_1 = \cdots = a_{n+1} = 0$ is a solution, but this solution cannot be the only solution: that would mean that fewer equations than unknowns determined the values of the unknowns.

To see this rigorously, row reduce the matrix $[\mathbf{v}_1, \ldots, \mathbf{v}_{n+1}, \mathbf{0}]$. At least one of the $n+1$ columns must not contain a leading 1, since there is at most one per row and there are fewer rows than columns. By Theorem L1.9, there cannot be a unique solution to the system of equations.

For the second part, let $\mathbf{v}_1, \ldots, \mathbf{v}_{n-1}$ be vectors in $\mathbb{R}^n$. To say they do not span is to say that there exists a vector $\mathbf{b}$ such that the equation

$$a_1\mathbf{v}_1 + \cdots + a_{n-1}\mathbf{v}_{n-1} = \mathbf{b}$$

has no solutions. Again write this out in full, to get

$$a_1 v_{1,1} + \cdots + a_{n-1}v_{1,n-1} = b_1$$
$$\vdots \quad \cdots \quad \vdots \quad = \vdots \quad,$$
$$a_1 v_{n,1} + \cdots + a_{n-1}v_{n,n-1} = b_n$$

i.e., n equations in the $n-1$ unknowns $a_1, \ldots, a_{n-1}$, with $v_{1,1}, \ldots, v_{n,n-1}$ as coefficients. We must see that the b_i's can be chosen so that the equations are incompatible. Write the matrix

$$[\mathbf{v}_1, \ldots, \mathbf{v}_{n-1}, \quad],$$

temporarily leaving the last column blank. Row-reduce it, at least the first $n-1$ columns. There must then be a row starting with $n-1$ zeroes, since any row must either begin with a leading one or be all zeroes; furthermore, there is at most one leading 1 per column. Now put a 1 in the nth, or last, position of that row and fill up the nth column by making all the other entries 0. By Theorem L1.9, case (b), such an echelon form of the matrix represents a system with no solution. Since any row operation can be undone, we can bring the echelon matrix back to where it started, with the matrix A on the left; the last column will then be a vector $\mathbf{b}$, making the system incompatible. $\square$

In a vector space, an ordered set of vectors $\mathbf{v}_1, \ldots, \mathbf{v}_n$ is called a *basis* if it satisfies one of the three following equivalent conditions (We will see below that they are indeed equivalent):

(1) The set is a maximal linearly independent set.

(2) The set is a minimal spanning set.

(3) The set is a linearly independent spanning set.

Example L2.7. The most fundamental example of basis is the *standard basis* of $\mathbb{R}^n$:

$$\mathbf{e}_1 = \begin{bmatrix} 1 \\ 0 \\ 0 \\ \vdots \\ 0 \end{bmatrix}, \ \mathbf{e}_2 = \begin{bmatrix} 0 \\ 1 \\ 0 \\ \vdots \\ 0 \end{bmatrix}, \dots, \mathbf{e}_n = \begin{bmatrix} 0 \\ 0 \\ 0 \\ \vdots \\ 1 \end{bmatrix}.$$

Clearly, every vector is in the span of $\mathbf{e}_1, \dots, \mathbf{e}_n$:

$$\begin{bmatrix} a_1 \\ \vdots \\ \vdots \\ a_n \end{bmatrix} = a_1\mathbf{e}_1 + \cdots + a_n\mathbf{e}_n;$$

it is equally clear that $\mathbf{e}_1, \dots, \mathbf{e}_n$ are linearly independent (Exercise L2#4).
▲

Example L2.8. The standard basis is not in any sense the only one. Quite the contrary: in general, any old n vectors in $\mathbb{R}^n$ form a basis. For instance,

$$\begin{bmatrix} 1 \\ 1 \end{bmatrix}, \ \begin{bmatrix} 1 \\ -1 \end{bmatrix} \text{ form a basis, as do } \begin{bmatrix} 2 \\ 0 \end{bmatrix}, \ \begin{bmatrix} 0.5 \\ -3 \end{bmatrix}. \ ▲$$

We need to show that the three conditions for a basis are indeed equivalent:

If a set $\mathbf{v}_1, \dots, \mathbf{v}_n$ is a maximal linearly independent set, then for any other vector $\mathbf{w}$, the set $\{\mathbf{v}_1, \dots, \mathbf{v}_n, \mathbf{w}\}$ is linearly dependent, and there exists a nontrivial relation

$$a_1\mathbf{v}_1 + \cdots + a_n\mathbf{v}_n + b\mathbf{w} = 0.$$

The coeffcient b is not zero, because the relation would then involve only the $\mathbf{v}$'s, which are linearly independent by hypothesis; so $\mathbf{w}$ can be expressed as a linear combination of the $\mathbf{v}$'s and we see that the $\mathbf{v}$'s do span.

But $\{\mathbf{v}_1, \dots, \mathbf{v}_n\}$ is a minimal spanning set: if one of the $\mathbf{v}_i$'s is omitted, the set no longer spans, since the omitted $\mathbf{v}_i$ is linearly independent of the others and hence cannot be in the span of the others. This shows that (1) and (2) are equivalent; the other equivalences are similar and left as Exercise L2#6.

Now Theorem L2.6 above can be restated:

Theorem L2.9. *Every basis of $\mathbb{R}^n$ has exactly n elements.*

Indeed, a set of vectors never spans if it has fewer than n elements, and is never linearly independent when it has more than n elements.

A vector space is *finite-dimensional* if it is spanned by finitely many elements. It then has a basis (in fact, many bases): find any finite spanning set and discard elements of it until what is left still spans, but if another vector is discarded, it no longer does.

Theorem L2.10. *Any two bases of a finite-dimensional vector space have the same number of elements.*

Proof. Let $v_1, \ldots, v_n$ be the first basis and $w_1, \ldots, w_m$ be the second. Then we can write

$$w_j = \sum_{i=1}^{n} a_{i,j} v_i.$$

Now any vector v can be written uniquely $v = \sum_{i=1}^{n} b_i v_i = \sum_{j=1}^{m} c_j w_j$, so

$$\sum_{i=1}^{n} b_i v_i = \sum_{j=1}^{m} c_j w_j = \sum_{j=1}^{m} c_j \sum_{i=1}^{n} a_{i,j} v_i = \sum_{i=1}^{n} \sum_{j=1}^{m} c_j a_{i,j} v_i,$$

so that $b_i = \sum_{j=1}^{m} c_j a_{i,j}$. Consider the b_i's known and this expression a system of n linear equations for the m unknowns $c_1, \ldots, c_m$. The statement that there exist such c_i's and that they are unique is precisely the statement that this system of equations has a unique solution. We have seen in Theorem L1.9 that this requires that there be as many unknowns as equations, i.e., that $n = m$. □

If a vector space is *finite-dimensional*, then the number of elements in a basis is called the *dimension* of the vector space. The dimension of a space is the same as the number of parameters used to describe it. Therefore, we can now more clearly state Theorem L1.9, case (c): the dimension of the solution space to a system of linear equations with infinitely many solutions is the same as the number of columns of $\tilde{A}$ not containing a leading 1.

Theorem L2.10 says precisely that $\mathbb{R}^n$ has dimension n. Therefore, we have

Principle L2.11. *An n-dimensional vector space with a basis is essentially the same as $\mathbb{R}^n$.*

More precisely, if $v_1, \ldots, v_n$ is a basis of a vector space V, then V can be identified to $\mathbb{R}^n$ by identifying v with its coefficients with respect to the basis:

$$a = \begin{bmatrix} a_1 \\ \vdots \\ a_n \end{bmatrix} \text{ can be identified with } a_1 v_1 + \cdots + a_n v_n = v.$$

One way to understand the proof of Theorem L2.10 is to think that we identified V with $\mathbb{R}^n$, via the basis $v_1, \ldots, v_n$; then in that identification,

$\mathbf{w}_1, \ldots, \mathbf{w}_m$ became m vectors of $\mathbb{R}^n$, namely the columns of the matrix $(a_{i,j})$. (Think about this, as Exercises L2#10, 11.) The identification preserves all "linear features"; in particular, these m vectors are a basis of $\mathbb{R}^n$, hence there are n of them. We shall use this fact in Section L2.4.

Remark. Not every vector space comes with an outstanding basis like $\mathbb{R}^n$ does.

Example L2.12. Consider the subspace $V \subset \mathbb{R}^3$ given by

$$V = \left\{ \begin{bmatrix} x \\ y \\ z \end{bmatrix} \, \middle| \, x + y + z = 0 \right\}.$$

This V does not (in the authors' opinion) have any *obvious* basis, but, of course, it does have lots of bases; for instance,

$$\begin{bmatrix} 1 \\ 0 \\ -1 \end{bmatrix} \begin{bmatrix} 0 \\ 1 \\ -1 \end{bmatrix} \text{ or } \begin{bmatrix} 1 \\ 1 \\ -2 \end{bmatrix} \begin{bmatrix} 1 \\ -1 \\ 0 \end{bmatrix}.$$

Each of these bases induces an identification of $\mathbb{R}^2$ with V as above. For instance, the first pair of basis vectors induces the identification of

$$\begin{bmatrix} a \\ b \end{bmatrix} \text{ with } a \begin{bmatrix} 1 \\ 0 \\ -1 \end{bmatrix} + b \begin{bmatrix} 0 \\ 1 \\ -1 \end{bmatrix} = \begin{bmatrix} a \\ b \\ -(a+b) \end{bmatrix}.$$

Exercise L2#8 asks you to write the identification corresponding to the second pair of basis vectors. ▲

More generally, any theorem about $\mathbb{R}^n$ is also a statement about every n-dimensional vector space, and any question about finite-dimensional vector spaces can be replaced by a question about $\mathbb{R}^n$. We will see many instances of this sort of thinking, both in these appendices and in the body of the book.

Remark. Once you have understood and absorbed the philosophy above, another question naturally comes up: if all vector spaces (at least finite dimensional) are just like $\mathbb{R}^n$, then why study abstract vector spaces? Why not just stick to $\mathbb{R}^n$? The answer is a bit subtle: $\mathbb{R}^n$ has more structure than just any old vector space—it has a distinguished basis. When you prove something about $\mathbb{R}^n$, you then need to check that your proof was "basis independent" before you can extend it to an arbitrary vector space. Even this answer is not really honest; proving things basis independent is usually quite easy. But mathematicians do not really like such proofs: if you can prove something by adding some structure and then showing you did not need it, it should be possible to prove the same thing without ever mentioning the extra structure. This aesthetic–philosophical consideration is also part of why abstract vector spaces are studied.

Let us turn briefly to the idea of a vector space that is *infinite-dimensional*.

Example L2.13. The vector space $C[0,1]$ of continuous functions on $[0,1]$, as in Example L2.3, is not finite-dimensional. This may be seen as follows:

Suppose functions $f_1,\ldots,f_n$ were a basis and pick $n+1$ distinct points $x_1,\ldots,x_{n+1}$ in $[0,1]$. Then given any values $c_1,\ldots,c_{n+1}$, there certainly exists a continuous function $f(x)$ with $f(x_i) = c_i$, for instance, the piecewise linear one whose graph consists of the line segments joining up the points (x_i, c_i).

We can write $f = \sum a_k f_k$ and, evaluating at the x_i, we get

$$f(x_i) = c_i = \sum a_k f_k(x_i), \qquad i = 1,\ldots,n+1.$$

This, for given c_i's, is a system of $n+1$ equations for the n unknowns $a_1,\ldots,a_n$; we know by Theorem L1.9 that for appropriate c_i's the equations will be incompatible. This is a contradiction to the hypothesis that $f_1,\ldots,f_n$ spanned. ▲

Remark. In infinite-dimensional vector spaces, bases still make sense but tend to be useless. The interesting notion is not *finite* linear combinations but *infinite* linear combinations, i.e., infinite series $\sum_{i=0}^{\infty} a_i \mathbf{v}_i$. This introduces questions of convergence, which are very interesting indeed, but a bit foreign to the spirit of linear algebra. We will examine such questions, however, when dealing with Fourier series.

L2.3. LINEAR TRANSFORMATIONS AND MATRICES

In "concrete" linear algebra, the central actors were "column vectors," inhabiting $\mathbb{R}^n$; these became "vectors" inhabiting "vector spaces" in "abstract" linear algegra. The other major actors of "concrete" linear algebra are "matrices"; these correspond to "linear transformations" in the abstract language.

If V and W are vector spaces, a *linear transformation* $T: V \to W$ is a *mapping* satisfying

$$T(a\mathbf{v}_1 + b\mathbf{v}_2) = aT(\mathbf{v}_1) + bT(\mathbf{v}_2)$$

for all scalars a, b and all $\mathbf{v}_1, \mathbf{v}_2 \in V$.

Example L2.14. Matrices give linear transformations: let $A = (a_{i,j})$ be an $m \times n$ matrix (that means m high and n wide). Then A defines a linear transformation $T: \mathbb{R}^n \to \mathbb{R}^m$ by matrix multiplication:

$$T(\mathbf{v}) = A\mathbf{v}.$$

Such mappings are indeed linear, because $A(\mathbf{v} + \mathbf{w}) = A\mathbf{v} + A\mathbf{w}$ and $A(c\mathbf{v}) = cA\mathbf{v}.$ ▲

Remark. In practice, a matrix and its associated linear transformation are usually identified, and the linear transformation $T : \mathbb{R}^n \to \mathbb{R}^m$ is denoted simply by its matrix A.

For an infinite-dimensional V, a linear transformation T is often given as a *differential operator* on the functions that comprise V:

Example L2.15. Let V be the vector space P_2 of polynomials $p(x)$ of degree at most 2; then an example of a linear transformation T is given by the formula

$$T(p)(x) = (x^2 + 1)p''(x) - xp'(x) + 2p(x).$$

The notation $T(p)(x)$ emphasizes that T acts on the polynomials p that comprise the vector space P_2, never on a number $p(x)$ that might describe p for a particular x.

We leave it to the reader as Exercise L2#13 to verify linearity. ▲

Example L2.16. A differential operator like that in Example L2.15 can be used on a more general function $f(x)$ to define a linear transformation such as

$$T(f)(x) = (x^2 + 1)f''(x) - xf'(x) + 2f(x), \qquad T: \mathcal{C}^2[0, 1] \to \mathcal{C}[0, 1]$$

from the space of twice continuously differentiable functions on $[0, 1]$ to the space of continuous functions on $[0, 1]$. ▲

Alternatively, for an infinite-dimensional V, a linear transformation T is sometimes given by an *integral* involving the functions of V, corresponding to the matrix A of the finite-dimensional case.

Example L2.17. There are analogs of matrices in $\mathcal{C}[0, 1]$. Let $K(x, y)$ be a continuous function of x and y defined for $0 \le x, y \le 1$, and consider the mapping

$$T: \mathcal{C}[0, 1] \to \mathcal{C}[0, 1]$$

given by

$$T(f)(x) = \int_0^1 K(x, y) f(y) dy.$$

This is analogous to Example L2.14, which could be written

$$T(\mathbf{v})_i = \sum_{j=1}^{n} a_{i,j}\mathbf{v}_j,$$

but here the discrete indices i, j have become "continuous indices" x, y.
▲

For finite-dimensional matrices, the following result is crucial:

Theorem L2.18. *Every linear transformation $T: \mathbb{R}^n \to \mathbb{R}^m$ is given by an $m \times n$ matrix M_T, the ith column of which is $T(\mathbf{e}_i)$.*

This means that Example L2.14 is "the general" linear transformation in finite-dimensional vector spaces.

Proof. Start with the linear transformation T of Example L2.14 and manufacture the matrix A according to the given rule. Then given any $\mathbf{v} \in \mathbb{R}^n$, we may write

$$\mathbf{v} = \sum v_i \mathbf{e}_i;$$

then, by linearity,

$$T(\mathbf{v}) = \sum v_i T(\mathbf{e}_i),$$

which is precisely the column vector $A\mathbf{v}$.   □

Example L2.19. Consider S_θ, the rotation about the origin by an angle θ in the plane, as shown in Figure L2.2. Since

$$S_\theta \begin{bmatrix} 1 \\ 0 \end{bmatrix} = \begin{bmatrix} \cos\theta \\ \sin\theta \end{bmatrix}, \qquad S_\theta \begin{bmatrix} 0 \\ 1 \end{bmatrix} = \begin{bmatrix} -\sin\theta \\ \cos\theta \end{bmatrix},$$

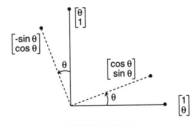

FIGURE L2.2.

Theorem L2.18 tells us that S_θ is given by the matrix

$$\begin{bmatrix} \cos\theta & -\sin\theta \\ \sin\theta & \cos\theta \end{bmatrix}.$$

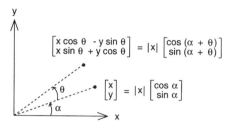

FIGURE L2.3.

Exercise L2#7a,b asks you to confirm the geometric picture shown in Figure L2.3 and compute some examples of rotations by matrices. ▲

Remark. In order for a rotation to be a *linear* transformation, it *must* be about the origin, which remains fixed. Rotation about any point other than the origin is *not* linear. (See Exercise L2#7c.)

Example L2.20. Consider Example L2.15. We will identify the polynomials of degree at most 2 with $\mathbb{R}^3$ by identifying

$$a_0 + a_1 x + a_2 x^2 \quad \text{with} \quad \begin{bmatrix} a_0 \\ a_1 \\ a_2 \end{bmatrix}.$$

We can find the matrix A that describes this linear transformation by the following observations:

$$T(p)(x) = (x^2 + 1)p''(x) - xp'(x) + 2p(x).$$

$$\begin{aligned} \text{if } p(x) &= 1, & T(1) &= 2; \\ \text{if } p(x) &= x, & T(x) &= x; \\ \text{if } p(x) &= x^2, & T(x^2) &= 2x^2 + 2. \end{aligned}$$

From this, you can show (Exercise L2#13) that T is given by the matrix

$$\begin{bmatrix} 2 & 0 & 2 \\ 0 & 1 & 0 \\ 0 & 0 & 2 \end{bmatrix}.$$

Exercise L2#13 also asks you to compute the matrices of the same differential operator, operating on polynomials of degree at most 3, 4, n. ▲

If V_1, V_2, and V_3 are vector spaces and if $S: V_1 \to V_2$ and $T: V_2 \to V_3$ are linear transformations, then the *composition* $T \circ S: V_1 \to V_3$ is again a linear transformation. In particular, if $V_1 = \mathbb{R}^{n_1}$, $V_2 = \mathbb{R}^{n_2}$, and $V_3 = \mathbb{R}^{n_3}$, then the matrix $M_{T \circ S}$ should be computable in terms of M_S and M_T.

Theorem L2.21. *Composition corresponds to matrix multiplication:*

$$M_{T \circ S} = M_T M_S.$$

Proof. Take $\mathbf{e}_i \in \mathbb{R}^{n_1}$, then $S(\mathbf{e}_i) \in \mathbb{R}^{n_2}$ is the ith column of M_S, and $T(S(\mathbf{e}_i)) = M_T S(\mathbf{e}_i) \in \mathbb{R}^{n_3}$ is the ith column of $M_T M_S$. $\square$

Remark. Many mathematicians would claim that this proposition "explains and justifies" the definition of matrix multiplication. This may seem odd to the novice, who probably feels that composition of linear mappings is far more baroque than matrix multiplication.

One immediate consequence of Theorem L2.21 is:

Corollary L2.22. *Matrix multiplication is associative: if A, B, and C are matrices such that the matrix multiplication $(AB)C$ is allowed, then so is $A(BC)$, and they are equal.*

Proof. Composition of mappings is associative. $\square$

L2.4. MATRICES OF LINEAR TRANSFORMATIONS WITH RESPECT TO A BASIS

So far, matrices have corresponded to linear transformations $\mathbb{R}^n \to \mathbb{R}^n$ or $\mathbb{C}^n \to \mathbb{C}^n$. What about linear transformations defined on more general vector spaces?

In accordance with Principle L2.11 that a vector space *with a basis* is $\mathbb{R}^n$, we will not be surprised to hear that if vector spaces V and W have bases $\{\mathbf{v}_1, \ldots, \mathbf{v}_n\}$ and $\{\mathbf{w}_1, \ldots, \mathbf{w}_m\}$, respectively, then any linear transformation $T: V \to W$ has a matrix $(t_{i,j})$ *with respect to the bases*. One way to see this is to say that the bases allow us to identify V and W to $\mathbb{R}^n$ and $\mathbb{R}^m$, respectively, via the coefficients of the basis vectors. Then the coefficients of $T(\mathbf{v})$ with respect to the $\mathbf{w}_i$ must depend linearly on the coefficients of $\mathbf{v}$ with respect to the $\mathbf{v}_i$. What this boils down to, in practice, is the following formula:

$$T(\mathbf{v}_i) = \sum t_{j,i} \mathbf{w}_j,$$

which can be taken as the definition of the matrix. Note the order of the indices: the ith column of $t_{i,j}$ is the vector of coefficients of $T(\mathbf{v}_i)$ with respect to $\{\mathbf{w}_1, \ldots, \mathbf{w}_m\}$. A look at Theorem L2.18 will show that this is indeed the correct definition.

In practical applications, the vector spaces V and W are often $\mathbb{R}^n$ and $\mathbb{R}^m$ themselves, but the bases $\{\mathbf{v}_1, \ldots, \mathbf{v}_n\}$ and $\{\mathbf{w}_1, \ldots, \mathbf{w}_m\}$ are not the standard bases. Is it possible to find the matrix of a linear transformation in terms of the new bases in terms of the old matrix? Of course, it is possible, and the important *change of basis* theorem is the answer:

Theorem L2.23. *If $\{\mathbf{v}_1, \ldots, \mathbf{v}_n\}$ and $\{\mathbf{w}_1, \ldots, \mathbf{w}_m\}$ are bases of $\mathbb{R}^n$ and $\mathbb{R}^m$ respectively, and if we form the matrices*

$$P = [\mathbf{v}_1, \ldots, \mathbf{v}_n] \quad and \quad Q = [\mathbf{w}_1, \ldots, \mathbf{w}_m],$$

then the linear transformation $T : \mathbb{R}^n \to \mathbb{R}^m$ having matrix A with respect to the standard basis has matrix $Q^{-1}AP$ with respect to the new bases.

The matrices P and Q are called the *change of basis matrices*, and $Q^{-1}AP$ is the *change of basis formula*.

Before proving Theorem L2.23 formally, let us see exactly what the change of basis matrices actually do.

If $\mathbf{a} \in \mathbb{R}^n$, then $P\mathbf{a} = a_1\mathbf{v}_1 + \cdots + a_n\mathbf{v}_n$, so that P takes a column of numbers and uses them as coefficients of the $\mathbf{v}_i$:

$$\begin{bmatrix} a_1 \\ a_2 \\ \vdots \\ a_n \end{bmatrix}$$

$$\begin{bmatrix} | & | & & | \\ \mathbf{v}_1 & \mathbf{v}_2 & \cdots & \mathbf{v}_n \\ | & | & & | \end{bmatrix} \begin{bmatrix} | \\ a_1\mathbf{v}_1 + a_2\mathbf{v}_2 + \cdots + a_n\mathbf{v}_n \\ | \end{bmatrix} = P\mathbf{a}.$$

Thus, P^{-1} must do the opposite; it takes vectors in $\mathbb{R}^n$ and gives you their coordinates with respect to the $\mathbf{v}_i$. You might think of P as synthesis and P^{-1} as analysis.

Proof of Theorem L2.23. The ith column of a matrix is the result of applying the matrix to $\mathbf{e}_i$. So let us compute $Q^{-1}AP(\mathbf{e}_i)$. Well, $P(\mathbf{e}_i) = \mathbf{v}_i$, so $AP(\mathbf{e}_i) = A(\mathbf{v}_i)$. Finally, Q^{-1} takes a vector and gives its coordinates with respect to the basis $\{\mathbf{w}_1, \ldots, \mathbf{w}_m\}$. So the ith column of $Q^{-1}AP$ is formed of the coordinates of $\mathbf{v}_i$ with respect to $\{\mathbf{w}_1, \ldots, \mathbf{w}_m\}$; this is precisely what was to be proved. $\square$

Of particular importance will be the case where $n = m$, and the bases in the domain and the range coincide. In that case, there is only one change of basis matrix P, and the matrix of the linear transformation with respect to the new basis is $P^{-1}AP$. The matrices A and $P^{-1}AP$ are said to be *conjugate*: they are different descriptions of the same linear transformation.

They will share many properties: determinants, characteristic polynomials, eigenvalues, dimensions of eigenspaces, etc. Over and over, we will say that some property of linear transformations is *"basis independent"*; this means that if a matrix has the property, then all its conjugates do too. (See Appendix L5, Theorems L5.6, L5.8, and L5.10.)

Note: *Any* invertible matrix is a change of basis matrix if you use its columns for the basis.

L2.5. KERNELS AND IMAGES

The *kernel* and the *image* of a linear transformation are the "abstract" notions that allow a precise treatment of uniqueness and existence of solutions to linear equations.

If V and W are vector spaces and $T: V \to W$ is a linear transformation, then the *kernel* of T is the subspace $\ker(T) \subset V$ given by

$$\ker(T) = \{\mathbf{v} \in V \mid T(\mathbf{v}) = 0\}.$$

For more or less obvious reasons, this space is also often called the *null space* of T.

The *image* of T is the set of vectors $\mathbf{w}$ such that there is a vector $\mathbf{v}$ with $T(\mathbf{v}) = \mathbf{w}$. That is,

$$\mathrm{Im}(T) = \{\mathbf{w} \in W \mid \text{ there exists } \mathbf{v} \in V \text{ such that } T(\mathbf{v}) = \mathbf{w}\}.$$

(Sometimes the image is also called the "range": this usage is a source of confusion since most authors, including ourselves, call the whole vector space W the range of T; consequently, we shall not use the word "range" to refer to the image.)

If V and W are finite-dimensional, we may choose bases for V and W; i.e., we may assume that $V = \mathbb{R}^n$, $W = \mathbb{R}^m$ and that T has an $m \times n$ matrix A. If we row-reduce A to echelon form $\tilde{A}$, we can read off a basis for the kernel and the image by the following two neat tricks:

    (i) The *image* is a subspace of $W = \mathbb{R}^m$. Hence any basis will consist of m-vectors. Such a basis is provided by those columns of the original matrix A that after row-reduction will contain a leading 1; the row-reduced matrix $\tilde{A}$ shows that these columns of A are linearly independent.

    (ii) The *kernel* is a subspace of $V = \mathbb{R}^n$. Hence, any basis will consist of n-vectors. We will find one such basis vector for each column of $\tilde{A}$ that does not contain a leading 1, as follows:

If the jth column of $\tilde{A}$ does not contain a leading 1, then it is a linear combination of those that do, giving a vector

$$
\mathbf{c}_j = \begin{bmatrix} c_{j1} \\ \vdots \\ c_{jj} \\ \vdots \\ c_{jn} \end{bmatrix},
$$

such that $\tilde{A}\mathbf{c}_j = \mathbf{0}$; hence, $A\mathbf{c}_j = \mathbf{0}$.

The vectors $\mathbf{c}_j$ found this way are linearly independent, since exactly one has a 0 sitting in each position corresponding to a column without leading 1's.

Example L2.24. Consider the matrix

$$
A = \begin{bmatrix} 1 & 2 & 3 & 7 & 0 \\ -1 & 1 & 0 & -1 & 1 \\ 1 & -1 & 0 & 1 & 0 \\ 0 & 2 & 2 & 4 & 1 \end{bmatrix}
$$

which row-reduces to

$$
\tilde{A} = \begin{bmatrix} \underline{1} & 0 & 1 & 3 & 0 \\ 0 & \underline{1} & 1 & 2 & 0 \\ 0 & 0 & 0 & 0 & \underline{1} \\ 0 & 0 & 0 & 0 & 0 \end{bmatrix}
$$

and describes a linear transformation from $\mathbb{R}^5$ to $\mathbb{R}^4$. According to the prescription above, we find bases for the image and for the kernel as follows:

(i) The leading 1's of the row-reduced matrix $\tilde{A}$ are in columns 1, 2, and 5, so columns 1, 2, and 5 of the original matrix A are a *basis for the image*:

$$
\begin{bmatrix} 1 \\ -1 \\ 1 \\ 0 \end{bmatrix} \begin{bmatrix} 2 \\ 1 \\ -1 \\ 2 \end{bmatrix} \begin{bmatrix} 0 \\ 1 \\ 0 \\ 1 \end{bmatrix}.
$$

(ii) A *basis for the kernel* is provided by the columns not containing leading 1's. We have in $\tilde{A}$

$$
\text{col } 3 = \text{col } 1 + \text{col } 2; \qquad \text{col } 4 = 3 \text{ col } 1 + 2 \text{ col } 2. \tag{1}
$$

The second equation of (1) can be rewritten as

$$
3(\text{col } 1) + 2(\text{col } 2) + 0(\text{col } 3) - 1(\text{col } 4) + 0(\text{col } 5) = \tilde{A}\mathbf{x} = \mathbf{0}.
$$

A solution for $\tilde{A}\mathbf{x} = \mathbf{0}$, and hence for $A\mathbf{x} = \mathbf{0}$, is found from the coefficients of the columns in the last line:

$$x = \begin{bmatrix} 3 \\ 2 \\ 0 \\ -1 \\ 0 \end{bmatrix}.$$

In like manner, the first equation of (1) can be rewritten and you can find another 5-vector, linearly independent of the one from the fourth column (just looking at the third and fourth entries shows that one is not a multiple of the other) and we now have *a basis for the kernel*:

$$\begin{bmatrix} 1 \\ 1 \\ -1 \\ 0 \\ 0 \end{bmatrix}, \begin{bmatrix} 3 \\ 2 \\ 0 \\ -1 \\ 0 \end{bmatrix}. \quad \blacktriangle$$

Many students find kernels and images difficult concepts to absorb and use comfortably. These important but rather abstract concepts are best understood in terms of linear equations, where the linear transformation T is represented by a matrix A:

> The kernel of T is the set of solutions of the homogeneous equation $A\mathbf{x} = \mathbf{0}$.

> The image of T is the set of vectors $\mathbf{b}$ for which there exists a solution of $A\mathbf{x} = \mathbf{b}$.

Kernels are related to uniqueness of solutions of linear equations, and images are related to their existence. The latter fact is stated directly above, and the former fact results from the following statement, which is easy to prove:

> Vectors $\mathbf{x}_1$ and $\mathbf{x}_2$ are both solutions of $A\mathbf{x} = \mathbf{b}$ if and only if $\mathbf{x}_1 - \mathbf{x}_2 \in \ker(A)$; in particular, solutions are unique if and only if $\ker(A) = \{\mathbf{0}\}$.

Much of the power of linear algebra comes from the following theorem, known as the *dimension formula*. It says that there is a conservation law concerning the kernel and the image, and that saying something about uniqueness *ipso facto* says something about existence.

Theorem L2.25. *Let V and W be finite-dimensional vector spaces, of dimensions n and m, respectively, and let $T: V \to W$ be a linear transformation. Then*

$$\dim(\ker(T)) + \dim(\mathrm{Im}(T)) = \dim V.$$

Proof. This is clear from tricks (i) and (ii) above. We find one basis vector for the kernel for each column of $\tilde{A}$ containing a leading 1, and one basis vector for the kernel for each column of $\tilde{A}$ containing a leading 1, so in all we find

$$\dim(\ker(T)) + \dim(\mathrm{Im}(T)) = \text{number of columns of} A = \dim V.$$

$\square$

The most important case of Theorem L2.25 is when the dimensions of V and W are equal.

Corollary L2.26. *If V and W are finite-dimensional vector spaces with the same dimension, and $T: V \to W$ is a linear transformation, then*

the equation $T(\mathbf{x}) = \mathbf{b}$ has a solution for any $\mathbf{b}$

if and only if

the equation $T(\mathbf{x}) = \mathbf{0}$ has only the solution $\mathbf{x} = \mathbf{0}$.

Proof. Let n be the dimension of V and W. The first statement above is that W is the image of T, so $\dim(\mathrm{Im}(T)) = n$; hence, $\dim(\ker(T)) = 0$, which is the second statement. $\square$

The *power* of linear algebra comes from Corollary L2.26. See the following example and Exercises L2#16 and L2#17.

Example L2.27. Partial fractions. This example is rather extensive but gives a good idea of the power of Corollary L2.26. Let

$$P(x) = (x - x_1)^{n_1} \cdots (x - x_k)^{n_k}$$

be a polynomial of degree $n = n_1 + \cdots + n_k$, with the x_i distinct. The claim of partial fractions is that for any polynomial q of degree $< n$, the rational function $q(x)/p(x)$ can be written uniquely as

$$\frac{q(x)}{p(x)} = \frac{q_1(x)}{(x - x_1)^{n_1}} + \cdots + \frac{q_k(x)}{(x - x_k)^{n_k}}, \tag{2}$$

with each q_i a polynomial of degree $< n_i$.

The space V of all such "vectors of polynomials" $(q_1(x), \ldots, q_k(x))$ is of dimension n, since the ith polynomial has n_i coefficients, and similarly the space W of polynomials q of degree $< n$ has dimension n. The process of multiplying out the right-hand side of (2) to get the left-hand side is a linear transformation $T: V \to W$, which can be written "explicitly" as

$$T(q_1, \ldots, q_k) = q_1(x)(x - x_2)^{n_2} \cdots (x - x_k)^{n_k} + \cdots + q_k(x)(x - x_1)^{n_1}$$
$$\cdots (x - x_{k-1})^{n_{k-1}}.$$

Except in the simplest cases, however, computing the matrix of T would be a big job.

Saying that q/p can be decomposed into partial fractions is precisely saying that W is the image of T, i.e., the first condition of Corollary L2.26. Since it is equivalent to the second alternative, we see that "partial fractions work" if and only if

$$\text{the only solution of } T(q_1, \ldots, q_k) = \mathbf{0} \text{ is } q_1 = \cdots = q_k = 0.$$

Lemma L2.28. *If $q_i \neq 0$ is a polynomial of degree $< n_i$, then*

$$\lim_{x \to x_i} |q_i(x)/(x - x_i)^{n_1}| = \infty \text{ and } T(q_1, \ldots, q_k) \neq \mathbf{0}.$$

Proof. By making the change of variables $u = x - x_i$, we can suppose $x_i = 0$. Since $q_i \neq 0$, $q_i(x) = a_m x^m +$ higher terms, with $a_m \neq 0$ and $m < n_i$. By Taylor's theorem, for any ε there exists δ such that

$$|q_i(x)| > (|a_m| - \varepsilon)|x|^m$$

if $|x| < \delta$. So

$$|q_i(x)/x^{n_i}| > (|a_m| - \varepsilon)|x|^{m - n_i}$$

which tends to ∞ as x tends to 0.

It follows from this that $T(q_1, \ldots, q_k) \neq \mathbf{0}$ if some $q_i \neq 0$, since for all $j \neq i$, the rational functions

$$q_j(x)/(x - x_j)^{n_j}$$

have the finite limits $q_j(x_i)/(x_i - x_j)^{n_j}$ as $x \to x_i$, and therefore the sum

$$\frac{q(x)}{p(x)} = \frac{q_1(x)}{(x - x_1)^{n_1}} + \cdots + \frac{q_k(x)}{(x - x_k)^{n_k}}$$

has infinite limit as $x \to x_i$ and therefore q cannot vanish identically. $\square$

Examine Example L2.27 carefully. It really put linear algebra to work. Even after translating the problem into linear algebra, via the linear transformation T, the answer was not clear, and only after using dimensions and the dimension formula is the result apparent. Still, all of this is *nothing more* than the intuitively obvious statement that either n equations in n unknowns are independent, the good case, or everything goes wrong at once.

L3 Vector Spaces and Inner Products

An inner product on a vector space is that extra structure that makes metric statements meaningful: such notions as angles, orthogonality, and length are only meaningful in *inner product spaces*. The archetypal examples are the dot product on $\mathbb{R}^n$ and $\mathbb{C}^n$. Just as in the case of vector spaces, we will see that these examples are not just the most obvious ones, but that all others are just like them.

However, also just as for vector spaces, there are frequent occasions where stripping away the extra structure of $\mathbb{R}^n$ and $\mathbb{C}^n$ is helpful, and we will begin with the general definition.

In this appendix, we will describe spaces with inner products; then in Appendix L4 we will deal with the relations between linear transformations and inner products.

L3.1. REAL INNER PRODUCTS

An *inner product* on a real vector space V is a rule that takes two vectors $\mathbf{a}$ and $\mathbf{b}$ and gives a number $\langle \mathbf{a}, \mathbf{b} \rangle \in \mathbb{R}$, satisfying the following three rules:

  (i) $\langle \mathbf{a}, \mathbf{b} \rangle = \langle \mathbf{b}, \mathbf{a} \rangle$ *symmetric*

  (ii) $\langle \alpha \mathbf{a}_1 + \beta \mathbf{a}_2, \mathbf{b} \rangle = \alpha \langle \mathbf{a}_1, \mathbf{b} \rangle + \beta \langle \mathbf{a}_2, \mathbf{b} \rangle$ *linear*

  (iii) $\langle \mathbf{a}, \mathbf{a} \rangle > 0$ if $\mathbf{a} \neq 0$ *positive definite.*

The *norm* of a vector $\mathbf{a}$ with respect to an inner product is

$$\|\mathbf{a}\| = \sqrt{\langle \mathbf{a}, \mathbf{a} \rangle}.$$

Example L3.1. The *standard inner product* or *dot product* on $\mathbb{R}^n$,

$$\mathbf{a} \cdot \mathbf{b} = \sum_{i=1}^{n} a_i b_i = \mathbf{a}^\top \mathbf{b},$$

satisfies all these properties. Furthermore,

$$\|\mathbf{a}\| = \sqrt{\mathbf{a} \cdot \mathbf{a}} = \sqrt{a_1^2 + a_2^2 + \cdots + a_n^2}$$

gives the *standard*, or *Euclidean norm*.   ▲

Example L3.2. On $\mathbb{R}^2$, consider the function

$$\langle \mathbf{a}, \mathbf{b} \rangle \equiv 2a_1 a_2 + 2b_1 b_2 + a_1 b_2 + a_2 b_1.$$

This function easily satisfies properties (i) and (ii). For property (iii), observe that

$$\langle \mathbf{a}, \mathbf{a} \rangle = 2a_1^2 + 2a_2^2 + 2a_1 a_2 = (a_1 + a_2)^2 + a_1^2 + a_2^2$$

is a sum of squares and is certainly strictly positive unless $\mathbf{a} = 0$. Hence, $\langle \mathbf{a}, \mathbf{b} \rangle$ is also an inner product.   ▲

The next example will probably be less familiar; it is, however, of great importance, particularly in trying to understand what Fourier series are all about.

Example L3.3. Consider V the vector space of continuous functions on an interval $[a, b]$, and $\rho(x)$ a positive continuous function on $[a, b]$. Then the *integral*

$$\langle f, g \rangle = \int_a^b \rho(x) f(x) g(x) dx$$

defines an inner product on V. The case $\rho = 1$ is the continuous analog of the standard inner product.   ▲

L3.2. COMPLEX INNER PRODUCTS

A *complex inner product* (also called a *Hermitian inner product*) on a complex vector space V is a complex-valued function on two vectors $\mathbf{a}$ and $\mathbf{b}$ giving a number $\langle \mathbf{a}, \mathbf{b} \rangle \in \mathbb{C}$, satisfying the following three rules:

(i) $\langle \mathbf{a}, \mathbf{b} \rangle = \overline{\langle \mathbf{b}, \mathbf{a} \rangle}$ *conjugate-symmetric*

(ii) $\langle \alpha \mathbf{a}_1 + \beta \mathbf{a}_2, \mathbf{b} \rangle = \alpha \langle \mathbf{a}_1, \mathbf{b} \rangle + \beta \langle \mathbf{a}_2, \mathbf{b} \rangle$ *linear with respect to first argument*

(iii) $\langle \mathbf{a}, \mathbf{a} \rangle > 0$ if $\mathbf{a} \neq 0$ *positive definite.*

where the bar denotes the *complex conjugate*. Note that property (ii) is unchanged from the real case, but, in combination with (i), it says that the complex inner product is linear with respect to the first variable but not the second. Note also that property (iii) makes sense because, from (i), $\langle \mathbf{a}, \mathbf{a} \rangle$ is real.

Example L3.4. In $\mathbb{C}^n$, the *standard inner product* is $\langle \mathbf{a}, \mathbf{b} \rangle = \sum_i a_i \bar{b}_i$, and the *norm* is $\|\mathbf{a}\| = \sqrt{\langle \mathbf{a}, \mathbf{a} \rangle}$.   ▲

Example L3.5. Let V be the vector space of complex-valued continuous functions on $[-\pi, \pi]$. Then the formula

$$\langle f, g \rangle = \int_{-\pi}^{\pi} f(x) \overline{g(x)} dx$$

defines an inner product on V [where $\overline{g(x)}$ is the complex conjugate of $g(x)$]. ▲

L3.3. BASIC THEOREMS AND DEFINITIONS USING INNER PRODUCTS

The following rather surprising theorem concerning the transpose of a matrix (defined at the end of Section L1.4) is very useful in manipulating and proving other theorems throughout the rest of linear algebra.

Theorem L3.6. *For any two vectors* $\mathbf{v}$ *and* $\mathbf{w}$ *in a real or complex inner product space and any real or complex-valued matrix A,*

$$\langle A^{\top}\mathbf{v}, \mathbf{w} \rangle = \langle \mathbf{v}, \bar{A}\mathbf{w} \rangle.$$

The proof of Theorem L3.6 is straightforward and is left as Exercise L3#1.

Remark. Theorem L3.6 says that the transpose depends on the inner product, not the basis; this is why it is important.

Suppose V and W are vector spaces with inner products (but without chosen bases). If $T: V \to W$ is a linear transformation, then there is an abstract transposed linear transformation $T^{\top}: W \to V$, often called the *adjoint*, defined by the formula

$$\langle T^{\top}\mathbf{w}, \mathbf{v} \rangle = \langle \mathbf{w}, T\mathbf{v} \rangle.$$

If $V = \mathbb{R}^n$ and $W = \mathbb{R}^m$, with inner products the standard dot product, the matrix for $T^{\top}$ is $(M_T)^{\top}$.

Another fact we shall want to have handy is the following:

Theorem L3.7. *The derivative of an inner product follows the ordinary product rule for derivatives:* $\langle \mathbf{v}(t), \mathbf{w}(t) \rangle' = \langle \mathbf{v}'(t), \mathbf{w}(t) \rangle + \langle \mathbf{v}(t), \mathbf{w}'(t) \rangle.$

The proof of Theorem L3.7 is straightforward and is left as Exercise L3#2.

In the case of a complex vector space, Theorem L3.7, in conjunction with the definition of inner product, leads to interesting results like the following:

Example L3.8. If a vector $\mathbf{v} \in \mathbb{C}^n$,

$$\frac{d}{dt}\|\mathbf{v}\|^2 = \langle \mathbf{v}', \mathbf{v} \rangle + \langle \mathbf{v}, \mathbf{v}' \rangle = \langle \mathbf{v}', \mathbf{v} \rangle + \overline{\langle \mathbf{v}', \mathbf{v} \rangle} = 2\,\mathrm{Re}\,\langle \mathbf{v}', \mathbf{v} \rangle,$$

where Re refers to the "real part" of the complex inner product $\langle \mathbf{v}', \mathbf{v} \rangle$.
▲

Another result, absolutely fundamental, about inner products is the following:

Theorem L3.9 (Schwarz's Inequality). *For any two vectors* **a** *and* **b** *in a real or complex inner product space, the inequality*

$$|\langle \mathbf{a}, \mathbf{b} \rangle| \leq \|\mathbf{a}\| \, \|\mathbf{b}\| \tag{1}$$

holds, and equality holds if and only if one of **a** *and* **b** *is a multiple of the other by a scalar.*

Proof. This proof is not simply crank-turning and really requires some thought. The inequality is no easier to prove for the standard dot product than it is in the general case. We will prove the real case only; the complex case is in Exercise L3#3.

For the first part, consider the function $\|\mathbf{a} + k\mathbf{b}\|^2$ as a function of k. It is a second degree polynomial, in fact

$$\|\mathbf{a} + k\mathbf{b}\|^2 = k^2 \|\mathbf{b}\|^2 + 2k\langle \mathbf{a}, \mathbf{b} \rangle + \|\mathbf{a}\|^2,$$

and it also only takes on values ≥ 0. But if the *discriminant*

$$4\langle \mathbf{a}, \mathbf{b} \rangle^2 - 4\|\mathbf{a}\|^2 \|\mathbf{b}\|^2$$

of the polynomial were positive, the polynomial would have two distinct roots and certainly would change sign. So the discriminant is nonpositive, as we wanted to show.

For the second part, assume $\mathbf{b} = c\mathbf{a}$ with $c > 0$; equality now drops out of inequality (1) when the discriminant is zero. Moreover, if **a** and **b** are linearly independent, the polynomial $\|\mathbf{a} + k\mathbf{b}\|^2$ never vanishes; so the discriminant is strictly negative, and equality cannot occur.   □

Having Schwarz's inequality, we can make the following definition:

The *angle* between two vectors **v** and **w** of an inner product space V is that angle α satisfying $0 \leq \alpha \leq \pi$ such that

$$\cos \alpha = \frac{\langle \mathbf{v}, \mathbf{w} \rangle}{\|\mathbf{v}\| \, \|\mathbf{w}\|}.$$

Schwarz's inequality was needed to make sure that there is such an angle, since cosines always have absolute value at most 1.

L3.4. Orthogonal Sets and Bases

Two vectors are called *orthogonal* (with respect to an inner product) if their inner product is zero; this means that the angle between them (just defined) is a right angle, as it should be. A set of vectors is orthogonal if each pair is orthogonal. A set of vectors is further said to be *orthonormal* if all the vectors are of unit length.

In general, you cannot say whether a set of vectors is independent just by looking at pairs, but if they are orthogonal, you can.

Theorem L3.10. *If a set of nonzero vectors is orthogonal, then the vectors are linearly independent.*

Proof. Let $\mathbf{a}_1, \ldots, \mathbf{a}_n$ be an orthogonal set of nonzero vectors and suppose $\sum k_i \mathbf{a}_i = 0$. Then

$$0 = \left\| \sum k_i \mathbf{a}_i \right\|^2 = \sum_{i,j} k_i k_j \langle \mathbf{a}_i, \mathbf{a}_j \rangle = \sum k_i^2 \|\mathbf{a}_i\|^2,$$

and the only way this can occur is for all the k_i to be zero. □

Orthogonal bases have a remarkable property. In general, if you wish to express a vector with respect to a basis, you must write it as a linear combination of the basis vectors, using unknowns as coefficients, and expand. You then obtain a system of linear equations, with as many equations and unknowns as the dimension of the space. Solving such a system is cumbersome when there are many equations. But *if the basis is orthogonal, each coefficient can be computed with one inner product*; this result is essential to motivating the definition of Fourier coefficients.

Theorem L3.11. *If $\mathbf{u}_1, \mathbf{u}_2, \ldots, \mathbf{u}_n$ is an orthogonal basis of V, then for any $\mathbf{a} \in V$, we have*

$$\mathbf{a} = \sum_{i=1}^{n} \frac{\langle \mathbf{a}, \mathbf{u}_i \rangle}{\langle \mathbf{u}_i, \mathbf{u}_i \rangle} \mathbf{u}_i.$$

Proof. Certainly, we can write $\mathbf{a} = \sum a_j \mathbf{u}_j$, since the $\mathbf{u}_j$ form a basis. Take the inner product of both sides with $\mathbf{u}_i$, to get $\langle \mathbf{a}, \mathbf{u}_i \rangle = a_i \langle \mathbf{u}_i, \mathbf{u}_i \rangle$, which can be solved for a_i. □

Remark. The formula of Theorem L3.11 will find its main use in calculating the coefficients of Fourier series. In that case, the basis in use is often orthogonal but not orthonormal.

Theorem L3.12. *If $\mathbf{w}_1, \ldots, \mathbf{w}_n$ is an orthonormal basis of a real vector space V, and if*

$$\mathbf{a} = \sum a_i \mathbf{w}_i \quad \text{and} \quad \mathbf{b} = \sum b_i \mathbf{w}_i,$$

then $\langle \mathbf{a}, \mathbf{b} \rangle = \sum a_i b_i$.

Proof. Just plug in: $\langle \mathbf{a}, \mathbf{b} \rangle = \sum_{i,j} a_i b_j \langle \mathbf{w}_i, \mathbf{w}_j \rangle = \sum_i a_i b_i$. $\square$

Simple though the proof might be, this result is important; it says that if you identify an inner product space V with $\mathbb{R}^n$ *using an orthonormal basis* of V, then the inner product becomes the ordinary dot product on the coordinates. This says that vector spaces with inner products have no individuality; they all look just like $\mathbb{R}^n$ with the ordinary dot product, at least if they have orthonormal bases, which they do, as we will see below.

L3.5. THE GRAM–SCHMIDT ALGORITHM

The Gram–Schmidt algorithm is one of the fundamental tools of linear algebra, used in innumerable settings.

Theorem L3.13 (Gram–Schmidt Orthogonalization). *Let V be a vector space with an inner product and $\mathbf{u}_1, \mathbf{u}_2, \ldots, \mathbf{u}_m$ be m linearly independent vectors of V. Then the algorithm below constructs an orthonormal set of vectors $\mathbf{w}_1, \ldots, \mathbf{w}_m$ having the same span.*

Proof. Define new vectors $\mathbf{a}_i$ and $\mathbf{w}_i$ inductively as follows:

$$
\begin{aligned}
\mathbf{a}_1 &= \mathbf{u}_1, & \mathbf{w}_1 &= \mathbf{a}_1 / \|\mathbf{a}_1\|, \\
\mathbf{a}_2 &= \mathbf{u}_2 - \langle \mathbf{u}_2, \mathbf{w}_1 \rangle \mathbf{w}_1, & \mathbf{w}_2 &= \mathbf{a}_2 / \|\mathbf{a}_2\|, \\
\mathbf{a}_3 &= \mathbf{u}_3 - \langle \mathbf{u}_3, \mathbf{w}_1 \rangle \mathbf{w}_1 - \langle \mathbf{u}_3, \mathbf{w}_2 \rangle \mathbf{w}_2, & \mathbf{w}_3 &= \mathbf{a}_3 / \|\mathbf{a}_3\|, \\
&\ \ \vdots & & \\
\mathbf{a}_m &= \mathbf{u}_m - \sum_{1 \le j < n-1} \langle \mathbf{u}_m, \mathbf{w}_j \rangle \mathbf{w}_j, & \mathbf{w}_m &= \mathbf{a}_m / \|\mathbf{a}_m\|.
\end{aligned}
$$

The $\mathbf{w}_1, \ldots, \mathbf{w}_k$ are clearly unit vectors, so the normal aspect is covered; showing algebraically that they are orthogonal is left as Exercise L3–4#6a.

To finish proving the theorem, assume by induction that the span of $\mathbf{w}_1, \ldots, \mathbf{w}_{k-1}$ is the same as the span of $\mathbf{u}_1, \ldots, \mathbf{u}_{k-1}$. Writing $\mathbf{a}_k = 0$ is precisely writing $\mathbf{u}_k$ as a linear combination of $\mathbf{w}_1, \ldots, \mathbf{w}_{k-1}$, which is impossible, so $\mathbf{a}_k \ne 0$ and the division by $\|\mathbf{a}_k\|$ is possible. Then in the kth line above, $\mathbf{u}_k$ is a linear combination of $\mathbf{w}_1, \ldots, \mathbf{w}_k$, and that $\mathbf{w}_k$ is a linear combination of $\mathbf{u}_k, \mathbf{w}_{k-1}, \ldots, \mathbf{w}_1$, verifying the inductive hypothesis for k. $\square$

Example L3.14. Figure L3.1 shows the Gram-Schmidt algorithm in action in $\mathbb{R}^2$.

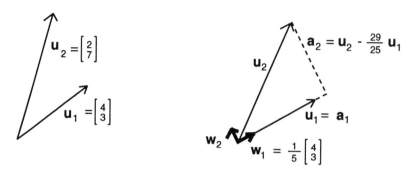

FIGURE L3.1. Gram–Schmidt orthogonalization.

$$\mathbf{a}_1 = \mathbf{u}_1 = \begin{bmatrix} 4 \\ 3 \end{bmatrix}, \qquad \mathbf{w}_1 = \frac{\mathbf{a}_1}{\|\mathbf{a}_1\|} = \frac{1}{5}\begin{bmatrix} 4 \\ 3 \end{bmatrix} = \begin{bmatrix} 0.8 \\ 0.6 \end{bmatrix},$$

$$\mathbf{a}_2 = \mathbf{u}_2 - \langle \mathbf{u}_2 \cdot \mathbf{w}_1 \rangle \mathbf{w}_1, \qquad \mathbf{w}_2 = \frac{\mathbf{a}_2}{\|\mathbf{a}_2\|} = \frac{1}{4.4}\begin{bmatrix} -2.64 \\ 3.52 \end{bmatrix} \begin{bmatrix} -0.6 \\ 0.8 \end{bmatrix},$$

$$= \begin{bmatrix} 2 \\ 7 \end{bmatrix} - \left\{ \frac{1}{5}\begin{bmatrix} 2 \\ 7 \end{bmatrix} \cdot \begin{bmatrix} 4 \\ 3 \end{bmatrix} \right\} \frac{1}{5}\begin{bmatrix} 4 \\ 3 \end{bmatrix}$$

$$= \begin{bmatrix} 2 \\ 7 \end{bmatrix} - \frac{29}{25}\begin{bmatrix} 4 \\ 3 \end{bmatrix} = \begin{bmatrix} -2.64 \\ 3.52 \end{bmatrix}. \quad \blacktriangle$$

Theorem L3.15. *Any finite-dimensional vector space with an inner product has an orthonormal basis.*

Proof. Apply Gram–Schmidt to any basis.   □

L3.6. ORTHOGONAL COMPLEMENTS AND PROJECTIONS

Let V be an inner product space and W be a subspace of V. Then the *orthogonal projection* $\pi_W(\mathbf{v})$ of $\mathbf{v}$ onto W is the unique vector in W closest to V. If W is finite dimensional, there is a formula for $\pi_W(\mathbf{v})$ in terms of an orthonormal basis for W.

Theorem L3.16. *If V is an inner product space and W a finite-dimensional subspace with an orthonormal basis $\mathbf{w}_1, \mathbf{w}_2, \ldots, \mathbf{w}_k$, then*

$$\pi_W(\mathbf{v}) = \sum_i \langle \mathbf{v}, \mathbf{w}_i \rangle \mathbf{w}_i. \tag{2}$$

Proof. We can express $\pi_W(\mathbf{v})$ as $\sum_i a_i \mathbf{w}_i$, and then the job is to find the a_i's. Since we want to minimize the distance from $\mathbf{v}$ to $\pi_W(\mathbf{v})$, this means we want to minimize

$$\left\| \mathbf{v} - \sum_i a_i \mathbf{w}_i \right\|^2 = \|\mathbf{v}\|^2 - 2 \sum_i a_i \langle \mathbf{v}, \mathbf{w}_i \rangle + \sum_i a_i^2,$$

which can be rewritten

$$= \|\mathbf{v}\|^2 + \sum_i (\langle \mathbf{v}, \mathbf{w}_i \rangle - a_i)^2 - \sum_i \langle \mathbf{v}, \mathbf{w}_i \rangle^2.$$

In this final form, the first and last terms do not depend at all on the a_i's, so the only term that can be adjusted by the a_i's is the middle term. When

$$a_i = \langle \mathbf{v}, \mathbf{w}_i \rangle,$$

the middle term is zero, producing a minimum of the distance from $\pi_W(\mathbf{v})$ to V. This proves uniqueness and existence of expression (2) representing $\pi_W(\mathbf{v})$ in terms of the orthonormal basis of W. $\square$

Example L3.17. Consider in $\mathbb{R}^3$ a vector $\mathbf{x}$ and a subspace S that is a plane through the origin. Then $\pi_S(\mathbf{x})$ is the vector lying in S obtained by dropping perpendiculars from the ends of the vector $\mathbf{x}$ to the plane S, as shown in Figure L3.2.

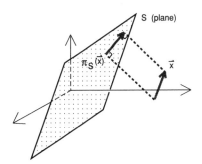

FIGURE L3.2. Orthogonal projection.

Remark. If W is infinite-dimensional, you can still try to use the argument of Theorem L3.16. The finite sum $\sum_i a_i \mathbf{w}_i$ now becomes a series; you need to worry about its convergence. We will see, when dealing with Fourier series, that such questions can be quite complicated. In particular, it is quite possible, when W is infinite dimensional, that the orthogonal projection does not exist: sometimes there will not be a vector in W closest to $\mathbf{v}$.

There is now one more term to define: The *orthogonal complement* of W is the subspace

$$W^\perp = \{\mathbf{v} \in V \mid \langle \mathbf{v}, \mathbf{w} \rangle = 0 \text{ for all } \mathbf{w} \in W\}.$$

Example L3.18. The orthogonal complement of the plane S in Example L3.17 is $S^\perp$, the line through the origin that is perpendicular to S, as shown in Figure L3.3. The projection $\pi_{S^\perp}(\mathbf{x})$ is the vector obtained by dropping perpendiculars from the ends of $\mathbf{x}$ to the line $S^\perp$.

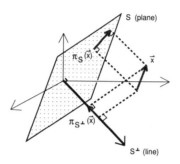

FIGURE L3.3. Orthogonal complement $S^\perp$ and projection $\pi_{S^\perp}(\mathbf{x})$.

We can now decompose the vector $\mathbf{x}$ into the two components shown in Figures L3.3 and L3.4; that is, we can write $\mathbf{x}$ as the sum of its orthogonal projection and its projection on the orthogonal complement:

FIGURE L3.4. Decomposition: $\mathbf{x} = \pi_S(\mathbf{x}) + \pi_{S^\perp}(\mathbf{x})$. ▲

Example L3.18 illustrates the following important theorem.

Theorem L3.19. *Any vector* $\mathbf{v} \in V$ *can be written uniquely as*

$$\mathbf{v} = \pi_W(\mathbf{v}) + \pi_{W^\perp}(\mathbf{v}).$$

Furthermore, $\|\mathbf{v}\|^2 = \|\pi_W(\mathbf{v})\|^2 + \|\pi_{W^\perp}(\mathbf{v})\|^2$; *in particular,* $\|\mathbf{v}\|^2 \geq \|\pi_W(\mathbf{v})\|^2$, *with equality if and only if* $\mathbf{v} \in W$.

Proof. Let $\mathbf{w} = \pi_W \mathbf{v}$ and $\tilde{\mathbf{w}} = \mathbf{v} - \pi_W \mathbf{v}$. First, let us check that $\tilde{\mathbf{w}} \in W^\perp$. It is enough to show that $\tilde{\mathbf{w}}$ is orthogonal to all vectors $\mathbf{w}_1, \ldots, \mathbf{w}_k$ of some orthonormal basis of W. We see this as follows:

$$\langle \tilde{\mathbf{w}}, \mathbf{w}_j \rangle = \left\langle \left(\mathbf{v} - \sum_i \langle \mathbf{v}, \mathbf{w}_i \rangle \mathbf{w}_i \right), \mathbf{w}_j \right\rangle$$

$$= \langle \mathbf{v}, \mathbf{w}_j \rangle - \sum_i \langle \mathbf{v}, \mathbf{w}_i \rangle \langle \mathbf{w}_i, \mathbf{w}_j \rangle = \langle \mathbf{v}, \mathbf{w}_j \rangle - \langle \mathbf{v}, \mathbf{w}_j \rangle = 0.$$

Next observe that

$$\|\mathbf{v}\|^2 = \|\mathbf{w} + \tilde{\mathbf{w}}\|^2 = \|\mathbf{w}\|^2 + 2\langle \mathbf{w}, \tilde{\mathbf{w}} \rangle + \|\tilde{\mathbf{w}}\|^2 = \|\mathbf{w}\|^2 + \|\tilde{\mathbf{w}}\|^2.$$

Remark. If we had been in a complex vector space, the cross-term above would have been replaced by $\langle \mathbf{w}, \tilde{\mathbf{w}} \rangle + \langle \tilde{\mathbf{w}}, \mathbf{w} \rangle = 2 \operatorname{Re} \langle \mathbf{w}, \tilde{\mathbf{w}} \rangle$.

Finally, we need to show that $\tilde{\mathbf{w}} = \pi_{W^\perp}(\mathbf{v})$, i.e., that if $\mathbf{u} \in W^\perp$, then the minimum of $\|\mathbf{v} - \mathbf{u}\|$ is realized by $\mathbf{u} = \tilde{\mathbf{w}}$. We see this as

$$\|\mathbf{v} - \mathbf{u}\|^2 = \|\mathbf{w} + \tilde{\mathbf{w}} - \mathbf{u}\|^2 = \|\mathbf{w}\|^2 + \|\tilde{\mathbf{w}} - \mathbf{u}\|^2,$$

which is clearly minimal when $\tilde{\mathbf{w}} = \mathbf{u}$. $\square$

L4 Linear Transformations and Inner Products

We shall explain in this section the three kinds of linear transformations that are intimately linked to real inner products: the *orthogonal* transformations, the *antisymmetric* ones, and the *symmetric* ones. All these are special cases of the corresponding complex linear transformations: *unitary*, *anti-Hermitian*, and *Hermitian*, respectively. We will deal here with the real case; the complex case will largely be left as exercises but will appear in Appendix L6 for characterization of eigenvalues.

L4.1. ORTHOGONAL TRANSFORMATIONS

A linear transformation $T_Q: V \to V$ from an inner product space to itself is called *orthogonal* (for that inner product) if for any vectors $\mathbf{v}_1$ and $\mathbf{v}_2$ of V and matrix Q for T, we have $\langle Q\mathbf{v}_1, Q\mathbf{v}_2 \rangle = \langle \mathbf{v}_1, \mathbf{v}_2 \rangle$. That is, an orthogonal transformation preserves inner products.

Remark. The name orthogonal is a bit unfortunate; it would be much better to call such linear transformations *orthonormal*.

Theorem L4.1. *The three following conditions are equivalent:*

(1) *A transformation $T_Q \colon \mathbb{R}^n \to \mathbb{R}^n$ is orthogonal for the dot product.*

(2) *The column vectors of the matrix Q of an orthogonal linear transformation form an orthonormal basis.*

(3) *The inverse of the matrix Q of a linear transformation is its transpose; i.e., $Q^\top = Q^{-1}$.*

Proof. (1) implies (2) follows from the fact that the columns of $\mathbf{w}_1, \ldots, \mathbf{w}_n$ of Q are the vectors $T_Q(\mathbf{e}_i)$; from the definition of an orthogonal transformation, we see that

$$\mathbf{w}_i \cdot \mathbf{w}_j = T_Q(\mathbf{e}_i) \cdot T_Q(\mathbf{e}_j) = \mathbf{e}_i \cdot \mathbf{e}_j.$$

The equivalence of (2) and (3) follows from the following computations: To see the equivalence, suppose that the columns of Q are $\mathbf{w}_1, \ldots, \mathbf{w}_n$. Then writing $Q^\top Q = I$, i.e.,

$$
\begin{bmatrix} \rule[0.5ex]{2em}{0.4pt} & \mathbf{w}_1^\top & \rule[0.5ex]{2em}{0.4pt} \\ & \vdots & \\ \rule[0.5ex]{2em}{0.4pt} & \mathbf{w}_n^\top & \rule[0.5ex]{2em}{0.4pt} \end{bmatrix}
\begin{bmatrix} | & \cdots & | \\ \mathbf{w}_1 & \cdots & \mathbf{w}_n \\ | & \cdots & | \end{bmatrix}
=
\begin{bmatrix} 1 & 0 & & & \\ 0 & 1 & & & \\ & & \cdots & & \\ & & \cdots & 1 & 0 \\ & & & 0 & 1 \end{bmatrix},
$$

we see that the dot product of each column with itself is 1 (the normal part) and with another is 0 (the orthogonal part) because the $\mathbf{w}_i$ form an orthonormal basis.

Finally, (3) implies (1) is seen as follows: pick any vectors $\mathbf{v}, \mathbf{w} \in \mathbb{R}^n$. Then, by the definition of dot product and Theorem L1.17 about transposes,

$$T_Q(\mathbf{v}) \cdot T_Q(\mathbf{w}) = (Q\mathbf{v})^\top (Q\mathbf{w}) = \mathbf{v}^\top Q^\top Q \mathbf{w} = \mathbf{v} \cdot \mathbf{w},$$

because we are now back to the definition of orthogonal transformation. $\square$

It follows immediately from the definition and Theorem L4.1 that the *orthogonal transformations form a group, denoted $O(n)$.* This means that the orthogonal transformations that compose the group must and do satisfy the following properties:

(i) the product of two orthogonal transformations is orthogonal (this was clear right from the definition);

(ii) an orthogonal transformation is invertible;

(iii) its inverse is orthogonal.

Example L4.2. As we saw in Example L2.19, the orthogonal matrix giving rotation in $\mathbb{R}^2$ by angle θ is

$$Q(\theta) = \begin{bmatrix} \cos\theta & \sin\theta \\ -\sin\theta & \cos\theta \end{bmatrix}.$$

You can confirm that the columns of $Q(\theta)$ form an orthonormal basis. The inverse of $Q(\theta)$ is rotation by $-\theta$, and since $\sin(-\theta) = -\sin\theta$, we see that

$$Q^{-1}(\theta) = Q(-\theta) = Q^\top(\theta), \tag{1}$$

the transpose, as it should. These are not the only elements of $O(2)$; the matrix

$$\begin{bmatrix} 1 & 0 \\ 0 & -1 \end{bmatrix}$$

is also orthogonal; geometrically, it represents reflection with respect to the x-axis. More generally, the product

$$\begin{bmatrix} \cos\theta & -\sin\theta \\ \sin\theta & \cos\theta \end{bmatrix} \begin{bmatrix} 1 & 0 \\ 0 & -1 \end{bmatrix} = \begin{bmatrix} \cos\theta & \sin\theta \\ \sin\theta & -\cos\theta \end{bmatrix}$$

is an element of $O(2)$, and in Exercise L3-4#16 you are asked to show that these and the matrices $Q(\theta)$ are all elements of $O(2)$. ▲

There is a way of expressing the Gram–Schmidt orthogonalization process in terms of orthogonal matrices; this formulation will be used twice in the remainder of these notes, once in the section on determinants and once in the section on the QR method.

Theorem L4.3. *Any matrix M can be written as the product $M = QR$, where Q is orthogonal and R is upper-triangular.*

Proof. This is just a way of condensing the Gram–Schmidt process. Let the columns of M be $\mathbf{u}_1, \ldots, \mathbf{u}_n$. First, we will assume that the $\mathbf{u}_i$ are linearly independent. Then we can apply Gram–Schmidt to them, and using the notation of Theorem L3.13 we see that all the Gram–Schmidt formulas can be condensed into the following matrix multiplication:

$$\begin{bmatrix} \|\mathbf{a}_1\| & \mathbf{u}_2 \cdot \mathbf{w}_1 & \mathbf{u}_3 \cdot \mathbf{w}_1 & \cdots & \mathbf{u}_n \cdot \mathbf{w}_1 \\ 0 & \|\mathbf{a}_2\| & \mathbf{u}_3 \cdot \mathbf{w}_2 & \cdots & \mathbf{u}_n \cdot \mathbf{w}_2 \\ \vdots & \vdots & & & \vdots \\ 0 & 0 & \cdots & & \|\mathbf{a}_n\| \end{bmatrix}$$

$$\begin{bmatrix} \mathbf{w}_1 & \cdots & \mathbf{w}_n \end{bmatrix} \begin{bmatrix} \mathbf{u}_1 & \mathbf{u}_2 & \cdots & \mathbf{u}_n \end{bmatrix},$$

which is precisely what we need, since the matrix $[\mathbf{w}_1, \ldots, \mathbf{w}_n]$ is orthogonal (Theorem L4.1, part 2). □

L4.2. ANTISYMMETRIC TRANSFORMATIONS

A linear transformation T from a vector space V to itself is called *antisymmetric* with respect to an inner product if for any vectors $\mathbf{v}_1$ and $\mathbf{v}_2$ we have

$$\langle T\mathbf{v}_1, \mathbf{v}_2 \rangle = -\langle \mathbf{v}_1, T\mathbf{v}_2 \rangle \qquad \text{(antisymmetric } T\text{)}.$$

The appropriate definition for the complex case is an *anti-Hermitian* transformation, which takes the complex conjugate of the right-hand side:

$$\langle T\mathbf{v}_1, \mathbf{v}_2 \rangle = -\overline{\langle \mathbf{v}_1, T\mathbf{v}_2 \rangle} \qquad \text{(anti-Hermitian } T\text{)}.$$

The reason why antisymmetry is an important property is that

> *Antisymmetric linear transformations are infinitesimal orthogonal transformations.*

More precisely,

Theorem L4.4. *Let $Q(t)$ be a family of orthogonal matrices depending on the parameter t. Then $Q^{-1}(t)Q'(t)$ is antisymmetric.*

Proof. Start with the definition of orthogonal matrices

$$\langle Q(t)\mathbf{v}_1, Q(t)\mathbf{v}_2 \rangle = \langle \mathbf{v}_1, \mathbf{v}_2 \rangle$$

and differentiate with respect to t to get

$$\langle Q'(t)\mathbf{v}_1, Q(t)\mathbf{v}_2 \rangle + \langle Q(t)\mathbf{v}_1, Q'(t)\mathbf{v}_2 \rangle = 0.$$

Since Q is orthogonal, we can apply Q^{-1} to both terms in each inner product, so the equation can be rewritten

$$\langle Q^{-1}(t)Q'(t)\mathbf{v}_1, \mathbf{v}_2 \rangle + \langle \mathbf{v}_1, Q^{-1}(t)Q'(t)\mathbf{v}_2 \rangle = 0. \quad \square$$

Theorem L4.4 will have the consequence that if T is antisymmetric with matrix A, then a solution $\mathbf{x}(t)$ of $\mathbf{x}' = A\mathbf{x}$ will have constant length; i.e., the vector $\mathbf{x}(t)$ will move on a sphere of constant radius. Such examples will occur and be of central importance in classical mechanics.

The reason for the word "antisymmetric" is the following.

Theorem L4.5. *If $V = \mathbb{R}^n$ with the standard inner product, then a linear transformation is antisymmetric if and only if its matrix A is antisymmetric, that is, if $A^\top = -A$.*

The proof is easy and left to the reader.

L4.3. Symmetric Linear Transformations

A linear transformation $T: V \to V$ is *symmetric* with respect to an inner product if for any vectors $\mathbf{v}_1, \mathbf{v}_2 \in V$ we have

$$\langle T\mathbf{v}_1, \mathbf{v}_2 \rangle = \langle \mathbf{v}_1, T\mathbf{v}_2 \rangle \qquad \text{(symmetric } T\text{)}.$$

The appropriate definition for the complex case is a *Hermitian* transformation, which takes the complex conjugate of the right-hand side:

$$\langle T\mathbf{v}_1, \mathbf{v}_2 \rangle = \overline{\langle \mathbf{v}_1, T\mathbf{v}_2 \rangle} \qquad \text{(Hermitian } T\text{)}.$$

Remark. Linear transformations symmetric with respect to an inner product are often called *self-adjoint* (and antisymmetric ones are called anti-self-adjoint).

The matrix interpretation is as follows, and the easy proof is left to the reader:

Theorem L4.6. *If $V = \mathbb{R}^n$ with the standard inner product, then a linear transformation $T: V \to V$ is symmetric if and only if its matrix A is symmetric, that is, if $A^\top = A$. Similarly, if $V = \mathbb{C}^n$ and T is Hermitian, then the matrix A of T is Hermitian; i.e., it satisfies $\overline{A} = A^\top$.*

Orthogonal linear transformations are more or less obviously important. Antisymmetric and anti-Hermitian transformations are important because of their intimate relation with orthogonal and unitary transformations. Symmetric and Hermitian transformations are rather harder to motivate, even though they are studied far more frequently than the anti-Hermitian ones.

Remark. One might look for motivation in quantum mechanics, where Hermitian matrices show up constantly. For instance, finding the energy levels of atoms or molecules boils down to computing the eigenvalues of a Hermitian matrix. Actually, *anti*-Hermitian matrices are the ones important to quantum mechanics, and the presence of Hermitian matrices is due to the following trick: If H is a Hermitian matrix, then iH is anti-Hermitian.

Indeed, the basic equation of quantum mechanics is the Schrödinger equation

$$\frac{\partial \psi}{dt} = \frac{h}{2\pi i}(\Delta \psi + V\psi) = \frac{h}{2\pi i}H(\psi),$$

where h is Planck's constant. As we will see in Example L4.7 for a one-dimensional case, the Laplace operator Δ (in that case, just the second derivative) is a symmetric operator, if appropriate boundary conditions are imposed, and so is the multiplication operator V. Thus, H is symmetric, and it is usually H, the *Hamiltonian operator*, which is studied.

But a glance at the equation shows that the really fundamental operator is $(h/2\pi i)H$, which is anti-Hermitian. One consequence of the principle "anti-Hermitian operators are infinitesimal unitary operators" is that $\|\psi\|^2$ is constant in time; physically, this just says that the probability of finding the system in *some* state is always 1.

Symmetric matrices show up in their own right when studying quadratic forms; in fact, they correspond precisely to quadratic forms. Our treatment will emphasize this relationship, which will be elaborated in Subsection L4.5 of this appendix and in Appendix L8.

There is not really much point in writing down finite-dimensional examples of Theorem L4.6; they are simply given by symmetric matrices.

The next example is the real reason for Fourier series. It is a bit less straightforward but well worth poring over a bit.

Example L4.7. Let V be the space of twice continuously differentiable functions on $[0, \pi]$ which vanish at 0 and π, and W the space of continuous functions on $[0, \pi]$. Both spaces are endowed with the inner product

$$\langle f, g \rangle = \int_0^\pi f(x)g(x)dx.$$

Consider the *second derivative* operator $D^2 \colon V \to W$. We claim that D^2 is symmetric.

Remark. The attentive reader may well complain: since V and W are *different* vector spaces, how can such an operator be symmetric? But V is a subspace of W, so the equation $\langle D^2 f, g \rangle = \langle f, D^2 g \rangle$ makes perfectly good sense.

The proof that this is so is an exercise in integration by parts:

$$\langle D^2 f, g \rangle = \int_0^\pi f''(x)g(x)dx = [f'(x)g(x)]_0^\pi - \int_0^\pi f'(x)g'(x)dx = \langle f, D^2 g \rangle.$$

The second equality is integration by parts; the third comes from the fact that the boundary term drops out since $g(0) = g(\pi) = 0$, and only the integral, which is obviously symmetric in f and g (in the ordinary sense of the word), is left. ▲

Several more examples of a similar style are given in Exercise L3–4#17.

L4.4. The Inner Products on $\mathbb{R}^n$

We can finally come full circle and describe all inner products on $\mathbb{R}^n$.

Theorem L4.8. *Let $\langle \ , \ \rangle$ be an inner product on $\mathbb{R}^n$, and let G be the symmetric matrix with entries*

$$G_{i,j} = \langle \mathbf{e}_i, \mathbf{e}_j \rangle.$$

Then the given inner product can be written in terms of the standard dot product as
$$\langle \mathbf{a}, \mathbf{b} \rangle = \mathbf{a} \cdot G\mathbf{b}. \qquad (2)$$

Conversely, if G is a symmetric matrix and $\mathbf{a} \cdot G\mathbf{a} \neq 0$ for all $\mathbf{a} \neq \mathbf{0}$, then (2) defines an inner product on $\mathbb{R}^n$.

Remark. We shall see in Section L4.5 that the property $\mathbf{a} \cdot G\mathbf{a} \neq 0$ for all $\mathbf{a} \neq \mathbf{0}$ makes G a *"positive definite"* or *"negative definite"* matrix.

Example L4.9. $G =$ the identity matrix gives the standard inner product.
▲

But Theorem L4.8 holds true for an arbitrary inner product, not just the standard one, as we show in the following:

Example L4.10. In Example L3.1, $\langle \mathbf{a}, \mathbf{b} \rangle \equiv 2a_1b_1 + 2a_2b_2 + a_1b_2 + a_2b_1$ is given by the matrix $\begin{bmatrix} 2 & 1 \\ 1 & 2 \end{bmatrix}$ because

$$\langle \mathbf{a}, \mathbf{b} \rangle = \begin{bmatrix} a_1 \\ a_2 \end{bmatrix} \cdot \begin{bmatrix} 2 & 1 \\ 1 & 2 \end{bmatrix} \begin{bmatrix} b_1 \\ b_2 \end{bmatrix}. \qquad ▲$$

Proof of Proposition L4.8. The first part follows from the fact that if you know an inner product on all pairs of basis vectors, then you know it on any pair of vectors by writing the vectors in terms of the basis and multiplying them out.

The second part, the converse, is easy to prove by referring to the symmetric, linear, and positive definite properties listed for an inner product in Section L3.1; this is left to the reader.   □

L4.5. QUADRATIC FORMS

After linear functions, the next simplest to study are the quadratics, and they are important in any mathematician's vocabulary. We often want to know the character of a surface at a critical point, for instance.

Consider the function $f(\mathbf{x})$ that describes the surface in $\mathbb{R}^{n+1}$ for a vector $\mathbf{x}$ in $\mathbb{R}^n$. Expanding about a critical point $\mathbf{a}$ by Taylor series gives

$$
\begin{aligned}
f(x_1, x_2, \ldots, x_n) = {} & f(a_1, a_2, \ldots, a_n) \\
& + \sum_i \frac{\partial f}{\partial x_i}(\mathbf{a})(x_i - a_i) \\
& + \sum_{i,j} \frac{\partial^2 f}{\partial x_i \partial x_j}(\mathbf{a})(x_i - a_i)(x_j - a_i) + \text{higher order terms.}
\end{aligned}
$$

The second (linear) term is zero at a critical point, so the shape of the surface near the critical point is dominated by the quadratic term (unless the quadratic terms have rank $< n$; see below). This quadratic term is a *homogeneous quadratic function* (meaning there are no nonquadratic terms).

What becomes important is to be able to tell whether at a critical point a given quadratic function represents a maximum, a minimum, or a saddle. The wonderful news is that we shall be able to use a correspondence between quadratic functions and symmetric linear transformations to answer this question easily for any example, using only the high school algebra technique of "completing the square." So let us begin.

A *quadratic form* on $\mathbb{R}^n$ is a function $q(\mathbf{v})\colon \mathbb{R}^n \to \mathbb{R}$ that is a *homogeneous quadratic function* of the coordinates $x_1, \ldots, x_n$ comprising the vector $\mathbf{v}$. That is, a quadratic form is a sum of terms of the form $a_{ii}x_i^2$ or $a_{ij}x_ix_j$.

Examples L4.11.

(i) $(4x^2 + xy - y^2)$ and xy are quadratic forms on $\mathbb{R}^2$. As you can see from the graphs of these functions in Figure L4.1, at the origin both these examples represent saddles.

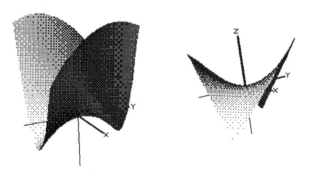

FIGURE L4.1. On the left: $q(\mathbf{x}) = 4x^2 + xy - y^2$; on the right: $q(\mathbf{x}) = xy$.

(ii) $(x^2 + xz - yz - z^2)$ and $(xy + xz + yz)$ are quadratic forms on $\mathbb{R}^3$.

(iii) $q(\mathbf{v}) = \langle \mathbf{v}, \mathbf{v} \rangle = \|\mathbf{v}\|^2$ is one outstanding quadratic form on $\mathbb{R}^n$. ▲

More generally, a quadratic form on a finite-dimensional vector space V is a function $q\colon V \to \mathbb{R}$ that is a homogeneous quadratic function in the coordinate functions of V with respect to any basis. Note: the authors do not know why quadratic forms are not simply called quadratic functions.

A very important aspect of quadratic forms is the following:

Any quadratic form can be represented by a symmetric linear transformation L_q, in the following way:

$$q(\mathbf{v}) = [x_1, \ldots, x_n] L_q \begin{bmatrix} x_1 \\ \vdots \\ x_n \end{bmatrix}.$$

Example L4.12.

$$x^2 + xz - yz - z^2 = [x,\ y,\ z] \begin{bmatrix} 1 & 0 & 1/2 \\ 0 & 0 & -1/2 \\ 1/2 & -1/2 & -1 \end{bmatrix} \begin{bmatrix} x \\ y \\ z \end{bmatrix}. \quad \blacktriangle$$

The recipe is to use the coefficients of x_i^2 for the diagonal terms a_{ii} and half the coefficients of $x_i x_j$ for the off-diagonal terms a_{ij}. This procedure justifies the following statements:

Theorem L4.13. *Let V be a finite-dimensional vector space with an inner product. Then, for any quadratic form q there exists a unique symmetric linear transformation $L_q\colon V \to V$ such that for any $\mathbf{v} \in V$,*

$$q(\mathbf{v}) = \langle \mathbf{v}, L_q(\mathbf{v}) \rangle.$$

Something useful to do with quadratic forms is a process called *decomposition as a sum of squares of linear functions* of the variables; these will turn out, by Theorem L4.15, to form a linearly independent set.

Example L4.14. The quadradic form shown in Figure L4.2 can be decomposed as follows:

$$x^2 + xy = x^2 + xy + \left(\frac{1}{4}\right) y^2 - \left(\frac{1}{4}\right) y^2 = \left(x + \frac{1}{2}y\right)^2 - \left(\frac{1}{2}y\right)^2.$$

FIGURE L4.2. $q(\mathbf{x}) = x^2 + xy.$ $\quad \blacktriangle$

Theorem L4.15. *Given a quadratic form $q(\mathbf{x})$ on $\mathbb{R}^n$, there exist linearly independent linear functions $\alpha_1(\mathbf{x}), \ldots, \alpha_m(\mathbf{x})$ such that*

$$q(\mathbf{x}) = \pm(\alpha_1(\mathbf{x}))^2 \pm \cdots \pm (\alpha_m(\mathbf{x}))^2.$$

Theorem L4.15 can be proved by "completing squares"; an example will show how this can be done, working from left to right on one variable at a time.

Example L4.16. Consider the quadratic form

$$q(\mathbf{x}) = x^2 + 2xy - 4xz + 2yz - 4z^2.$$

Taking all terms in which x appears and completing the square yields

$$q(\mathbf{x}) = [(x + y - 2z)^2 - (y^2 + 4z^2 - 4yz)] + 2yz - 4z^2.$$

Collecting all remaining terms in which y appears and completing the square gives:

$$q(\mathbf{x}) = (x + y - 2z)^2 - (y - 3z)^2 + z^2.$$

The linear functions that appear in the decomposition,

$$\alpha_1(x, y, z) = x + y - 2z, \quad \alpha_2(x, y, z) = y - 3z, \quad \alpha_3(x, y, z) = z,$$

are certainly independent since x appears only in α_1, y only in α_2, and z only in α_3. You can also, of course, check by considering $A\mathbf{x} = \boldsymbol{\alpha}$ and showing $\det A \neq 0$.

The decomposition just derived is not unique. For example, Exercise L3–4#18 asks you to start with the terms in z first and derive the following alternative decomposition:

$$q(\mathbf{x}) = x^2 + 2xy - 4xz + 2yz - 4z^2 = x^2 + (x + y/2)^2 - (x - y/2 + 2z)^2. \quad \blacktriangle$$

The algorithm for completing the squares should be pretty clear: as long as the square of some coordinate function actually figures in the expression, every appearance of that variable can be incorporated into a perfect square; by subtracting off that perfect square, you are left with a quadratic form in fewer variables. This works as long as there is at least one square, but what should you do with something like the following?

Example L4.17. Consider the quadratic form

$$q(\mathbf{x}) = xy - xz + yz.$$

One possibility is to trade x for the new variable $u = x - y$, i.e., $x = u + y$, to give

$$
\begin{aligned}
(u + y)y - uz &= y^2 + uy - uz \\
&= (y + u/2)^2 - u^2/4 - uz - z^2 + z^2 \\
&= (y + u/2)^2 - (u/2 + z)^2 + z^2 \\
&= (x/2 + y/2)^2 - (x/2 - y/2 + z)^2 + z^2.
\end{aligned}
$$

There was not anything magical about the choice of u; almost anything would have done.

We leave it as Exercise L3–4#20 to verify that the linear functions found in this way are indeed linearly independent. ▲

One more example elaborates on the crucial hypothesis of Theorem L4.15:

Example L4.18. Consider the quadratic form from Example L3.2:

$$
q(\mathbf{x}) = \langle \mathbf{x}, \mathbf{x} \rangle = 2x_1^2 + 2x_2^2 + 2x_1x_2.
$$

We can write

$$
\begin{aligned}
q(\mathbf{x}) = 2x_1^2 + 2x_2^2 + 2x_1x_2 &= (2x_1^2 + 2x_1x_2 + x_2^2/2) + 2x_2^2 - x_2^2/2 \\
&= (\sqrt{2}\,x_1 + x_2/\sqrt{2})^2 + (\sqrt{3/2}\,x_2)^2.
\end{aligned}
$$

So the quadratic form shown in Figure L4.3 has been decomposed into the sum of *two* squared linear functions.

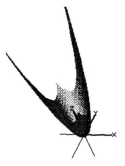

FIGURE L4.3. $q(\mathbf{x}) = 2x_1^2 + 2x_2^2 + 2x_1x_2$.

But, how about writing $q(\mathbf{x}) = x_1^2 + x_2^2 + (x_1 + x_2)^2$, as we did in Example L3.2, with *three* squared linear functions? This alternative "decomposition" does not reflect Theorem L4.15 because the three linear functions x_1, x_2, and $x_1 + x_2$ are not linearly independent. ▲

There are many ways of decomposing a quadratic form as a sum of linearly independent squares. We will see in Theorem L4.19 that the number m of squares which appear does not depend on the decomposition.

In order to use decomposition of a quadratic form to identify the character of a quadric surface, we must introduce a few more terms.

The number m of linearly independent squares in a representation as above is called the *rank* of the quadratic form; if $m = n$, then the form is called *nondegenerate*. The examples in this section are all nondegenerate; some degenerate cases are illustrated in Exercise L3–4#21.

If there are k plus signs appearing and ℓ minus signs, then the form has *signature* (k, ℓ); a form of signature $(n, 0)$ is called *positive definite*, and one of signature $(0, n)$ is called *negative definite*; the others are called *indefinite*. The symmetric matrices corresponding to the forms are also called positive or negative definite. Note: Without loss of generality we may assume that the first k have plus signs and the last ℓ have minus signs.

According to these definitions, the form of Example L4.14 has signature $(1, 1)$ and rank 2; Examples L4.16 and L4.17 both have signature $(2, 1)$ and rank 3; Example L4.18 is positive definite with rank 2; all four examples are nondegenerate.

Now we are ready for the following:

Theorem L4.19. *In any two representations of a quadratic form as the sums of squares of linearly independent linear functions*

$$q(\mathbf{x}) = \alpha_1(\mathbf{x})^2 + \cdots + \alpha_k(\mathbf{x})^2 - \alpha_{k+1}(\mathbf{x})^2 - \cdots - \alpha_{k+\ell}(\mathbf{x})^2,$$

the number (k) of plus and the number (ℓ) of minus signs appearing will be the same; i.e., the signature (k, ℓ) is the same in every representation of a given quadratic form.

There will usually be many ways of representing a quadratic form as a sum of squares of linearly independent linear functions; why should the signs be independent of the representation? In the case of definite matrices, this is clear, as follows:

> A quadratic form is positive (resp. negative) definite if and only if it takes positive (resp. negative) values on all nonzero vectors.

This second characterization of positive definiteness says nothing about any decomposition. Similarly, the following statement describes the signature in a third way, also independently of any decomposition:

> The numbers (k, ℓ) of the signature are respectively the largest dimensions of the subspaces on which a quadratic form is positive and negative definite.

Proof of Theorem L4.19. Let us make vectors $\boldsymbol{\alpha}_i$ from the coefficients of the variables in the α_i; e.g., for

$$\alpha_1 = x + 2y - z, \qquad \alpha_2 = -x + z,$$

then

$$\boldsymbol{\alpha}_1 = \langle 1, 2, -1 \rangle; \qquad \boldsymbol{\alpha}_2 = \langle -1, 0, 1 \rangle.$$

Next add to the $k + \ell$ vectors formed this way $\boldsymbol{\alpha}_i$ independent vectors $\boldsymbol{\alpha}_{k+\ell+1}, \ldots, \boldsymbol{\alpha}_n$ to make a basis for $\mathbb{R}^n$.

If we now define the positive definite subspace

$$P \equiv \{\mathbf{x} \mid \alpha_{k+1}(\mathbf{x}) = \cdots = \alpha_n(\mathbf{x}) = 0\},$$

then P is a k-dimensional subspace of vectors in $\mathbb{R}^n$, by the dimension formula of Theorem L2.9. Likewise, we can define the negative definite subspace which has dimension ℓ

$$N \equiv \{\mathbf{x} \mid \alpha_1(\mathbf{x}) = \cdots = \alpha_k(\mathbf{x}) = \alpha_{k+\ell+1}(\mathbf{x}) = \cdots = \alpha_n(\mathbf{x}) = 0\},$$

omitting the ℓ terms from $k + 1$ to $k + \ell + 1$.

If $W \subset \mathbb{R}^n$ is a subspace with $\dim(W) = k + 1$, then there must be $\mathbf{x} \neq \mathbf{0}$ in W such that

$$\alpha_1(\mathbf{x}) = \cdots = \alpha_k(\mathbf{x}) = \mathbf{0},$$

again by the dimension formula. On such a vector the quadratic form must take a nonpositive value, so the form is not positive definite on W. This proves that the number k of positive α_i cannot change; likewise, we can show that the number ℓ of negative α_i cannot change. ▲

Remark. The number $n - m$ is *not* the dimension of a maximal subspace on which q vanishes. For instance, the form $xy = ((x+y)/2)^2 - ((x-y)/2)^2$ has signature $(1, 1)$ and is nondegenerate, but it vanishes on the x-axis and on the y-axis.

The following example shows the use of the quadratic form representation in terms of the x_i's to determine whether a function has a maximum, minimum, or neither.

Example L4.20. Consider the function

$$f(x, y, z) = 3 + (x^2 + 2xy - 4yz + 5z^2) + \text{terms of order three or higher.}$$

This function has a critical point at $(x, y, z) = (0, 0, 0)$ since $\partial f/\partial x$, $\partial f/\partial y$, and $\partial f/\partial z$ are all zero there. We want to see whether it is a maximum, a minimum, or neither.

For small x, y, and z the quadratic terms are larger than the higher order terms, so we look at

$$x^2 + 2xy - 4yz + 5z^2 = \underbrace{(x+y)^2}_{\alpha_1} - \underbrace{(y+2z)^2}_{\alpha_2} + \underbrace{(3z)^2}_{\alpha_3}.$$

Since the signature is $(k, \ell) = (2, 1)$, $f(x, y, z)$ has *neither* a maximum nor a minimum.

If you move away from $(0,0,0)$ in a direction where $y = -2z$ (i.e., where $\alpha_2 = 0$), then the function will *increase*.

$$\text{Try } f(0, -2, 1) > 3 = f(0, 0, 0).$$

If you move away from $(0,0,0)$ in a direction where $x = -y$ and $z = 0$ (i.e., $\alpha_1 = \alpha_3 = 0$), then the function will *decrease*.

$$\text{Try } f(1, -1, 0) < 3 = f(0, 0, 0).$$

This illustrates how the α_i give you planar directions along which the function increases (if the α is preceded by a $+$) or decreases (if the α is preceded by a $-$). Different planes can be chosen (by the α_i), but in every case, the *number* of linearly independent planes along which the function increases/decreases stays the same. ▲

L5 Determinants and Volumes

The *determinant* is a function of square matrices. It is given by various formulas, all of which are far too messy for hand computation once the matrix gets larger than 3×3. It can also be computed (much more reasonably) by row reduction. The determinant has a geometric interpretation as a signed volume, and this really makes it important.

In order not to favor any one of the formulas and to obtain the volume interpretation most easily, we shall define the determinant by three properties that characterize it. We will think of the determinant as a function of n vectors rather than as a function of a matrix; this is a minor point, since whenever you have n vectors in $\mathbb{R}^n$, you can always place them side by side to make an $n \times n$ matrix.

L5.1. DEFINITION, EXISTENCE, AND UNIQUENESS OF DETERMINANT FUNCTION

The *determinant*

$$\det A = \det \begin{bmatrix} | & | & & | \\ \mathbf{a}_1, & \mathbf{a}_2, & \ldots, & \mathbf{a}_n \\ | & | & & | \end{bmatrix} = \det(\mathbf{a}_1, \mathbf{a}_2, \ldots, \mathbf{a}_n)$$

is the unique real-valued function of n vectors in $\mathbb{R}^n$ such that

(1) $\det A$ is *linear* with respect to *each* of its arguments, i.e.,

(*multilinearity*)

$$\det(\mathbf{a}_1, \ldots, \mathbf{a}_{i-1}, (\alpha\mathbf{u} + \beta\mathbf{w}), \mathbf{a}_{i+1}, \ldots, \mathbf{a}_n)$$
$$= \alpha \det(\mathbf{a}_1, \ldots, \mathbf{a}_{i-1}, \mathbf{u}, \mathbf{a}_{i+1}, \ldots, \mathbf{a}_n)$$
$$+ \beta \det(\mathbf{a}_1, \ldots, \mathbf{a}_{i-1}, \mathbf{w}, \mathbf{a}_{i+1}, \ldots, \mathbf{a}_n).$$

(2) $\det A$ is *antisymmetric*, i.e.,   (*antisymmetry*)

$$\det(\mathbf{a}_1, \ldots, \mathbf{a}_i, \ldots, \mathbf{a}_j, \ldots, \mathbf{a}_n) = -\det(\mathbf{a}_1, \ldots, \mathbf{a}_j, \ldots, \mathbf{a}_i, \ldots, \mathbf{a}_n).$$

(3) $\det I$ of the *identity* is 1, i.e.,   (scaling, or *normalization*)

$$\det(\mathbf{e}_1, \mathbf{e}_2, \ldots, \mathbf{e}_n) = 1, \text{ where } \mathbf{e}_i \text{ is the } i\text{th standard unit basis vector.}$$

In order to see that this definition of determinant is reasonable, we will want the following theorem:

Theorem L5.1. *There indeed exists a function satisfying the three properties of determinant,* $\det A$, *and it is unique.*

The proofs of these two parts are quite different, with a somewhat lengthy but necessary construction for each. The outline for the next several pages is as follows:

> First, we shall use a computer program to construct a function $D(A)$ by a process called "development according to the first row." Of course, this could be developed differently, e.g., according to the first column, but you can show in Exercise L5#4 that the result is equivalent to this definition.
>
> Then, we shall prove that $D(A)$ satisfies the properties of $\det A$, thus establishing *existence* of a function appropriate to the definition of determinant.
>
> Finally, we shall proceed by "column operations" to evaluate this function $D(A)$ and show that it is unique, which will prove *uniqueness* of the determinant function.

Development according to the first row. Consider the function

$$D(A) = \sum_{i=1}^{n} (-1)^{1+i} a_{1,i} \det A_{1,i}, \tag{1}$$

where $A_{1,i}$ is the $(n-1) \times (n-1)$ matrix obtained from A by erasing the first line and the ith column.

Example L5.2. If

$$A = \begin{bmatrix} 1 & 3 & 4 \\ 0 & 1 & 1 \\ 1 & 2 & 0 \end{bmatrix}, \text{ then } A_{1,3} = \begin{bmatrix} 1 & 3 & 4 \\ 0 & 1 & 1 \\ 1 & 2 & 0 \end{bmatrix} = \begin{bmatrix} 0 & 1 \\ 1 & 2 \end{bmatrix},$$

so equation (1) corresponds to

$$D(A) = 1\det\begin{bmatrix} 1 & 1 \\ 2 & 0 \end{bmatrix} - 3\det\begin{bmatrix} 0 & 1 \\ 1 & 0 \end{bmatrix} + 4\det\begin{bmatrix} 0 & 1 \\ 1 & 2 \end{bmatrix}$$
$$= 1(-2) - 3(-1) + 4(-1) = -3. \quad \blacktriangle$$

Of course, the definition (1) is recursive, and to start it out we must say that the determinant of a 1×1 matrix, i.e., a number, is the number itself. The following Pascal program implements this definition and will compute $D(A)$ for any square matrix of side at most 10; this program will actually run on a personal computer and compute a determinant of the maximum size in about an hour.

```
Program determinant;

Const maxsize = 10;

Type matrix = record
                        size:integer;
                        array[1..maxsize,1..maxsize] of real;
            end;

        submatrix = record
                        size:integer;
                        rows,cols:array[1..maxsize,1..maxsize] of boolean;
                end;
Var    M:Matrix;
        S:submatrix;
        d:real;

function det(S:submatrix):real;
Var  tempdet:real;
            i,sign:integer;
            S1:submatrix;
Procedure erase(S:SubMatrix; i,j:integer; var S1:SubMatrix);
Var k:integer;
begin {erase}
  S1.size := S.size-1;
  for k = S.size-1 downto i do S1.cols[k] := S.cols[k+1];
  for k = i-1 downto 1 do S1.cols[k] := S.cols[k];
  for k = S.size-1 downto j do S1.rows[k] := S.rows[k+1];
  for k = j-1 downto 1 do S1.rows[k] := S.rows[k+1];
end;
```

```
begin {function det}
     If  S.size = 1 then det := M.coeffs[row[1],col[1]]
     else
     begin   tempdet := 0; sign := 1;
             for i := 1 to S.size do
                 begin
                       erase(S,i,1,S1);
                       tempdet := tempdet + sign*M.coeffs[S.rows[1],S.cols[i]]*det(S1);
                       sign := -sign;
                 end;
                    det := tempdet
          end;
end;

Procedure InitSubMatrix(Var S:SubMatrix);
Var k:integer;
begin
   S.size := M.size;
   for k := 1 to S.size do begin S.rows[k] := k; S.cols[k] := k end;
end;

Procedure InitMatrix;
begin {define M.size and  M.coeffs any way you like) end;

Begin {main program}
   InitMatrix;
   InitSubMatrix(S);
   d := det(S);
   writeln('determinant = 'd);
end.
```

This program embodies the recursive nature of the determinant as defined above: the key point is that the function det calls itself. It would be quite a bit more difficult to write this program in a language like Basic, which does not allow that sort of thing.

Please note that this program is very time-consuming. This can be seen as follows: suppose that the function det takes time $T(k)$ to compute the determinant of a $k \times k$ matrix. Then, since it makes k "calls" of det for a $(k-1) \times (k-1)$ matrix, as well as k multiplications, $k-1$ additions, and k calls of the subroutine erase, we see that

$$T(k) > kT(k-1),$$

so that $T(k) > k!\, T(1)$. On a personal computer one call or operation takes about 10^{-6} seconds; on a large mainframe computer that would take more like 10^{-8} seconds. The time to compute determinants by this method is at least the factorial of the size of the matrix; if you were to try for a larger matrix, say 15×15, this would take $15! \approx 1.3 \times 10^{12}$ calls or operations, which translates into several minutes on a large mainframe. So if this program were the only way to actually compute determinants, they would be of theoretical interest only. But as we shall soon show, determinants can also be computed by row or column reduction, and this is much more efficient as soon as the matrix is even moderately large.

However, the construction of the function $D(A)$ is most convenient for establishing the following:

Proof of existence for Theorem L5.1. We shall verify that the function $D(A)$, the development along the first row, does indeed satisfy properties (1), (2); and (3) for the determinant $\det A$. This is a messy and uninspiring exercise in the use of induction:

(1) First, let us check multilinearity: Let $\mathbf{a}_1, \ldots, \mathbf{a}_n$ be n vectors in $\mathbb{R}^n$, and write $\mathbf{a}_i = \beta\mathbf{b} + \gamma\mathbf{c}$. Let

$$A = [\mathbf{a}_1, \ldots, \mathbf{a}_i, \ldots, \mathbf{a}_n],$$
$$B = [\mathbf{a}_1, \ldots, \mathbf{b}, \ldots, \mathbf{a}_n],$$
$$C = [\mathbf{a}_1, \ldots, \mathbf{c}, \ldots, \mathbf{a}_n].$$

We need to check $D(A) = \beta D(B) + \gamma D(C)$. Just write

$$
\begin{aligned}
D(A) &= \sum_{j=1}^{n} (-1)^{j+1} a_{1,j} D(A_{1,j}) \\
&= (-1)^{i+1}(\beta b_{1,i} + \gamma c_{1,i}) D(A_{1,i}) + \sum_{j \neq i} (-1)^{j+1} a_{1,j} D(A_{1,j}) \\
&= \beta(-1)^{i+1} b_{1,i} D(A_{1,i}) + \gamma(-1)^{i+1} c_{1,i} D(A_{1,i}) \\
&\quad + \beta \sum_{j \neq i} (-1)^{j+1} b_{1,j} D(B_{1,j}) + \gamma \sum_{j \neq i} (-1)^{j+1} c_{1,j} D(C_{1,j}) \\
&= \beta D(B) + \gamma D(C).
\end{aligned}
$$

The third and fourth terms on the third line are by induction, using the fact that all the $A_{1,i}$ are of size $n-1$, whereas one passes from the third to the fourth line by observing that, except for the ith column, the matrices A, B, and C coincide. This gives property (1).

(2) Next we come to antisymmetry: Let $A = [\mathbf{a}_1, \ldots, \mathbf{a}_i, \ldots, \mathbf{a}_j, \ldots, \mathbf{a}_n]$ with $j > i$, and let A' be the same matrix with the ith and jth columns exchanged. Then

$$
\begin{aligned}
D(A) &= \sum_{k=1}^{n} (-1)^{k+1} a_{1,k} D(A_{1,k}) \\
&= (-1)^{i+1} a_{1,i} D(A_{1,i}) + (-1)^{j+1} a_{1,j} D(A_{1,j}) \\
&\quad + \sum_{k \neq i,j} (-1)^{k+1} a_{1,k} D(A_{1,k}).
\end{aligned}
$$

The last term above can be dealt with by induction; if $k \neq i, j$, then $A'_{1,k}$ is obtained from $A_{1,k}$ by exchanging the ith and jth columns, and since these matrices are of size $n-1$, we can assume by induction that $D(A'_{1,k}) = -D(A_{1,k})$, so

$$\sum_{k \neq i,j} (-1)^{k+1} a_{1,k} D(A_{1,k}) = - \sum_{k \neq i,j} (-1)^{k+1} a'_{1,k} D(A'_{1,k}).$$

We will be done if we can show that

$$(-1)^i a_{1,i} D(A_{1,i}) = -(-1)^j a'_{1,j} D(A'_{1,j}),$$
$$(-1)^i a_{1,j} D(A_{1,j}) = -(-1)^j a'_{1,i} D(A'_{1,i}).$$

We will show the first; the second is identical. First, $a_{1,i} = a'_{1,j}$. The relation between $A_{1,i}$ and $A'_{1,j}$ is as follows: $A'_{1,j}$ is obtained from $A_{1,i}$ by moving the jth column over into the ith position. This can be done by exchanging it with its neighbor $j - i - 1$ times. Again using induction, since these matrices are of size $n - 1$, each of these exchanges changes the sign of D, so

$$(-1)^i a_{1,i} D(A_{1,i}) = (-1)^i (-1)^{j-i-1} a'_{1,j} D(A'_{1,j})$$
$$= -(-1)^j a'_{1,j} D(A'_{1,j}).$$

This proves property (2).

(3) The normalization condition is much simpler: If $A = [e_1, \ldots, e_n]$, then only the first entry $a_{1,1}$ on the top row is nonzero, and $A_{1,1}$ is the identity matrix one size smaller, so that D of it is 1 by induction. So

$$D(A) = a_{1,1} D(A_{1,1}) = 1,$$

and we have also proved property (3). □

Now we know there is at least one function of a matrix A satisfying properties (1), (2), and (3), namely $D(A)$. Of course, there might be others; but we will now show that in the course of row-reducing (or rather column-reducing) a matrix, we simultaneously compute the determinant. Our discussion will use only properties (1), (2) and (3), without the function $D(A)$. We will get out of it an effective method of computing determinants, and at the same time a proof of uniqueness.

The effect of column operations on the determinant. We will use column operations (defined just like the row operations of Appendix L1) rather than row operations in our construction, because we defined the determinant as a function of the n column vectors. This convention will make the interpretation in terms of volumes simpler, and in any case, you will be able to show in Exercise L5#4 that row operations could have been used just as well.

Let us check how each of the three column operations affects the determinant. It turns out that each multiplies the determinant by an appropriate factor μ:

(a) *Multiply a column through by a number m.* Clearly, by property (1), this will have the effect of multiplying the determinant by the same number, so

$$\mu = m.$$

(b) *Exchange two columns.* By property (2), this changes the sign of the determinant, so

$$\mu = -1.$$

(c) *Add a multiple of one column onto another.* By property (1), this does not change the determinant, because

$$\det(\mathbf{a}_1, \ldots, \mathbf{a}_j, \ldots, \mathbf{a}_j + \beta \mathbf{a}_i, \ldots, \mathbf{a}_n)$$
$$= \det(\mathbf{a}_1, \ldots, \mathbf{a}_i, \ldots, \mathbf{a}_j, \ldots, \mathbf{a}_n)$$
$$+ \beta \det(\mathbf{a}_1, \ldots, \mathbf{a}_i, \ldots, \mathbf{a}_i, \ldots, \mathbf{a}_n)$$

and the last term is zero (since two columns are equal, so exchanging them both changes the sign of the determinant and leaves it unchanged). Therefore,

$$\mu = 1.$$

Now any square matrix can be column-reduced until, at the end, you either get the identity or you get a matrix with a column of zeroes (by a corollary of Theorem L1.13). A sequence of matrices resulting from column operations can be denoted as follows, with the multipliers μ_i atop arrows for each operation:

$$A \xrightarrow{\mu_1} A_1 \xrightarrow{\mu_2} A_2 \xrightarrow{\mu_3} \cdots \xrightarrow{\mu_{n-1}} A_{n-1} \xrightarrow{\mu_n} A_n.$$

Then, working backward by property (1),

$$\det A_{n-1} = \left(\frac{1}{\mu_n}\right) \det A_n, \tag{2a}$$

$$\det A_{n-2} = \left(\frac{1}{\mu_{n-1}\mu_n}\right) \det A_n, \tag{2b}$$

$$\downarrow$$

$$\det A = \left(\frac{1}{\mu_1\mu_2 \cdots \mu_{n-1}\mu_n}\right) \det A_n. \tag{2c}$$

Therefore, from equation (2c),

if $A_n = I$, det $A_n = 1$ by property (3)

$$\det A = \frac{1}{\mu_1 \mu_2 \ldots \mu_{n-1} \mu_n};$$ (3)

if $A_n \neq I$, det $A_n = 0$ by property (1)

$$\det A = 0.$$ (4)

Proof of uniqueness of determinant. Suppose that the function $D_1(A)$ obeys properties (1), (2), and (3). Then for any matrix A,

$$D_1(A) = \left(\frac{1}{\mu_1 \mu_2 \ldots \mu_n} \right) \det(A_n) = D(A);$$

i.e., $D_1 = D$. You may object that a different sequence of row operations might lead to a different sequence of μ_i's. If that were the case, it would show that the axioms for the determinant were inconsistent; we know they are consistent because of Theorem L5.1.

We have finally finished proving Theorem L5.1.   □

L5.2. THEOREMS RELATING MATRICES AND DETERMINANTS

There are some matrices whose determinant is easy to compute: the triangular matrices, both upper and lower.

Theorem L5.3. *If a matrix is triangular, then its determinant is the product of the entries along the diagonal.*

Proof. Suppose the matrix is upper-triangular. If the entries along the diagonal are nonzero, then by adding multiples of the first columns to the others, the first row can be cleared beyond the diagonal, then adding multiples of the second to the third and later columns, the second row can be cleared, etc. We are left (Exercise L5#2) with a diagonal matrix with the same diagonal entries as the original matrix; such a matrix can be turned into the identity by dividing each column through by its entry on the diagonal.

If any of the diagonal entries are zero, then the same clearing operation as above, carried out until you come to such a column, leaves a matrix with a column of zeroes and, hence, has determinant 0 (like the product of the diagonal entries).

If the original matrix had been lower-triangular, the same argument would work, only starting with the last column, and first clearing the bottom row.   □

Remark. It is fairly easy to prove Theorem L5.3 from the recursive definition of D, at least if the matrix is lower-triangular. Exercise L5#3 asks you to do so and to adapt the proof to the upper-triangular case.

Appendix L2.5 gives several characterizations of invertible matrices; here are some more:

Theorem L5.4. *A matrix A is invertible if and only if its determinant is not zero.*

Proof. This follows immediately from the column-reduction algorithm and the uniqueness proof, since along the way we showed, in equations (3) and (4), that a matrix has a nonzero determinant if and only if it can be column-reduced to the identity. (See Section L1.4.)   □

A key property of the determinant, for which we will see a geometric interpretation later, is

Theorem L5.5. *If A and B are $n \times n$ matrices, then*

$$\det(A)\det(B) = \det(AB).$$

Proof. If A is invertible, consider the function $\det(AB)/\det(A)$, as a function of the matrix B. As the reader will easily check, it has properties (1), (2), and (3), which characterize the determinant function. If A is not invertible, the left-hand side of the theorem is zero; the right-hand side must be zero also, because

$$\mathrm{Image}(AB) \subset \mathrm{Image}(A).$$

Therefore, $\dim \mathrm{Image}(AB) \leq \dim \mathrm{Image}(A) < n$, so AB is not invertible either.   □

Theorem L5.6. *If a matrix A is invertible, then $\det A^{-1} = 1/\det A$.*

Proof. This is a simple consequence of Theorem L5.5.   □

Theorem L5.7. *The determinant function is basis independent, that is,*

$$\det A = \det(P^{-1}AP).$$

Again this follows immediately from Theorem L5.5.   □

These facts will be very important in Appendix L6 on eigenvalues and eigenvectors, as will be the next theorem and the next section.

Theorem L5.8. *If A is an $n \times n$ matrix and B is an $m \times m$ matrix, then for the $(n+m) \times (n+m)$ matrix formed with these as diagonal elements,*

$$\det \begin{bmatrix} A & 0 \\ 0 & B \end{bmatrix} = \det A \det B.$$

The proof of Theorem L5.8 is left to the reader.

L5.3. THE CHARACTERISTIC POLYNOMIAL OF A MATRIX AND ITS COEFFICIENTS

The *characteristic polynomial* of a square $n \times n$ matrix A is the polynomial $\det(\lambda I - A)$. Such λ's will play a crucial role in the next Appendix, L6.

Example L5.9. Computing characteristic polynomials is usually just a grind:
The characteristic polynomial of

$$\begin{bmatrix} -1 & 0 & 1 \\ -2 & 1 & -2 \\ 0 & -1 & -1 \end{bmatrix}$$

is

$$\det \begin{bmatrix} \lambda+1 & 0 & -1 \\ 2 & \lambda-1 & 2 \\ 0 & 1 & \lambda+1 \end{bmatrix} = (\lambda+1)^2(\lambda-1) - 2 - 2(\lambda+1)$$

$$= \lambda^3 + \lambda^2 - 3\lambda - 5. \quad \blacktriangle$$

Remark. You may encounter elsewhere the definition of the characteristic polynomial as $\det(A - \lambda I)$, but you will still get essentially the same results. These expressions match exactly if n is even, and are of opposite sign if n is odd.

The characteristic polynomial is an important application of the determinant, which will really come into its own when we deal with eigenvectors and eigenvalues. However, in this section, we want to define and prove some of the properties of the *trace*, another important function of a matrix, and finding the characteristic polynomial is an easy way of going about it.

Until now, we have behaved as if all the matrices we had were matrices of *numbers*. But $\lambda I - A$ is a matrix of *polynomials*; does our theory of determinants still apply? Certainly, the development according to the first row defined in Section L5.1 makes perfectly good sense even for a matrix of polynomials. Actually, the entire development works for matrices whose entries are any *rational function*. The only thing we use in row-reduction is the ability to do arithmetic, i.e., to add, subtract, multiply, and divide by everything except 0, and these operations do hold for rational functions.

Theorem L5.10. *The characteristic polynomial is basis independent.*

Proof. Since $\lambda I - P^{-1}AP = P^{-1}(\lambda I - A)P$, by Theorem L5.7 we see that

$$\det(\lambda I - A) = \det(\lambda I - P^{-1}AP)$$

for any invertible matrix P.   □

Theorem L5.10 has some interesting consequences:

Theorem L5.11. *The characteristic polynomial of an $n \times n$ matrix is a polynomial of degree n, with leading term λ^n, negative trace as coefficient of λ^{n-1}, and $(-1)^n \det A$ as constant term, i.e.,*

$$\det(\lambda I - A) = \lambda^n - (\operatorname{tr} A)\lambda^{n-1} + \cdots + (-1)^n \det A.$$

Furthermore, each coefficient of the powers of λ is individually basis independent.

Proof. The fact that the constant term is $\det(-A) = (-1)^n \det A$ is seen by setting $\lambda = 0$ in the formula for the characteristic polynomial.

The coefficient of λ^{n-1} is actually minus the sum of the diagonal entries: we use this fact to define the *trace* of a matrix as

$$\operatorname{tr} A = \sum a_{i,i}.$$

Directly from this definition it is not at all clear that $\operatorname{tr} A = \operatorname{tr} P^{-1}AP$, i.e., that the trace is basis independent. In fact, it is *not true* that $\operatorname{tr} AB = \operatorname{tr} A \operatorname{tr} B$. But the basis independence does follow from the description of the trace in terms of the characteristic polynomial, using Theorem L5.10.

The other coefficients of the characteristic polynomial are important also but do not seem to have names. They are all basis-independent functions of the matrix, and it can be shown conversely that any basis-independent function of matrices is a function of these coefficients. All of this again comes from Theorem L5.10.   □

Example L5.12. A very useful example for constructing exercises is the *companion matrix*. Given any polynomial

$$p(\lambda) = \lambda^n + a_{n-1}\lambda^n + \cdots + a_0,$$

there is a way of manufacturing a matrix with precisely $p(\lambda)$ as its characteristic polynomial:

$$\begin{bmatrix} 0 & 1 & 0 & \cdots & 0 & 0 \\ 0 & 0 & 1 & \cdots & 0 & 0 \\ 0 & 0 & 0 & \cdots & \cdots & \\ \vdots & \vdots & & \vdots & \vdots & \\ \vdots & \vdots & & \vdots & \vdots & \\ -a_0 & -a_1 & -a_2 & & \cdots & -a_{n-1} \end{bmatrix}.$$

The computation showing that this is true is Exercise L5#6. An illustrative case is the polynomial $p(\lambda) = \lambda^3 + 2\lambda^2 + 3\lambda + 4$, which is the characteristic polynomial of

$$\begin{bmatrix} 0 & 1 & 0 \\ 0 & 0 & 1 \\ -4 & -3 & -2 \end{bmatrix},$$

because

$$\det(\lambda I - A) = \det \begin{bmatrix} \lambda & -1 & 0 \\ 0 & \lambda & -1 \\ 4 & +3 & \lambda+2 \end{bmatrix}$$

$$= \lambda^2(\lambda+2) + 4 + 3\lambda = \lambda^3 + 2\lambda^2 + 3\lambda + 4. \quad \blacktriangle$$

For some matrices, the characteristic polynomial is easy to compute:

Theorem L5.13. *The characteristic polynomial of a triangular matrix* $A = (a_{i,j})$ *is the product* $\prod(\lambda - a_{i,i})$.

Proof. If A is triangular, then so is $\lambda I - A$; hence, the characteristic polynomial is the product of the diagonal terms. $\quad \square$

Generally speaking, the characteristic polynomial cannot effectively be computed for large matrices. Using the explicit formula as in Example L5.9, it is already time-consuming for a 10×10 matrix, even on a large computer. There are other ways of computing it (for instance, by row-reduction); but we will not go into them as there are more effective ways (like the QR method and Jacobi's method, described in some detail in Appendices L7 and L8, respectively) of dealing with the prime goal of the characteristic polynomial: eigenvalues and eigenvectors.

L5.4. VOLUMES

Determinants have a geometric interpretation (to be further detailed in Section L5.5) that really makes them important:

Determinants measure n-dimensional volumes.

This requires a definition of "volume." Here we will only deal with volumes of k-dimensional parallelepipeds in $\mathbb{R}^n$, and leave more arbitrary shapes to a course in multiple integrals.

Let $\mathbf{v}_1, \ldots, \mathbf{v}_k$ be k vectors in $\mathbb{R}^n$; the k-dimensional parallelepiped which they span is the space of points

$$P_{\mathbf{v}_1,\ldots,\mathbf{v}_k} = \{a_1\mathbf{v}_1 + \cdots + a_k\mathbf{v}_k \mid 0 \le a_i \le 1 \text{ for all } i = 1, \ldots, k\}.$$

Example L5.14.

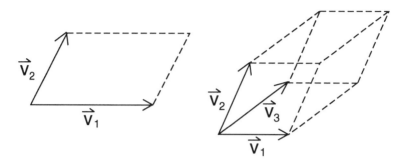

FIGURE L5.1. Right-handed systems of vectors in $\mathbb{R}^2$ and $\mathbb{R}^3$ with the "volumes" they span. ▲

The *k-dimensional volume* of $P_{\mathbf{v}_1,\ldots\mathbf{v}_k}$ is defined by induction:

if $k = 1$, then the volume is simply the length, $\mathrm{vol}_1(P_{\mathbf{v}}) = \|\mathbf{v}\|$;

if $k > 1$, then

$$\mathrm{vol}_k(P_{\mathbf{v}_1,\ldots,\mathbf{v}_k}) = \mathrm{vol}_{k-1}(P_{\mathbf{v}_1,\ldots,\mathbf{v}_{k-1}}) \cdot \|\pi_{V_{k-1}^{\perp}}(\mathbf{v}_k)\|,$$

where V_{k-1} is the subspace spanned by $\mathbf{v}_1, \ldots, \mathbf{v}_{k-1}$, and $\pi_{V_{k-1}^{\perp}}$ is the projection on the orthogonal complement, as defined in Subsection L3.6.

Note that this last formula is precisely "volume = base·height."

Before this definition is of much use, we will need to know a few of its properties. The first and hardest is

Theorem L5.15. *The volume of $P_{\mathbf{v}_1,\ldots,\mathbf{v}_k}$ does not depend on the order of $\mathbf{v}_1, \ldots, \mathbf{v}_k$.*

Proof. The heart of the proof is the case $k = 2$; i.e., the statement that the area of a parallelogram is given by base·height, and that *you can choose either side as the base*. You are asked in Exercise L5#8 to prove this for vectors in the plane, by classical geometric methods; even this is not absolutely obvious. The notation gets messy, but we are just "algebraicizing" high school plane geometry.

Let us verify by computation:

$$(\mathrm{vol}_2(\mathbf{v}_1, \mathbf{v}_2))^2 = \left(\|\mathbf{v}_1\| \left\|\mathbf{v}_2 - \frac{(\mathbf{v}_1 \cdot \mathbf{v}_2)}{(\mathbf{v}_1 \cdot \mathbf{v}_1)}\mathbf{v}_1\right\|\right)^2 = \|\mathbf{v}_1\|^2\|\mathbf{v}_2\|^2 - (\mathbf{v}_1, \mathbf{v}_2)^2.$$

This last expression is symmetric in $\mathbf{v}_1$ and $\mathbf{v}_2$.

Now let us try the case $k > 2$. It is enough to show that the volume is unchanged if the last two vectors are exchanged. Indeed, by induction, we may assume that the volume is unchanged if the first $k - 1$ vectors are permuted (that is, written in a different order), and an arbitrary permutation (or ordering) which moves the kth vector can be achieved by first permuting the first $k - 1$ so as to bring into the $k - 1$ position whatever is supposed to go in the kth position, then exchanging the last two, then permuting the first $k - 1$ again to achieve whatever permutation is desired.

Using the definition of volume twice, we see that

$$\mathrm{vol}_k(P_{\mathbf{v}_1,\ldots,\mathbf{v}_k}) = \mathrm{vol}_{k-2}(P_{\mathbf{v}_1,\ldots,\mathbf{v}_k}) \|\pi_{V^\perp_{k-2}}(\mathbf{v}_{k-1})\| \, \|\pi_{V^\perp_{k-1}}(\mathbf{v}_k)\|. \qquad (5)$$

Since $V_{k-2} \subset V_{k-1}$, we have $V^\perp_{k-1} \subset V^\perp_{k-2}$, and so

$$\pi_{V^\perp_{k-1}} = \pi_{V^\perp_{k-1}} \circ \pi_{V^\perp_{k-2}}.$$

For instance, if you have a line contained in a plane in $\mathbb{R}^3$, then if you project first onto the plane and then onto the line, or directly onto the line, you get the same thing.

The last two factors in (5) above can therefore be rewritten

$$\|\pi_{V^\perp_{k-2}}(\mathbf{v}_{k-1})\| \, \|\pi_{V^\perp_{k-1}} \circ \pi_{V^\perp_{k-2}}(\mathbf{v}_k)\|,$$

which is the volume (area, really, since $\mathrm{vol}_2 = $ area) of the parallelogram spanned by

$$\pi_{V^\perp_{k-2}}(\mathbf{v}_{k-1}) \quad \text{and} \quad \pi_{V^\perp_{k-2}}(\mathbf{v}_k).$$

By the case for two vectors above, this does not depend on the order of $\mathbf{v}_{k-1}$ and $\mathbf{v}_k$. Hence, the theorem is proved for all k. $\square$

The next proposition is easier and intuitive.

Theorem L5.16. *If* $\mathbf{v}_1, \ldots, \mathbf{v}_k$ *are orthogonal, then the volume of* $P_{\mathbf{v}_1,\ldots,\mathbf{v}_k}$ *is the product of the lengths of* $\mathbf{v}_1, \ldots, \mathbf{v}_k$:

$$\mathrm{vol}_k(P_{\mathbf{v}_1,\ldots,\mathbf{v}_k}) = \|\mathbf{v}_1\| \cdot \ldots \cdot \|\mathbf{v}_k\|.$$

Proof. This is proved by induction. It is certainly true if $k = 1$; from the definition and by induction,

$$\mathrm{vol}_k(P_{\mathbf{v}_1,\ldots,\mathbf{v}_k}) = \mathrm{vol}_{k-1}(P_{\mathbf{v}_1,\ldots,\mathbf{v}_{k-1}}) \cdot \|\pi_{V^\perp_{k-1}}(\mathbf{v}_k)\|$$

$$= (\|\mathbf{v}_1\| \cdot \ldots \cdot \|\mathbf{v}_{k-1}\|) \cdot \|\mathbf{v}_k\|.$$

The first factor in the last expression above is equal to the first term in the next to last by induction, and the last terms are equal since $\mathbf{v}_k$ is already orthogonal to V_{k-1}, so $\pi_{V^\perp_{k-1}}(\mathbf{v}_k) = \mathbf{v}_k$. $\square$

One more intuitively desirable property of volumes:

Theorem L5.17. *If Q is an orthogonal matrix, then the image under Q of a parallelepiped has the same volume as the original parallelepiped:*

$$\text{vol}_k(P_{Q(\mathbf{v}_1),\ldots Q(\mathbf{v}_k)}) = \text{vol}_k(P_{\mathbf{v}_1,\ldots,\mathbf{v}_k}).$$

Proof. Again, this is proved by induction on k. For $k = 1$, it is immediate since orthogonal transformations preserve length. By the definition of volume,

$$\text{vol}_k(P_{\mathbf{v}_1,\ldots,\mathbf{v}_k}) = \text{vol}_{k-1}(P_{\mathbf{v}_1,\ldots,\mathbf{v}_{k-1}}) \cdot \|\pi_{V_{k-1}^\perp}(\mathbf{v}_k)\|,$$

and we may assume by induction that

$$\text{vol}_{k-1}(P_{\mathbf{v}_1,\ldots,\mathbf{v}_{k-1}}) = \text{vol}_{k-1}(P_{Q(\mathbf{v}_1),\ldots,Q(\mathbf{v}_{k-1})});$$

the proposition will follow if we know $Q(\pi_{V_{k-1}^\perp}(\mathbf{v}_k)) = \pi_{Q(V_{k-1})^\perp}(Q(\mathbf{v}_k))$. This formula should be clear. $\square$

L5.5. THE RELATION BETWEEN DETERMINANTS AND VOLUMES

Theorem L5.18. *The determinant measures volume, in the sense that its absolute value*

$$|\det[\mathbf{a}_1,\ldots,\mathbf{a}_n]| = \text{vol}(P_{\mathbf{a}_1,\ldots,\mathbf{a}_n}).$$

Proof. Let us see how $|\det|$ is affected by column operations.

(i) Multiplying a column through by μ multiplies $|\det|$ by $|\mu|$.

(ii) Exchanging two columns leaves $|\det|$ unchanged.

(iii) Adding a multiple of one column onto another leaves $|\det|$ unchanged.

Moreover,

(iv) $|\det(I)| = 1$, $|\det(A)| = 0$ if a column of A is 0,

and exactly as in Section L5.1, these four properties characterize $|\det|$, since any matrix can be column-reduced.

> However, *these four properties are also true of the function that takes a matrix and gives the volume of the parallelepiped spanned by the column vectors.*

Property (i) is clear, from the definition of volume, if the vector chosen is the last and is therefore true for any column by Theorem L5.15. Property (ii) is Theorem L5.15. Finally, Property (iii) is clear if the column being added to is the last, since if $\mathbf{w}$ is obtained by adding an arbitrary linear combination of $\mathbf{v}_1, \ldots, \mathbf{v}_{k-1}$ to $\mathbf{v}_k$, then $\pi_{V_{k-1}^\perp} \mathbf{w} = \pi_{V_{k-1}^\perp}(\mathbf{v}_k)$. That is, (iii) is true of any column by (ii). As for (iv), the first part follows from the definition of volume, and the second part is obvious if the column is the last, hence once more is true for any column by (ii). $\square$

Corollary L5.19. *The determinant of an orthogonal matrix is ± 1.*

Proof. The columns of an orthogonal matrix form an orthnormal basis, so the parallelepiped which they span has volume 1, by either Theorem L5.15 or Theorem L5.17. $\square$

We are finally in a position to prove the following result. It is surprising how hard it is to prove; we will be using just about everything developed up to now.

Theorem L5.20. *For any $n \times n$ matrix A, we have $\det(A) = \det(A^\top)$.*

Proof. By Theorem L4.3, any matrix can be written $A = QR$ with Q orthogonal and R upper-triangular. Thus, $A^\top = R^\top Q^\top$ and $\det(A^\top) = \det(R^\top)\det(Q^\top)$. By Theorem L5.3, $\det(R^\top) = \det(R)$, since both are the product of their diagonal entries. Moreover, $\det(Q^\top) = \det(Q)$, since they are both of absolute value 1 and their product is 1. $\square$

One important consequence of Theorem L5.20 is that throughout this text, whenever we were speaking of column operations, we could just as well have spoken of row operations.

Now for the promised geometric explanation of Theorem L5.5, that

$$\det(AB) = \det(A)\det(B).$$

As we have just seen, $\det(A)$ is the volume of the image under A of the unit cube, i.e., the image of the parallelepiped spanned by $\mathbf{e}_1, \ldots, \mathbf{e}_n$, which, of course, has volume 1. Actually, for any "body" X in $\mathbb{R}^n$, $\det(A)$ measures the ratio of the volume of the image $A(X)$ to the volume of X; that is, $\det(A)$ measures the *expansion factor on volumes.*

This fact can be heuristically seen as follows: Fill X up with little disjoint cubes whose faces are aligned with the coordinate axes; it seems reasonable to define the volume of X to be the maximum of the sum of the volumes of such little cubes. Then the images of these little cubes under A will fill up $A(X)$, but the volume of each of the image cubes is precisely the volume of the original little cube multiplied by $|\det(A)|$.

Now it should be clear that $|\det(AB)| = |\det(A)\det(B)|$, since under AB the unit cube is first taken to B(unit cube) with volume $|\det(B)|$, and then to $A(B$(unit cube)), with volume $|\det(A)|\,|\det(B)|$.

In particular, if A is not invertible, then the column vectors are linearly dependent, and hence A squashes the unit cube down onto a space of lower dimension, by making one or more sides the zero vector. Of course, the volume of the image of the unit cube is then zero.

Signed volumes. As anybody who has integrated functions knows, the area under the curve has to be taken negative when the curve is beneath the x-axis. In integration theory, it is usually much easier to deal with signed areas, volumes, etc., than with unsigned ones. The first step in doing this in n dimensions is the following definition:

The *signed volume* of the parallelepiped P_A spanned by the column vectors of A is given by $\det A$.

Of course, the signed volume does depend on the order of the spanning vectors. The meaning of signed volume is best explained by the following two examples.

Example L5.21. In $\mathbb{R}^2$, two vectors $\mathbf{a}_1$ and $\mathbf{a}_2$ *in that order* span a parallelogram of positive area if and only if the smallest angle from $\mathbf{a}_1$ to $\mathbf{a}_2$ is counterclockwise, as shown in Example L5.14.   ▲

Example L5.22. In $\mathbb{R}^3$, three vectors $\mathbf{a}_1$, $\mathbf{a}_2$, and $\mathbf{a}_3$ *in that order* span a parallelepiped of positive signed volume if and only if they form a right-handed coordinate system, as shown in Example L5.14.   ▲

L6 Eigenvalues and Eigenvectors

L6.1. Eigenvalues and Eigenvectors

Let V be a complex vector space and $T\colon V \to V$ be a linear transformation with matrix A. If for some number $\lambda \in \mathbb{C}$ and some vector $\mathbf{v} \in V$, $\mathbf{v} \neq \mathbf{0}$,

$$A\mathbf{v} = \lambda\mathbf{v},$$

then the number λ is called an *eigenvalue* of A, and every nonzero vector $\mathbf{v}$ for which $A\mathbf{v} = \lambda\mathbf{v}$ is called a corresponding *eigenvector* of A.

Note that eigenvectors corresponding to a given eigenvalue are never unique: nonzero multiples of an eigenvector are eigenvectors. Note also that the zero vector is never, by definition, an eigen*vector*. However, nothing prevents zero from being an eigen*value*; in fact, the eigenvectors for the eigenvalue 0 form the *kernel* of A, and 0 is an eigenvalue if and only if $\ker A \neq \{\mathbf{0}\}$.

Example L6.1. If $A = \begin{bmatrix} 0 & -1 \\ 1 & 0 \end{bmatrix}$, the number $-i$ is an eigenvalue and

$\mathbf{v} = \begin{bmatrix} 1 \\ i \end{bmatrix}$ is a corresponding eigenvector of A, since

$$\begin{bmatrix} 0 & -1 \\ 1 & 0 \end{bmatrix} \begin{bmatrix} 1 \\ i \end{bmatrix} = \begin{bmatrix} -i \\ 1 \end{bmatrix} = -i \begin{bmatrix} 1 \\ i \end{bmatrix}. \qquad \blacktriangle$$

The next example is the real reason for Fourier series. It is a bit less straightforward but well worth poring over a bit.

Example L6.2. Recalling Example L4.7, let V be the space of infinitely differentiable functions on $[0, \pi]$ which vanish at 0 and at π. Consider the *second derivative operator* $D^2 : V \to V$. We will describe the eigenvectors (often called *eigenfunctions*, since they live in a function space) of D^2.

> *The functions $f_k(x) = \sin kx$ are eigenvectors of D^2, with eigenvalue $-k^2$.*

This is nothing more or less than the fact that $d^2/dx^2(\sin kx) = -k^2 \sin kx$, which everyone learned in elementary calculus. $\qquad \blacktriangle$

L6.2. Eigenvalues and the Characteristic Polynomial

From a theoretical point of view, the characteristic polynomial is the main actor in the theory of eigenvectors and eigenvalues, at least in finite-dimensional vector spaces.

Theorem L6.3. *The eigenvalues of a square matrix A are the roots of its characteristic polynomial.*

Proof. From Appendix L2.5 and Theorem L5.4, we can state (Exercise L6#1) that the following four conditions are equivalent:

(a) A has no inverse.

(b) There exists a vector $\mathbf{x} \neq \mathbf{0}$ such that $A\mathbf{x} = \mathbf{0}$.

(c) There exists a vector $\mathbf{y}$ that cannot be written as $A\mathbf{x}$.

(d) $\det A = 0$.

By part (b) of these conditions for noninvertibility, $\det(\lambda I - A) = 0$ if and only if there exists $\mathbf{v} \neq \mathbf{0}$ with $(\lambda I - A)\mathbf{v} = \mathbf{0}$, i.e., $A\mathbf{v} = \lambda\mathbf{v}$. $\quad \square$

Theorem L6.3 has many corollaries:

Corollary L6.4. *An $n \times n$ matrix has at most n eigenvalues, and at least one.*

Proof. The first part is immediate from Theorem L6.3; the second part is the fundamental theorem of algebra (see below). □

The *fundamental theorem of algebra* is the following statement:

> *Every polynomial of degree > 0 with complex coefficients has at least one complex root.*

This theorem is what guarantees that an nth degree polynomial equation $p_n(x) = 0$ has n roots, although some of them may coincide. Naming the root promised as r_1, you divide the polynomial p_n by $(x - r_1)$ and reapply the theorem to the resulting new polynomial of degree $n - 1$. As you might guess from the name, the fundamental theorem of algebra is an important theorem. It is also hard to prove, and, more seriously, the proofs tend not to be constructive: they do not give an algorithm for finding the root. We need the statement now in this section and not before, because the subject of eigenvectors and eigenvalues is the first time we must solve polynomial equations, which are *not* linear; nonlinear equations cannot be treated using row reduction or anything else which uses only arithmetic (addition, subtraction, multiplication, division).

Corollary L6.5. *The trace of a matrix is the sum of the eigenvalues and the determinant is their product, each one being taken with its multiplicity.*

Proof. This follows directly from Theorem L5.11. □

Remark. It is perfectly common for real polynomials to have complex roots, so there is no reason to expect that real matrices will have real eigenvalues. If the eigenvalues of a real matrix are complex, then so are the corresponding eigenvectors, and the vector space we are dealing with is $\mathbb{C}^n$ rather than $\mathbb{R}^n$. Furthermore, we see that an $n \times n$ matrix has at most n distinct eigenvalues, and "in general" will have exactly that many, since a polynomial will "in general" have distinct roots.

Example L6.6. If $A = \begin{bmatrix} 0 & -1 \\ 1 & 0 \end{bmatrix}$, then the eigenvalues are the roots of $\det \begin{bmatrix} \lambda & 1 \\ -1 & \lambda \end{bmatrix}$, or $\lambda^2 + 1$, so the eigenvalues are $\pm i$. Since

$$(\lambda I - A)\mathbf{v} = \begin{bmatrix} \lambda & 1 \\ -1 & \lambda \end{bmatrix} \begin{bmatrix} x \\ y \end{bmatrix} = \begin{bmatrix} 0 \\ 0 \end{bmatrix},$$

we can write the following system of linear equations:

$$\lambda x + y = 0,$$
$$-x + \lambda y = 0,$$

which is linearly *dependent.* The eigenvectors are the solutions to this system, but they are determined only up to a multiplicative constant. Solving the system for each value of λ, we find that

$$\mathbf{v}_1 = \begin{bmatrix} x \\ y \end{bmatrix} = \begin{bmatrix} 1 \\ -i \end{bmatrix} \quad \text{is an eigenvector with eigenvalue } i;$$

$$\mathbf{v}_2 = \begin{bmatrix} x \\ y \end{bmatrix} = \begin{bmatrix} 1 \\ i \end{bmatrix} \quad \text{is an eigenvector with eigenvalue } -i. \quad \blacktriangle$$

We need one more theorem to complete the general theory:

Theorem L6.7. *If A is a square matrix and $\mathbf{v}_1, \ldots, \mathbf{v}_m$ are eigenvectors with distinct eigenvalues, then $\mathbf{v}_1, \ldots, \mathbf{v}_m$ are linearly independent.*

Proof. This result is a good deal harder to prove than the earlier results. We shall do it by contradiction. If $\mathbf{v}_1, \ldots, \mathbf{v}_m$ are not linearly independent, then there is a first vector $\mathbf{v}_j$ which is a linear combination of the earlier linearly independent ones. Thus, we can write

$$\mathbf{v}_j = a_1 \mathbf{v}_1 + \cdots + a_{j-1} \mathbf{v}_{j-1},$$

where at least one of the a_i is different from zero. Apply $\lambda_j I - A$ to both sides to get

$$\mathbf{0} = (\lambda_j I - A)\mathbf{v}_j = a_1(\lambda_j - \lambda_1)\mathbf{v}_1 + \cdots + a_{j-1}(\lambda_j - \lambda_{j-1})\mathbf{v}_{j-1}.$$

Since the λ's are all different, the coefficients $a_i(\lambda_i - \lambda_j)$ cannot all be zero. Move the ith such term to the other side of the equality and divide out by the (nonzero) coefficient. This expresses an even earlier $\mathbf{v}_i$ with $i < j$ as a linear combination of the $\mathbf{v}_1, \ldots, \mathbf{v}_{j-1}$, which is a contradiction of the assumption that $\mathbf{v}_j$ is the earliest. $\square$

Putting Theorems L6.3 and L6.7 together yields the following result:

Theorem L6.8. *If A is a square $n \times n$ matrix whose characteristic polynomial has n distinct roots, then $\mathbb{C}^n$ has a basis $\mathbf{v}_1, \ldots, \mathbf{v}_n$ made of eigenvectors of A.*

If all the roots of the characteristic polynomial are real, then the basis vectors in Theorem L6.8 can also be chosen to be real, providing a basis of $\mathbb{R}^n$, but if any eigenvalues are not real, then there definitely will not be a basis of $\mathbb{R}^n$ made up of eigenvectors of A.

Theorem L6.8 describes the nicest linear transformations, which are also the most common ones (except in textbooks). Bases of eigenvectors are very important, so we invent a new word to describe them.

Definition L6.9. Let $T: V \to V$ be a linear transformation. A basis $\mathbf{v}_1, \ldots, \mathbf{v}_n$ will be called an *eigenbasis for* T if each vector $\mathbf{v}_i$ is an eigenvector of T.

Note that Theorem L6.8 does *not* say that if the characteristic polynomial of a matrix A has multiple roots, then there is no eigenbasis for A. For instance, the identity has only 1 as an eigenvalue, but every basis is an eigenbasis. The correct statement, of which Theorem L6.8 is a special case, is the following:

Theorem L6.10. *If A is a square $n \times n$-matrix whose characteristic polynomial has $m \le n$ distinct roots, then $\mathbb{C}^n$ has an eigenbasis for A if and only if, for each root λ of the characteristic polynomial, the dimension of the eigenspace E_λ for λ,*

$$E_\lambda = \{\mathbf{v} \in \mathbb{C}^n \mid A\mathbf{v} = \lambda\mathbf{v}\},$$

is equal to the multiplicity of λ as a root of the characteristic polynomial.

Proof: The proof is very similar to that of Theorem L6.8 and is left to the reader. $\square$

L6.3. DIAGONALIZATION

There is another way of saying that a vector space has an eigenbasis for a matrix:

Theorem L6.11. *Suppose that $\mathbf{v}_1, \ldots, \mathbf{v}_n$ is an eigenbasis of $\mathbb{C}^n$ for a linear transformation with matrix A; let λ_i be the eigenvalue of $\mathbf{v}_i$. Let P be the matrix whose i^{th} column is $\mathbf{v}_i$, i.e., the change of basis matrix. Then*

$$P^{-1}AP = \begin{bmatrix} \lambda_1 & & \\ & \ddots & 0 \\ 0 & & \\ & & \lambda_n \end{bmatrix}.$$

The result of Theorem L6.11 is the reason for which the theory of eigenvalues and eigenvectors is frequently called *diagonalization* of matrices. A matrix A is diagonalizable if and only if there is an eigenbasis for A.

Proof of Theorem L6.11. Let us see what $P^{-1}AP$ does to a standard basis vector $\mathbf{e}_i$. Well, $P(\mathbf{e}_i) = \mathbf{v}_i$; so $AP(\mathbf{e}_i) = \lambda_i\mathbf{v}_i$, by definition of eigenvector and eigenvalue. Finally, $P^{-1}AP(\mathbf{e}_i) = \lambda_i\mathbf{e}_i$, by undoing the first step. However, the diagonal matrix with λ_i in the ith place on the diagonal is precisely the matrix that sends $\mathbf{e}_i$ to $\lambda_i\mathbf{e}_i$. $\square$

Definition L6.12 A matrix P such that $P^{-1}AP$ is diagonal will be said to *diagonalize A*.

Most square matrices can be diagonalized; the one place where this is not true is in the exercise sections of textbooks. More precisely, among all $n \times n$ matrices, those that cannot be diagonalized form a subset of a locus defined by a polynomial equation in the entries.

The 2×2 case can be worked out explicitly. A matrix $A = \begin{bmatrix} a & b \\ c & d \end{bmatrix}$ is definitely diagonalizable if its characteristic polynomial has distinct roots, which happens except when its discriminant vanishes, i.e., except if

$$\text{tr}^2 A = 4 \det A \quad \text{or} \quad (a+d)^2 = 4(ad - bc).$$

It may happen that $\text{tr}^2 A = 4 \det A$ and A is diagonalizable anyway; this occurs precisely if A is a multiple of the identity. If A satisfies $\text{tr}^2 A = 4 \det A$ and if A is not already diagonal, then A is not diagonalizable, as you can prove in Exercise L6#2 So the set of nondiagonalizable 2×2 matrices is precisely those whose entries satisfy $(a+d)^2 = 4(ad - bc)$ but for which b and c do not both vanish.

Example L6.13. The matrix $A = \begin{bmatrix} 1 & 1 \\ 0 & 1 \end{bmatrix}$ is not diagonalizable.   ▲

The 3×3 case has a similar but messier analysis, which you are invited to provide in Exercise L6#3.

PSEUDO-DIAGONAL MATRICES

All the above applies to complex matrices, and even if A is real, it can usually not be diagonalized over the reals. There are bases $\mathbf{w}_1, \dots, \mathbf{w}_n$ of $\mathbb{R}^n$ such that if $P = [\mathbf{w}_1, \dots, \mathbf{w}_n]$ is the change of basis matrix, then $P^{-1}AP$ has many of the pleasant features of diagonal matrices.

Theorem L6.14. *Let A be the matrix of a real linear transformation with eigenvalues*

$$\lambda_1 = \xi_1 + i\eta_1, \lambda_2 = \overline{\lambda_1} = \xi_1 - i\eta_1, \dots, \lambda_{2k-1}, \quad \lambda_{2k} = \overline{\lambda_{2k-1}}, \lambda_{2k+1}, \dots, \lambda_n$$

with the first $2k$ eigenvalues not real and the last $n - 2k$ real. Let

$$\mathbf{v}_1, \mathbf{v}_2 = \overline{\mathbf{v}_1}, \dots, \mathbf{v}_{2k-1}, \mathbf{v}_{2k} = \overline{\mathbf{v}_{2k-1}}, \mathbf{w}_{2k+1}, \dots, \mathbf{w}_n$$

be a corresponding eigenbasis of $\mathbb{C}^n$, and we may assume that the vectors $\mathbf{w}_{2k+1}, \dots, \mathbf{w}_n$ are real. For $j = 1, \dots, k$ set

$$\mathbf{w}_{2j-1} = \text{Re}\,\mathbf{v}_{2j-1} \quad \text{and} \quad \mathbf{w}_{2j} = \text{Im}\,\mathbf{v}_{2j-1}.$$

Let P be the matrix whose ith column is $\mathbf{w}_i$, *i.e., the change of basis matrix. Then*

$$
P^{-1}AP = \begin{bmatrix}
\xi_1 & -\eta_1 & & & & & \\
\eta_1 & \xi_1 & & & & 0 & \\
& & \ddots & & & & \\
& & & \xi_k & -\eta_k & & \\
& & & \eta_k & \xi_k & & \\
& 0 & & & & \lambda_{2k+1} & \\
& & & & & & \ddots \\
& & & & & & & \lambda_n
\end{bmatrix}.
$$

Again we leave the proof to the reader. We shall say that a matrix as above is *k-pseudo-diagonal*. More specifically, a matrix will be k-pseudo-diagonal if it is the sum of a diagonal matrix and of antisymmetric matrices in the k leading 2×2 blocks along the diagonal.

L6.4. TRIANGULARIZATION

We have seen in the last section that most, but not all, square matrices are diagonalizable. That is, *some matrices do not have enough eigenvectors to make a basis*; this fact is the main nuisance of linear algebra. There are many ways of dealing with this nuisance, all more or less unsatisfactory. Most use words like "generalized eigenvector" and "Jordan canonical form" (commonly abbreviated as JCF).

We have two related reasons for omitting Jordan canonical form (usually the centerpiece of a course in linear algebra). First, the JCF is not constructive; if you know a matrix only up to a certain precision, then you never *know* that it is not diagonalizable. Thus, the JCF is not computable.

The second reason is actually more important. JCF is useful for studying *one matrix at a time*, especially exceptional nondiagonalizable matrices. It tends to put the spotlight on the behavior of these matrices for their own sake. We view these matrices as mainly important because you have to go through them to get from one generic sort to another (most often when going from a matrix with real eigenvalues to one with complex eigenvalues). We want to see these exceptional nondiagonalizable matrices in *context*; this is just what JCF does not encourage.

We shall take a different approach, perhaps "truer to life," certainly a good deal easier, and correspondingly less precise: *triangularization*.

Definition L6.15. A basis $\mathbf{v}_1, \dots, \mathbf{v}_n$ of a vector space V *triangularizes* a linear transformation $T: V \to V$ if $T(\mathbf{v}_i)$ is a linear combination of $\mathbf{v}_1, \dots, \mathbf{v}_i$ (but not $\mathbf{v}_{i+1}, \dots, \mathbf{v}_n$) for all i with $1 \le i \le n$.

If $V = \mathbb{R}^n$ or $\mathbb{C}^n$, and T has matrix A, then T is triangularizable if there exists a matrix P such that $P^{-1}AP$ is upper-triangular; indeed, simply take the vectors $\mathbf{v}_i$ as the columns of P, i.e., take P as the appropriate change of basis matrix. We will say that P *triangularizes* A.

Triangularization is similar to diagonalization. In particular, recall from Theorem L5.13 and Exercise L6#4 or L6#5 that *the diagonal terms of a triangular matrix are the eigenvalues.* Therefore, triangularizing a matrix is at least as hard as finding the eigenvalues.

The following theorem is our main tool:

Theorem L6.16. *Given any complex matrix A, there exists a unitary matrix M such that $M^{-1}AM$ is upper-triangular.*

Equivalently, given any complex n-dimensional vector space V with an inner product and any linear transformation $T\colon V \to V$, there exists an orthonormal basis $\mathbf{v}_1, \mathbf{v}_2, \ldots, \mathbf{v}_n$ of V such that for any i, $1 \leq i \leq n$, the vector $T(\mathbf{v}_i)$ is a linear combination of $\mathbf{v}_1, \ldots, \mathbf{v}_i$ and not the further vectors.

The proof we will give is fairly simple but not very constructive: it depends on the fact that any linear transformation from a vector space to itself has at least one (perhaps complex) eigenvector (Theorem L6.4). Recall that this requires the fundamental theorem of algebra, as applied to the characteristic polynomial; both are rather hard to compute. Moreover, the fundamental theorem of algebra is a result about *complex numbers*, which explains why unitary matrices, the complex analog of orthogonal matrices, as defined in Section L4.1, make their way into the statement. The corresponding statement to Theorem L6.16 with "complex" replaced by "real" and "unitary" replaced by "orthogonal" is simply false; we will see in Theorem L6.17 what replaces it.

The QR method (Appendix L7) gives precisely the orthonormal basis promised by Theorem L6.16, when it converges; so, in fact, exhibiting such a basis is usually possible in practice.

Proof. The equivalence of the two statements is simply the matrix characterization of triangularizability; we will prove the second statement. The proof is by induction on n; so assume that the theorem is true for all vector spaces of dimension $n - 1$.

Choose $\mathbf{v}_1$ a unit eigenvector of T, let V_1 be the space orthogonal to $\mathbf{v}_1$, and let $\pi\colon V \to V_1$ be the orthogonal projection. Now let $T_1\colon V_1 \to V_1$ be given by $T_1(\mathbf{v}) = \pi(T(\mathbf{v}))$. Since V_1 has dimension $n - 1$, we can apply the inductive hypothesis and find an orthonormal basis $\mathbf{v}_2, \ldots, \mathbf{v}_n$ of V_1 such that $T_1(\mathbf{v}_i)$ is a linear combination of $\mathbf{v}_2, \ldots, \mathbf{v}_i$ for all $i = 2, \ldots, n$.

We claim that the vectors $\mathbf{v}_1, \ldots, \mathbf{v}_n$ do the trick. It is an orthonormal set, with n elements, hence a basis. For $2 \leq i \leq n$, the vector $T(\mathbf{v}_i)$ differs from $T_1(\mathbf{v}_i)$ only by some multiple of $\mathbf{v}_1$, so it is a linear combination of $\mathbf{v}_1, \ldots, \mathbf{v}_i$. Finally, $\mathbf{v}_1$ is an eigenvector. $\square$

When we must work with real matrices, Theorem L6.17 is often an adequate substitute. Let us call a matrix k-pseudo-upper-triangular if it is the sum of a pseudo-diagonal matrix and a matrix with nonzero entries only above the k-pseudo-diagonal. In Exercise L6#6, you are asked to show that the characteristic polynomial of a real k-pseudo-upper-triangular matrix has at most k conjugate pairs of nonreal eigenvalues.

Theorem L6.17 *If A is a real $n \times n$ matrix with k conjugate pairs of nonreal eigenvalues, there exists a real invertible matrix M such that $M^{-1}AM$ is k-pseudo-upper-triangular.*

Equivalently, given a real vector space V with an inner product and a linear transformation $T: V \to V$ with k conjugate pairs of nonreal eigenvalues, there exists a basis $\mathbf{w}_1, \mathbf{w}_2, \ldots, \mathbf{w}_n$ of V such that in that basis the matrix of T is k-pseudo-upper-triangular.

Proof. The proof is similar to that of Theorem L6.16. Suppose the statement is true for all vector spaces of dimension less than n; we may also suppose $k > 0$: if all the eigenvalues are real the proof of Theorem L6.16 actually shows that there is a real orthonormal basis in which the matrix of T is upper-triangular.

So let $\lambda = \xi + i\eta$ be a nonreal eigenvalue, and $\mathbf{v}$ a corresponding eigenvector (necessarily complex). Let $\mathbf{w}_1$ and $\mathbf{w}_2$ be the real and imaginary parts of $\mathbf{v}$; note that $\mathbf{w}_1$ and $\mathbf{w}_2$ are linearly independent, since their complex span is the space spanned by $\mathbf{v}$ and $\overline{\mathbf{v}}$, which are linearly independent since they are eigenvectors for the distinct eigenvalues λ and $\overline{\lambda}$. This is where we use that λ is not real.

Let W be the orthogonal of the plane spanned of $\mathbf{w}_1$ and $\mathbf{w}_2$, and consider, as above, $T_1 = \pi_W \circ T|_W$. By induction, we can assume that there exists a basis $\mathbf{w}_3, \ldots, \mathbf{w}_n$ of W such that the matrix of T_1 in that basis is l-pseudo-upper-triangular for some l. The proof that in the basis $\mathbf{w}_1, \ldots, \mathbf{w}_n$ the matrix of T is $(l+1)$-pseudo-upper-triangular is the same as in Theorem L6.16

Since the determinant of a pseudo-upper-triangular matrix is the same as the determinant of the pseudo-diagonal part, we see that the eigenvalues of T are exactly those of T_1 and λ, with multiplicity. Thus, $l = k - 1$. $\square$

As we have seen, the eigenvalues of A are precisely the diagonal entries of the triangular matrix. One consequence of this result is that, as we have been claiming all along, *any matrix can be perturbed arbitrarily little so that its eigenvalues become distinct*. Indeed, write any matrix A as MDM^{-1} with D triangular, and let D_1 be a slight perturbation of D, where only the diagonal entries have been modified to make the entries distinct. Then MD_1M^{-1} is close to A and has distinct eigenvalues.

Unfortunately, triangular matrices are not diagonal. But, as we shall see in Chapter 8, if we go beyond Theorem L6.16 and add a little trick of allowing bases that are not orthonormal, the triangular matrix can be further modified to make the off-diagonal terms as small (but not zero) as one wishes.

It is easy to see how to do this. Let us look at the 2×2 case: suppose that with respect to some basis $\mathbf{v}_1$, $\mathbf{v}_2$, a linear transformation T has an upper-triangular matrix. This means that

$$T(\mathbf{v}_1) = \lambda_1 \mathbf{v}_1 + a\mathbf{v}_1,$$
$$T(\mathbf{v}_2) = \lambda_2 \mathbf{v}_2 \tag{1}$$

so that the matrix for T is

$$T\begin{bmatrix} \mathbf{v}_1 \\ \mathbf{v}_2 \end{bmatrix} = \begin{bmatrix} \lambda_1 & a \\ 0 & \lambda_2 \end{bmatrix} \begin{bmatrix} \mathbf{v}_1 \\ \mathbf{v}_2 \end{bmatrix}.$$

Note that $\mathbf{v}_2$ is an eigenvector and a is the off-diagonal term of the matrix.

Now replace $\mathbf{v}_1$ by $w_1 = \varepsilon \mathbf{v}_1$. Equations (1) can now be written

$$T(\mathbf{v}_1) = \lambda_1 \mathbf{w}_1 + \varepsilon a \mathbf{v}_2 \quad \text{(we haven't touched $\mathbf{v}_2$) and}$$
$$T(\mathbf{v}_2) = \lambda_2 \mathbf{v}_2.$$

so that

$$T\begin{bmatrix} \mathbf{w}_1 \\ \mathbf{v}_2 \end{bmatrix} = \begin{bmatrix} \lambda_1 & \epsilon a \\ 0 & \lambda_2 \end{bmatrix} \begin{bmatrix} \mathbf{w}_1 \\ \mathbf{v}_2 \end{bmatrix}.$$

In the basis $\mathbf{w}_1$, $\mathbf{v}_2$, the matrix of T has as its only off-diagonal term εa, which by taking ε small can be made as small as you like. (Yes, that also makes $\mathbf{w}_1$ very small, but we care only about its direction, not its size.)

Theorem L6.18. *Given any $n \times n$ complex matrix A and any $\varepsilon > 0$, there exists an orthogonal basis $\mathbf{w}_1, \mathbf{w}_2, \ldots, \mathbf{w}_n$ of $\mathbb{C}^n$ such that if $P = [\mathbf{w}_1, \mathbf{w}_2, \ldots, \mathbf{w}_n]$, then $P^{-1}AP$ is upper-triangular, with off-diagonal terms of absolute value smaller than ε.*

Proof. First find an orthonormal basis $\mathbf{v}_1, \ldots, \mathbf{v}_n$ such that if $P = [\mathbf{v}_1, \ldots, \mathbf{v}_n]$ then $P^{-1}AP$ is upper-triangular, as above. Next let $\mathbf{w}_k = \varepsilon^k \mathbf{v}_k$. Since

$$A\mathbf{v}_k = a_{1,k}\mathbf{v}_1 + \cdots + a_{k,k}\mathbf{v}_k,$$

we find

$$A\mathbf{w}_k = \varepsilon^k(a_{1,k}\mathbf{v}_1 + \cdots + a_{k,k}\mathbf{v}_k)$$
$$= a_{1,k}\varepsilon^{k-1}\mathbf{w}_1 + a_{2,k}\varepsilon^{k-2}\mathbf{w}_2 + \cdots + a_{k,k}\mathbf{w}_k.$$

We see that in the basis $\mathbf{w}_1, \ldots, \mathbf{w}_n$ all the off-diagonal terms are multiplied by a positive power of ε and hence can be made as small as one wishes by taking ε sufficiently small. $\square$

There is a real variant of this result.

Theorem L6.19. *Given any $n \times n$ real matrix A with k pairs of nonreal eigenvalues and any $\varepsilon > 0$, there exists a basis $\mathbf{w}_1, \mathbf{w}_2, \ldots, \mathbf{w}_n$ of $\mathbb{R}^n$ such that if $P = [\mathbf{w}_1, \mathbf{w}_2, \ldots, \mathbf{w}_n]$, then $P^{-1}AP$ is k-pseudo-upper-triangular, with off-pseudo-diagonal terms of absolute value smaller than ε.*

The proof is analogous to the proof of L6.18 and is left to the reader (Exercise L6#7).

L6.5. EIGENVALUES AND INNER PRODUCTS

The three classes (symmetric, antisymmetric, and orthogonal) of linear transformations that are closely related to inner products (Sections L4.1, L4.2, and L4.3) have rather special eigenvalues and eigenvectors.

We have strived so far to treat the real and the complex cases separately; at this point it becomes misleading to do so; so please remember that real antisymmetric is a special case of complex anti-Hermitian and that orthogonal is the real case of unitary.

Theorem L6.20.

(a) *If A is a real symmetric or Hermitian linear transformation, then the eigenvalues of A are real.*

(b) *If A is a real antisymmetric or anti-Hermitian linear transformation, then the eigenvalues of A are purely imaginary.*

(c) *If A is an orthogonal or unitary linear transformation, then the eigenvalues of A are of absolute value 1.*

In all three cases, any two eigenvectors of A with distinct eigenvalues are orthogonal.

Note that (c) says in particular that the only real eigenvalues an orthogonal matrix can have are ± 1.

Proof. Theorem L6.20 is an easy computation, outlined in the exercises. The proof of Theorem L6.22 in any case will also prove Theorem L6.20 in the finite-dimensional case.   □

Example L6.21. One consequence of Examples L4.7, and L6.2 and Exercise L3–4#17 is that if $n \neq m$, then

$$\int_{-\pi}^{\pi} \sin nx \sin mx \, dx = \int_{-\pi}^{\pi} \cos nx \sin mx \, dx$$
$$= \int_{-\pi}^{\pi} \cos nx \cos mx \, dx = 0.$$

Indeed, the functions $\sin nx$, $\sin mx$, $\cos nx$, and $\cos mx$ are eigenvectors of the second derivative operator with distinct eigenvalues $-n^2$ and $-m^2$ as appropriate, so they are hereby shown to be orthogonal.

You will meet these integrals when Fourier series are introduced in partial differential equations. We do not want to claim that the proof given here of their being zero is especially simple, but it is more revealing than just computing the integrals using trigonometric formulas. ▲

The next result is really of very great importance. It is central to classical mechanics and to quantum mechanics; it is surely the most important mathematical contribution to chemistry, and probably to the engineering of structures as well. The proof we will give here, essentially a corollary of Theorem L6.16, is short but suffers from being nonconstructive, since it relies on the characteristic polynomial and the fundamental theorem of algebra. In application, one needs not only to know that an eigenvector exists but also how to compute it. Jacobi's method (Appendix L8) provides such a method, as well as a constructive proof of the theorem.

Theorem L6.22 (Spectral Theorem).

(a) *If A is a Hermitian, anti-Hermitian, or unitary matrix, then $\mathbb{C}^n$ has an orthonormal eigenbasis $\mathbf{v}_1, \mathbf{v}_2, \ldots, \mathbf{v}_n$ for A.*

(b) *If A is real symmetric, then the eigenvalues are real and the basis vectors can be chosen real so that $\mathbb{R}^n$ has an orthonormal eigenbasis for A.*

Proof. By Theorem L6.16, there exists a unitary matrix P such that $P^{-1}AP$ is upper-triangular. However, since P is unitary, $P^{-1}AP$ is still Hermitian, anti-Hermitian, or unitary. Clearly, for a Hermitian matrix to be upper-triangular, it must be diagonal *with real entries on the diagonal*; for an anti-Hermitian matrix to be upper-triangular, it must be diagonal *with purely imaginary entries on the diagonal*. Furthermore, since the inverse of an upper-triangular matrix is also upper-triangular, if $P^{-1}AP$ is unitary, then it must be diagonal, this time with entries of absolute value 1 on the diagonal (Exercise L6#8).

Since in all three cases $P^{-1}AP$ turned out to be diagonal, the column vectors of P form an eigenbasis for A, which is orthonormal since P is unitary. The italicized statements, together with the fact that P is unitary, prove Theorem L6.22, part (a).

Part (b) follows from the fact, just proved, that the eigenvalues of a Hermitian matrix, hence of a real symmetric matrix, are real; hence, the eigenvectors can be chosen real.   □

In Exercise L6#9 we will define and prove appropriate properties of *normal* linear transformations. These include all three of the classes above—

unitary, Hermitian, and anti-Hermitian—and allow a unified proof of Theorem L6.22.

L6.6. FACTORING THE CHARACTERISTIC POLYNOMIAL

The characteristic polynomial has already had many uses in this section. The next ones are rather deeper, and techniques along these lines are the most powerful that linear algebra has at its disposal. On the other hand, the characteristic polynomial is more or less uncomputable, and so the results are mainly of theoretical interest.

A first statement is the following:

Theorem L6.23. *Any matrix A satisfies its characteristic polynomial p_A, i.e., $p_A(A) = 0$.*

Example L6.24. The characteristic polynomial of the matrix

$$A = \begin{bmatrix} 1 & 0 & 2 \\ -1 & 0 & 1 \\ 0 & 1 & 0 \end{bmatrix}$$

is $p_A(\lambda) = \lambda^3 - \lambda^2 - \lambda + 3$. If we compute $p_A(A) = A^3 - A^2 - A + 3I$, we find

$$\underbrace{\begin{bmatrix} -1 & 2 & 4 \\ -2 & -2 & -1 \\ -1 & 1 & -2 \end{bmatrix}}_{A^3} - \underbrace{\begin{bmatrix} 1 & 2 & 2 \\ -1 & 1 & -2 \\ -1 & 0 & 1 \end{bmatrix}}_{A^2} - \underbrace{\begin{bmatrix} 1 & 0 & 2 \\ -1 & 0 & 1 \\ 0 & 1 & 0 \end{bmatrix}}_{A} + \underbrace{\begin{bmatrix} 3 & 0 & 0 \\ 0 & 3 & 0 \\ 0 & 0 & 3 \end{bmatrix}}_{3I}$$

$$= \begin{bmatrix} 0 & 0 & 0 \\ 0 & 0 & 0 \\ 0 & 0 & 0 \end{bmatrix} = 0. \quad \blacktriangle$$

Proof of Theorem L6.23. The method of proof is typical of a large class of proofs in this part of linear algebra. First, do it for the diagonal matrices, then the diagonalizable matrices, then the others by approximation.

(i) Suppose A is diagonal, with the eigenvalues λ_i along the diagonal. Then all A^k are also diagonal, with powers of the eigenvalues on the diagonal, so $P_A(A)$ is diagonal, with $P_A(\lambda_i)$ in the ith spot on the diagonal. Since the eigenvalues are the roots of the characteristic polynomial, the result is true.

(ii) Now suppose A is diagonalizable; say, $P^{-1}AP = A_1$ is diagonal. Since by Theorem L5.10 $p_{A_1} = p_A$, we have

$$P^{-1}p_A(A)P = p_A(A_1) = p_{A_1}(A_1) = 0.$$

(iii) Now suppose A is anything. In any case, there exists a sequence of matrices A_i converging to A with all the A_i diagonalizable. Since $A \mapsto P_A(A)$ is a continuous function, $P_A(A) = \lim_{i \to \infty} p_{A_i}(A_i) = 0$. $\quad\square$

Now we come to the main idea of this section: factorizations of the characteristic polynomial. We will be mainly concerned with factorizations into *relatively prime* factors, i.e., factors that have no common roots. Such a factorization corresponds to a *grouping of the roots* of the characteristic polynomial. This will allow the astute reader to isolate the part of the behavior coming from various kinds of eigenvalues.

Theorem L6.25. *Let A be an $n \times n$ matrix, with characteristic polynomial p_A, and suppose that p_A is factored: $p_A = p_1 p_2$, with p_1 and p_2 relatively prime with leading coefficient 1. Let $V_1 = \ker p_1(A)$ and $V_2 = \ker p_2(A)$. Then*

(i) *Any vector $\mathbf{x} \in \mathbb{R}^n$ can be written uniquely as $\mathbf{x} = \mathbf{x}_1 + \mathbf{x}_2$, with $\mathbf{x}_1 \in V_1$ and $\mathbf{x}_2 \in V_2$.*

(ii) *For any vector $\mathbf{x} \in V_i$ we have $A(\mathbf{x}) \in V_i$.*

(iii) *The eigenvectors of A with corresponding eigenvalues roots of p_i are in V_i.*

(iv) *The characteristic polynomial of $A: V_i \to V_i$ is p_i.*

Before proceeding to the proof of this result, we make a few comments, insert an example, and prove a lemma.

What does (iv) mean? After all, $A: V_i \to V_i$ is not a matrix, at least not until a basis has been chosen for V_i. But recall that characteristic polynomials do not depend on the basis, all they require is a linear transformation mapping a vector space to itself; (ii) says that $A: V_i \to V_i$ is such a thing.

Let's move to the example:

Example L6.26. The characteristic polynomial of the matrix

$$A = \begin{bmatrix} 0 & 1 & 0 \\ 0 & 0 & 1 \\ 2 & -1 & 2 \end{bmatrix}$$

is

$$p_A(\lambda) = \lambda^3 - 2\lambda^2 + \lambda - 2 = (\lambda^2 + 1)(\lambda - 2).$$

So we want to examine $V_1 = \ker(A - 2I)$ and $V_2 = \ker(A^2 + I)$. The space $V_1: \ker(A - 2I)$ contains all vectors $\mathbf{x} \in \mathbb{R}^3$ such that $(A - 2I)\mathbf{x} = 0$;

i.e.,

$$(A - 2I)\mathbf{x} = \begin{bmatrix} -2 & 1 & 0 \\ 0 & -2 & 1 \\ 2 & -1 & 0 \end{bmatrix} \begin{bmatrix} x_1 \\ x_2 \\ x_3 \end{bmatrix} = \begin{bmatrix} 0 \\ 0 \\ 0 \end{bmatrix} \Rightarrow \begin{matrix} -2x_1 + x_2 = 0 \\ -2x_2 + x_3 = 0 \\ 2x_1 - x_2 = 0 \end{matrix}$$

$$\Rightarrow \begin{matrix} x_2 = 2x_1 \\ x_3 = 2x_2 \end{matrix} \Rightarrow \begin{bmatrix} x_1 \\ x_2 \\ x_3 \end{bmatrix} = \begin{bmatrix} c \\ 2c \\ 4c \end{bmatrix}.$$

The result for the vector $\mathbf{x}$ shows that with eigenvalue 2, we have but one parameter, a *one-dimensional* space of eigenvectors of A. We have shown that V_1 is a *line*.

$V_2 \colon \ker(A^2 + I)$ contains all vectors $\mathbf{u} \in \mathbb{R}^3$ such that $(A^2 + I)\mathbf{u} = \mathbf{0}$; i.e.,

$$(A^2 + i)\mathbf{u} = \left(\begin{bmatrix} 0 & 1 & 0 \\ 0 & 0 & 1 \\ 2 & -1 & 2 \end{bmatrix} \begin{bmatrix} 0 & 1 & 0 \\ 0 & 0 & 1 \\ 2 & -1 & 2 \end{bmatrix} + \begin{bmatrix} 1 & 0 & 0 \\ 0 & 1 & 0 \\ 0 & 0 & 1 \end{bmatrix} \right) \begin{bmatrix} u_1 \\ u_2 \\ u_3 \end{bmatrix}$$

$$= \begin{bmatrix} 1 & 0 & 1 \\ 2 & 0 & 2 \\ 4 & 0 & 4 \end{bmatrix} \begin{bmatrix} u_1 \\ u_2 \\ u_3 \end{bmatrix} = \begin{bmatrix} 0 \\ 0 \\ 0 \end{bmatrix} \Rightarrow u_1 + u_3 = \mathbf{0},$$

the *only* condition, so vectors in V_2 are of the form

$$\begin{bmatrix} u_1 \\ u_2 \\ u_3 \end{bmatrix} = \begin{bmatrix} b \\ a \\ -b \end{bmatrix}.$$

The fact that we have *two* arbitrary constants a, b for the vector $\mathbf{u}$ shows that with eigenvalue $\pm i$ we have a *two-dimensional* space of eigenvectors of A. We have shown that V_2 is a *plane*, described, for example, by vectors

$$b \begin{bmatrix} 1 \\ 0 \\ -1 \end{bmatrix} \quad \text{and} \quad a \begin{bmatrix} 0 \\ 1 \\ 0 \end{bmatrix}.$$

Of course, there are other ways to derive these results; the above method was chosen to most clearly show what is happening; a more efficient purely matrix method uses row reduction and the dimension Theorem L2.25 (see Exercise L6#10a).

Now we are ready to illustrate the conclusions of Theorem L6.25:

(i) Any vector $\begin{bmatrix} x \\ y \\ z \end{bmatrix}$ in $\mathbb{R}^3$ can be written uniquely as

$$\begin{bmatrix} b \\ a \\ -b \end{bmatrix} + \begin{bmatrix} c \\ 2c \\ 4c \end{bmatrix}$$

with a correct choice of a, b, and c (from three equations in these three unknowns).

(ii)

$$A \begin{bmatrix} c \\ 2c \\ 4c \end{bmatrix} = \begin{bmatrix} 0 & 1 & 0 \\ 0 & 0 & 1 \\ 2 & -1 & 2 \end{bmatrix} \begin{bmatrix} c \\ 2c \\ 4c \end{bmatrix} = \begin{bmatrix} 2c \\ 4c \\ 8c \end{bmatrix} = \begin{bmatrix} c^* \\ 2c^* \\ 4c^* \end{bmatrix},$$

where $c^* = 2c$; hence, the resulting vector belongs to $\ker(A - 2I)$, i.e., A maps $V_1 \to V_1$.

$$A \begin{bmatrix} b \\ a \\ -b \end{bmatrix} = \begin{bmatrix} 0 & 1 & 0 \\ 0 & 0 & 1 \\ 2 & -1 & 2 \end{bmatrix} \begin{bmatrix} b \\ a \\ -b \end{bmatrix} = \begin{bmatrix} a \\ -b \\ -a \end{bmatrix} = \begin{bmatrix} b^* \\ a^* \\ -b^* \end{bmatrix},$$

where $b^* = a$, and $a^* = -b$; hence, the resulting vector belongs to $\ker(A^2 + I)$, i.e., A maps $V_2 \to V_2$.

Exercise L6#10b asks you to verify properties (iii), and (iv) in similar fashion. ▲

The lemma we need is the following, which appears to have nothing to do with linear algebra at all. It is known as Bezout's theorem and is an important result in its own right. Another proof appears as Exercise L2#17.

Lemma L6.27 (Bezout's Theorem). *Two polynomials p_1 and p_2 are relatively prime if and only if there exist polynomials q_1 and q_2 such that $p_1 q_1 + p_2 q_2 = 1$.*

Proof. In one direction, this is clear: if q divides both p_1 and p_2, then it will also divide $p_1 q_1 + p_2 q_2$ for any q_1 and q_2; so q must divide 1, which implies that q must be a constant. This exactly means that p_1 and p_2 are relatively prime.

The converse is more interesting and involves an algorithm that is important in itself: the *Euclidean Algorithm*. Let $\deg(p_1) \geq \deg(p_2)$, and define a finite sequence of polynomials $p_3, p_4, \ldots, p_k$ by division with remainder as follows:

$$p_1 = q_2 p_2 + p_3 \quad \text{with} \quad \deg(p_3) < \deg(p_2),$$
$$p_2 = q_3 p_3 + p_4 \quad \text{with} \quad \deg(p_4) < \deg(p_3),$$
$$\vdots$$

Since the degrees of the p_i are strictly decreasing, this process must end, in fact, in at most $\deg(p_1)$ steps. The only way in which it can end is with $p_{k+1} = 0$ for some k, since 0 is the only polynomial by which you cannot divide.

Consider p_k; it divides p_{k-1}, and hence p_{k-2}, etc., and finally both p_1 and p_2. As such, it must be a constant. On the other hand, all the p_i can be written in the form $s_1 p_1 + s_2 p_2$ for appropriate s_1 and s_2:

$$p_3 = p_1 - q_2 p_2,$$
$$p_4 = p_2 - q_3 p_3 = p_2 - q_3(p_1 - q_2 p_2) = -q_3 p_1 + (1 + q_2 q_3) p_2,$$
$$\vdots$$

This ends the proof: p_k is a constant c which can be written $c = s_1 p_1 + s_2 p_2$; set $q_2 = s_2/c$, $q_2 = s_2/c$ to find $1 = q_1 p_1 + q_2 p_2$. $\quad\square$

Remark. Note that the proof of Lemma L6.27 actually gave a recipe for computing q_1 and q_2 (as well as the greatest common divisor of two polynomials). It can also be used to compute greatest common divisors of integers in precisely the same way.

Proof of Theorem L6.25. (i) Suppose $q_1(\lambda)$ and $q_2(\lambda)$ are two polynomials such that $p_1 q_1 + p_2 q_2 = 1$. Then since $p_1(A)q_1(A) + p_2(A)q_2(A) = I$, we have for any vector $\mathbf{x}$ that

$$\mathbf{x} = p_2(A)q_2(A)\mathbf{x} + p_1(A)q_1(A)\mathbf{x}.$$

Now $p_2(A)q_2(A)\mathbf{x} \in V_1$ since $p_1(A)p_2(A)q_2(A) = p_A(A)q_2(A) = 0$ by Theorem L6.23 and, similarly, $p_1(A)q_1(A)\mathbf{x} \in V_2$. This shows that there is a decomposition as in $\mathbf{x} = \mathbf{x}_1 + \mathbf{x}_2$ with $\mathbf{x}_1 \in V_1$ and $\mathbf{x}_2 \in V_2$.

For uniqueness of the decomposition, it is enough to show $V_1 \cap V_2 = \{0\}$; indeed, if

$$\mathbf{x} = \mathbf{x}_1 + \mathbf{x}_2 = \mathbf{y}_1 + \mathbf{y}_2$$

with $\mathbf{x}_1, \mathbf{y}_1 \in V_1$ and $\mathbf{x}_2, \mathbf{y}_2 \in V_2$, then $\mathbf{x}_1 - \mathbf{y}_1 = \mathbf{y}_2 - \mathbf{x}_2$ is in both V_1 and in V_2. Therefore, if the intersection is 0, we have $\mathbf{x}_1 = \mathbf{y}_1$ and $\mathbf{x}_2 = \mathbf{y}_2$. If $\mathbf{x} \in V_1 \cap V_2$, then

$$\mathbf{x} \in \ker(p_1(A)q_1(A) + p_2(A)q_2(A)) = \ker I = \mathbf{0},$$

so $\mathbf{x} = \mathbf{0}$. This proves (i).

(ii) If $\mathbf{x} \in V_1$, then $p_1(A)\mathbf{x} = \mathbf{0}$ and $p_1(A)A\mathbf{x} = Ap_1(A)\mathbf{x} = \mathbf{0}$.

(iii) Suppose $p_1(\lambda) = 0$, and $A\mathbf{v} = \lambda\mathbf{v}$. Then $p_1(A)\mathbf{v} = p_1(\lambda)\mathbf{v} = \mathbf{0}$, so $\mathbf{v} \in V_1$.

(iv) This part is a bit more unpleasant. If you choose a basis of V_1 and a basis of V_2, then together these form a basis of V, and since A sends each of these spaces into itself, in this basis A has a matrix of the form

$$\begin{bmatrix} A_1 & 0 \\ 0 & A_2 \end{bmatrix}.$$

By Theorem L5.8, the characteristic polynomial p_A can be written q_1q_2, where q_1 and q_2 are the characteristic polynomials of A_1 and A_2, respectively, which in turn are simply the matrices of $A: V_i \to V_i$ with respect to the chosen bases of these subspaces. So we have $p_A = q_1q_2 = p_1p_2$, and we need to show $p_1 = q_1$, $p_2 = q_2$. If we show that no root of q_1 is also a root of p_2 and no root of q_2 is also a root of p_1, we will be done. But any root λ of q_1 is an eigenvalue of $A: V_1 \to V_1$. On the other hand, any eigenvalue of A with corresponding eigenvector in V_1 is a root of p_1, hence not a root of p_2. $\square$

L7   Finding Eigenvalues: The QR Method

As indicated in Chapter 7 and subsequent chapters, the main task in solving linear differential equations with constant coefficients,

$$\mathbf{x}' = A\mathbf{x},$$

is to find the *eigenvectors* and *eigenvalues* of A.

At the end of Section 7.3, we remarked that computing the characteristic polynomial of A, finding the eigenvalues as its roots, and solving for the eigenvectors by solving linear equations turns out to be impractical as soon as the matrix A is at all large. The QR method (and its variants) is a different algorithm, and it is probably the one most often used in applications, at least for nonsymmetric matrices. Although it is much more practical than the characteristic polynomial approach, it still has its problems. For one thing, the explanation and proof are especially lengthy and complicated. Finding eigenvectors and eigenvalues really is hard, and nothing will make it quite simple. Our presentation was inspired by a lecture by Michael Shub.

We will begin by a description of the *power method*, which is a simpler cousin of the QR method.

L7.1. THE "POWER" METHOD

Suppose an $n \times n$ real matrix A has eigenvalues $\lambda_1, \ldots, \lambda_n$ with distinct absolute values, and suppose $|\lambda_1| > |\lambda_i|$, $i = 2, \ldots, n$. We then call λ_1 the dominant eigenvalue.

Take from $\mathbb{R}^n$ any vector $\mathbf{x} \neq 0$, and consider the sequence of unit vectors

$$\mathbf{x}_0 = \frac{\mathbf{x}}{\|\mathbf{x}\|}, \quad \mathbf{x}_1 = \frac{A\mathbf{x}_0}{\|A\mathbf{x}_0\|}, \ldots, \mathbf{x}_{k+1} = \frac{A\mathbf{x}_k}{\|A\mathbf{x}_k\|}.$$

It is trivial (for a computer) to compute the sequence $\mathbf{x}_k$. The main point of the "power" method is the following result.

Theorem L7.1. *In general, the sequence of vectors $\mathbf{x}_{2k}$ converges to a unit eigenvector of A with dominant eigenvalue λ_1.*

Proof. We might as well work in a basis $v_1, \ldots, v_n$ of unit eigenvectors of A. Suppose that $x = c_1 v_1 + \cdots + c_n v_n$, with $c_1 \neq 0$. (This last condition is what "in general" means.) Then

$$A^k x = c_1 \lambda_1^k v_1 + \cdots + c_n \lambda_n^k v_n,$$

and the various terms have very different magnitudes. In fact, if we divide through by $c_1 \lambda_1^k$, we get

$$\frac{A^k x}{c_1 \lambda_1^k} = v_1 + \cdots + \frac{c_n \lambda_n^k}{c_1 \lambda_1^k} v_n,$$

and all the coefficients except the first tend to zero as $k \to \infty$.

Now x_k is a multiple of $A^k x$, so for k large, x_k is close to one of the vectors $+v_1$ or $-v_1$. If $\lambda_1 > 0$, then x_{k+1} is close to the same vector and the sequence x_k converges. If $\lambda_1 < 0$, then x_k or x_{k+1} is close to the opposite vector; in any case, the even terms x_{2k} converge. □

You may wonder why we normalized x_i at each step rather than computing $A^k x$ directly. The proof of Theorem L7.1 shows that the entries of $A^k x$ grow like λ_1^k and will rapidly get out of hand.

The QR method is essentially a way of teasing the other eigenvectors away from the voracious attraction of v_1.

Remark. It may seem strange to find the eigenvectors first and then the eigenvalues, but we shall see in Theorem L7.4 how it works.

L7.2. A DESCRIPTION OF THE QR METHOD

Begin with an invertible matrix A.

(i) Find Q_1 and R_1 so that $A = Q_1 R_1$, where Q_1 is *orthogonal*; R_1 is *upper-triangular*; this is done by the Gram-Schmidt algorithm, Theorem L3.13.

(ii) Now, *reverse the order of the factors* and write $A_1 = R_1 Q_1$; then repeat step (i) using A_1 instead of A.

Note: $A = Q_1 R_1$ implies $Q_1^\top A = R_1$, which implies $A_1 = Q_1^\top A Q_1 = Q_1^{-1} A Q_1$, so A_1 and A have the same eigenvalues, by Theorem L5.7. We continue, with

$$A_{k-1} = Q_k R_k, \quad A_k \equiv R_k Q_k.$$

As we will see in Section L7.7, the matrices A_k miraculously converge under appropriate circumstances to an upper-triangular matrix with the eigenvalues on the diagonal.

Furthermore, a computer program like *MacMath's Eigenfinder* will automatically perform the iterations, so the actual process can be rather painless. The following example illustrates how it works; the iterations continue until all the entries below the diagonal are zero, to within whatever tolerance you choose.

Example L7.2. For the matrix $\begin{bmatrix} -3 & 1 & 2 \\ 0 & 0 & 1 \\ 1 & 0 & 0 \end{bmatrix}$, the iterations of the QR method are shown in Figure L7.1.

Matrix after 1 iterations:

-3.6000	-1.2294	0.16903
0.37416	0.74285	-0.8583
0.00000	-0.2258	-0.1428

Matrix after 5 iterations:

-3.4912	-1.6154	-0.1448
0.00098	0.82984	-0.6142
0.00000	-0.0090	-0.3386

Matrix after 2 iterations:

-3.4656	1.65260	0.33635
0.06741	0.86308	-0.4648
0.00000	0.14673	-0.3974

Matrix after 6 iterations:

-3.4907	1.61496	0.16243
0.00023	0.83608	-0.6014
0.00000	0.00378	-0.3453

Matrix after 3 iterations:

-3.4974	-1.6029	-0.0871
0.01764	0.81133	-0.6563
0.00000	-0.0514	-0.3138

Matrix after 7 iterations:

-3.4908	-1.6153	-0.1551
0.00005	0.83346	-0.6067
0.00000	-0.0015	-0.3425

Matrix after 4 iterations:

-3.4893	1.61600	0.18752
0.00405	0.84385	-0.5826
0.00000	0.02271	-0.3545

Eigenvalues:
-3.4909
0.8343
-0.3434

FIGURE L7.1. QR iteration of a matrix.

The program stops after eight iterations because the subdiagonal elements are all less than the chosen tolerance of 0.0001.

The eigenvalues have been boxed on the main diagonal and have rounded values (using more decimal places than shown on the diagonal) of -3.4909, 0.8343, and -0.3434. ▲

L7.3. THE MAIN CONVERGENCE RESULT

Theorem L7.3. *Suppose A has n real eigenvalues, with distinct absolute values. Then the matrices A_k converge as $k \to \infty$ to an upper-triangular matrix with the eigenvalues of A on the diagonal.*

The proof of Theorem L7.3 requires laying quite a bit of groundwork, to which we shall shortly proceed. We will be able to explain why this result is true in Section L7.7. Meanwhile, we shall first make some observations.

There is more to be said than Theorem L7.3 gives. First, the requirement that the eigenvalues, or for that matter the matrix A, be real is unnecessary. If A is complex, then Q should be chosen unitary, and Gram–Schmidt works exactly the same way; the reference is Appendix L3.

Unfortunately, if A is real and has complex eigenvalues, then they never have distinct absolute values since they come in conjugate pairs. It turns out that all is still not lost: a pair of complex conjugate eigenvectors leads to matrices A_k with a 2×2 submatrix M along the diagonal. The *entries* in M do not converge as the algorithm proceeds; however, the pair of complex eigenvalues are then the roots of the characteristic polynomial of M. It turns out that this *characteristic polynomial* does converge even though the entries of the submatrix do not, and characteristic polynomials for 2×2 matrices are easy to compute: they always have the form

$$\lambda^2 - (\operatorname{tr} M)\lambda + (\det M), \tag{1}$$

as discussed in Chapter 7, Section 5. We do not here prove these results for the complex case, but for completeness we state the theorem that explains how the QR algorithm extends to complex eigenvalues.

Theorem L7.4. *Suppose A has n complex (including real) eigenvalues, with distinct absolute values, except for complex conjugate pairs. Then the matrices A_k tend as $k \to \infty$ to almost upper-triangular matrices. The exceptions are 2×2 submatrices along the diagonal, which yield the nonreal complex eigenvalues of A in conjugate pairs, as solutions to the characteristic polynomials of these 2×2 submatrices. The real eigenvalues of A lie on the diagonal.*

Example L7.5. Find the eigenvalues for the matrix

$$\begin{bmatrix} 0 & 1 & 0 & 0 & 0 & 0 \\ 0 & 0 & 1 & 0 & 0 & 0 \\ 0 & 0 & 0 & 1 & 0 & 0 \\ 0 & 0 & 0 & 0 & 1 & 0 \\ 0 & 0 & 0 & 0 & 0 & 1 \\ 1 & 1 & 4.25 & 4 & .75 & 0 \end{bmatrix},$$

corresponding to the characteristic polynomial

$$\lambda^6 = .75\lambda^4 + 4\lambda^3 + 4.25\lambda^2 + \lambda + 1.$$

The results of the computer program *Eigenfinder* are shown in Figure L7.2.

Original matrix:

0.00000	1.00000	0.00000	0.00000	0.00000	0.00000
0.00000	0.00000	1.00000	0.00000	0.00000	0.00000
0.00000	0.00000	0.00000	1.00000	0.00000	0.00000
0.00000	0.00000	0.00000	0.00000	1.00000	0.00000
0.00000	0.00000	0.00000	0.00000	0.00000	1.00000
1.00000	1.00000	4.25000	4.00000	0.75000	0.00000

Matrix after 20 iterations:

1.99962	-0.9375	2.89401	-1.3520	1.28155	3.54095
0.00557	-0.0794	-1.6945	0.48031	-0.6236	-1.3997
0.00000	1.13375	-0.9200	0.42265	-0.3620	-1.0748
0.00000	0.00000	0.00148	-1.0001	0.49814	0.41483
0.00000	0.00000	0.00000	0.00000	-0.0715	-1.3578
0.00000	0.00000	0.00000	0.00000	0.18788	0.07151

Matrix after 10 iterations:

2.04994	-2.1280	2.31694	-1.2635	-1.3134	-3.6190
0.15048	-0.1326	-1.1163	0.11214	0.30364	0.54856
0.00000	1.55062	-0.9219	0.51184	0.57989	1.50513
0.00000	0.00000	0.04116	-0.9960	-0.4967	-0.4104
0.00000	0.00000	0.00000	0.00201	-0.0705	-1.3571
0.00000	0.00000	0.00000	0.00000	0.18771	0.07126

Matrix after 30 iterations: (almost done)

1.99992	0.13775	3.03348	-1.3534	-1.2800	-3.5373
0.00019	-0.3692	-1.9079	0.60336	0.71417	1.69643
0.00000	0.92608	-0.6306	0.22650	0.11869	0.51248
0.00000	0.00000	0.00005	-1.0000	-0.4978	-0.4144
0.00000	0.00000	0.00000	0.00000	-0.0715	-1.3578
0.00000	0.00000	0.00000	0.00000	0.18788	0.07151

FIGURE L7.2.

Working down the boxes of *Eigenfinder's* final matrix, we see, with the help of the trace and determinant expression (1), that the matrix A has

a real eigenvalue $1.9999 \approx 2$;
complex eigenvalues satisfying $\lambda^2 + 0.9998\lambda + 2.0000 \approx \lambda^2 + \lambda + 2 = 0$;
a real eigenvalue $-0.9999 \approx -1$;
complex eigenvalues satisfying $\lambda^2 - 0\lambda + .2499 \approx \lambda^2 + \frac{1}{4} = 0$.

Therefore, the eigenvalues are 2, $\frac{-1 \pm \sqrt{7}i}{2}$, -1, $\pm \frac{1}{2}i$. These calculations, in fact, match Eigenfinder's numerical listing:

$$2, \quad -0.5 \pm 1.3229i, \quad -1, \quad \pm 0.5i.$$

L7.4. Flags

Unfortunately, although the QR method is quite easy to implement, it is fairly complicated to understand why it works. The key notion is the following:

A *flag* of $\mathbb{R}^n$ is an ascending sequence of subspaces

$$F = (U_1 \subset U_2 \subset \cdots \subset U_n),$$

with each subspace U_i of dimension i, and therefore $U_n = \mathbb{R}^n$.

A flag may appear to be a nebulous notion, but it is quite easy to work with in practice. For instance, if the columns $\mathbf{a}_1 \ldots, \mathbf{a}_n$ of a matrix A are linearly independent, they form a basis of $\mathbb{R}^n$; we may consider the flag F_A defined as above with the subspaces defined as follows: each U_i is spanned by $\mathbf{a}_1, \ldots, \mathbf{a}_i$; this flag is called the *flag associated to the basis*.

$$F_A = \{\mathrm{sp}\{\mathbf{a}_1\} \subset \mathrm{sp}\{\mathbf{a}_1, \mathbf{a}_2\} \subset \cdots \subset \mathrm{sp}\{\mathbf{a}_1, \mathbf{a}_2, \ldots, \mathbf{a}_n\}\}.$$

Of course, many bases have the same associated flag.

Example L7.6. Consider $A = \begin{pmatrix} 1 & 2 & 3 \\ 4 & 5 & 6 \\ 7 & 8 & 9 \end{pmatrix}$ and $B = \begin{pmatrix} 1 & 10 & 31 \\ 4 & 28 & 73 \\ 7 & 46 & 115 \end{pmatrix}$. We can write

$$F_A = \left\{ \mathrm{sp}\left\{ \begin{pmatrix} 1 \\ 4 \\ 7 \end{pmatrix} \right\} \subset \mathrm{sp}\left\{ \begin{pmatrix} 1 \\ 4 \\ 7 \end{pmatrix}, \begin{pmatrix} 2 \\ 5 \\ 8 \end{pmatrix} \right\} \subset \mathbb{R}^3 \right\},$$

$$F_B = \left\{ \mathrm{sp}\left\{ \begin{pmatrix} 1 \\ 4 \\ 7 \end{pmatrix} \right\} \subset \mathrm{sp}\left\{ \begin{pmatrix} 1 \\ 4 \\ 7 \end{pmatrix}, \begin{pmatrix} 10 \\ 28 \\ 46 \end{pmatrix} \right\} \subset \mathbb{R}^3 \right\}.$$

The first and last entries in each flag are the same; if we can show that the middle entries each span the same plane, we will have shown that the flags are the same. That is, can we find numbers a and b such that

$$a \begin{pmatrix} 1 \\ 4 \\ 7 \end{pmatrix} + b \begin{pmatrix} 2 \\ 5 \\ 8 \end{pmatrix} = \begin{pmatrix} 10 \\ 28 \\ 46 \end{pmatrix}?$$

Yes: $a = 2$, $b = 4$. Therefore, $F_A = F_B$. ▲

In fact, if A is the matrix with ith column $\mathbf{a}_i$, then B has the same flag associated to its columns as A *if and only if* $B = AR$, with R an invertible upper-triangular matrix. The proof of this fact involves a straightforward calculation in Exercise L7#2, which you should consider compulsory and central to getting a feel for the heart of the QR method. An essential implication is the following: *If you apply Gram–Schmidt to any basis, the vectors of the new orthonormal basis in the same order produce the same flag.*

We will need the following special cases of flags:

(1) The flag associated to the standard basis is called the *base flag* F_0.

(2) Let A be a square diagonalizable $n \times n$ matrix and let $\mathbf{v}_1, \ldots, \mathbf{v}_n$ be an eigenbasis of $\mathbb{R}^n$ for A. Then the flag associated to $\mathbf{v}_1, \ldots, \mathbf{v}_n$ is called an *eigenflag* of A. There are lots $(n!)$ of eigenflags, depending on the order in which the eigenvectors are arranged.

(3) If the eigenvalues $\lambda_1, \ldots, \lambda_n$ corresponding to $\mathbf{v}_1, \ldots, \mathbf{v}_n$ have distinct absolute values and are ordered from largest to smallest, i.e.,

$$|\lambda_1| > |\lambda_2| > \cdots > |\lambda_n|,$$

then the associated flag Λ_A is called the *dominant eigenflag* of A. If they are ordered from smallest to largest, we get the *recessive eigenflag* Λ_A^* of A.

L7.5. OPERATIONS OF LINEAR TRANSFORMATIONS ON FLAGS

If $F = (U_1 \subset \cdots \subset U_n)$ is a flag of $\mathbb{R}^n$ and A is an invertible matrix, then $A(U_i)$ is the subspace generated by the vectors $A\mathbf{v}_1, A\mathbf{v}_2, \ldots, A\mathbf{v}_i$, and

$$A(U_1) \subset A(U_2) \subset \cdots \subset A(U_n)$$

is another flag, which we will write AF.

Although it does not look as if it has anything to do with it, the following statement really explains the QR method.

Theorem L7.7. *Let A be an invertible diagonalizable matrix whose eigenvalues have distinct absolute values. If*

$$F = (U_1 \subset U_2 \subset \cdots \subset U_n)$$

is any flag, then, in general, $A^k F$ converges to Λ_A as $k \to \infty$.

Remarks. The phrase "in general" means that if

$$\Lambda_A^* = (W_1^* \subset W_2^* \subset \cdots)$$

is the recessive flag, then $U_i \cap W_{n-i}^* = 0$ for all $i = 1, \ldots, n$. If any flag is jiggled slightly, it will satisfy this "general" requirement. Example L7.8 shows the geometric significance of the phrase "in general" if $n = 3$.

Example L7.8. For a 3×3 invertible diagonalizable matrix with distinct eigenvalues, the geometry of the "general" requirement is illustrated in Figure L7.3.

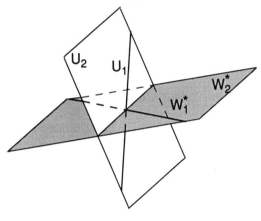

FIGURE L7.3.

In the case $n = 3$, the "general" requirement amounts to only two sentences:

$$U_1 \cap W_2^* = 0,$$

which means that the line U_1 intersects the plane W_2^* only at the point 0; similarly,

$$U_2 \cap W_1^* = 0$$

means that the plane U_2 can only intersect W_1^* at the point 0. That is, the plane U_2 can only intersect W_2^* in a line $\neq W_1^*$.

Although we will not show this, as long as the eigenvalues of A have distinct absolute values, then for any flag F, even without the "general" requirement, $A^k F$ *will still converge.* However, the limit will not be the dominant eigenflag if F fails to satisfy the condition above.

Proof of Theorem L7.7. We might as well work in a basis $\mathbf{v}_1, \ldots, \mathbf{v}_n$ of eigenvectors ordered so as to give the dominant eigenflag. Choose a basis $\mathbf{x}_1, \ldots, \mathbf{x}_n$ whose associated flag is F; in coordinates,

$$\mathbf{x}_i = \begin{bmatrix} x_{1,i} \\ \vdots \\ x_{n,i} \end{bmatrix}.$$

The "in general" requirement says that $x_{i,i} \neq 0$, $i = 1, \ldots, n$. In the basis of eigenvectors, it is easy to see what $A^k F$ is: it is the flag associated to the basis

$$\begin{bmatrix} \lambda_1^k x_{1,1} \\ \lambda_2^k x_{2,1} \\ \vdots \\ \lambda_n^k x_{n,1} \end{bmatrix}, \begin{bmatrix} \lambda_1^k x_{1,2} \\ \lambda_2^k x_{2,2} \\ \vdots \\ \lambda_n^k x_{n,2} \end{bmatrix}, \ldots, \begin{bmatrix} \lambda_1^k x_{1,n} \\ \lambda_2^k x_{2,n} \\ \vdots \\ \lambda_n^k x_{n,n} \end{bmatrix}.$$

The coordinates of the vectors above have very different magnitudes; in order to see this more clearly, divide the ith vector by $\lambda_i^k x_{i,i}$. These new vectors give the same flag, but this time they look like

$$\begin{bmatrix} 1 \\ * \\ * \\ \vdots \\ * \end{bmatrix}, \begin{bmatrix} \# \\ 1 \\ * \\ \vdots \\ * \end{bmatrix}, \ldots, \begin{bmatrix} \# \\ \# \\ \# \\ \vdots \\ 1 \end{bmatrix},$$

where $*$ represents an entry that becomes arbitrarily small as $k \to \infty$, and $\#$ represents an entry that is probably very large but its magnitude is irrelevant.

The flag associated to these vectors is very nearly the dominant eigenflag of A. The first vector nearly points in the direction of $\mathbf{v}_1$. The second vector is a large multiple of $\mathbf{v}_1$ plus a vector which nearly points in the direction of $\mathbf{v}_2$, so the first and second vectors nearly span the subspace spanned by $\mathbf{v}_1$ and $\mathbf{v}_2$, and so forth. □

We shall see in Section L7.7 that while Theorem L7.7 is true theoretically, *numerically* it does not lead to a productive algorithm unless the vectors are orthogonal. Without orthogonality, the vectors can get too wild; e.g., if the first $\#$ above the diagonal is huge, you could divide the second vector by that number and it would be very close to the first vector.

L7.6. THE RELATION BETWEEN EIGENFLAGS AND EIGENVALUES

There is a good reason for wanting to find an eigenflag for a matrix A.

Theorem L7.9. *If $\mathbf{w}_1, \ldots, \mathbf{w}_n$ is a basis of $\mathbb{R}^n$ with the associated flag an eigenflag of A, then the matrix of A in that basis is upper-triangular with the eigenvalues on the diagonal.*

Proof. Let $\mathbf{v}_1, \ldots, \mathbf{v}_n$ be eigenvectors such that $\mathbf{w}_i$ is a linear combination of $\mathbf{v}_1, \ldots, \mathbf{v}_i$, say

$$\mathbf{w}_i = b_1 \mathbf{v}_1 + \cdots + b_i \mathbf{v}_i. \tag{2}$$

Then

$$A\mathbf{w}_i = b_1\lambda_1\mathbf{v}_1 + \cdots + b_i\lambda_i\mathbf{v}_i$$

from equation (2) for the ith term

$$= b_1\lambda_1\mathbf{v}_1 + \cdots + b_{i-1}\lambda_{i-1}\mathbf{v}_{i-1} - \overbrace{\lambda_i(b_1\mathbf{v}_1 + \cdots + b_{i-1}\mathbf{v}_{i-1})} + \lambda_i\mathbf{w}_i \ .$$

In this last expression, all the terms except the last are linear combinations of $\mathbf{v}_1, \ldots, \mathbf{v}_{i-1}$, i.e., of $\mathbf{w}_1, \ldots, \mathbf{w}_{i-1}$, so the ith coefficient of $A\mathbf{v}_i$ is λ_i and the jth coefficient is zero for $j > i$. Therefore,

$$\begin{bmatrix} \uparrow & \uparrow & & \uparrow \\ \mathbf{w}_1 & \mathbf{w}_2 & \cdots & \mathbf{w}_i \\ \downarrow & \downarrow & & \downarrow \end{bmatrix}^{-1} A \begin{bmatrix} \uparrow & \uparrow & & \uparrow \\ \mathbf{w}_1 & \mathbf{w}_2 & \cdots & \mathbf{w}_i \\ \downarrow & \downarrow & & \downarrow \end{bmatrix}$$

$$= \begin{bmatrix} \lambda_1 & - & - & - & - \\ 0 & \lambda_2 & - & - & - \\ \vdots & 0 & \lambda_3 & - & - \\ \vdots & \vdots & \vdots & \ddots & - \\ 0 & 0 & 0 & 0 & \lambda_n \end{bmatrix} . \quad \square$$

L7.7. THE QR METHOD

The strategy of the QR method is as follows: For a matrix A, take the base flag F_0, and define $F_k \equiv A^k F_0$.

For example,

$$F_0 = \{\mathrm{sp}\{\mathbf{e}_1\} \subset \mathrm{sp}\{\mathbf{e}_1, \mathbf{e}_2\} \subset \cdots \subset \mathrm{sp}\{\mathbf{e}_1, \mathbf{e}_2, \ldots, \mathbf{e}_n\}\},$$
$$F_1 = \{\mathrm{sp}\{A\mathbf{e}_1\} \subset \mathrm{sp}\{A\mathbf{e}_1, A\mathbf{e}_2\} \subset \cdots \subset \mathrm{sp}\{A\mathbf{e}_1, A\mathbf{e}_2, \ldots, A\mathbf{e}_n\}\} \quad (3)$$
$$= \{\mathrm{sp}\{\mathbf{a}_1\} \subset \mathrm{sp}\{\mathbf{a}_1, \mathbf{a}_2\} \subset \cdots \subset \mathrm{sp}\{\mathbf{a}_1, \mathbf{a}_2, \ldots, \mathbf{a}_n\}\}.$$

This means that F_0 and F_1 *both serve as flags for A, but with different bases.* By Theorem L7.7, F_k converges to the dominant eigenflag in general.

There is an obvious way to find the eigenflag: F_k is the flag associated to the columns of A^k. Unfortunately, this approach (from Theorem L7.7) does not work numerically, because, as we saw in Section L7.1, the entries of A^k get out of hand. Even if we scale the columns to be of unit length (this still has the same associated flag), it still does not work. All columns tend to multiples of the dominant eigenvector, so that although it is true that the plane spanned by the first two columns does tend to the plane of the dominant eigenflag, the spanning vectors are separated by a smaller and smaller angle, and it becomes difficult to tell them apart with finite precision.

A more reasonable idea is to take the base flag F_0, compute

$$F_1 = AF_0,$$

and then describe F_1 by an orthogonal basis (from Gram–Schmidt). Continue this way, each time computing F_k by applying A to an *orthonormal* basis whose associated flag is F_{k-1} and reencoding F_k by an orthonormal basis. This is *exactly* what the QR method does, but we have to prove it:

Theorem L7.10. *Let* $A = Q_1 R_1$, $A_1 = R_1 Q_1$, $A_1 = Q_2 R_2, \ldots$. *The flag associated to the orthogonal matrix*

$$P_k = Q_1 Q_2 \cdots Q_k$$

is $F_k = A^k F_0$, *and* $A_k = P_k^{-1} A P_k$ *is the linear transformation A written in the basis formed by the columns of P_k. In particular, this means that A_k has the same eigenvalues as A.*

Proof. The proof is by induction on k. The base step is for $k = 1$ with $P_1 = Q_1$. By Gram–Schmidt, we can write $A = Q_1 R_1$, and then we see that

$$A_1 = R_1 Q_1 = Q_1^{-1} A Q_1$$

is actually A in a new orthonormal basis. We showed in (3) that the flags F_0 and $A F_0 = F_1$ both serve as flags (by different bases) for A. The matrices A and Q_1 have the same flag, by Exercise L7#2d. Therefore, $A F_0$ indeed is a flag in the orthonormal basis of Q_1.

Next suppose that the theorem is true for $k - 1$, that the flag for Q_{k-1} is $A^{k-1} F_0$, and $A_{k-1} = P_{k-1}^{-1} A P_{k-1}$. Writing, by the definition of Q_k and R_k,

$$A_{k-1} = Q_k R_k$$

means that Q_k has the same associated flag as A_{k-1}; by expressions similar to (3), we can show that the flags F_k and $A(F_{k-1})$ both serve (by different bases) for A_k. On the other hand,

$$A_k = R_k Q_k = Q_k^{-1} A_{k-1} Q_k$$

shows that A_k is A_{k-1} written in the orthonormal basis of columns of Q_k, so for k we indeed have that the flag $F_k = A^k F_0$. So when we get done, by Theorem L7.9,

$$A_k \to \begin{bmatrix} \lambda_1 & - & - & - & - \\ 0 & \lambda_2 & - & - & - \\ \vdots & 0 & \lambda_3 & - & - \\ \vdots & \vdots & \vdots & \ddots & - \\ 0 & 0 & 0 & 0 & \lambda_n \end{bmatrix}. \quad \square$$

The combination of Theorems L7.10 and L7.9 proves Theorem L7.3 at least if F_0 satisfies the "general" condition about the recessive eigenflag of A, the general case.

L7.8. ACTUAL COMPUTATIONS WITH QR

We have proved in Section L7.7 that the QR method will find the eigenvalues of A, at least for the usual case of distinct eigenvalues. The flags, which are essential for the proofs, are not used in the actual calculations. The computational algorithm simply is as follows:

(i) Find Q_k by Gram–Schmidt on A_{k-1}, so that $A_{k-1} = Q_k R_k$.

(ii) Find $A_k = Q_k^\top A_{k-1} Q_k$, because $A_k = R_k Q_k = Q_k^{-1} A_{k-1} Q_k = Q_k^\top A_{k-1} Q_k$.

(iii) Repeat steps (i) and (ii) until A_k is *sufficiently upper-triangular*.

The phrase "sufficiently upper-triangular" means the following: First, "zero" in practice means below a preset tolerance (like 0.001), and we are aiming at that for all elements below the diagonal. But we must also allow for the possibility of complex eigenvalues, which leave 2×2 submatrices on the diagonal rather than in strictly upper-triangular form. In those cases, as we have already noted in Section L7.3 and Example L7.5, it is simple to compute the eigenvalues of a submatrix from its characteristic polynomial; we note here the same truth even if the eigenvalues are real. So, in practice, we simply continue QR until the subdiagonal of A_k has no two consecutive nonzero elements. Then for each 2×2 submatrix, we form the characteristic polynomial from the trace and the determinant and find its eigenvalues; any remaining eigenvalues will be real and already lying on the diagonal of the overall matrix.

There are matrices for which QR does not work. Some examples are given in Exercise L7#8. One is the case of repeated eigenvalues (which need more eigenvectors than the method can find). Another possibility is a matrix with distinct eigenvalues that nevertheless have the same absolute value (which has the effect that none can dominate).

Nevertheless, "in general," the QR method does its job well. Moreover, there is yet an extra twist that enormously improves the computational efficiency of the QR algorithm—the use of *Hessenberg matrices*. This is what the program *Eigenfinder* actually does, and the remaining sections of this appendix will explain it.

L7.9. HESSENBERG MATRICES IN QR

Each iteration of the QR method involves doing Gram–Schmidt to the columns of a square $n \times n$ matrix, and multiplying two $n \times n$ matrices. A bit of checking will show that each of these computations takes Cn^3 operations, where C is a constant not depending on n (and actually only well defined after a unit has been chosen). In this section, we will describe a trick to reduce each of these operations to be of order n^2. The underlying mathematics is not actually very different, but the efficiency of the calculation is so much greater that many numerical analysts would say this is the heart of the QR method.

As we described the QR method, we started with an invertible matrix A and the standard flag F_0, and considered the images $F_1 = A(F_0)$, $F_2 = A(F_1), \ldots$, claiming that it converged in general to the dominant eigenflag. The idea behind the Hessenberg calculation is to start with a flag H_A better adapted to the situation.

A *Hessenberg matrix* is a matrix that has nonzero entries only on or above the *subdiagonal*, the diagonal line beneath the main diagonal.

The neat thing is that a Hessenberg matrix can be reduced to an upper-triangular matrix R efficiently. You can remove the subdiagonal elements one at a time by carefully chosen rotations; the product of these rotations gives the Q in QR.

Furthermore, everything we do with a Hessenberg matrix will remain Hessenberg: QR and RQ, so if we can just get H_A to start, we can continue to find all Q_k by the simple product-of-rotations process just mentioned.

Product-Rotation Algorithm

Given a Hessenberg matrix H on which to perform the QR method, we can focus on $R = Q^{-1}H = Q^\top H$ and find $Q^\top$ by a product of rotations that gradually lead to the required upper-triangular form for R. We shall find that

$$Q^\top H = S(n-1, \theta_{n-1}) \cdot \ldots \cdot S(1, \theta_1)H,$$

where

$$S(i, \theta_i) \equiv \begin{pmatrix} 1 & 0 & \cdots & 0 & 0 & 0 & \cdots & 0 \\ 0 & 1 & \cdots & 0 & 0 & 0 & \cdots & 0 \\ \cdots & \cdots & \cdots & \cdots & & \cdots & \cdots & 0 \\ 0 & 0 & \cdots & \cos\theta & -\sin\theta & \cdots & \cdots & 0 \\ 0 & 0 & \cdots & \sin\theta & \cos\theta & \cdots & \cdots & 0 \\ \cdots & \cdots & \cdots & 0 & 0 & 1 & \cdots & 0 \\ \vdots & \vdots & \vdots & \vdots & \vdots & \vdots & \vdots & \vdots \\ 0 & 0 & \cdots & \cdots & & \cdots & 0 & \cdots & 1 \end{pmatrix} \begin{matrix} \\ \\ \\ \cdots i \\ \cdots i+1 \\ \\ \\ \\ \end{matrix}$$

with columns i and $i+1$ indicated above.

That is $S(i, \theta_i)$ represents the rotation by θ_i in the $(i, i+1)$ plane. In particular, $S(i, \theta_i)$ is orthogonal, and $(S(i, \theta_i))^{-1} = (S(i, \theta_i))^\top = S(i, -\theta_i)$.

The product-rotation algorithm begins with $i = 1$ and proceeds through $i = n - 1$, at which point all the subdiagonal elements will have been cleared.

The idea is to use $S(i, \theta_i)$ to cancel the ith term on the subdiagonal, after the first $i - 1$ terms on the subdiagonal of H have vanished. You can accomplish this by solving for θ_i whatever equation results from the multiplication for that particular ith entry on the subdiagonal (and you can choose either of the two solutions that occur), or by more formally setting

$$\sin\theta = \frac{b_{i+1,i}}{\sqrt{b_{i,i}^2 + b_{i+1,i}^2}}, \quad \cos\theta = \frac{-b_{i,i}}{\sqrt{b_{i,i}^2 + b_{i+1,i}^2}}.$$

A careful look at Figure L7.4 should convince the reader that the product $S(i,\theta_i)H$ is indeed one step closer to upper-triangular form, with the first i terms on the subdiagonal vanishing.

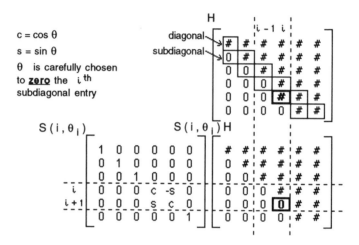

FIGURE L7.4. A step in the middle of QR for a Hessenberg matrix H.

Theorem L7.11. *Given H an $n \times n$ matrix in Hessenberg form, there exists a sequence of rotations $Q^{\top} = S(n-1, \theta_{n-1}) \cdot \ldots \cdot S(1,\theta_1)$ as defined above such that $Q^{\top} H$ is an upper-triangular matrix R and the matrix $H = RQ = Q^{\top} H Q$ is also Hessenberg.*

Proof. The first half of the theorem is proven by the construction of the rotations $S(i,\theta_i)$ as described above; the second half is left to the reader. See Exercises L7#5 and L7#7. □

Note that we never actually find Q; we keep the $S(i,\theta_i)$. It is faster to multiply by all the $S(i,\theta_i)$ than it is to multiply them together.

Thus, once we start with a Hessenberg matrix, the entire QR method gives a sequence H_k of Hessenberg matrices, for which it is very much easier to find Q's. The product-rotation algorithm is far more efficient than Gram–Schmidt, which we have already noted will take an order of n^3 calculations to compute for each Q_k.

Theorem L7.12. *Multiplying an $n \times n$ matrix by a matrix of the form $S(i, \theta_i)$, on the right or on the left, requires $4n$ multiplications.*

The proof is left to the reader. Note how much smaller $4n$ is than n^3.

Theorem L7.13. *Given H an $n \times n$ matrix in Hessenberg form, the sequence of rotations $S(n-1, \theta_{n-1}) \cdot \ldots \cdot S(1, \theta_1)$ as defined above such that $Q^\top H$ is an upper-triangular matrix R can be found using on the order of n^2 operations.*

We leave the proof as Exercise L7#6.

The final thing we need to do is to get a Hessenberg matrix to start.

Hessenberg Algorithm

We must find a basis $\mathbf{h}_1, \ldots, \mathbf{h}_n$ such that $A(\mathbf{h}_i)$ is in the span of $\mathbf{h}_1, \ldots, \mathbf{h}_{i+1}$ for all $i = 1, \ldots, n$. We will, in fact, find an orthonormal basis with this property. The Hessenberg algorithm to do this is very similar to the Gram–Schmidt process, Theorem L3.13.

(1) Start with any unit vector $\mathbf{h}_1$, for instance the standard basis vector $\mathbf{e}_1$

$$\mathbf{u}_2 = A(\mathbf{h}_1) - (A(\mathbf{h}_1) \cdot \mathbf{h}_1)\mathbf{h}_1, \qquad\qquad \mathbf{h}_2 = \mathbf{u}_2/\|\mathbf{u}_2\|,$$
$$\mathbf{u}_3 = A(\mathbf{h}_2) - (A(\mathbf{h}_2) \cdot \mathbf{h}_1)\mathbf{h}_1 - (A(\mathbf{h}_2) \cdot \mathbf{h}_2)\mathbf{h}_2, \qquad \mathbf{h}_3 = \mathbf{u}_3/\|\mathbf{u}_3\|,$$
$$\vdots$$

until either you have a basis, which will usually be the case, or until some $\mathbf{u}_i = \mathbf{0}$. If this happens, actually something rather nice has happened, namely, the subspace spanned by $\mathbf{h}_1, \ldots, \mathbf{h}_{i-1}$ is sent into itself by A and that part, which has lower dimension, can be studied separately. In any case, we then get to freely choose another vector, say $\mathbf{e}_2$, which we must then make orthogonal to all those chosen so far. If this leads to $\mathbf{0}$ again (i.e., if $\mathbf{e}_2$ was also in the span of $h_1, \ldots, h_{i-1}$), then try $\mathbf{e}_3$, and so forth. You will certainly find a vector not in this span eventually, since any i vectors in the span of $\mathbf{h}_1, \ldots, \mathbf{h}_{i-1}$ are linearly dependent, and the standard basis vectors are linearly independent.

(2) Let P be the change-of-basis matrix with the $\mathbf{h}_i$ as columns and set

$$H_A = P^{-1}AP.$$

You now have a Hessenberg matrix H_A, which has the same eigenvalues as A. Performing QR on H_A is efficient. The summary of Section L7.8 becomes (in *Eigenfinder*) as follows:

(i) Find H_A by the Hessenberg algorithm giving $P = (\mathbf{h}_1, \mathbf{h}_2, \ldots, \mathbf{h}_n)$, $H_A = P^{-1}AP$.

(ii) Find $Q_k^\top = S(n - 1, \theta_{n-1}) \cdot \ldots \cdot S(1, \theta_1)$ by the product-rotation algorithm; this makes an upper-triangular $R_k = Q_k^\top H_{k-1}$, so that $H_{k-1} = Q_k R_k$. (Start with $R_1 = Q_1^\top H_A$.)

(iii) Find $H_k = Q_k^\top H_{k-1} Q_k = R_k Q_k$,

(iv) Repeat steps (ii) and (iii) until H_k is *sufficiently upper-triangular.*

L8 Finding Eigenvalues: Jacobi's Method

Jacobi's method is a reasonably efficient algorithm for finding eigenvalues for a symmetric matrix, it is not as hard to understand as the QR algorithm, and it provides a constructive proof of the *Spectral Theorem* L6.22. We restate it here for the case we shall discuss.

Theorem L8.1 (Spectral Theorem). *If A is a real symmetric matrix, the eigenvalues are real. Furthermore, $\mathbb{R}^n$ has an orthonormal basis $\mathbf{v}_1, \mathbf{v}_2, \ldots, \mathbf{v}_n$ of eigenvectors of A.*

The algorithm is not everybody else's method of choice at present. We choose it because it is fairly easy to analyze, and it corresponds to operations that can be geometrically motivated. We suspect also that with the advent of parallel computation, it will regain favor in large-scale computations and may become the method of choice in the not too distant future.

L8.1. JACOBI'S METHOD: THE 2×2 CASE

Lemma L8.2. *If A is the symmetric 2×2 matrix $\begin{bmatrix} a & b \\ b & d \end{bmatrix}$ and*

$$\tan 2\theta = \frac{2b}{a - d},$$

then

$$Q(-\theta) A Q(\theta) = \begin{bmatrix} \lambda_1 & 0 \\ 0 & \lambda_2 \end{bmatrix},$$

that is, a diagonal matrix, with the eigenvalues λ_1, λ_2 along the diagonal.

The proof of Lemma L8.2 consists of carrying out the multiplication $Q(-\theta) A Q(\theta)$ and observing that if θ satisfies the condition of the theorem, the off-diagonal terms of the product vanish. Exercise L8#1 shows how the computer deals with this calculation.

L8.2. JACOBI'S METHOD: THE $n \times n$ CASE

1. *The basic step.*

Lemma L8.3. *Let A be now any $n \times n$ symmetric matrix, with entries $a_{i,j}$, and let $Q(i, \theta_i) =$*

$$
\begin{matrix}
& & & i & & & & j & & \\
& & & \vdots & & & & \vdots & & \\
\end{matrix}
$$

$$
\begin{pmatrix}
1 & 0 & \cdots & 0 & 0 & \cdots & 0 & \cdots & 0 & 0 & \cdots & 0 \\
0 & 1 & \cdots & 0 & 0 & \cdots & 0 & \cdots & 0 & 0 & \cdots & 0 \\
\vdots & \vdots & \ddots & 0 & 0 & \cdots & 0 & \cdots & 0 & 0 & \cdots & 0 \\
0 & 0 & \cdots & 1 & 0 & \cdots & 0 & \cdots & 0 & 0 & \cdots & 0 \\
0 & 0 & \cdots & 0 & \cos\theta & \cdots & 0 & \cdots & -\sin\theta & 0 & \cdots & 0 \\
\vdots & \vdots & \vdots & \vdots & \vdots & \ddots & 0 & \cdots & 0 & 0 & \cdots & 0 \\
0 & 0 & \cdots & 0 & 0 & 0 & 1 & \cdots & 0 & 0 & \cdots & 0 \\
\vdots & \vdots & \vdots & \vdots & \vdots & \vdots & \vdots & \ddots & 0 & 0 & \cdots & 0 \\
0 & 0 & \cdots & 0 & \sin\theta & 0 & 0 & 0 & \cos\theta & 0 & \cdots & 0 \\
0 & 0 & \cdots & 0 & 0 & 0 & 0 & 0 & 0 & 1 & \cdots & 0 \\
\vdots & \vdots & \vdots & \vdots & \vdots & \vdots & \vdots & \vdots & \vdots & \vdots & \ddots & 0 \\
0 & 0 & \cdots & 0 & 0 & 0 & 0 & 0 & 0 & 0 & \cdots & 1
\end{pmatrix}
\begin{matrix}
\\ \\ \\ \\ \cdots\ i \\ \\ \\ \\ \cdots\ j \\ \\ \\ \\
\end{matrix}
$$

be the matrix of rotation by angle θ in the plane containing the ith and jth standard basis vectors.

Choose any i, j with $i < j$, and choose θ so that $\tan 2\theta = 2a_{i,j}/(a_{i,i}-a_{j,j})$. If

$$B = Q_{i,j}(-\theta)AQ_{i,j}(\theta),$$

then

(a) *B is symmetric;*

(b) *if $b_{k,\ell}$ is the (k,ℓ)-entry of B, then $b_{i,j} = b_{j,i} = 0$;*

(c) *the sum of the squares of the off-diagonal terms of B is the sum of the squares of the off-diagonal terms of A, decreased by the squares of the terms just killed; i.e.,*

$$\sum_{k \neq \ell} b_{k,\ell}^2 = \sum_{k \neq \ell} a_{k,\ell}^2 - 2a_{i,j}^2.$$

Proof.

(a) For this part, $Q_{i,j}(\theta)$ could be any orthogonal matrix Q:

$$B^\top = (Q^\top AQ)^\top = Q^\top A^\top Q = Q^\top AQ = B.$$

(b) This is exactly the computation in the 2×2 case of Lemma L8.2. You can show as Exercise L8#2 that for $n > 2$ the resulting relevant computations involve just the ith and jth rows and columns, which is just like the 2×2 case.

(c) This is where the real content of Jacobi's method is. Let $\mathbf{e}_k$ be the standard basis vectors and $\mathbf{w}_k$ be the orthonormal basis vectors forming the columns of $Q_{i,j}(\theta)$; in particular, $\mathbf{e}_k = \mathbf{w}_k$ if $k \neq i, j$. Note that

$$a_{k,\ell} = \underbrace{\mathbf{e}_k \cdot \underbrace{A\mathbf{e}_\ell}_{\ell\text{th column of }A}}_{k\text{th component of }\ell\text{th column of }A}$$

and

$$b_{k,\ell} = \mathbf{e}_k \cdot B\mathbf{e}_\ell = \mathbf{e}_k \cdot Q^{-1}AQ\mathbf{e}_\ell = \mathbf{w}_k \cdot A\mathbf{w}_\ell.$$

This restates that B is the matrix of A in the basis of the $\mathbf{w}_i$'s.

Now we will look at $\sum_{k \neq \ell} b_{k,\ell}^2$ and $\sum_{k \neq \ell} a_{k,\ell}^2$, considering the three different possibilities for k and ℓ:

(i) If neither index is an i or a j, then $b_{k,\ell} = a_{k,\ell}$, so the corresponding squares are certainly equal ($k \neq i, j$; $\ell \neq i, j$; $k \neq \ell$).

ii) On the other hand, if exactly one index is an i or a j, we will handle the entries in pairs ($k \neq i, j$): The quantity

$$a_{k,i}^2 + a_{k,j}^2 = (A\mathbf{e}_k \cdot \mathbf{e}_i)^2 + (A\mathbf{e}_k \cdot \mathbf{e}_j)^2$$

is the squared length of the projection of $A\mathbf{e}_k$ onto the plane spanned by $\mathbf{e}_i$ and $\mathbf{e}_j$. Since the same plane is spanned by $\mathbf{w}_i$ and $\mathbf{w}_j$, we see that

$$b_{k,i}^2 + b_{k,j}^2$$

is also the squared length of the projection of $A\mathbf{w}_k = A\mathbf{e}_k$ onto that plane, so that

$$a_{k,i}^2 + a_{k,j}^2 = b_{k,i}^2 + b_{k,j}^2.$$

The same argument applies to indices (i, k) and (j, k) with $k \neq i, j$.

(iii) Lastly, if both indices are from the off-diagonal term i, j or j, i, then

$$b_{i,j}^2 + b_{j,i}^2 = 2b_{i,j}^2 = 0,$$

though

$$a_{i,j}^2 + a_{j,i}^2 = 2a_{i,j}^2 \neq 0.$$

Summing up the contributions from (i), (ii), and (iii), we have

$$\sum_{k \neq \ell} b_{k,\ell}^2 = \sum_{k \neq i,j; \ell \neq i,j; k \neq \ell} b_{k,\ell}^2 + 2\sum_{k \neq i,j}(b_{k,i}^2 + b_{k,j}^2) + 0$$

$$= \sum_{k \neq i,j, \ell \neq i,j} a_{k,\ell}^2 + 2 \sum_{k \neq i,j} (a_{k,i}^2 + a_{k,j}^2)$$

$$= \sum_{k,\ell} a_{k,\ell}^2 - (a_{i,j}^2 + a_{j,i}^2) \quad \square$$

2. *Completion of Jacobi's method.* In conjunction with Lemmas L8.2 and L8.3, the following argument completes the proof of the Spectral Theorem.

Proof of Theorem 8.1, continued. Let A be a symmetric matrix. Choose i, j with $i \neq j$ so that $|a_{i,j}|$ is maximal, and θ so that if

$$A_1 = Q_{i,j}(-\theta) A Q_{i,j}(\theta),$$

then the (i, j)th term of A_1 is 0. Define $P_1 = Q_{i,j}(\theta)$.

Continue this way, killing on the next move the largest off-diagonal term of A_1 by conjugating A_1 by $Q_{i_1,j_1}(\theta_1)$, setting

$$A_2 = Q_{i_1,j_1}(-\theta_1) A_1 Q_{i_1,j_1}(\theta_1) \quad \text{and} \quad P_2 = P_1 Q_{i_1,j_1}(\theta_1),$$

and so on. Note that the (i, j) term, which had been killed in A_1, may come back to life in A_2. However, *at each stage, the sum of the squares of the off-diagonal terms is decreased by the square of the largest off-diagonal term*; in that sense, the A_n are becoming more and more diagonal.

At this point, a rather unpleasant point needs to be brought up. In general, the sequence of A_n and the sequence of P_n will converge; in fact, we have never seen a case where they do not. But there is trouble in showing that this is so. However, it is easy to show that a subsequence of the P_n converges, because these form a sequence in the orthogonal group, which is compact. Let Q_m be such a subsequence, so that the sequence

$$A_m = Q_m^{-1} A Q_m$$

converges also. Then $Q = \lim_{m \to \infty} Q_m$ is orthogonal and $Q^{-1} A Q$ is diagonal. $\square$

L8.3. Geometric Significance of Jacobi's Method in $\mathbb{R}^3$

A symmetric 3×3 matrix A represents a *quadratic form* (or quadratic function) $\mathbf{x}^\top A \mathbf{x}$ on $\mathbb{R}^3$, as described in Appendix L4.5. The equation $\mathbf{x}^\top A \mathbf{x} = \alpha$ represents a quadric surface. We shall be particularly interested in the cases where $\alpha = 1$ or $\alpha = -1$. Some are illustrated in Examples L8.4 for the cases where the axes of the surface coincide with the coordinate axes. In $\mathbb{R}^2$, the basic types are ellipses or hyperbolas; in $\mathbb{R}^3$, the basic types are ellipsoids or hyperboloids (of one or two sheets); in both cases, there are also degenerate examples.

Examples L8.4. In $\mathbb{R}^2$ we see conic sections such as in Figure L8.1.

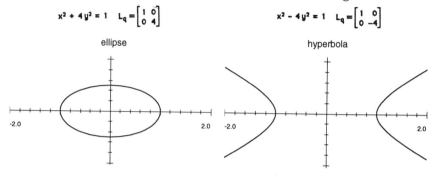

FIGURE L8.1. Conic sections in $\mathbb{R}^2$.

In $\mathbb{R}^3$ we see quadric surfaces such as in Figure L8.2.

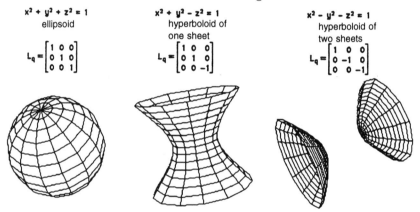

FIGURE L8.2. Quadric surfaces in $\mathbb{R}^3$. ▲

Notice that all the matrices of Examples L8.4 are diagonal matrices; this is no coincidence, as you can tell from the equations. If a symmetric matrix is not diagonal, there will be cross-terms in the equation, and the quadric surface it represents will be rotated, with respect to the x, y, and z axes. ▲

Example L8.5. For $3x^2 + 2y^2 + 5z^2 - 2xy + 4xz - 4yz = 1$, or

$$[x\ y\ z] \begin{bmatrix} 3 & -1 & 2 \\ -1 & 2 & -2 \\ 2 & -2 & 5 \end{bmatrix} \begin{bmatrix} x \\ y \\ z \end{bmatrix} = 1,$$

the quadric surface is an ellipsoid. Figure L8.3 shows it by the elliptical intersections of each coordinate plane with the surface; admittedly, this picture is not so easy to interpret.

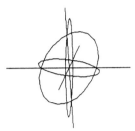

FIGURE L8.3.

Figure L8.4 shows the same ellipsoid, in the same position, but with the coordinate axes rotated to lie along the axes of the ellipsoid; now you can see it more easily.

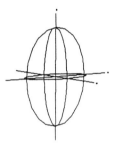

FIGURE L8.4.

Remark. Without the arrows on the axes it can be hard to interpret what is in front and what is behind. However, if you remember that we draw right-handed systems, you can see that the arrow on the z-axis is in the *front* of the picture (but high).   ▲

The question now is "how do we get from a drawing like Figure L8.3 to one like Figure L8.4?" The answer is, by Jacobi's method! Each step of Jacobi's method represents a rotation of the axes in one coordinate plane. The program *JacobiDraw* does this for you.

Example L8.6. The sequence of pictures from *JacobiDraw* in Figure L8.5 shows all the steps used in going from Figure L8.3 to Figure L8.4:

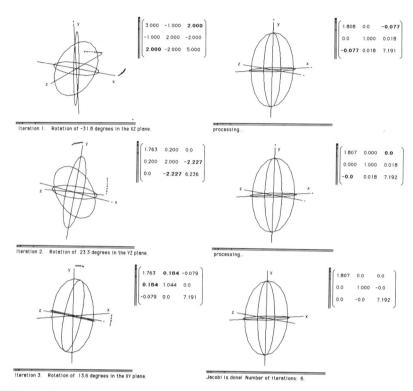

FIGURE L8.5. Action of Jacob's method on a quadratic form representing an ellipsoid.

The matrix at the right of each picture represents the *start* of each step (and therefore the *end* of the last step), highlighting the off-diagonal element to be zeroed in the current step.

The tolerance for this series of iterations was set at 0.0001 for determining whether an off-diagonal entry has become zero; you can change that tolerance.

The eigenvalues are 1.807, 1.000, and 7.192.

$$\text{The eigenvectors are } \begin{bmatrix} .882 \\ .210 \\ -.420 \end{bmatrix}, \begin{bmatrix} 0 \\ .894 \\ .447 \end{bmatrix}, \text{ and } \begin{bmatrix} .470 \\ -.394 \\ .789 \end{bmatrix},$$

as can be read off the eigenvector matrix displayed by *JacobiDraw*. In Exercise L8#5 you can confirm that the eigenvectors form an orthonormal basis as stated in the Spectral Theorem L8.1.   ▲

A line through the origin is a *principal axis* for a quadratic form if it intersects the level surfaces for that quadratic form *orthogonally*. When a symmetric matrix A becomes diagonal in an orthonormal basis, then the lines through the basis vectors appear to be principal axes for the quadratic form. This is, in fact, true in general and explains the alternate name *Principal Axis Theorem* for the *Spectral Theorem* L6.22. Jacobi's method tries to approximate the basis of principal axes by successively rotating in one plane at a time toward the principal axis in that plane.

Example L8.7 shows *JacobiDraw* in action for a hyperboloid of one sheet.

Example L8.7. For the matrix

$$\begin{bmatrix} 3 & -1 & 2 \\ -1 & 0 & -2 \\ 2 & -2 & 5 \end{bmatrix}$$

Jacobi's method is illustrated geometrically as shown in Figure L8.6.

The eigenvalues are 1.777, -0.728, and 6.950.

$$\text{The eigenvectors are } \begin{bmatrix} .886 \\ .069 \\ -.494 \end{bmatrix}, \begin{bmatrix} .094 \\ .949 \\ .298 \end{bmatrix}, \text{ and } \begin{bmatrix} .490 \\ -.305 \\ .816 \end{bmatrix}.$$

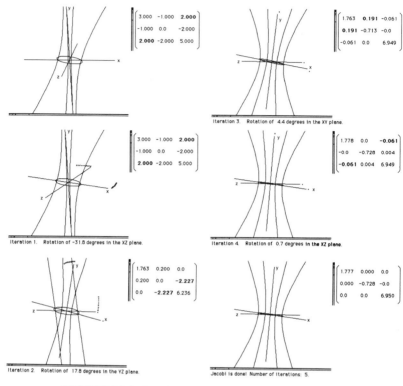

FIGURE L8.6. Jacobi's method on a hyperboloid. ▲

We show in Example L8.8 what happens in a degenerate case, where one of the eigenvalues is zero.

Example L8.8. The matrix $\begin{bmatrix} 2 & 0 & 3 \\ 0 & 0 & 0 \\ 3 & 0 & -1 \end{bmatrix}$ represents a degenerate case of a quadric surface, because, as you can see, one of the eigenvalues is zero. *JacobiDraw* gives the representation of the resulting *hyperbolic cylinder* in Figure L8.7.

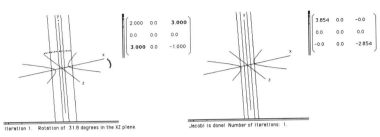

FIGURE L8.7. Jacobi's method on a degenerate case, a hyperbolic cylinder.

The eigenvalues are 3.854, 0, −2.854.

$$\text{The eigenvectors are } \begin{bmatrix} .850 \\ 0 \\ .525 \end{bmatrix}, \begin{bmatrix} 0 \\ 1 \\ 0 \end{bmatrix}, \begin{bmatrix} -.525 \\ 0 \\ .850 \end{bmatrix}. \quad \blacktriangle$$

L8.4. RELATIONSHIP BETWEEN EIGENVALUES AND SIGNATURES

Finally, we close this appendix with a reprise of another aspect of quadratic forms—their *signatures*, as discussed in Theorem L4.19. If a symmetric matrix A has a signature (k, ℓ), then its quadratic form $\mathbf{x}^\top A\mathbf{x}$ can be written as the sum of the squares of linearly independent linear expressions, k of them with plus signs and ℓ of them with minus signs.

Geometrically the signature tells whether a form is elliptic (all plus signs, so $\ell = 0$), hyperbolic (a mixture of plus and minus signs, so $k \neq 0$, $\ell \neq 0$), or not graphable at all (all negative signs, so $k = 0$). For $\mathbb{R}^3$,

signature $(3, 0)$ represents an ellipsoid;
signature $(2, 1)$ represents an hyperboloid of one sheet;
signature $(1, 2)$ represents an hyperboloid of two sheets;
signature $(0, 3)$ has no graph at all.

Theorem L8.9. *If a symmetric matrix A has signature (k, ℓ), then k eigenvalues are positive, and ℓ eigenvalues are negative.*

Proof. A diagonal matrix represents a quadratic form for which the simplest decomposition is $\lambda_1 x_1^2 + \lambda_2 x_2^2 + \cdots + \lambda_n x_n^2$, making the eigenvalues the coefficients of the squared terms. Furthermore, Theorem L4.19 says that for *any* decomposition of a quadratic form into a sum of squares of linearly independent linear functions, k and ℓ remain unchanged. So the numbers k

and ℓ of the signature correspond to the numbers of positive and negative eigenvalues, respectively. □

Recall that a positive definite quadratic form is one for which the signature is $(n, 0)$; we now see that this corresponds to all the eigenvalues of the associated matrix being positive.

Appendix L   Exercises

Exercises L1   Theory of Linear Equations: in Practice

L1#1. Perform the following matrix multiplications:

(a) $\begin{bmatrix} 1 & 2 & 3 \\ 4 & 5 & 6 \end{bmatrix} \begin{bmatrix} 7 & 8 \\ 9 & 0 \\ 1 & 2 \end{bmatrix}$

(b) $\begin{bmatrix} 1 & 2 \\ 0 & 3 \end{bmatrix} \begin{bmatrix} 1 & 4 \\ -1 & 3 \end{bmatrix} \begin{bmatrix} 0 & 1 \\ -1 & 3 \end{bmatrix}$

(c) $\begin{bmatrix} 1 & -1 & 1 \\ -1 & 0 & 2 \\ -1 & 1 & 1 \end{bmatrix} \begin{bmatrix} 0 & 1 & -1 \\ -1 & 1 & 2 \\ 2 & 0 & -2 \end{bmatrix}$

(d) $\begin{bmatrix} 7 & 1 \\ -1 & 0 \\ 2 & 3 \end{bmatrix} \begin{bmatrix} 5 \\ -4 \end{bmatrix}$

L1#2. Show that any row operation can be undone by another row operation. Note the importance of the word "nonzero" in the algorithm for row reduction.

L1#3. For each of the three matrices in Examples L1.5, find (and label) row operations that will bring them to echelon form.

L1#4. For each of the following matrices, find (and label) row operations that will bring them to echelon form:

(a) $\begin{bmatrix} 1 & 2 & 3 \\ 4 & 5 & 6 \end{bmatrix}$

(d) $\begin{bmatrix} 1 & 3 & -1 & 4 \\ 1 & 2 & 1 & 2 \\ 3 & 7 & 1 & 9 \end{bmatrix}$

(b) $\begin{bmatrix} 1 & -1 & 1 \\ -1 & 0 & 2 \\ -1 & 1 & 1 \end{bmatrix}$

(e) $\begin{bmatrix} 1 & 1 & 1 & 1 \\ 2 & -3 & 3 & 3 \\ 1 & -4 & 2 & 2 \end{bmatrix}$

(c) $\begin{bmatrix} 1 & 2 & 3 & 5 \\ 2 & 3 & 0 & -1 \\ 0 & 1 & 2 & 3 \end{bmatrix}$

L1#5. For Example L1.7, analyze precisely where the troublesome errors occur.

L1#6. Solve the following systems of equations using row operations:

(a)
$$2x + 13y - 3z = -7$$
$$x + y = 1$$
$$x + 7z = 22$$

(d)
$$x + 3y + z = 4$$
$$-x - y + z = -1$$
$$2x + 4y = 0$$

(b)
$$x - 2y - 12z = 12$$
$$2x + 2y + 2z = 4$$
$$2x + 3y + 4z = 3$$

(e)
$$x + 2y + z - 4w + v = 0$$
$$x + 2y - z + 2w - v = 0$$
$$2x + 4y + z - 5w + v = 0$$
$$x + 2y + 3z - 10w + 2v = 0$$

(c)
$$x + y + z = 5$$
$$x - y - z = 4$$
$$2x_6 y + 6z = 12$$

L1#7. Draw a sketch to verify each of the following geometric interpretations of systems of linear equations. Then for each of the parts (a)–(d) of Exercise L1#6, give the geometric interpretation of the solutions.

i. An equation $ax + by = d$ represents a *line* in $\mathbb{R}^2$ (and a vertical plane in $\mathbb{R}^3$).

An equation $ax + by + cz = d$ represents a *plane* in $\mathbb{R}^3$.

ii. A system of two linear equations in $\mathbb{R}^2$ represents the intersection of two lines, $a_i x + b_i y = d_i$, for $i = 1, 2$.

The equations are incompatible if and only if one line is parallel (but not equal) to the other.

The equations have infinitely many solutions if and only if the lines coincide.

iii. A system of three linear equations in $\mathbb{R}^3$ represents the intersection of three planes, $a_i x + b_i y + c_i z = d_i$, for $i = 1, 2, 3$.

The equations are incompatible if either at least two of the planes are parallel, *or* the line of intersection of any two planes is parallel to the third.

The equations have infinitely many solutions if either at least two of the planes coincide and the third is not parallel, *or* all three planes intersect in a common line.

L1#8.

(a) Derive from Theorem L1.3 the fact that only square matrices can have inverses.

(b) Construct an example where $AB = I$, but $BA \neq I$.

L1#9. For Example L1.12:

(a) Confirm by matrix multiplication that $A^{-1}A = AA^{-1} = I$.

(b) Use A^{-1} to solve the system of Example L1.3.

L1#10. Find the inverse, or show it does not exist, for each of the following matrices:

(a) $\begin{bmatrix} 1 & -5 \\ 9 & 9 \end{bmatrix}$

(d) $\begin{bmatrix} 3 & 2 & -1 \\ 0 & 1 & 1 \\ 8 & 3 & 9 \end{bmatrix}$

(b) $\begin{bmatrix} 1 & 3 \\ 3 & 9 \end{bmatrix}$

(e) $\begin{bmatrix} 1 & 0 & 1 \\ 2 & 1 & -1 \\ 1 & 1 & -1 \end{bmatrix}$

(c) $\begin{bmatrix} 1 & 2 & 3 \\ 2 & 3 & 0 \\ 0 & 1 & 2 \end{bmatrix}$

L1#11. Prove Theorem L1.17, that $(AB)^\top = B^\top A^\top$.

Exercises L2 Theory of Linear Equations: Vocabulary

L2#1. For Example L2.2, verify that the restricted set of vectors $\begin{bmatrix} x \\ y \\ z \end{bmatrix}$ such that $2x - 3y + 2z = 0$ satisfies all ten of the rules for a vector space.

L2#2. Show that a subset W of a vector space V is a subspace if the following two statements are true for any elements $\mathbf{w}_i$, $\mathbf{w}_j$ of W:

$$\mathbf{w}_i + \mathbf{w}_j \in W; \quad \alpha \mathbf{w}_i \in W \quad \text{for any real number } \alpha.$$

That is, show that this short two rule checklist automatically invokes the longer list of ten rules for a vector space when you know W is a subset of a vector space.

L2#3. Show that the following two statements are equivalent to saying that a set of vectors $\mathbf{v}_1, \ldots, \mathbf{v}_k$ is linearly independent:

i. The only way to write the zero vector 0 as a linear combination of the $\mathbf{v}_i$ is to use only zero coefficients.

ii. None of the $\mathbf{v}_i$ is a linear combination of the others.

L2#4. Show that the standard basis vectors $\mathbf{e}_1, \ldots, \mathbf{e}_k \in \mathbb{R}^k$ are linearly independent.

L2#5. Consider the following vectors: $\begin{bmatrix} 1 \\ 1 \\ 0 \end{bmatrix}$, $\begin{bmatrix} 1 \\ 2 \\ 1 \end{bmatrix}$, and $\begin{bmatrix} 0 \\ 1 \\ \alpha \end{bmatrix}$.

(a) For what values of α are these three vectors linearly dependent?

(b) Show that for each such α the three vectors lie in the same plane and give an equation of the plane.

L2#6. Let $\{\mathbf{v}_1, \ldots, \mathbf{v}_k\}$ be a set of vectors of a vector space V. Show that the following three conditions are equivalent:

i. The set is a maximal linearly independent set.

ii. The set is a minimal spanning set.

iii. The set is a linearly independent spanning set.

L2#7.

(a) For Figure L2.3 in Example L2.19, confirm both geometrically and trigonometrically the vector expressions for the higher point.

(b) Take the triangle with vertices at $(1,1)$, $(2,4)$, and $(5,3)$ and find the new vertices when the triangle is rotated by $30°$ about the origin.

(c) Confirm the necessity of rotating about the origin in order to make a *linear* transformation. Hint: look at what would happen to the origin.

L2#8. In Example L2.12, write the identification with $\mathbb{R}^2$ induced by the vectors $\begin{bmatrix} 1 \\ 1 \\ -2 \end{bmatrix}$ and $\begin{bmatrix} 1 \\ -1 \\ 0 \end{bmatrix}$ as a basis of the plane $x + y + z = 0$.

L2#9. Let $v_1 = \begin{bmatrix} 1 \\ 1 \end{bmatrix}$ and $v_2 = \begin{bmatrix} 1 \\ 3 \end{bmatrix}$. Let x and y be the coordinates with respect to the standard basis $\{e_1, e_2\}$ and let u and v be the coordinates with respect to $\{v_1, v_2\}$. Write the equations to translate from (x, y) to (u, v) and back. Use these equations to write the vector $\begin{bmatrix} 3 \\ -5 \end{bmatrix}$ in terms of v_1 and v_2.

L2#10. Let $v_1, \ldots, v_n$ be vectors of a vector space V and let $P_{[v]} : \mathbb{R}^n \to V$ be given by

$$P_{[v]} \left(\begin{bmatrix} a_1 \\ \vdots \\ a_n \end{bmatrix} \right) = \sum a_i v_i.$$

(a) Show that $v_1, \ldots, v_n$ are linearly independent if and only if the map $P[v]$ is one-to-one.

(b) Show that $v_1, \ldots, v_n$ span V if and only if $P[v]$ is onto.

(c) Show that $v_1, \ldots, v_n$ is a basis of V if and only if $P_{[v]}$ is one-to-one and onto.

L2#11. Suppose $v_1, \ldots, v_n$ is a basis of V and that $w_1, \ldots, w_m$ is another basis of V. Using the quantity $P_{[v]}$ defined in Exercise L2#10, show that $P_{[v]}^{-1}(w_1), \ldots, P_{[v]}^{-1}(w_n)$ is then a basis of $\mathbb{R}^n$. Use this to show that $n = m$.

L2#12. Show that the mapping from $\mathbb{R}^n$ to $\mathbb{R}^m$ described by the product Av (multiplying a vector v in $\mathbb{R}^n$ by an $m \times n$ matrix A, as in Example L2.9) is indeed linear.

L2#13. For the mapping from the vector space P_2 of polynomials of degree at most two, given by Example L2.15,

$$T(p)(x) = (x^2 + 1)p''(x) - xp'(x) + 2p(x)$$

(a) Verify linearity.

(b) Show that in the basis $\begin{bmatrix} 1 \\ x \\ x^2 \end{bmatrix}$, the matrix $\begin{bmatrix} 2 & 0 & 2 \\ 0 & 1 & 0 \\ 0 & 0 & 2 \end{bmatrix}$ gives T.

(c) Compute the matrices of the same differential operator, on $P_3, P_4, \ldots, P_n$ (polynomials of degree at most 3, 4, and n).

L2#14. Confirm that in general the procedure suggested in Section L2.5 for finding a basis of the kernel from n linearly dependent columns of an $m \times n$ matrix A leads to n automatically linearly independent m-vectors.

L2#15. For each of the following matrices, find a basis for the kernel and a basis for the image, using Theorem L2.9:

(a) $\begin{bmatrix} 1 & 1 & 3 \\ 2 & 2 & 6 \end{bmatrix}$

(b) $\begin{bmatrix} 1 & 2 & 3 \\ -1 & 1 & 1 \\ -1 & 4 & 5 \end{bmatrix}$

(c) $\begin{bmatrix} 1 & 1 & 1 \\ 1 & 2 & 3 \\ 2 & 3 & 4 \end{bmatrix}$

L2#16. Decompose the following into partial fractions, as requested, being explicit in each case about the system of linear equations involved and showing that its matrix is invertible:

(a) Write

$$\frac{x + x^2}{(x+1)(x+2)(x+3)} \quad \text{as} \quad \frac{A}{x+1} + \frac{B}{x+2} + \frac{C}{x+3}.$$

(b) Write

$$\frac{x + x^3}{(x+1)^2(x-1)^3} \quad \text{as} \quad \frac{Ax + B}{(x+1)^2} + \frac{Cx^2 + Dx + F}{(x-1)^3}.$$

L2#17. Given p_1 and p_2 polynomials of degree k_1 and k_2 respectively, consider the mapping

$$T: (q_1, q_2) \to p_1 q_1 + p_2 q_2,$$

where q_1 and q_2 are polynomials of degrees $k_2 - 1$ and $k_1 - 1$, respectively, so that

$$p_1 q_1 + p_2 q_2 \quad \text{is of degree} \leq k_1 + k_2 - 1.$$

Note that the space of such (q_1, q_2) is of dimension $k_1 + k_2$, and the space of polynomials of degree $k_1 + k_2 - 1$ is also of dimension $k_1 + k_2$.

(a) Show that $\ker T = \{0\}$ if and only if p_1 and p_2 are relatively prime.

(b) Use Theorem L2.10 to show that if p_1 and p_2 are relatively prime, then there exist unique q_1 and q_2 as above such that

$$p_1 q_1 + p_2 q_2 = 1.$$

This procedure gives a proof of Bezout's Theorem, which you will see as Theorem L6.14.

Exercises L3–4 The Inner Product

L3–4#1. Prove Theorem L3.1 and its extension to $\mathbb{C}^n$, to show that

$$\overline{A^T}\,\mathbf{v}\cdot\mathbf{w} = \mathbf{v}\cdot A\mathbf{w}.$$

L3–4#2. Prove Theorem L3.2.

L3–4#3.

(a) Prove Schwarz's inequality, Theorem L3.3, in $\mathbb{R}^2$ for the standard inner product by direct computation, i.e., show that for any numbers x_1, x_2, y_1, and y_2, we have

$$|x_1y_1 + x_2y_2| \le \sqrt{x_1^2 + x_2^2}\,\sqrt{y_1^2 + y_2^2}.$$

(b) Prove Schwarz's inequality for Hermitian inner products. (Hint: the proof for real inner products almost goes over to the complex case.)

L3–4#4. Calculate the angles between the following pairs of vectors:

(a) $\begin{bmatrix} 1 \\ 0 \\ 0 \end{bmatrix}$, $\begin{bmatrix} 1 \\ 1 \\ 1 \end{bmatrix}$

(b) $\begin{bmatrix} 1 \\ 0 \\ -1 \\ 0 \end{bmatrix}$, $\begin{bmatrix} 1 \\ 1 \\ 1 \\ 1 \end{bmatrix}$

(c) $\displaystyle\lim_{n\to\infty}$ (angle between $\begin{bmatrix} 1 \\ 0 \\ 0 \\ 0 \\ \vdots \end{bmatrix}$, $\begin{bmatrix} 1 \\ 1 \\ 1 \\ 1 \\ \vdots \end{bmatrix}$ as vectors in $\mathbb{R}^n$).

L3–4#5.

(a) Show that $\begin{bmatrix} 1 \\ 1 \\ 1 \end{bmatrix}$, $\begin{bmatrix} 1 \\ -1 \\ 0 \end{bmatrix}$, $\begin{bmatrix} 1 \\ 1 \\ -2 \end{bmatrix}$ is an orthogonal basis of $\mathbb{R}^3$.

(b) Use Theorem L3.11 to express $\begin{bmatrix} 1 \\ 0 \\ 0 \end{bmatrix}$ as a linear combination of these vectors.

L3–4#6.

(a) Prove algebraically that the Gram–Schmidt construction gives orthogonal vectors, to complete the proof of Theorem L3.13.

(b) Describe geometrically what the Gram–Schmidt algorithm does. Hint: Use Example L3.14 to begin.

L3–4#7.

(a) Apply Gram–Schmidt to the vectors

$$\mathbf{w}_1 = \begin{bmatrix} 1 \\ 0 \end{bmatrix}, \ \mathbf{w}_2 = \begin{bmatrix} 0 \\ 1 \end{bmatrix} \text{ where } \langle \mathbf{x}, \mathbf{y} \rangle = \mathbf{x}^T K \mathbf{y} \text{ for } K = \begin{bmatrix} 1 & 1 \\ 1 & 2 \end{bmatrix}.$$

(b) Let $\{\mathbf{u}_1, \mathbf{u}_2\}$ be the resulting orthonormal basis. Let T be defined by $T(\mathbf{x}) = K^{-1} H \mathbf{x}$ for

$$H = \begin{bmatrix} 1 & 5 \\ 5 & 7 \end{bmatrix}.$$

Show that the matrix of T with respect to $\{\mathbf{u}_1, \mathbf{u}_2\}$ is symmetric.

L3–4#8. Using the Gram–Schmidt process, obtain an orthonormal basis for $\mathbb{R}^3$ with the standard inner product from the vectors

$$\beta_1 = \begin{bmatrix} 1 \\ 1 \\ 1 \end{bmatrix} \quad \beta_2 = \begin{bmatrix} 1 \\ 0 \\ 1 \end{bmatrix} \quad \beta_3 = \begin{bmatrix} 0 \\ 2 \\ 3 \end{bmatrix}.$$

L3–4#9. Let V be the space of polynomials of degree at most three. Let

$$\langle f, g \rangle = \int_0^1 f(t) g(t) dt.$$

Apply the Gram–Schmidt process to the basis $\{1, t, t^2, t^3\}$.

L3–4#10. For the inner product $\langle p, q \rangle = \int_0^\infty p(t) \overline{q(t)} \, e^{-t} dt$

(a) Find an orthonormal basis of the space P_3 of polynomials degree at most three, that is, with elements $p_0(t)$, $p_1(t)$, $p_2(t)$, and $p_3(t)$, by applying Gram–Schmidt to $\{1, x, x^2, x^3\}$.

(b) Write $t^2 + t^3$ as a linear combination of $p_0(t), \ldots, p_3(t)$, using Lemma L3.11.

L3–4#11.

(a) What is the angle between the vectors $\begin{bmatrix} a_1 \\ b_1 \end{bmatrix}$ and $\begin{bmatrix} a_2 \\ b_2 \end{bmatrix}$ for the inner product $\langle \mathbf{a}, \mathbf{b} \rangle = a_1 b_1 + 2 a_2 b_2$?

(b) Find a geometric construction that will change the angle to a standard one.

L3–4#12. Show that the projection operator π_W gives a decomposition of $\mathbb{R}^n$ into the kernel and image of π_W. That is, show that

$$W^\perp = \ker \pi_W$$

and that

$$\pi_W(\mathbb{R}^n) = \operatorname{Im} \pi_W.$$

L3–4#13. Prove Theorem L3.9 for the complex case. That is, for vectors $\mathbf{v} \in \mathbb{C}^n$, with vectors $\mathbf{w}_1, \mathbf{w}_2, \ldots, \mathbf{w}_n$ forming a basis, show that

$$\left\| \mathbf{v} - \sum_i a_i \mathbf{w}_i \right\|^2 = \|\mathbf{v}\|^2 - 2\operatorname{Re}\sum_i \bar{a}_i \langle \mathbf{v}, \mathbf{w}_i \rangle + \sum_i |a_i|^2,$$

and then finish the proof.

L3–4#14. For each of the following matrices, determine whether it is Hermitian, anti-Hermitian, unitary, or none of these:

(a) $\begin{bmatrix} 1 & i \\ -i & 0 \end{bmatrix}$

(d) $\begin{bmatrix} \cos\theta & \sin\theta \\ \sin\theta & \cos\theta \end{bmatrix}$

(b) $\begin{bmatrix} 1 & 1 \\ -1 & 1 \end{bmatrix}$

(e) $\begin{bmatrix} e^{i\theta_1} & 0 \\ 0 & e^{i\theta_2} \end{bmatrix}$

(c) $\begin{bmatrix} i & 1 \\ -1 & 2i \end{bmatrix}$

(f) $\begin{bmatrix} \cos\theta_1 & 0 & 0 & \sin\theta_1 \\ 0 & \cos\theta_2 & -\sin\theta_2 & 0 \\ 0 & \sin\theta_2 & \cos\theta_2 & 0 \\ -\sin\theta & 0 & 0 & \cos\theta_1 \end{bmatrix}$

L3–4#15. Write each of the following matrices in the form QR, where Q is orthogonal and R is upper-triangular (Theorem L4.3):

(a) $\begin{bmatrix} 1 & 2 \\ 1 & 1 \end{bmatrix}$

(b) $\begin{bmatrix} 1 & 0 & -1 \\ 0 & 1 & 0 \\ 1 & 1 & 1 \end{bmatrix}$

L3–4#16. Show that the matrices of Example L4.2 are exactly the vertices forming $O(2)$, the orthogonal group in two dimensions.

L3–4#17. Consider Example L4.7.

(a) Show that the derivative operater D operating on the space $\mathcal{C}_0^1[a, b]$ of continuously differentiable functions on $[a, b]$, vanishing at a and b, is antisymmetric or anti-Hermitian, as a result of the fact that the eigenvalues are purely imaginary. Find the eigenvectors. Then show that D^2 is symmetric.

(b) Do the same for the space of functions satisfying $f(a) = f(b)$, $f'(a) = f'(b)$.

(c) Show that D^2 is symmetric, operating on the space of functions which vanish at a and whose derivatives vanish at b.

(d) Do the same as in part (c) for the space of functions such that $f(a) = f'(a)$ and $f(b) = f'(b)$.

(e) For any numbers α and β not both zero, and any numbers γ and δ not both zero, do the same as in part (c) for the space of functions f such that

$$\alpha f(0) + \beta f'(0) = 0$$
$$\gamma f(\pi) + \delta f'(\pi) = 0.$$

L3–4#18. For the quadratic form of Example L4.16,

$$x^2 + 2xy - 4xz + 2yz - 4z^2$$

(a) Show that completing the square starting with the z terms, then the y term, and finally the x terms leads to $x^2 + (x+y/2)^2 - (x-y/2+2z)^2$, as stated in the example.

(b) Find the decomposition by completing the square if you start with the y terms, then do the x terms, and finally the z terms.

L3–4#19. Decompose each of the following quadratic forms by completing squares, and determine its signature.

(a) $x^2 + xy - y^2$

(b) $x^2 + 2xy - y^2$

(c) $x^2 + xy + yz$

(d) $xy + yz$

L3–L4#20. Consider Example L4.17

$$q(\mathbf{x}) = xy - xz + yz.$$

(a) Verify that the decomposition

$$(x/2 + y/2)^2 - (x/2 - y/2 + z)^2 + z^2$$

is indeed composed of linearly independent functions.

(b) Decompose $q(\mathbf{x})$ with a different choice of u, to support the statement that $u = x - y$ was not a magical choice.

L3–4#21. Check that the following quadratic forms are degenerate.

(a) $x^2 + 4xy + 4y^2$ on $\mathbb{R}^2$.

(b) $x^2 + 2xy + 2y^2 + 2yz + z^2$ on $\mathbb{R}^3$.

(c) $2x^2 + 2y^2 + z^2 + w^2 + 4xy + 2xz - 2xw - 2yw$ on $\mathbb{R}^4$.

L3–4#22. On R^4 as described by $M = \begin{bmatrix} a & c \\ b & d \end{bmatrix}$, consider the quadratic form $q(M) = \det M$. What is its signature?

L3–4#23. On R^4 as described by $H = \begin{bmatrix} a & b+ic \\ b-ic & d \end{bmatrix}$, the space of Hermitian 2×2 matrices, consider the quadratic form $q(H) = \det H$. What is its signature?

L3–4#24. For any real $n \times n$ matrix M

(a) Show that M can be written uniquely as a sum of a symmetric matrix and an antisymmetric matrix. (Hint: $M + M^\top$ is symmetric.)

(b) Consider the quadratic form on the vector space of symmetric matrices M_S given by $q_S(M_S) = \mathrm{tr}(M_s^2)$. Show that q_S is positive definite.

(c) Consider the quadratic form on the vector space of antisymmetric matrices M_A given by $q_A(M_A) = \mathrm{tr}(M_A^2)$. Show that q_A is negative definite.

(d) Find the signature of q, defined by $q(M) = \mathrm{tr}(M^2)$.

L3–4#25. For any complex $n \times n$ matrix M, find the signature of q, the Hermitian quadratic form defined by $q(M) = \mathrm{tr}(M \overline{M})$.

L3–4#26. Consider again $q(M) = \mathrm{tr}(M^2)$, operating on the space of upper triangular matrices described by $M = \begin{bmatrix} a & b \\ 0 & d \end{bmatrix}$.

(a) What kind of surface in $\mathbb{R}^3$ do you get by setting $q(M^2) = 1$?

(b) What kind of surface in $\mathbb{R}^3$ do you get by setting $q(MM^T) = 1$?

L3–4#27.

(a) Show that the function

$$\left\langle \begin{bmatrix} a_1 \\ a_2 \end{bmatrix}, \begin{bmatrix} b_1 \\ b_2 \end{bmatrix} \right\rangle = \int_0^1 (a_1 + a_2 x)(b_1 + b_2 x)dx$$

defines an inner product on $\mathbb{R}^2$, and compute its matrix.

(b) There is an obvious generalization of part a to $\mathbb{R}^3$, using second degree polynomials. Do the corresponding problem.

L3–4#28. Show that a 2×2 matrix $G = \begin{bmatrix} a & b \\ c & d \end{bmatrix}$ represents an inner product on $\mathbb{R}^2$ if and only if G is symmetric, $\det G > 0$, and $\operatorname{tr} G > 0$.

Exercises L5 Determinants and Volumes

L5#1. Compute the determinants of the following matrices according to development by the first row:

$$\text{(a)} \begin{bmatrix} 1 & 2 & 3 \\ 4 & 0 & 1 \\ 5 & -1 & 2 \end{bmatrix} \quad \text{(b)} \begin{bmatrix} 1 & -1 & 2 \\ 0 & 3 & 4 \\ 2 & 1 & 0 \end{bmatrix} \quad \text{(c)} \begin{bmatrix} 1 & 2 & 3 & 4 \\ 0 & 1 & -1 & 2 \\ 3 & 0 & 1 & -1 \\ 1 & -1 & 2 & 0 \end{bmatrix}$$

L5#2. In the proof of Theorem L5.3, show that the specified row operations give a diagonal matrix with the same diagonal entries as the original matrix.

L5#3. For Theorem L5.3,

(a) Give an alternate proof direct from the recursive definition of D for the case where the matrix is lower triangular.

(b) Adapt the proof in part (a) to the case where the matrix is upper-triangular.

L5#4. For a matrix A, we defined the determinant $D(A)$ recursively by development according to the first row. Show that it could have equally well been defined, with the same result, as development according to the first column.

L5#5. For the matrix $A = \begin{bmatrix} 1 & 0 \\ 1 & 2 \end{bmatrix}$

(a) Compute the characteristic polynomial and find its roots.

(b) With the change of basis matrix $P = \begin{bmatrix} 2 & 1 \\ 0 & 1 \end{bmatrix}$, compute the characteristic polynomial of $P^{-1}AP$ and find its roots, thus confirming the basis independence of the characteristic polynomial.

L5#6. As stated in Example L5.12, prove that given a polynomial $p(\lambda)$, you can easily construct a matrix that has $p(\lambda)$ as its characteristic polynomial. You simply need to show that the stated general "companion" matrix M (or its transpose) indeed has the stated $p(\lambda)$ as $\det|\lambda I - M|$.

L5#7. Referring to change-of-basis as introduced in Section L2.4, prove the part of Theorem L5.15 that states the trace is also basis independent.

L5#8. Using classical geometry, show that for a parallelogram you can use either side as a base and still get the same area, confirming Theorem L5.15 for $k = 2$.

L5#9. To get a better feeling for the relation between volumes and determinants, make a drawing in the plane illustrating that property (iii) of Theorem L5.18 is true for areas of parallelograms. That is, show that the "before" and "after" parallelograms have the same $area = base \times height$. (Hint: use a dissection argument.)

Remark: The general case is analogous, using $volume = base \times height$.

L5#10. Find and identify geometrically the volumes of the solids spanned by the following sets of vectors:

(a) $\begin{bmatrix} 5 \\ 2 \end{bmatrix}, \begin{bmatrix} -2 \\ -1 \end{bmatrix}$ (c) $\begin{bmatrix} 1 \\ 3 \\ -1 \end{bmatrix}, \begin{bmatrix} 2 \\ -1 \\ -1 \end{bmatrix}, \begin{bmatrix} 3 \\ 6 \\ 0 \end{bmatrix}$

(b) $\begin{bmatrix} 6 \\ 4 \end{bmatrix}, \begin{bmatrix} 3 \\ 2 \end{bmatrix}$ (d) $\begin{bmatrix} 2 \\ 3 \\ 4 \end{bmatrix}, \begin{bmatrix} -1 \\ 2 \\ 3 \end{bmatrix}, \begin{bmatrix} 0 \\ 1 \\ 2 \end{bmatrix}$

Exercises L6 Eigenvalues and Eigenvectors

L6#1. Show that for a square matrix A the following four statements are equivalent, thus confirming the proof of Theorem L6.1:

  i. A has no inverse.

  ii. There exists a vector $\mathbf{x} \neq 0$ such that $A\mathbf{x} = \mathbf{0}$.

  iii. There exists a vector $\mathbf{y}$ that cannot be written as $A\mathbf{x}$.

  iv. $\det A = 0$.

L6#2. Prove that if a 2×2 matrix A has only one eigenvalue and is not already diagonal, then A is not diagonalizable.

L6#3. For a 3×3 matrix A, derive the more complicated conditions and results for diagonalizability. You might refer to Exercise 9.6#14.

L6#4. Prove that if A is a triangular matrix, the eigenvalues of A are the diagonal entries, by applying Theorem L5.3 to $\lambda I - A$.

L6#5. Prove that if A is a triangular matrix with entries a_{ij}, the eigenvalues of A are the diagonal entries, by showing that there exists a nontrivial solution to the equation $A\mathbf{v} = \lambda\mathbf{v}$ if and only if $\lambda = a_{ii}$ for some $i = 1, 2, \ldots, n$.

L6#6. Show that the characteristic polynomial of a real k-pseudo-upper-triangular matrix has at most $k/2$ conjugate pairs of nonreal eigenvalues.

L6#7. Prove Theorem L6.19. This may be done analogously to the proof given for Theorem L6.18.

L6#8. Show that an upper-triangular orthogonal (or unitary) matrix *must* be diagonal.

L6#9. A *normal* linear transformation is one for which $A\overline{A^\top} = \overline{A^\top}A$.

(a) Show that Hermitian, anti-Hermitian, and unitary matrices are all normal.

(b) Show that if P is a unitary matrix and A is a normal matrix, then $P^{-1}AP = B$ is a normal matrix.

(c) Show by Theorem L6.16 that there exists (for A normal) a P such that $B = P^{-1}AP$ is normal and upper-triangular.

(d) Show that if a matrix B is normal and upper-triangular, then B is diagonal.

L6#10. Consider Example L6.26 where the matrix $A = \begin{bmatrix} 0 & 1 & 0 \\ 0 & 0 & 1 \\ 2 & -1 & 2 \end{bmatrix}$, and

we want to examine $V_1 = \ker(A - 2I)$ and $V_2 = \ker(A^2 + I)$. An alternative and faster way to find the V_i is by row reduction as follows:

$$A - 2I = \begin{bmatrix} -2 & 1 & 0 \\ 0 & -2 & 1 \\ 2 & -1 & 0 \end{bmatrix} \rightarrow \underbrace{\begin{bmatrix} 1 & 0 & -1/4 \\ 0 & 1 & -1/2 \\ 0 & 0 & 0 \end{bmatrix}}_{\text{2-dimensional image}} \Rightarrow \underbrace{\begin{array}{l} x = +\frac{1}{4}z \\ y = \frac{1}{2}z \end{array}}_{\text{a line}}$$

2-dimensional image

$\Downarrow$

1-dimensional kernel
by Theorem L2.25

so V_1 is one-dimensional and $x \in V_1$ looks like $\begin{bmatrix} c \\ 2c \\ 4c \end{bmatrix}$.

(a) By a similar analysis, find V_2.

(b) Verify properties (iii) and (iv) of Theorem L6.25 to finish Example L6.26.

L6#11. Apply the Euclidean algorithm to the following pairs of relatively prime numbers p_1 and p_2 to verify Bezout's Theorem L6.27. That is, find a pair of numbers a and b such that $p_1 a + p_2 b = 1$:

(a) $21, 16$.   (b) $537, 791$.

Exercises L7   Finding Eigenvalues, Eigenvectors: QR Method

L7#1. Using *Eigenfinder* (or applying QR by hand), find the eigenvalues

of the matrix $\begin{bmatrix} 0 & 1 & 0 \\ 0 & 0 & 1 \\ 6 & -11 & 6 \end{bmatrix}$. Then verify that if you find the characteristic

polynomial, its roots are indeed the eigenvalues found by QR.

L7#2. For the heart of the QR method, you should provide the following proofs and examples:

(a) Verify the following lemma: For any $n \times n$ matrix

$$
A = \begin{bmatrix} \uparrow & \uparrow & & \uparrow \\ \mathbf{a}_1 & \mathbf{a}_2 & \cdots & \mathbf{a}_n \\ \downarrow & \downarrow & & \downarrow \end{bmatrix},
$$

where we label the ith column as a vector $\mathbf{a}_i$, and any $n \times n$ *upper-triangular* matrix R, the product AR is given by

$$
\begin{bmatrix} \uparrow & \uparrow & \uparrow & & \uparrow \\ r_{1,1}\mathbf{a}_1 & (r_{1,2}\mathbf{a}_1 + r_{2,2}\mathbf{a}_2) & (r_{1,3}\mathbf{a}_1 + r_{2,3}\mathbf{a}_2 + r_{3,3}\mathbf{a}_3) & \cdots & \left(\sum_{j=1}^{n}\sum_{i=1}^{j} r_{i,j}\mathbf{a}_j\right) \\ \downarrow & \downarrow & \downarrow & & \downarrow \end{bmatrix}.
$$

(b) Use the lemma in part (a) to prove the following theorem:

> *Invertible matrices B and A have the same flag if and only if there exists an R such that B = AR.*

(c) Use the theorem in part (b) to show the core statement:

> *If you apply Gram-Schmidt to any basis, the vectors of the new orthonormal basis in the same order produce the same flag.*

(d) As an example, consider

$$A = \begin{bmatrix} 1 & 1 & -1 \\ 0 & 0 & 2 \\ 0 & 1 & 0 \end{bmatrix} \text{ and } R = \begin{bmatrix} 1 & 2 & 3 \\ 0 & 4 & 5 \\ 0 & 0 & 6 \end{bmatrix}.$$

For $B = AR$, show that the flag F_A associated with the column vectors of A as a basis is the same flag F_B associated with the column vectors of B as a basis.

L7#3. Confirm the second step of the induction proof for Theorem L7.10 by showing explicitly that $A^2 F_0$ is a flag associated with $Q_1 Q_2$.

L7#4. Explain the reasons for each step in the proof of Theorem L7.12.

L7#5. For Figure L7.4 in the Product-Rotation Algorithm (Section L7.4)

(a) Verify that each zero is indeed zero.

(b) Show that this procedure works if you begin with $\begin{bmatrix} C & -S \\ S & C \end{bmatrix}$ in the upper left corner and end with $\begin{bmatrix} C & -S \\ S & C \end{bmatrix}$ in the lower right corner, but *not* if you try to cancel nonzero entries by working *up* the matrix starting at the lower right. In other words, the order is important.

L7#6. Show that the count of operations in Theorem L7.13 is indeed of order n^2.

L7#7. Prove Theorem L7.11 using a manner similar to Figure L7.4. That is, show that $RQ_i^\top = RS(1, \theta_1)^\top S(2, \theta_2)^\top \cdots S(n-1, \theta_{n-1})^\top$ works to *fill* the subdiagonal with nonzero terms, from the top down but leaves all zeroes below the subdiagonal.

L7#8. Confirm that for the following matrices, QR gets hung up. The reason is that the eigenvalues do not have distinct absolute values (for different reasons, which you should explain).

(a) $\begin{bmatrix} 0 & 1 & 0 & 0 \\ 0 & 0 & 1 & 0 \\ 0 & 0 & 0 & 1 \\ 1 & 0 & 0 & 0 \end{bmatrix}$ (b) $\begin{bmatrix} 0 & 1 & 0 \\ 0 & 0 & 1 \\ 2 & -5 & 4 \end{bmatrix}$

Exercises L8 Finding Eigenvalues, Eigenvectors: Jacobi's Method

L8#1. Let A be the symmetric 2×2 matrix $\begin{bmatrix} a & b \\ b & d \end{bmatrix}$. Then Jacobi's method will multiply by a rotation matrix $Q(\theta)$, where $\tan 2\theta = 2b/(a - d)$, in order to eliminate the off-diagonal terms. Find formulas for $\cos\theta$ and $\sin\theta$ directly from a, b, and d, so as not to have to calculate θ. Note that there are four choices for θ (two for 2θ, and from each of those, two for θ); take the smallest.

L8#2. Show that for $n > 2$, the multiplication $Q_{ij}^\top A Q_{ij}$ involves just the ith and jth rows and columns.

L8#3. Sketch the quadric surfaces for the quadratic forms $\mathbf{x}^\top A\mathbf{x}$ represented by the following matrices:

(a) $\begin{bmatrix} -1 & 0 \\ 0 & 4 \end{bmatrix}$ (e) $\begin{bmatrix} 0 & 0 & 0 \\ 0 & 2 & 0 \\ 0 & 0 & 0 \end{bmatrix}$

(b) $\begin{bmatrix} 4 & 0 \\ 0 & 1 \end{bmatrix}$ (f) $\begin{bmatrix} 0 & 0 & 0 \\ 0 & 0 & 0 \\ 0 & 0 & -3 \end{bmatrix}$

(c) $\begin{bmatrix} 1 & 0 & 0 \\ 0 & -1 & 0 \\ 0 & 0 & 0 \end{bmatrix}$ (g) $\begin{bmatrix} 1 & 0 & 0 \\ 0 & 2 & 0 \\ 0 & 0 & 0 \end{bmatrix}$

(d) $\begin{bmatrix} -1 & 0 & 0 \\ 0 & 1 & 0 \\ 0 & 0 & -1 \end{bmatrix}$ (h) $\begin{bmatrix} -1 & 0 & 0 \\ 0 & 0 & 0 \\ 0 & 0 & -3 \end{bmatrix}$

L8#4. Either by hand or with the computer program *Eigenfinder*, find the eigenvalues for the following matrices, then the eigenvectors. Use these to sketch the quadric surfaces for the quadratic forms $\mathbf{x}^\top A\mathbf{x}$ represented by the following matrices. Use the program *JacobiDraw* to confirm your sketches, and label the axes in the computer printout.

(a) $\begin{bmatrix} 1 & -.5 & -.3 \\ -.5 & 2 & -.4 \\ -.3 & -.4 & 3 \end{bmatrix}$ (c) $\begin{bmatrix} 2 & 4 & -3 \\ 4 & 1 & 3 \\ -3 & 3 & -1 \end{bmatrix}$

(b) $\begin{bmatrix} -1 & 3 & 2 \\ 3 & 0 & -2 \\ 2 & -2 & 5 \end{bmatrix}$ (d) $\begin{bmatrix} 1 & -3 & -3 \\ -3 & -2 & 0 \\ -3 & 0 & 0 \end{bmatrix}$

L8#5. For the following examples, confirm that the eigenvectors found by the program *JacobiDraw* do indeed form an orthonormal basis of $\mathbb{R}^3$:
(a) Example L8.2;   (b) Example L8.3;   (c) Example L8.4.

L8#6. Identify and sketch the conic sections and quadric surfaces represented by the quadratic forms defined by the following matrices:

(a) $\begin{bmatrix} 2 & 1 \\ 1 & 3 \end{bmatrix}$

(d) $\begin{bmatrix} 2 & 1 & 0 \\ 1 & 2 & 1 \\ 0 & 1 & 2 \end{bmatrix}$

(b) $\begin{bmatrix} -1 & 0 \\ 1 & 4 \end{bmatrix}$

(e) $\begin{bmatrix} 2 & 4 & -3 \\ 4 & 1 & 3 \\ -3 & 3 & -1 \end{bmatrix}$

(c) $\begin{bmatrix} 1 & 2 \\ 2 & 4 \end{bmatrix}$

(f) $\begin{bmatrix} 2 & 0 & 3 \\ 0 & 0 & 0 \\ 3 & 0 & -1 \end{bmatrix}$

L8#7. Determine the signature of each of the following quadratic forms. Where possible, sketch the curve or surface represented by the equation.

(a) $x^2 + xy - y^2 = 1$ 　　　　　 (c) $x^2 + xy + yz = 1$

(b) $x^2 + 2xy - y^2 = 1$ 　　　　 (d) $xy + yz = 1$

Compare results with Exercise L3–4#19.

L8#8. Show that the property of being a principal axis is equivalent to being an eigenvector of the original matrix A.

Appendix L   Summary

L1 Summary: Basic Tools for Solving Systems of Linear Equations

Given *vectors*, as columns, and *matrices*, with multiplication, we can write

$$
\begin{array}{ccccc}
a_{1,1}x_1 & + & \cdots & + & a_{1,n}x_n & = & b_1 \\
\vdots & & \cdots & & \vdots & & \vdots \\
a_{m,1}x_1 & + & \cdots & + & a_{m,n}x_n & = & b_m
\end{array}
$$

as

$$
\begin{pmatrix} a_{1,1}x_1 & + & \cdots & + & a_{1,n}x_n \\ \vdots & & \cdots & & \vdots \\ a_{m,1}x_1 & + & \cdots & + & a_{m,n}x_n \end{pmatrix}
\begin{pmatrix} x_1 \\ \vdots \\ x_n \end{pmatrix}
=
\begin{pmatrix} b_1 \\ \vdots \\ b_m \end{pmatrix}
\quad \text{or, as } A\mathbf{x} = \mathbf{b}.
$$

Given a matrix A, a *row operation* on A is one of the following three operations:

(1) multiplying a row by a nonzero mumber;

(2) adding a multiple of a row onto another row;

(3) exchanging two rows.

A matrix is in *echelon form* if

(a) for every *row*, the first nonzero entry is a 1; called a *leading* 1;

(b) the leading 1 of a lower row is always to the right of that of a higher row;

(c) for every *column* containing a leading 1, all other entries are 0.

Given any matrix A, there exists a unique matrix A^* in echelon form which can be obtained from A by row operations.   (Th. L 1.6)

Row Reduction. The algorithm for bringing a matrix to echelon form is as follows:

(1) Look down the first column until you find a nonzero entry (called a *pivot*). If you don't find one, then look in the second column, etc.

(2) Move the row containing the pivot to the first row, divide that row by the pivot to make the leading entry 1.

(3) Add appropriate multiples of this row onto the other rows to cancel the entries in the first column of each of the other rows.

For a system of linear equations $Ax = b$, with the matrix $[A, b]$:

If the matrix $[A', b']$ is obtained fom $[A, b]$ by row operations, then the set of solutions of $a'x' = b'$ coincides with the set of solutions of $Ax = b$.   (Th. L 1.2)

If $[A, b]$ row reduces to echelon form $[A^*, b^*]$, then   (Th. L 1.9)

(a) If A^* is the identity matrix, then $x = b^*$ is the *unique* solution.

(b) If b^* contains a leading 1, then there are *no* solutions.

(c) If b^* does not contain a leading 1, and if A^* is not the identity, then there are *infinitely many* solutions.

A matrix A may have an *inverse* A^{-1} such that $AA^{-1} = A^{-1}A = I$; this inverse, if there is one, may be found by *row-reducing* $(A \mid I)$ to $(I \mid A^{-1})$. $Ax = b \Leftrightarrow x = A^{-1}b$.   (Th. L 1.13)

The *transpose* $A^\top$ of a matrix A interchanges the rows and the columns. $(AB)^\top = B^\top A^\top$.   (Th. L 1.17)

L2 Summary: Vocabulary for Theory of Linear Equations

One of the "hard" facts for the novice about linear algebra is simply that there are so many different ways to say any one thing. In the text of the Appendix, we've tried to mention a lot of them; here we try to be brief rather than comprehensive in that sense.

A *vector space* V is a set with two operations, *addition* and *multiplication by scalars*, satisfying ten basic axioms regarding identities, inverses, and algebraic laws.

For vectors $\mathbf{v}_1, \mathbf{v}_2, \ldots, \mathbf{v}_n \in V$, a *linear combination* $\mathbf{v} = \sum_{i=1}^{k} a_i \mathbf{v}_i$;

> the *span* is the set of linear combinations, $\mathrm{sp}\{\mathbf{v}_1, \mathbf{v}_2, \ldots, \mathbf{v}_k\}$;

> the $\mathbf{v}_i$ are *linearly independent* if no $\mathbf{v}_i$ can be written in terms of the others.

A *basis* is a set of $\mathbf{v}_i$ that spans *and* is linearly independent.

In $\mathbb{R}^n$, $n+1$ vectors are never linearly independent;   (Th. L 2.6)

> n linearly independent vectors always span;

> $n-1$ vectors never span, even if they are linearly independent;

> every basis has exactly n elements;   (Th. L 2.9)

> the *standard basis* is $\mathbf{e}_1, \mathbf{e}_2, \ldots, \mathbf{e}_n$, where $\mathbf{e}_i$ is an n-dimensional vector with 1 in the ith place and zeros elsewhere.

Any V with finite *dimension* n has n elements in any basis;   (Th. L 2.10)

> can be identified to $\mathbb{R}^n$.   (Princ. L 2.11)

Infinite-dimensional V's also exist (e.g., spaces of functions, as in Examples L 2.13, 2.15, 2.16).

A *linear transformation* $T : V \longrightarrow W$ is a *mapping* satisfying $T(a\mathbf{v}_1 + b\mathbf{v}_2) = aT(\mathbf{v}_1) + bT(\mathbf{v}_2)$.

For a finite-dimensional V, there is a *matrix* A corresponding to T.
                                                                         (Ex. L 2.14)

> $T : \mathbb{R}^n \longrightarrow \mathbb{R}^m$ is given by an $m \times n$ matrix A, with ith column $T(\mathbf{e}_i)$:
> $T(\mathbf{v}) = A\mathbf{v}$.   (Th. L 2.18)

Rotation by a counterclockwise angle θ in the x, y-plane is given by matrix   (Ex. L 2.19)

$$s_\theta = \begin{pmatrix} \cos\theta & -\sin\theta \\ \sin\theta & \cos\theta \end{pmatrix}$$

Change of basis: (Th. L 2.23)

If P = matrix whose columns are n basis vectors of $\mathbb{R}^n$,
Q = matrix whose columns are m basis vectors of $\mathbb{R}^m$,
A = matrix giving T with respect to the *old* basis, then
$Q^{-1}AP$ = matrix giving T with respect to the *new* basis.

$$
\begin{array}{ccc}
\mathbb{R}^n & \xrightarrow{Q^{-1}AP} & \mathbb{R}^m \\
\downarrow{\scriptstyle P} & & \downarrow{\scriptstyle Q} \\
\mathbb{R}^n & \xrightarrow{\;A\;} & \mathbb{R}^m
\end{array}
$$

Trace, determinant, and characteristic poynomial are all *basis-independent*.   (See Appendix L5, Theorems L 5.7, 5.10, 5.11)

Composition $T(S(\mathbf{v}))$ corresponds to matrix multiplication: $M_{T(S)} = M_T M_S$.   (Th. L 2.21)

Matrix multiplication is associative: $(AB)C = A(BC)$.   (Th. L 2.22)

For an infinite-dimensional V, some of the different possibilities for $T(f)(x)$ are as follows:

a *differential operator* acting on the functions $f(x) \in V$;

(Ex. L 2.15)

or

an *integral*, analogous to a matrix;   (Ex. L 2.17)

or

if a function can be expressed with a finite number of parameters, a finite-dimensional V.   (Ex. L 2.15, 2.20)

For $T : V \longrightarrow W$,

the *kernel* $\ker(T) = \{\mathbf{v} \mid T(\mathbf{v}) = \mathbf{0}\}$;

the *image* $\mathrm{Im}(T) = \{\mathbf{w} \mid T(\mathbf{v}) = \mathbf{w}\}$.

For linear equations:

$\ker(T)$ is a set of solutions $\mathbf{x}$ to $T(\mathbf{x}) = \mathbf{0}$;

$\mathrm{Im}(T)$ is a set of vectors $\mathbf{b}$ for which a solution exists to $T(\mathbf{x}) = \mathbf{b}$.

For finite-dimensional V, any W, and linear transformation $T : V \longrightarrow W$, then

$$\dim(\ker(T)) + \dim(\mathrm{Im}(T)) = \dim(V) \qquad \text{(Th. L 2.25)}$$

If V and W have the same dimension, then

equation $T(\mathbf{x}) = \mathbf{b}$ has solution for any $\mathbf{b}$ if and only if $T(\mathbf{x}) = \mathbf{0}$

has only the solution $\mathbf{x} = \mathbf{0}$. \qquad (Th. L 2.26)

L3–4 Summary: Inner Products

An *inner product* on a vector space V takes two vectors $\mathbf{a}$ and $\mathbf{b}$ and gives a number $\langle \mathbf{a}, \mathbf{b} \rangle$, satisfying the following rules (with bar for complex conjugate):

(i) $\langle \mathbf{a}, \mathbf{b} \rangle = \overline{\langle \mathbf{b}, \mathbf{a} \rangle}$ \hfill *symmetric*

(ii) $\langle \alpha \mathbf{a}_1 + \beta \mathbf{a}_2, \mathbf{b} \rangle = \alpha \langle \mathbf{a}_1, \mathbf{b} \rangle + \beta \langle \mathbf{a}_2, \mathbf{b} \rangle$ \hfill *linear*

(iii) $\langle \mathbf{a}, \mathbf{a} \rangle > 0$ if $\mathbf{a} \neq \mathbf{0}$. \hfill *positive definite*

$\langle A^\top \mathbf{v}, \mathbf{w} \rangle = \langle \mathbf{v}, A\mathbf{w} \rangle$ \hfill (Th. L 3.6)

$\langle \mathbf{v}, \mathbf{w} \rangle' = \langle \mathbf{v}', \mathbf{w} \rangle + \langle \mathbf{v}, \mathbf{w}' \rangle$ \hfill (Th. L 3.7)

standard inner product or *dot product* on $\mathbb{R}^n$, $\mathbf{a} \cdot \mathbf{b} = \sum_{i=1}^n a_i \overline{b_i}$

norm of a vector $\mathbf{a}$ with respect to an inner product is $\|\mathbf{a}\| = \sqrt{\langle \mathbf{a}, \mathbf{a} \rangle}$.

Schwarz's inequality: $|\langle \mathbf{a}, \mathbf{b} \rangle| \leq \|\mathbf{a}\| \|\mathbf{b}\|$, with equality iff $\mathbf{a} = c\mathbf{b}$, for some number c. \hfill (Th. L 3.9)

The *angle* α between two vectors $\mathbf{v}$ and $\mathbf{w}$ satisfies $\cos \alpha = \dfrac{\langle \mathbf{v}, \mathbf{w} \rangle}{(\|\mathbf{v}\| \|\mathbf{w}\|)}$, for $0 \leq \alpha \leq \pi$.

Two vectors $\mathbf{v}$ and $\mathbf{w}$ are *orthogonal* if $\langle \mathbf{v}, \mathbf{w} \rangle = 0$; *orthonormal* if also $\|\mathbf{v}\| = \|\mathbf{w}\| = 1$.

A set of nonzero orthogonal vectors is linearly independent.
$$\text{(Th. L 3.10)}$$

With an orthogonal basis $\mathbf{u}_1, \mathbf{u}_2, \ldots, \mathbf{u}_n$ of V, then any
$$\mathbf{a} = \sum_{i=1}^{n} \frac{\langle \mathbf{a}, \mathbf{u}_i \rangle}{\langle \mathbf{u}_i, \mathbf{u}_i \rangle} \mathbf{u}_i. \qquad \text{(Th. L 3.11)}$$

For an orthonormal basis $\mathbf{w}_1, \ldots, \mathbf{w}_n$ of V with $\mathbf{a} = \sum a_i \mathbf{w}_i$;
$\mathbf{b} = \sum b_i \mathbf{w}_i$, then $\langle a, b \rangle = \sum a_i \overline{b_i}$.
$$\text{(Th. L 3.12)}$$

Gram-Schmidt orthogonalization: For $\mathbf{u}_1, \mathbf{u}_2, \ldots, \mathbf{u}_m$ linearly independent vectors of V with inner product, this algorithm constructs an *orthonormal* set $\mathbf{w}_1, \ldots, \mathbf{w}_m$ with the same span:
$$\text{(Th. L 3.13)}$$

$$
\begin{aligned}
\mathbf{a}_1 &= \mathbf{u}_1 & \mathbf{w}_1 &= \mathbf{a}_1/\|\mathbf{a}_1\| \\
\mathbf{a}_2 &= \mathbf{u}_2 - \langle \mathbf{u}_2 - \mathbf{w}_1 \rangle \mathbf{w}_1 & \mathbf{w}_2 &= \mathbf{a}_2/\|\mathbf{a}_2\| \\
\mathbf{a}_3 &= \mathbf{u}_3 - \langle \mathbf{u}_3, \mathbf{w}_1 \rangle \mathbf{w}_1 - \langle \mathbf{u}_3, \mathbf{w}_2 \rangle \mathbf{w}_2 & \mathbf{w}_3 &= \mathbf{a}_3/\|\mathbf{a}_3\| \\
&\ \ \vdots & &\ \ \vdots \\
\mathbf{a}_m &= \mathbf{u}_1 - \sum_{1 \le j < n-1} \langle \mathbf{u}_m, \mathbf{w}_j \rangle \mathbf{w}_j & \mathbf{w}_m &= \mathbf{a}_m/\|\mathbf{a}_m\|
\end{aligned}
$$

Gram-Schmidt implies:

Any finite-dimensional V with inner product has an orthonormal basis.
$$\text{(Th. L 3.15)}$$

Any matrix A can be written as QR, with

Q orthogonal transformation, (defined below)
R upper-triangular.   (Th. L 4.3, App. L 7)

For W a finite-dimensional subspace of an inner product space V:

orthogonal projection $\pi_W(\mathbf{v}) =$ unique vector $\mathbf{w}$ in W that is closest to $\mathbf{v}$.

If $\mathbf{w}_i$ form an orthonormal basis of W, $\pi_W(\mathbf{v}) = \sum_i \langle \mathbf{v}, \mathbf{w}_i \rangle \mathbf{w}_i$
$$\text{(Th. L 3.16)}$$

orthogonal complement $W^\perp = \{ \mathbf{v} \in V \mid \langle \mathbf{v}, \mathbf{w} \rangle = 0 \text{ for all } \mathbf{w} \in W \}$.

Any vector $\mathbf{v} \in V$ can be written uniquely $\mathbf{v} = \pi_W(\mathbf{v}) + \pi_{W^\perp}(\mathbf{v})$
and $\|\mathbf{v}\|^2 = \|\pi_W(\mathbf{v})\|^2 + \|\pi_{W^\perp}(\mathbf{v})\|^2$.   (Th. L 3.19)

For linear transformations $T : V \longrightarrow V$ with matrix A and any vectors $\mathbf{v}_1, \mathbf{v}_2$ of V,
$$\text{(L 4)}$$

	for real A	for complex A
if $\langle A\mathbf{v}_1, A\mathbf{v}_2 \rangle = \langle \mathbf{v}_1, \mathbf{v}_2 \rangle$, transformation is	*orthogonal*	*unitary*
if $\langle A\mathbf{v}_1, \mathbf{v}_2 \rangle = -\langle \mathbf{v}_1, A\mathbf{v}_2 \rangle$, transformation is	*antisymmetric*	*anti-Hermitian*
if $\langle A\mathbf{v}_1, \mathbf{v}_2 \rangle = \langle \mathbf{v}_1, A\mathbf{v}_2 \rangle$, transformation is	*symmetric*	*Hermitian*

A linear transformation $T : \mathbb{R}^n \longrightarrow \mathbb{R}^n$ with standard inner product is *orthogonal* if and only if for matrix Q of T

the column vectors of Q form an orthonormal basis;

Q satisfies $Q^\top Q = I$, so $Q^\top = Q^{-1}$ and is orthogonal also.

(These last two lines are equivalent; the inner product is preserved.)
(Th. L 4.1)

A linear transformation $T : \mathbb{R}^n \longrightarrow \mathbb{R}^n$ with standard inner product is *antisymmetric* if and only if the matrix A of T is antisymmetric; that is, $A^\top = -A$.   (Th. L 4.5)

A linear transformation $T : \mathbb{R}^n \longrightarrow \mathbb{R}^n$ with standard inner product is *symmetric* if and only if the matrix A of T is symmetric; that is, $A^\top = A$.
(Th. L 4.6)

Antisymmetric linear transformations are infinitesimal orthogonal transformations. For $Q(t)$, a family of orthogonal matrices $Q^{-1}(t)Q'(t)$ is antisymmetric.   (Th. L 4.4)

Orthogonal transformations form a *group*.

A *quadratic form* on a finite-dimensional vector-space V is a *quadratic function* $q : V \longrightarrow \mathbb{R}$, *homogeneous* in the coordinates of V, with respect to any basis.

Symmetric linear transformations and quadratic forms correspond *bijectively*.

For any quadratic form q, there exists a unique symmetric linear transformation $L_q : V \longrightarrow V$ such that for any $\mathbf{v} \in V$, $q(\mathbf{v}) = \langle \mathbf{v}, L_q(\mathbf{v}) \rangle$.   (Th. L 4.13)

Any quadratic form $q(\mathbf{x}) : \mathbb{R}^n \longrightarrow \mathbb{R}$ can be decomposed into a sum of squares of linearly independent linear functions,

$$q(\mathbf{x}) = \pm\alpha_1^2(\mathbf{x}) \pm \alpha_2^2(\mathbf{x}) \pm \cdots \pm \alpha_m^2(\mathbf{x}),$$

and the number of plus and minus signs appearing will be the same for any two such decompositions of $q(\mathbf{x})$.   (Th. L 4.15, 4.19)

$m = rank$; if $m = n$, $q(\mathbf{x})$ is *nondegenerate*.

If n signs are $+$, $q(\mathbf{x})$ and its symmetric matrix are *positive definite*.

If n signs are $-$, $q(\mathbf{x})$ and its symmetric matrix are *negative definite*.

If k signs are $+$ and ℓ signs are $-$, $q(\mathbf{x})$ has *signature* (k, ℓ).

An often quicker way to determine the signature of a symmetric matrix is by the signs of its *eigenvalues* (Appendix L 6). In Appendix L 8.3 we discuss that, for a positive definite matrix, all the eigenvalues are positive, and, more generally, that a matrix of signature (k, ℓ) will have k positive eigenvalues and ℓ negative eigenvalues.

For $\langle\,,\,\rangle$ an inner product on $\mathbb{R}^n$, and G, the symmetric matrix with entries

$$G_{i,j} = \langle \mathbf{e}_i, \mathbf{e}_j \rangle,$$

the given inner product can be written in terms of the standard dot product:

$$\langle \mathbf{a}, \mathbf{b} \rangle = \mathbf{a} \cdot G\mathbf{b}. \qquad (*)$$

Conversely, if G is a positive definite symmetric matrix, then $(*)$ defines an inner product on $\mathbb{R}^n$.
(Th. L 4.8)

L5 Summary: Determinants

The *determinant* function $\det A$ satisfies three properties, exists, and is unique:
(Th. L 5.1)

(1) $\det(\mathbf{a}_1, \ldots, \mathbf{a}_{i-1}, (\alpha\mathbf{u} + \beta\mathbf{w}), \mathbf{a}_{i+1}, \ldots, \mathbf{a}_n)$
$= \alpha \det(\mathbf{a}_1, \ldots, \mathbf{a}_{i-1}, \mathbf{u}, \mathbf{a}_{i+1}, \ldots, \mathbf{a}_n)$
$\quad + \beta \det(\mathbf{a}_1, \ldots, \mathbf{a}_{i-1}, \mathbf{w}, \mathbf{a}_{i+1}, \ldots, \mathbf{a}_n).$
(multilinearity)

(2) $\det(\mathbf{a}_1, \ldots, \mathbf{a}_i, \ldots, \mathbf{a}_j, \ldots, \mathbf{a}_n)$
$= -\det(\mathbf{a}_1, \ldots, \mathbf{a}_j, \ldots, \mathbf{a}_i, \ldots, \mathbf{a}_n)$
(antisymmetry)

(3) $\det I = \det(\mathbf{e}_1, \mathbf{e}_2, \ldots, \mathbf{e}_n) = 1$, where $\mathbf{e}_i$ is the ith standard unit basis vector.
(normalization)

Construction of the determinant function can be done either by

(i) development according to the first row: $D(A)$ given by computer program;

(ii) column operations: Each of the three column operations multiplies the determinant by an appropriate factor μ:

(a) Multiply a column through by a number m. $\quad \mu = m$.
(b) Exchange two columns. $\quad \mu = -1$.
(c) Add a multiple of one column onto another. $\quad \mu = 1$.

Any square matrix can be column-reduced until

$$\det A = \mu_1 \mu_2 \ldots \mu_{n-1} \mu_n \det A_n,$$

where either $A_n = I$, and $\det A = \mu_1 \mu_2 \ldots \mu_{n-1} \mu_n$;
    or $A_n = 0$, and $\det A = 0$.

Theorems:

If A is triangular, $\det A = $ product of diagonal entries; (Th. L 5.3)

characteristic polynomial: $\det(\lambda I - A) = \prod(\lambda - a_{ii})$. (Th. L 5.13)

For A, B both $n \times n$, $\det A \det B = \det AB$. (Th. L 5.5)

For A, B both square matrices, $\det \left[\begin{array}{c|c} A & 0 \\ \hline 0 & B \end{array}\right] = \det A \det B$.

(Th. L 5.8)

A is invertible $\iff$ $\det A \neq 0$, and then $\det A^{-1} = (1/\det A)$.

(Th. L 5.6)

The following are all *basis independent*:

determinant, $\det A$ because $\det A = \det P^{-1}AP$ (Th. L 5.7)

trace, tr $A \equiv$ *sum* of diagonal entries, but tr A tr $B \neq$ tr AB

(Th. L 5.11)

characteristic polynomial, $\det(\lambda I - A)$ (Th. L 5.10)

The *k-dimensional volume* of $P_{\mathbf{v}_1,\ldots,\mathbf{v}_k}$ is defined by induction:

if $k = 1$, then the volume is simply the length: $\text{vol}_1(P_{\mathbf{v}} = \|\mathbf{v}\|$;

if $k > 1$, then $\text{vol}_k(P_{\mathbf{v}_1,\ldots,\mathbf{v}_k}) = \text{vol}_{k-1}(P_{\mathbf{v}_1,\ldots,\mathbf{v}_{k-1}}) \cdot \|\pi_{V_{k-1}^{\perp}}(\mathbf{v}_k)\|$,
where V_{k-1} is the subspace spanned by $\mathbf{v}_1, \ldots, \mathbf{v}_{k-1}$.

The volume of $P_{\mathbf{v}_1,\ldots,\mathbf{v}_k}$ does not depend on the order of $\mathbf{v}_1, \ldots, \mathbf{v}_k$.

(Th. L 5.15)

If $\mathbf{v}_1, \ldots, \mathbf{v}_k$ are orthogonal, then $\text{vol}_k(P_{\mathbf{v}_1,\ldots,\mathbf{v}_k}) = \|\mathbf{v}_1\| \cdot \ldots \cdot \|\mathbf{v}_k\|$.

(Th. L 5.16)

If Q is an orthogonal matrix, then the image under Q of a parallelepiped has the same volume as the original parallelepiped: (Th. L 5.17)

$$\text{vol}_k(P_{Q(\mathbf{v}_1),\ldots,Q(\mathbf{v}_k)}) = \text{vol}_k(P_{\mathbf{v}_1,\ldots,\mathbf{v}_k}).$$

The volume of $P_{\mathbf{v}_1,\ldots,\mathbf{v}_k}$ is measured by $\det(\mathbf{v}_1, \ldots, \mathbf{v}_k)$. (Th. L 5.18)

The determinant of an orthogonal matrix is ± 1. (Cor. L 5.19)

For any $n \times n$ matrix A, $\det A = \det A^{\top}$. (Th. L 5.20)

L6 Summary: Eigenvalues and Eigenvectors

For an $n \times n$ matrix A, if there exist $\lambda \in \mathbb{C}$ and nonzero $\mathbf{v} \in \mathbb{C}^n$ such that $\lambda \mathbf{v} = A\mathbf{v}$, then λ is an *eigenvalue* of A and $\mathbf{v}$ is its associated *eigenvector*.

λ is an eigenvalue if and only if it is a root of the *characteristic polynomial*:

$$\det(\lambda I - A) = 0, \qquad \text{(Th. L 6.3)}$$

or,

$$\lambda^n - \operatorname{tr} A \lambda^{n-1} + \cdots + (-1)^n \det A = 0. \qquad \text{(Th. L 5.11)}$$

The characteristic polynomial, and hence the set of eigenvalues, is independent of change of basis. Thus A and $P^{-1}AP$ have the same eigenvalues.   (Th. L 6.10)

The matrix A has at least one and at most n eigenvalues.
                                                                        (Th. L 6.4)

$\left(\begin{array}{l} \operatorname{tr} A = \text{sum} \\ \det A = \text{product} \end{array} \right\}$ of eigenvalues, each taken with its multiplicity.

                                                                        (Th. L 6.5)

Eigenvectors with distinct eigenvalues are linearly independent.
                                                                        (Th. L 6.7)

If n λ's are distinct, then $\mathbb{C}^n$ has an *eigenbasis* (basis of eigenvectors).
                                                                        (Th. L 6.8)

If m λ's are distinct, for $m \leq n$, then $\mathbb{C}^n$ has an eigenbasis for A if and only if dim(eigenspace E_λ of eigenvectors) = multiplicity of λ in the characteristic polynomial.   (Th. L 6.10)

If P is an invertible matrix whose columns are eigenvectors, then we can *diagonalize*:

$$P^{-1}AP = \begin{pmatrix} \lambda_1 & & 0 \\ & \ddots & \\ 0 & & \lambda_n \end{pmatrix},$$

with all eigenvalues of A along the diagonal.   (Th. L 6.11)

Most matrices can be diagonalized; all can be *triangularized*; the eigenvalues of a triangular matrix will be along the diagonal.

There exists unitary matrix M such that $M^{-1}AM$ is *upper-triangular*.
                                                                        (Th. L 6.16)

For any A and any $\varepsilon > 0$, there exists a matrix M with orthogonal columns (but not normalized) such that $M^{-1}AM$ is *upper-triangular* with off-diagonal terms of absolute value $< \varepsilon$.

                                                            (Th. L 6.17, 6.18, 6.19)

For a linear transformation with $n \times n$ matrix A: (Th. L 6.20)

(a) if A is real symmetric or Hermitian, then the eigenvalues of A are real;

(b) if A is real antisymmetric or anti-Hermitian, then the eigennvalues of A are purely imaginary;

(c) if A is real orthogonal or unitary, then the eigenvalues of A are of absolute value 1;

In all three cases, any two eigenvectors of A with distinct eigenvalues are orthogonal.

Spectral Theorem: (Th. L 6.22)

For A Hermitian, anti-Hermitian, or unitary, $\mathbb{C}^n$ has an orthonormal eigenbasis for A;

If A is real symmetric, the eigenvalues are real, and the basis vectors can be chosen real (meaning that $P^{-1}AP$ is diagonal and the entries are the eigenvalues.)

Appendix L 8, Jacobi's method, gives a constructive proof and discusses for quadratic forms (Appendix L 4.4) the fact that the *signature* (k, ℓ) is given by the signs of the eigenvalues of the symmetric matrix associated with the quadratic form: the matrix will have k positive eigenvalues and ℓ negative eigenvalues; a positive definite matrix has all positive eigenvalues.

Factoring the characteristic polynomial p_A for an $n \times n$ matrix A:

$$p_A(A) = 0.$$ (Th. L 6.23)

If $p_A = p_1 p_2$, where p_1 and p_2 are relatively prime with leading coefficient 1, and $V_i = \ker p_i(A)$, then   (Th. L 6.25)

(i) any vector $\mathbf{x} \in \mathbb{R}^n$ can be written uniquely as $\mathbf{x}_1 + \mathbf{x}_2$, where $\mathbf{x}_i \in V_i$;

(ii) for any vector $\mathbf{x}_i \in V_i$, $A\mathbf{x}_i \in V_i$;

(iii) if an eigenvalue λ_i is a root of p_i, its corresponding eigenvectors $\in V_i$;

(iv) the characteristic polynomial of $A : V_i \longrightarrow V_i$ is p_i.

This is proved using Bezout's Theorem: (Th. L 6.27)

Two polynomials p_1 and p_2 are relatively prime if and only if there exist polynomials q_1 and q_2 such that $p_1 q_1 + p_2 q_2 = 1$.

Appendices L7 and L8 each elaborate on a single result, and so are not summarized here.

Appendix T

Key Theorems and Definitions from Parts I and III

FOR A FIRST ORDER DIFFERENTIAL EQUATION

$$x' = f(t, x).$$

Definition 1.3.1 (Lower fence). For the differential equation $x' = f(t, x)$, we call a continuous and continuously differentiable function $\alpha(t)$ a *lower fence* if $\alpha'(t) \leq f(t, \alpha(t))$ for all $t \in \mathbf{I}$.

lower fence

Definition 1.3.2 (Upper fence). For the differential equation $x' = f(t, x)$, we call a continuous and continuously differentiable function $\beta(t)$ an *upper fence* if $f(t, \beta(t)) \leq \beta'(t)$ for all $t \in \mathbf{I}$.

upper fence

An intuitive idea is that *a lower fence pushes solutions up, an upper fence pushes solutions down.*

Definition 1.4.1 (Funnel). If for the differential equation $x' = f(t, x)$, over some t-interval $\mathbf{I}$, $\alpha(t)$ is a *lower fence* and $\beta(t)$ an *upper fence*, and if $\alpha(t) < \beta(t)$, then the set of points (t, x) for $t \in \mathbf{I}$ with $\alpha(t) \leq x \leq \beta(t)$ is called a *funnel.*

funnel

Once a solution enters a funnel, it stays there. (See Theorem 4.7.2.)

Definition 1.4.3 (Antifunnel). If for the differential equation $x' = f(t, x)$, over some t-interval **I**, $\alpha(t)$ is a *lower fence* and $\beta(t)$ a *upper fence*, and if $\alpha(t) > \beta(t)$, then the set of points (t, x) for $t \in \mathbf{I}$ with $\alpha(t) \geq x \geq \beta(t)$ is called an *antifunnel*.

Solutions are, in general, *leaving* an antifunnel. But *at least one solution is trapped inside the antifunnel*, as is guaranteed by Theorems 4.7.3, 4.7.4, and 4.7.5.

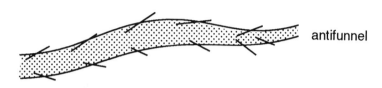

antifunnel

Sharpening the Theory

Definition 4.3.1 (Lipschitz Constant). A number K is a *Lipschitz constant* with respect to x for a function $f(t, x)$ defined on a region A of $\mathbb{R}^2$ (the t, x-plane) if

$$|f(t, x_1) - f(t, x_2)| \leq K|x_1 - x_2|, \tag{9}$$

for all (t, x_1), (t, x_2) in A. We call this inequality a *Lipschitz condition* in x.

A differential equation $x' = f(t, x)$ admits a Lipschitz condition if the function f admits a Lipschitz condition. What makes a Lipschitz condition important is that

> the Lipschitz constant K bounds the rate at which solutions can pull apart,

as the following computation shows. If in the region A, $u_1(t)$ and $u_2(t)$ are two solutions to the differential equation $x' = f(t, x)$, then they pull apart at a rate

$$|(u_1 - u_2)'(t)| = |f(t, u_1(t)) - f(t, u_2(t))| \leq K|u_1(t) - u_2(t)|.$$

So in practice we will want the smallest possible value for K.

As we will see, such a number K also controls for numerical solutions of a differential equation the rate at which errors compound.

The *existence* of a Lipschitz condition is often very easy to ascertain, if the function in question is *continuously differentiable*.

Theorem 4.3.2. *If on a rectangle $R = [a,b] \times [c,d]$ a function $f(t,x)$ is differentiable in x with continuous derivative $\partial f/\partial x$, then $f(t,x)$ satisfies a Lipschitz condition in x with the best possible Lipschitz constant equal to the maximum value of $|\partial f/\partial x|$ achieved in R. That is,*

$$K \equiv \sup_{(t,x) \in R} \left| \frac{\partial f}{\partial x} \right|.$$

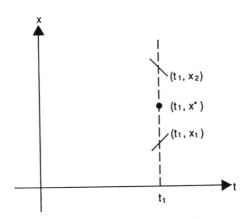

FIGURE 4.3.1. (From Part I.)

Theorem 4.4.1 (Fundamental Inequality). *If, on a rectangle $R = [a,b] \times [c,d]$, the differential equation $x' = f(t,x)$ satisfies a Lipschitz condition with respect to x, with Lipschitz constant $K \neq 0$, and if $u_1(t)$ and $u_2(t)$ are two approximate solutions, piecewise differentiable, satisfying*

$$|u_1'(t) - f(t, u_1(t))| \leq \varepsilon_1$$
$$|u_2'(t) - f(t, u_2(t))| \leq \varepsilon_2$$

for all $t \in [a,b]$ at which $u_1(t)$ and $u_2(t)$ are differentiable; and if for some $t_0 \in [a,b]$

$$|u_1(t_0) - u_2(t_0)| \leq \delta;$$

then for all $t \in [a,b]$,

$$\boxed{|u_1(t) - u_2(t)| \leq \delta e^{K|t-t_0|} + \left(\frac{\varepsilon}{K} \right) \left(e^{K|t-t_0|} - 1 \right),} \tag{10}$$

where $\varepsilon = \varepsilon_1 + \varepsilon_2$. See Figure 4.4.1 on the next page.

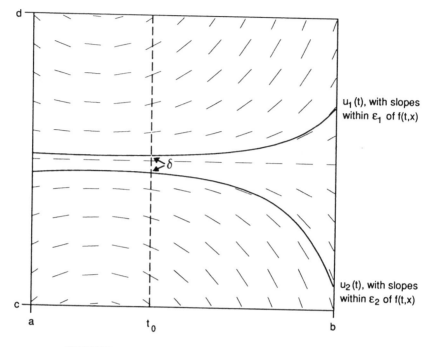

FIGURE 4.4.1. Slope field for $x' = f(t, x)$. (From Part I.)

Theorem 4.5.1 (Uniqueness). *Consider the differential equation $x' = f(t, x)$, where f is a function satisfying a Lipschitz condition with respect to x on a rectangle $R = [a, b] \times [c, d]$ in the tx-plane. Then for any given initial condition (t_0, x_0), if there exists a solution, there is exactly one solution $u(t)$ with $u(t_0) = x_0$.*

Theorem 4.7.2 (Funnel Theorem). *Let $\alpha(t)$ and $\beta(t)$, $\alpha(t) \leq \beta(t)$ be two fences defined for $t \in [a, b)$, where b might be infinite, defining a funnel for the differential equation $x' = f(t, x)$. Furthermore, let $f(t, x)$ satisfy a Lipschitz condition in the funnel.*

Then any solution $x = u(t)$ that starts in the funnel at $t = a$ remains in the funnel for all $t \in [a, b)$.

FIGURE 4.7.2. Funnel. (From Part I.)

Theorem 4.7.3 (Antifunnel Theorem; Existence). *Let $\alpha(t)$ and $\beta(t)$, $\beta(t) \leq \alpha(t)$, be two fences defined for $t \in [a, b)$, where b might be infinite, that bound an **antifunnel** for the differential equation $x' = f(t, x)$. Furthermore, let $f(t, x)$ satisfy a Lipschitz condition in the antifunnel.*

Then there exists a solution $x = u(t)$ that remains in the antifunnel for all $t \in [a, b)$ where $u(t)$ is defined.

FIGURE 4.7.3. Antifunnel. (From Part I).

The really interesting results about antifunnels are the ones which give properties which ensure that the solutions which stay in them are unique. We will give two such properties; the first is a special case of the second, but is so much easier to prove that it seems worthwhile to isolate it.

Theorem 4.7.4 (First uniqueness criterion for antifunnels). *Let $\alpha(t)$ and $\beta(t)$, $\beta(t) \leq \alpha(t)$, be two fences defined for $t \in [a, b)$ that bound an **antifunnel** for the differential equation $x' = f(t, x)$. Let $f(t, x)$ satisfy a Lipschitz condition in the antifunnel. Furthermore, let the antifunnel be* **narrowing,** *with*

$$\lim_{t \to b}(\alpha(t) - \beta(t)) = 0.$$

If $\partial f / \partial x \geq 0$ in the antifunnel, then there is a unique solution that stays in the antifunnel.

Theorem 4.7.5 (Second uniqueness criterion for antifunnels). *Let*
$\alpha(t)$ *and* $\beta(t)$, $\beta(t) \leq \alpha(t)$, *be two fences defined for* $t \in [a, b]$ *that bound*
an **antifunnel** *for the differential equation* $x' = f(t, x)$. *Let* $f(t, x)$ *satisfy*
a Lipschitz condition in the antifunnel. Furthermore, let the antifunnel be
narrowing, *with*

$$\lim_{t \to b} (\alpha(t) - \beta(t)) = 0.$$

If $(\partial f / \partial x)(t, x) \geq w(t)$ *in the antifunnel, where* $w(t)$ *is a function satis-*
fying

$$\int_a^b w(s)ds > -\infty,$$

then there is a unique solution which stays in the antifunnel.

Note that the first uniqueness criterion is a special case of the second,
with $w(t) = 0$, since $\int_a^b 0 \, ds = 0 > -\infty$.

NUMERICAL METHODS

Our numerical methods for solving a differential equation $x' = f(t, x)$ are
based on the same idea as Euler's method, in that using intervals of step
size h,

$$t_{i+1} = t_i + h \text{ and } x_{i+1} = x_i + hm, \text{ where } m = slope.$$

1. *Euler.* For *Euler's* method we simply use the slope, $f(t_i, x_i)$, available
at the point where we begin to "follow our noses," the left endpoint of the
interval.

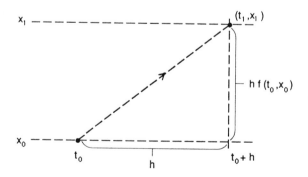

FIGURE 3.1.1. Euler's method. Single step, starting at (t_0, x_0). (From Part I.)

For fancier methods we "sniff ahead," and then can do a better job of "following."

2. *Midpoint Euler.* For the *midpoint Euler* method (also called *modified Euler*) we use the slope m_M at the *midpoint* of the segment we would have obtained with Euler's method, as shown in Figure 3.2.1.

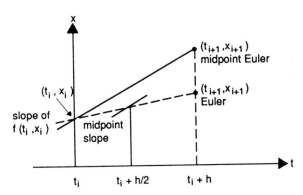

FIGURE 3.2.1. Midpoint slope $= m_M = f(t_i + \frac{h}{2}, x_i + \frac{h}{2}f(t_i, x_i))$. (From Part I.)

3. *Runge-Kutta.* For the *Runge–Kutta* method we use slope m_{RK}, a weighted average of beginning, midpoint, and ending slopes. The result is a much better convergence with fewer steps.

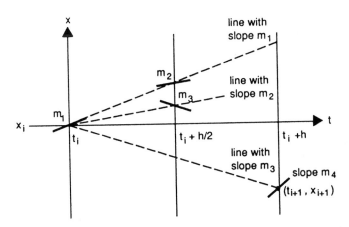

FIGURE 3.2.2. Runge–Kutta makes a linear combination of these four slopes using $m_{RK} = \left(\frac{1}{6}\right)(m_1 + 2m_2 + 2m_3 + m_4)$. (From Part I.)

$$m_1 = f(t_i, x_i) \qquad\qquad\qquad \text{slope at } \textit{beginning} \text{ of interval}$$

$$m_2 = f\left(t_i + \tfrac{h}{2}, x_i + \tfrac{h}{2}m_1\right)$$

slope at *midpoint* of a segment with slope m_1

$$m_3 = f\left(t_i + \tfrac{h}{2}, x_i + \tfrac{h}{2}m_2\right)$$

slope at *midpoint* of a segment with slope m_2

$$m_4 = f(t_i + h, x_i + hm_3)$$

slope at *end* of a segment with slope m_3

$$m_{RK} = \tfrac{1}{6}(m_1 + 2m_2 + 2m_3 + m_4).$$

Theorem 4.5.2 (Bound on slope error, Euler's method). *Consider the differential equation* $x' = f(t, x)$, *where* f *is a continuous function on a rectangle* $R = [a, b] \times [c, d]$ *in the tx-plane. Let* u_h *be the Euler approximate solution with step* h. *Then*

(i) *for every* h *there is an* ε_h *such that* u_h *satisfies*

$$|u_h'(t) - f(t, u_h(t))| \le \varepsilon_h$$

at any point where u_h *is differentiable (and the inequality holds for left- and right-hand derivatives elsewhere);*

(ii) $\varepsilon_h \to$ *as* $h \to 0$;

(iii) *if furthermore* f *is a function on* R *with continuous derivatives with respect to* x *and* t, *with the following bounds over* R:

$$\sup |f| \le M; \quad \sup\left|\frac{\partial f}{\partial t}\right| \le P; \quad \sup\left|\frac{\partial f}{\partial x}\right| \le K,$$

then there is a specific bound on ε_h:

$$|u_h'(t) - f(t, u_h(t))| \le h(P + KM).$$

Theorem 4.6.1 (Bound on slope error, midpoint Euler method). *Consider the differential equation* $x' = f(t, x)$, *where* f *is a continuously differentiable function on a rectangle* $R = [a, b] \times [c, d]$ *in the tx-plane. Consider also the midpoint Euler approximate solution* u_h, *with step* h. *Then there is an* ε_h *such that* u_h *satisfies*

$$|u_h'(t) - f(t, u_h(t))| \le \varepsilon_h$$

at any point where u_h *is differentiable (or has left- and right-hand derivatives elsewhere), and* $\varepsilon_h \to 0$ *as* $h \to 0$.

Furthermore, if f is a function on R with continuous derivatives up to order two with respect to x and t, then there is a constant B_M such that

$$|u'_h(t) - f(t, u_h(t))| \leq B_M h^2.$$

This computation is not too difficult if you don't insist on *knowing B_M*, which is a fairly elaborate combination of sup's of the second order partial derivatives.

FROM MULTIVARIABLE CALCULUS

Theorem 13.6.4 (Implicit Function Theorem). *Let $U = V \times W$ be an open set in $\mathbb{R}^n \times \mathbb{R}^m$, $\mathbf{f}: U \to \mathbb{R}^m$ a differentiable mapping, and $(\mathbf{x}_0, \mathbf{y}_0)$ a point in U such that $\mathbf{f}(\mathbf{x}_0, \mathbf{y}_0) = 0$. If the square $m \times m$ matrix of partial derivatives $\partial \mathbf{f} / \partial \mathbf{y}(\mathbf{x}_0, \mathbf{y}_0)$ is invertible, and if $\partial \mathbf{f} / \partial \mathbf{y}$ satisfies the Lipschitz condition*

$$\|\partial \mathbf{f} / \partial \mathbf{y}(\mathbf{x}, \mathbf{y}_1) - \partial \mathbf{f} / \partial \mathbf{y}(\mathbf{x}, \mathbf{y}_2)\| \leq K \mathbf{y}_1 - \mathbf{y}_2\|$$

on U for some K, then the equation $\mathbf{f}(\mathbf{x}, \mathbf{y}) = 0$ expresses $\mathbf{y}$ locally as an implicit function of $\mathbf{x}$ near x_0.

More precisely, there exists a neighborhood V_1 of $\mathbf{x}_0$ in $\mathbb{R}^n$ and a differentiable function $\mathbf{g}: V_1 \to \mathbb{R}^m$ with $\mathbf{g}(\mathbf{x}_0) = \mathbf{y}_0$ and $\mathbf{f}(\mathbf{x}, \mathbf{g}(\mathbf{x})) = 0$. This function can be computed as follows: Choose $\mathbf{x}$ and solve the equation $\mathbf{f}(\mathbf{x}, \mathbf{y}) = 0$ for $\mathbf{y}$ by Newton's method, starting at $\mathbf{y}_0$, i.e., set

$$\mathbf{y}_{n+1} = \mathbf{y}_n - [\partial \mathbf{f} / \partial \mathbf{y}(\mathbf{x}, \mathbf{y}_n)^{-1}] \mathbf{f}(\mathbf{x}, \mathbf{y}_n).$$

Then if $\mathbf{x}$ is chosen sufficiently close to $\mathbf{x}_0$, this sequence will converge to a limit which depends of course on $\mathbf{x}$; if you set $g(\mathbf{x})$ equal to this limit, then g will be the required implicit function.

Moreover, the function g computed this way is differentiable at $\mathbf{x}_0$, with derivative

$$d_{\mathbf{x}_0} \mathbf{g} = -\frac{\partial \mathbf{f}}{\partial \mathbf{y}}(\mathbf{x}_0, \mathbf{y}_0)^{-1} \frac{\partial \mathbf{f}}{\partial \mathbf{x}}(\mathbf{x}_0, \mathbf{y}_0).$$

Corollary 13.6.5 (Inverse Function Theorem). *Let $U \subset \mathbb{R}^m$ be an open neighborhood of $\mathbf{0}$, and $\mathbf{f}: U \to \mathbb{R}^m$ a continuously differentiable function, with $d_{\mathbf{x}} \mathbf{f}$ satisfying a Lipschitz condition*

$$|d_{\mathbf{x}_1} \mathbf{f} - d_{\mathbf{x}_1} \mathbf{f}| \leq K |\mathbf{x}_1 - \mathbf{x}_2|$$

for some constant K. If $d_0 \mathbf{f}$ is invertible, there exist neighborhoods $U_1 \subset U$ and V_1 of $\mathbf{0}$ in $\mathbb{R}^m$ and a continuously differentiable mapping $\mathbf{g}: V_1 \to U_1$ such that $\mathbf{g} \circ \mathbf{f}$ is the identity of U_1 and $\mathbf{f} \circ \mathbf{g}$ is the identity of V_1. The derivative of $\mathbf{g}$ at the origin is given by

$$d_0 \mathbf{g} = (d_0 \mathbf{f})^{-1}.$$

References

Arnold, V.I., *Ordinary Differential Equations*, M.I.T. Press, 1973 (Russian original, Moscow, 1971).

Artigue, Michèle and Gautheron, Véronique, *Systèmes Différentiels: Étude Graphique*, CEDIC, Paris, 1983.

Birkhoff, G. and Rota, G. C., *Ordinary Differential Equations*, Ginn, 1962.

Borrelli, R., Boyce, W. E., and Coleman, C., *Differential Equations Laboratory Workbook*, John Wiley, 1992.

Borrelli, R. and Coleman, C., *Differential Equations: A Modeling Approach*, Prentice-Hall 1987.

Boyce, W. E. and DiPrima, R. C., *Elementary Differential Equations*, John Wiley, 1969, 1994.

Dieudonne, Jean, *Calcul Infinitesimal*, Hermann, Paris, 1968.

Guckenheimer, J. and Holmes, P., *Nonlinear Oscillations, Dynamical Systems, and Bifurcations of Vector Fields*, Springer-Verlag, Applied Mathematical Sciences 42, 1983.

Hale, Jack, *Ordinary Differential Equations*, Wiley-Interscience, 1969.

Hale, Jack and Koçak, Huseyn, *Dynamics and Bifurcations*, Springer-Verlag, 1991.

Hirsch, M.W. and Smale, S., *Differential Equations, Dynamical Systems, and Linear Algebra*, Academic Press, 1974.

Kaplansky, I., *Differential Algebra*.

LaSalle, J.P. and Lefschetz, S., *Stability by Liapunov's Direct Method with Applications*, Academic Press, 1961.

Lorenz, E.N., "Deterministic Nonperiodic Flow," *J. Atmos. Sci.* **20**, 130, 1963.

Marsden, Jerrold, and McCracken, M., *The Hopf Bifurcation and Its Applications*, Springer-Verlag, New York 1976.

Peixoto, M.M., *Dynamical Systems*, Proceedings of 1971 Bahia conference, Academic Press, 1973.

Peixoto, M.M., "Structural Stability on Two Dimensional Manifolds," *Topology* **1**, 101, 1962.

Roessler, O.E., "Continuous Chaos — Four Prototype Equations," *Bifurcation Theory and Applications in Scientific Disciplines*, edited by O. Gurel and O.E. Rossler, *Ann. N.Y. Acad. Sci.* **316**, 376, 1979.

Sanchez, David A., Allen, Richard C., Jr., and Kyner, Walter T., *Differential Equations, An Introduction*, Addison-Wesley, 1983.

Simmons, George F., *Differential Equations, with Applications and Historical Notes*, McGraw-Hill, 1972.

Smale, Stephen, "Structurally Stable Systems Are Not Dense," *American Journal of Mathematics* **88**, 491, 1966.

Sparrow, C., *The Lorenz Equations: Bifurcations, Chaos, and Strange Attractors*, Springer-Verlag, Applied Mathematical Sciences 41, 1982.

Strogatz, Steven H., *Nonlinear Dynamics and Chaos, with Applictions to Physics, Biology, Chemistry, and Engineering*, Addison-Wesley, 1994.

Williams, R.F., "The Structure of Lorenz Attractors," *Publ. Math. IHES* **50**, 101, 1979.

Ye Yan-Qian, *The Theory of Limit Cycles*, American Mathematical Society, 1984.

Software

Hubbard, J.H. and West, B.H., *MacMath 9.2, A Dynamical Systems Software Package*, Springer-Verlag, 1994.

Hubbard, J.H., Hinkle, B. and West, B.H., *Extensions of MacMath*, Springer-Verlag (to appear).

Gollwitzer, Herman, *Differential Systems*, U'Betcha Publications, 1991.

Khibnik, Alexander and Levitin, V.V., *TraX: Analysis of Dynamical Systems*, Exeter Software, 1992.

Koçak, Huseyn, *Phaser*, Springer-Verlag, 1989.

Answers to Selected Problems

Solutions for Exercises 6.1

6.1#2. (b)

$$\frac{dx}{dt} = x + 2 - y$$

$$\frac{dy}{dt} = x^2 - y.$$

Vertical isocline: line $y = x+2$. Horizontal isocline: parabola $y = x^2$. Notice that in Example 6.1.5 the only difference is the sign of x'; however, the two phase planes do *not* look at all alike.

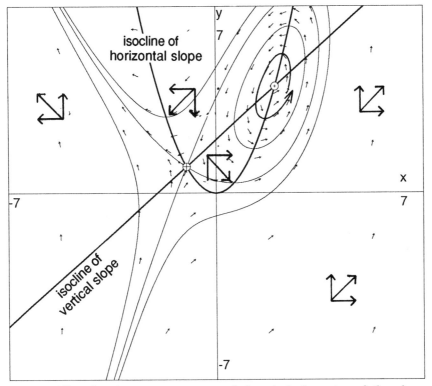

Note: This drawing captures most of the vital character of the phase plane for this system, which is all we can expect without the further study provided in Section 8.6. If you make a computer drawing, you will see that the "spirals" actually close up into ellipses. But without knowing about area-preserving systems, you could not predict that fine point in a hand sketch.

6.1#2. (g)

$$\frac{dx}{dt} = x\left(1 - \frac{1}{4}x - y\right)$$

$$\frac{dy}{dt} = y(2 - y - x).$$

Vertical isoclines: $x = 0$ and $y = 1 - \frac{1}{4}x$. Horizontal isoclines: $y = 0$ and $y = 2 - x$.

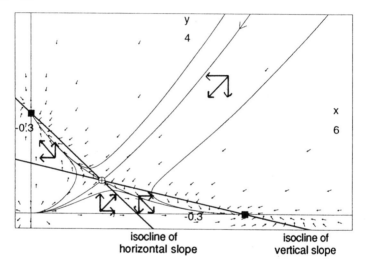

isocline of   isocline of
horizontal slope   vertical slope

6.1#4. (c) The *vertical* isoclines are the 2 circles: $(x - 2)^2 + y^2 = 1$ and $x^2 + y^2 = 9$. The *horizontal* isocline is the circle $(x - 1)^2 + y^2 = 4$.

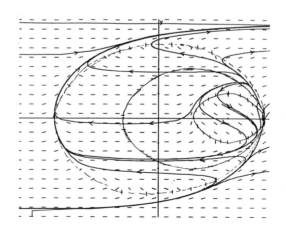

Solutions for Exercises 6.2

6.2#1. To prove the Fundamental Inequality, let

$$\gamma(t) = \|\mathbf{u}_1(t) - \mathbf{u}_2(t)\|.$$

Then

$$\gamma^2(t) = \langle \mathbf{u}_1(t) - \mathbf{u}_2(t), \mathbf{u}_1(t) - \mathbf{u}_2(t)\rangle$$

and differentiation with respect to t gives

$$2\gamma(t)\gamma'(t) = 2\langle \mathbf{u}_1(t) - \mathbf{u}_2(t), \mathbf{u}_1'(t) - \mathbf{u}_2'(t)\rangle$$

therefore, by the Schwarz Inequality,

$$|\gamma(t)\gamma'(t)| \le \|\mathbf{u}_1(t) - \mathbf{u}_2(t)\| \, \|\mathbf{u}_1'(t) - \mathbf{u}_2'(t)\|$$
$$\gamma(t)|\gamma'(t)| \le \gamma(t)\|\mathbf{u}_1'(t) - \mathbf{u}_2'(t)\|, \quad \text{since } \gamma(t) \ge 0 \text{ for all } t.$$

If $\gamma(t) \equiv 0$, on some t-interval, then $\mathbf{u}_1(t) = \mathbf{u}_2(t)$ and hence $\mathbf{u}_1' = \mathbf{u}_2'$ in that interval. Where $\gamma(t) \ne 0$, the above equation says

$$|\gamma'(t)| \le \|\mathbf{u}_1'(t) - \mathbf{u}_2'(t)\|$$
$$= \|\mathbf{f}(t, \mathbf{u}_1(t)) + \varepsilon_1 - \mathbf{f}(t, \mathbf{u}_2(t)) - \varepsilon_2\|$$
$$\le \|\mathbf{f}(t, \mathbf{u}_1(t)) - \mathbf{f}(t, \mathbf{u}_2(t))\| + \varepsilon$$
$$\le k\|\mathbf{u}_1(t) - \mathbf{u}_2(t)\| + \varepsilon, \quad \text{by the Lipschitz condition.}$$

Therefore, $\gamma'(t) \le k\gamma(t) + \varepsilon$, and the proof in Part I applies. Suppose $\mathbf{u}_1$ and $\mathbf{u}_2$ are piecewise linear functions, for example $\mathbf{u}_1(t) = \begin{bmatrix} 2t+1 \\ -t \end{bmatrix}$, $\mathbf{u}_2(t) = \begin{bmatrix} 2.1t + .9 \\ -1.1t \end{bmatrix}$. Then

$$\|\mathbf{u}_1(t) - \mathbf{u}_2(t)\| = \left\| \begin{bmatrix} -0.1t + 0.1 \\ 0.1t \end{bmatrix} \right\|$$
$$= \sqrt{(-0.1t + 0.1)^2 + (0.1t)^2} = \sqrt{0.02t^2 - 0.02t + 0.01}$$

is not a linear function, but the proof still works because of the Schwarz Inequality.

6.2#8. (a) The rate of change of the function F along a solution curve is

$$\frac{dF}{dt} = \frac{\partial F}{\partial x}\frac{dx}{dt} + \frac{\partial F}{\partial y}\frac{dy}{dt} = 2x(y) + 2y(-x) \equiv 0.$$

Therefore, the trajectories of the solutions in the phase plane are circles $x^2 + y^2 = C$.

(b) Solving the system by Euler's method is equivalent to iterating the equations

$$x_{n+1} = x_n + hx'_n = x_n + hy_n$$
$$y_{n+1} = y_n + hy'_n = y_n - hx_n.$$

In matrix form,

$$\begin{pmatrix} x \\ y \end{pmatrix}_{n+1} = \begin{pmatrix} 1 & h \\ -h & 1 \end{pmatrix} \begin{pmatrix} x \\ y \end{pmatrix}_n = H \begin{pmatrix} x \\ y \end{pmatrix}_n.$$

If the initial conditions are $x = x_0$, $y = y_0$, Euler's method will produce the sequence of points $H \begin{pmatrix} x_0 \\ y_0 \end{pmatrix}$, $H^2 \begin{pmatrix} x_0 \\ y_0 \end{pmatrix}, \ldots, H^n \begin{pmatrix} x_0 \\ y_0 \end{pmatrix}$. The matrix

$$H = \sqrt{1+h^2} \begin{pmatrix} \frac{1}{\sqrt{1+h^2}} & \frac{h}{\sqrt{1+h^2}} \\ -\frac{h}{\sqrt{1+h^2}} & \frac{1}{\sqrt{1+h^2}} \end{pmatrix} = \sqrt{1+h^2} \begin{pmatrix} \cos\theta & \sin\theta \\ -\sin\theta & \cos\theta \end{pmatrix},$$

where $\theta = \sin^{-1}\left(\frac{h}{\sqrt{1+h^2}}\right)$, and $\begin{pmatrix} \cos\theta & \sin\theta \\ -\sin\theta & \cos\theta \end{pmatrix}$ is rotation by an angle θ in the clockwise direction. Hence the nth power of H is

$$H^n = (1+h^2)^{n/2} \begin{pmatrix} \cos(n\theta) & \sin(n\theta) \\ -\sin(n\theta) & \cos(n\theta) \end{pmatrix}.$$

Thus, at each step the vector $\begin{pmatrix} x_n \\ y_n \end{pmatrix}$ is rotated by a fixed angle, but also multiplied by $\sqrt{1+h^2} > 1$, so the Euler solution will spiral as shown below:

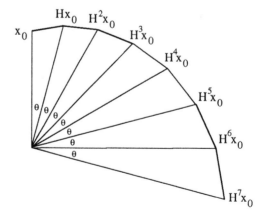

Solutions for Exercises 6.3

6.3#2. (a) For x and y both positive, $F(x, y) = |x|^b|y|^a e^{-(fx+cy)}$ is continuous and differentiable. Any maximum must occur either on the boundary (positive x or y axes) or at an interior point where $\frac{\partial F}{\partial x} = \frac{\partial F}{\partial y} = 0$. With a, b, c and f all positive, $F \to 0$ as $x \to 0$ or $y \to 0$, and the exponential approaches 0 if either x or $y \to \infty$. Since F is positive in the interior of the positive quadrant, its maximum must occur where

$$\begin{cases} \dfrac{\partial F}{\partial x} = 0 = bx^{b-1}y^a e^{-(fx+cy)} - fx^b y^a e^{-(fx+cy)} \\ \qquad \Rightarrow bx^{b-1}y^a = fx^b y^a \Rightarrow b = fx \\[2mm] \dfrac{\partial F}{\partial y} = 0 = ax^b y^{a-1} e^{-(fx+cy)} - cx^b y^a e^{-(fx+ey)} \\ \qquad \Rightarrow ax^b y^{a-1} = cx^b y^a \Rightarrow a = cy \end{cases}$$

That is, $x = b/f$ and $y = a/c$.

(b) If a level curve of F comes very close to the origin, it will also have to travel very far away. This would mean that if a population became very small, at some point in the cycle it would have to become very large. When the populations get away from their equilibrium value, they tend to oscillate wildly.

Solutions for Exercises 6.4

6.4#2. Equation (a) is $x'' + 3x' + 5x = 0$, which has solution

$$x = e^{-\frac{3}{2}t}\left(c_1 \cos\left(\frac{\sqrt{11}}{2}t\right) + c_2 \sin\left(\frac{\sqrt{11}}{2}t\right) \right),$$

as you will see, using the methods of Chapter 7. If you let $y = \frac{dx}{dt}$, $y' = y\frac{dy}{dx}$, then

$$x'' = -3x' - 5x$$
$$y\frac{dy}{dx} = -3y - 5x$$

can be solved by letting $v = \frac{y}{x}$ or $y = xv$ and $\frac{dy}{dx} = x\frac{dv}{dx} + v$. Then

$$\frac{dy}{dx} = x\frac{dv}{dx} + v = -3 - \frac{5}{v}$$

and you need to solve

$$\frac{v^2 + 3v + 5}{v} = -x\frac{dv}{dx} \quad \text{or} \quad \frac{v\,dv}{v^2 + 3v + 5} = -\frac{dx}{x}$$

for $v = \frac{y}{x}$, then solve the resulting equation with y replaced by $\frac{dx}{dt}$. This is definitely the *hard way*—you can forget it.

For equation (b), $x'' + 3tx' = t$, letting $y = x'$ and $y' = x''$, then $y' + 3ty = t$ and $\frac{dy}{dt} = t(1 - 3y)$. Therefore, $\int \frac{dy}{1-3y} = \int t\,dt$ by separation of variables, and

$$\frac{\ln|1 - 3y|}{-3} = \frac{t^2}{2} + C$$

$$1 - 3y = \alpha e^{-\frac{3}{2}t^2}$$

$$\text{or } y = \frac{1}{3} - \alpha' e^{-\frac{3}{2}t^2} \equiv \frac{dx}{dt}.$$

This leads to $x(t) = \frac{1}{3}t + c_1 \int_0^t e^{-\frac{3}{2}s^2}\,ds + c_2$, which is about as good a solution as you can get by any method.

For equation (c), $x'' - xx' = 0$, replacing x' by y and x'' by $y\frac{dy}{dx}$, $y\frac{dy}{dx} - xy = 0 \Rightarrow$ either $y = 0$ so that $x(t) = c_1$, or $\frac{dy}{dx} - x = 0$ so $y = \frac{x^2}{2} + c_2$, or $\frac{dx}{dt} = \frac{x^2}{2} + c_3$, which can be solved by separation of variables $\frac{2dx}{x^2+c} = dt$.

if $c > 0$: $\frac{2}{\sqrt{c}} \tan^{-1}\left(\frac{x}{\sqrt{c}}\right) = t + k \Rightarrow x = \sqrt{c}\tan\left(\frac{\sqrt{c}t}{2} + k_1\right)$

if $c < 0$: let $c = -\alpha^2$. Then $\displaystyle\int \frac{2dx}{x^2 - \alpha^2} = dt$

$$\Rightarrow \frac{1}{\alpha}\ln|x - \alpha| - \frac{1}{\alpha}\ln|x + \alpha| = t + k_1$$

and therefore

$$\ln\left|\frac{x - \alpha}{x + \alpha}\right| = \alpha t + k_2.$$

Solving this for $x(t)$ gives

$$x(t) = \alpha\frac{(1 + k_3 e^{\alpha t})}{1 - k_3 e^{\alpha t}}.$$

if $c = 0$: $\frac{dx}{dt} = \frac{x^2}{2}$ can be solved by writing $\frac{dx}{x^2} = \frac{1}{2}dt$

$$-\frac{1}{x} = \frac{1}{2}t + k$$

$$x = -\frac{2}{t + 2k}.$$

Solutions for Exercises 6.5

6.5#3. The graph shows the potential function $V(x) = x^4 - x^2 = x^2(x^2 - 1)$ and below it the phase plane of the system

$$x' = y$$
$$y' = -\frac{\partial V}{\partial x} = -4x^3 + 2x$$

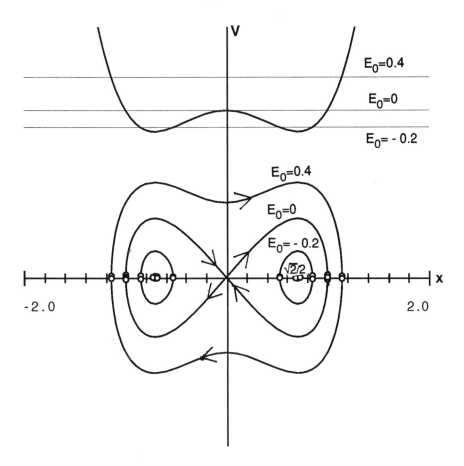

$E = \frac{1}{2}y^2 + x^4 - x^2 = $ constant E_0, so $y = \pm\sqrt{2(x^2 - x^4 + E_0)}$ gives curves in the phase plane.

Minimum of $V(x)$ occurs when

$$4x^3 - 2x = 2x(2x^2 - 1) = 0$$
$$\Rightarrow x = \pm\sqrt{\frac{1}{2}}.$$

Maximum of $V(x)$ is at $x = 0$.

Therefore the *equilibria* are at $(0,0)$, $\left(\pm\frac{\sqrt{2}}{2}, 0\right)$.

If $E_0 < 0$, the particle oscillates back and forth about one of the two equilibrium points $\left(\pm\frac{\sqrt{2}}{2}, 0\right)$.

If $E_0 = 0$, the particle either sits at $(0,0)$, or moves on a trajectory which takes infinite time to go between -1 and 0 or 0 and 1.

If $E_0 > 0$, the particle oscillates infinitely between $\pm x_{\max}$, where $|x_{\max}| > 1$. To get a good idea of the motion, imagine moving along a wire shaped like the function $V(x)$, rolling down to the first minimum, over the maximum (hump) at $x = 0$, back to the minimum and up the other side, and repeating infinitely often.

(b) The phase plane for $x' = y$, $y' = -4x^3 + 2x - 0.2y$ shows that any trajectory spirals in and eventually ends up at one of the two points $\left(\pm\frac{\sqrt{2}}{2}, 0\right)$. The origin is still a saddle point, so if the particle starts at $(0,0)$ it will remain there.

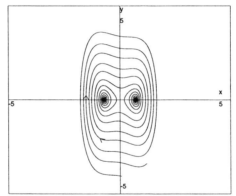

6.5#7. To have $V(x, y)$ constant along solutions of $x' = x^2 - 1$, $y' = x^2(3 - 2y)$, we need

$$\frac{dV}{dt} = \frac{\partial V}{\partial x} \cdot \frac{dx}{dt} + \frac{\partial V}{\partial y} \cdot \frac{dy}{dt} = V_x(x^2 - 1) + V_y x^2(3 - 2y) \equiv 0.$$

There is no unique way to find v from this equation, so try letting

$$V(x, y) = F(x) + G(y).$$

Then $V_x = F'(x)$ and $V_y = G'(y)$ are ordinary derivatives of functions of a single variable. Therefore,

$$(x^2 - 1)F'(x) + x^2(3 - 2y)G'(y) = 0$$

$$\left(\frac{x^2 - 1}{x^2}\right) F'(x) = -(3 - 2y)G'(y)$$

implies that both sides must be equal to the same constant, since each side depends on a different variable.

Therefore, assume $\left(\frac{x^2-1}{x^2}\right) F'(x) = 2 = (2y-3)G'(y)$. $F'(x) = \frac{2x^2}{x^2-1}$ can be
integrated to give $F(x) = 2x + \ln\left|\frac{x-1}{x+1}\right|$. $G'(y) = \frac{2}{2y-3} \Rightarrow G(y) = \ln|2y-3|$.
Then $V(x,y) = 2x + \ln\left|\frac{x-1}{x+1}\right| + \ln|2y-3|$ is constant along trajectories of
the system. (It goes to ∞ on the lines $y = \frac{3}{2}$ and $x = \pm 1$.)

6.5#10. (a) The equation is equivalent to the system

$$x' = y$$
$$y' = -\cos x - 2\alpha x$$

(b) The conserved quantity is

$$E = \frac{1}{2}(x')^2 + V(x) = \frac{1}{2}y^2 + \sin x + \alpha x^2 = \text{const}.$$

Phase planes are given for $\alpha = 0.01, 0.1, 0.5$ and 1.0.

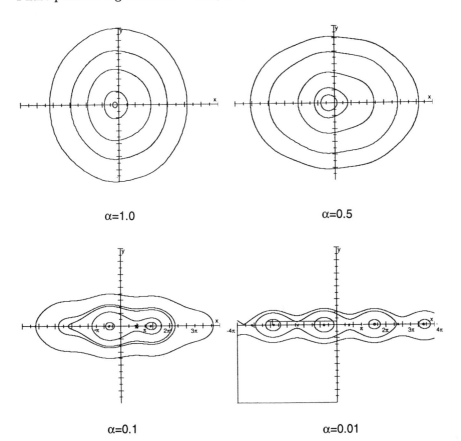

$\alpha = 1.0$ $\alpha = 0.5$

$\alpha = 0.1$ $\alpha = 0.01$

(c) The easiest way to describe the equilibria as functions of α is to look at the zeros of $y' = -\cos x - 2\alpha x$. The equilibria consist of all intersections of $f(x) = \cos x$ with $g(x) = -2\alpha x$.

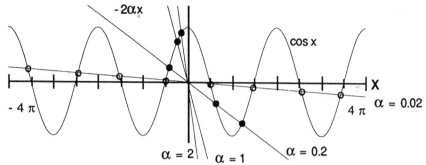

For $\alpha = 1$ and 0.5 there is *only one* equilibrium at $(x_0, 0)$ with x_0 small and negative. That will be true for any $\alpha < \bar{\alpha}$ where $\bar{\alpha}$ is the value of α such that $y = -2\alpha x$ is tangent to the graph of $\cos x$ at a value $\frac{\pi}{2} < x < \pi$.

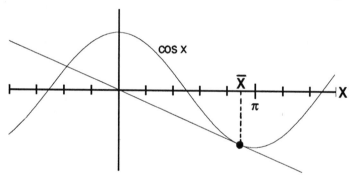

At $\bar{x}$, $-2\alpha = \frac{d}{dx}(\cos x) = -\sin x$ (slope of line and curve are equal) and $-2\alpha x = \cos x$ (two curves intersect).

$$\Rightarrow \frac{\cos \bar{x}}{\bar{x}} = -\sin \bar{x} \quad \text{or} \quad \tan \bar{x} = -\frac{1}{\bar{x}}.$$

Using Newton's method, $\bar{x} \approx 2.798$ and $\bar{\alpha} = \frac{\cos(\bar{x})}{-2\bar{x}} \approx 0.168$. Notice that for $\alpha = 0.1$ there are 3 equilibria, and for $\alpha = 0.01$ there are a lot. Using methods you will learn in Chapter 7, it can be shown that when there is more than one equilibrium, they will be alternately stable and unstable (as in Fig. 6.5.4).

Solutions for Exercises 6.6

6.6.#2. (a)

$\mathbf{r} \times \mathbf{r}' = (x\mathbf{i} + y\mathbf{j} + z\mathbf{k}) \times (x'\mathbf{i} + y'\mathbf{j} + z'\mathbf{k})$
$\equiv (yz' - zy')\mathbf{i} + (zx' - xz')\mathbf{j} + (xy' - yx')\mathbf{k}$ (def. of cross product)

$\dfrac{d}{dt}(\mathbf{r} \times \mathbf{r}') = (yz'' + y'z' - z'y' - zy'')\mathbf{i} + (zx'' + z'x' - x'z' - xz'')\mathbf{j}$
$\qquad + (xy'' + x'y' - y'x' - yx'')\mathbf{k})$
$\qquad\qquad\qquad\qquad$ (product rule for differentiation)
$= (yz'' - zy'')\mathbf{i} + (zx'' - xz'')\mathbf{j} + (xy'' - yx'')\mathbf{k}$
$\equiv \mathbf{r} \times \mathbf{r}''$ (def. of cross product).

Using equation (26), $\mathbf{r}'' = \dfrac{\|\mathbf{f}(r)\|}{mr}\mathbf{r}$ states that $\mathbf{r}''$ is a constant multiple of $\mathbf{r}$, hence parallel to $\mathbf{r}$.
Therefore, $\mathbf{r} \times \mathbf{r}'' = 0$ (Cross product of parallel vectors is 0).
(b) Let $\mathbf{n} = \mathbf{r}(t_0) \times \mathbf{r}'(t_0)$, the normal vector to the plane spanned by $\mathbf{r}(t_0)$ and $\mathbf{r}'(t_0)$. From part (a), since $\frac{d}{dt}(\mathbf{r}(t) \times \mathbf{r}'(t)) = 0$, the cross product of $\mathbf{r}(t)$ and $\mathbf{r}'(t)$ at any point along the trajectory will also be $\mathbf{n}$. That is, $\mathbf{r}(t)$ always lies in the plane with normal $\mathbf{n}$, the plane spanned by $\mathbf{r}(t_0)$ and $\mathbf{r}'(t_0)$.
(c) In polar coordinates, $\mathbf{r} = \begin{pmatrix} x(t) \\ y(t) \end{pmatrix} = \begin{pmatrix} r(t)\cos\theta(t) \\ r(t)\sin\theta(t) \end{pmatrix}$ and

$$\mathbf{r}' = \begin{pmatrix} r'(t)\cos\theta(t) - r(t)\theta'(t)\sin\theta(t) \\ r'(t)\sin\theta(t) + r(t)\theta'(t)\cos\theta(t) \end{pmatrix}.$$

$$\mathbf{r} \times \mathbf{r}' = \begin{pmatrix} \mathbf{i} & \mathbf{j} & \mathbf{k} \\ r\cos t & r\sin t & 0 \\ r'\cos\theta - r\theta'\sin\theta & r'\sin\theta + r\theta'\cos\theta & 0 \end{pmatrix}$$
$= \mathbf{i}(0) - \mathbf{j}(0) + \mathbf{k}[rr'\cos\theta\sin\theta + r^2\theta'\cos^2\theta$
$\qquad - r\sin\theta\, r'\cos\theta + r^2\theta'\sin^2\theta]$
$= r^2\theta'(\cos^2\theta + \sin^2\theta)\mathbf{k}$

and $\|\mathbf{r} \times \mathbf{r}'\| = \|r^2\theta'\mathbf{k}\| = |r^2\theta'| \equiv$ the angular momentum M.

Solutions for Exercises 6.7

6.7#2. Using Newton's law $\mathbf{F} = m\mathbf{a}$, the differential equations for $\mathbf{x}_1$, $\mathbf{x}_2$, and $\mathbf{x}_3$ are:

$$\mathbf{x}_1'' = \frac{Gm_2(\mathbf{x}_2 - \mathbf{x}_1)}{|\mathbf{x}_2 - \mathbf{x}_1|^3} + \frac{Gm_3(\mathbf{x}_3 - \mathbf{x}_1)}{|\mathbf{x}_3 - \mathbf{x}_1|^3}$$

$$\mathbf{x}_2'' = \frac{Gm_1(\mathbf{x}_1 - \mathbf{x}_2)}{|\mathbf{x}_1 - \mathbf{x}_2|^3} + \frac{Gm_3(\mathbf{x}_3 - \mathbf{x}_2)}{|\mathbf{x}_3 - \mathbf{x}_2|^3}$$

$$\mathbf{x}_3'' = \frac{Gm_1(\mathbf{x}_1 - \mathbf{x}_3)}{|\mathbf{x}_1 - \mathbf{x}_3|^3} + \frac{Gm_2(\mathbf{x}_2 - \mathbf{x}_3)}{|\mathbf{x}_2 - \mathbf{x}_3|^3}.$$

The center of gravity $\mathbf{X} = \frac{m_1\mathbf{x}_1 + m_2\mathbf{x}_2 + m_3\mathbf{x}_3}{m_1 + m_2 + m_3}$. Therefore,

$$\mathbf{X}'' = \frac{1}{m_1 + m_2 + m_3}(m_1\mathbf{x}_1'' + m_2\mathbf{x}_2'' + m_3\mathbf{x}_3'')$$

$$= \frac{1}{m_1 + m_2 + m_3}\left[\frac{Gm_1m_2(\mathbf{x}_2 - \mathbf{x}_1)}{|\mathbf{x}_2 - \mathbf{x}_1|^3} + \frac{Gm_1m_3(\mathbf{x}_3 - \mathbf{x}_1)}{|\mathbf{x}_3 - \mathbf{x}_1|^3}\right.$$

$$+ \frac{Gm_1m_2(\mathbf{x}_1 - \mathbf{x}_2)}{|\mathbf{x}_1 - \mathbf{x}_2|^3} + \frac{Gm_2m_3(\mathbf{x}_3 - \mathbf{x}_2)}{|\mathbf{x}_3 - \mathbf{x}_2|^3}$$

$$\left. + \frac{Gm_1m_3(\mathbf{x}_1 - \mathbf{x}_3)}{|\mathbf{x}_1 - \mathbf{x}_3|^3} + \frac{Gm_2m_3(\mathbf{x}_2 - \mathbf{x}_3)}{|\mathbf{x}_2 - \mathbf{x}_3|^3}\right] \equiv 0.$$

Since $\mathbf{X}''(t) \equiv 0$, $\mathbf{X}'(t) = \mathbf{b}$, a constant vector.

Therefore $\mathbf{X}(t) = \mathbf{a} + \mathbf{b}t$,

which says $\mathbf{X}$ moves in a straight line.

6.7#4. (a) The equations for a central force field (in polar coordinates) are:

$$\begin{cases} r'' - r(\theta')^2 = f(r) \\ 2r'\theta' + r\theta'' = 0 \end{cases} \tag{*}$$

and using the fact that $(r^2\theta')' = 2rr'\theta' + r^2\theta'' = r(2r'\theta' + r\theta'') = 0$, so that $M = r^2\theta'$ is constant, the first equation in $(*)$ becomes

$$r'' - \frac{M^2}{r^3} = f(r).$$

The force $f(r)$ was defined as the multiplier of $\mathbf{r}/r$ in the equation $\mathbf{x}'' = f(r)\begin{bmatrix} \cos\theta \\ \sin\theta \end{bmatrix}$, so for this problem $\boxed{f(r) \equiv -\frac{k}{r}}$.

(36) becomes $r'' - \dfrac{M^2}{r^3} = -\dfrac{k}{r}$

(37) becomes $r'' = \dfrac{M^2}{r^3} - \dfrac{k}{r} = -\dfrac{dW}{dr}$

so that

(38) becomes $W(r) = -\displaystyle\int \left(\frac{M^2}{r^3} - \frac{k}{r}\right)dr = -\left(\frac{M^2 r^{-2}}{-2} - k\ln r\right)$

$$\boxed{W(r) = \frac{M^2}{2r^2} + k\ln r}.$$

The necessary condition for bounded solutions is that $W(r)$ *have a local maximum or minimum* for $r > 0$.

But $\frac{dW}{dr} = \frac{k}{r} - \frac{M^2}{r^3} = 0$ has one solution $M^2 r - kr^3 = r(M^2 - kr^2) = 0$, so that $r = \frac{M}{\sqrt{k}}$ is a minimum. (M and k are both assumed to be positive constants.) Ex: If $M = k = 1$, the graph of $W(r)$ has a minimum at $r = 1$.

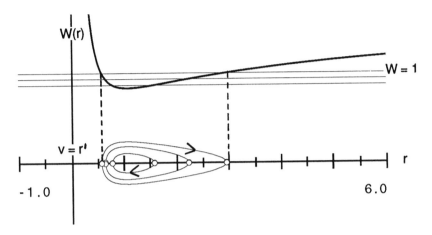

Solutions for Exercises 6.8

6.8#6. (a) $x' = x^3 = f(x)$. The solution, by separation of variables, is

$$x(t) = \sqrt{\frac{1}{c - 2t}}.$$

$$x(0) = \frac{1}{\sqrt{c}} \Rightarrow c = \frac{1}{(x(0))^2}.$$

Therefore $x(t) = x_0/\sqrt{1 - 2tx_0^2}$. The flow is $\phi_f(t, x) = x/\sqrt{1 - 2tx^2}$, with the condition $tx^2 < \frac{1}{2}$. Domain of ϕ_f:

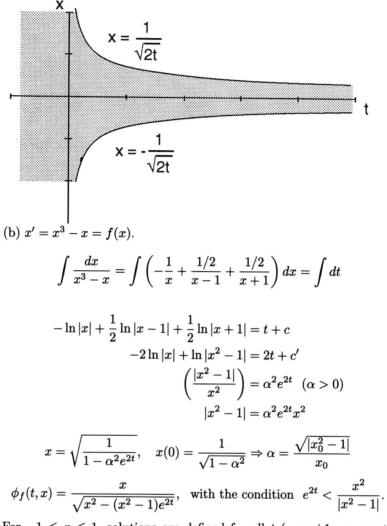

(b) $x' = x^3 - x = f(x)$.

$$\int \frac{dx}{x^3 - x} = \int \left(-\frac{1}{x} + \frac{1/2}{x-1} + \frac{1/2}{x+1} \right) dx = \int dt$$

$$-\ln|x| + \frac{1}{2}\ln|x-1| + \frac{1}{2}\ln|x+1| = t + c$$

$$-2\ln|x| + \ln|x^2 - 1| = 2t + c'$$

$$\left(\frac{|x^2 - 1|}{x^2} \right) = \alpha^2 e^{2t} \quad (\alpha > 0)$$

$$|x^2 - 1| = \alpha^2 e^{2t} x^2$$

$$x = \sqrt{\frac{1}{1 - \alpha^2 e^{2t}}}, \quad x(0) = \frac{1}{\sqrt{1 - \alpha^2}} \Rightarrow \alpha = \frac{\sqrt{|x_0^2 - 1|}}{x_0}$$

$$\phi_f(t, x) = \frac{x}{\sqrt{x^2 - (x^2 - 1)e^{2t}}}, \quad \text{with the condition} \quad e^{2t} < \frac{x^2}{|x^2 - 1|}.$$

For $-1 \le x \le 1$, solutions are defined for all t ($x \equiv \pm 1$ are constant solutions). If $|x| > 1$, the requirement is that $e^{2t} < \frac{x^2}{x^2 - 1}$, i.e., $|x| < \frac{e^t}{\sqrt{e^{2t} - 1}}$.

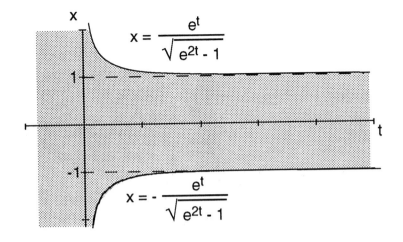

Solutions for Exercises 7.1–7.2

7.1–7.2#1. (i)

(a) $A(t) = \begin{pmatrix} 0 & t \\ 1 & 1 \end{pmatrix}$, $\quad g(t) = \begin{pmatrix} t \\ 0 \end{pmatrix}$.

(b) $A(t) = \begin{pmatrix} 0 & t & 1 \\ 1 & 1 & 0 \\ 2 & 0 & 0 \end{pmatrix}$, $\quad g(t) = \begin{pmatrix} t \\ 0 \\ 0 \end{pmatrix}$.

(ii)

(c) $\begin{pmatrix} x(t) \\ x'(t) \end{pmatrix}' = \begin{pmatrix} 0 & 1 \\ -5 & 3 \end{pmatrix} \begin{pmatrix} x(t) \\ x'(t) \end{pmatrix} + \begin{pmatrix} 0 \\ t^2 \end{pmatrix}$.

(d) $\begin{pmatrix} x(t) \\ x'(t) \end{pmatrix}' = \begin{pmatrix} 0 & 1 \\ \cos t & 0 \end{pmatrix} \begin{pmatrix} x(t) \\ x'(t) \end{pmatrix} + \begin{pmatrix} 0 \\ e^t \end{pmatrix}$.

7.1–7.2#2.

(a) $x(t) = c_1 e^{(2+\sqrt{3})t} + c_2 e^{(2-\sqrt{3})t}$.

(b) $x(t) = e^{2t}(c_1 \cos(2t) + c_2 \sin(2t))$.

(c) $x(t) = e^{-6t}(-2 - 15t)$.

(d) $x(t) = \left(\dfrac{1}{2} - \dfrac{1}{2\sqrt{5}} \right) e^{(\frac{1}{2}-\frac{\sqrt{5}}{2})t} + \left(\dfrac{1}{2} + \dfrac{1}{2\sqrt{5}} \right) e^{(\frac{1}{2}+\frac{\sqrt{5}}{2})t}$.

7.1–7.2#3.

(a) $x(t) = \sqrt{3}\, e^t \sin\left(\dfrac{t}{\sqrt{3}}\right)$.

(b) $x(t) = e^{t/2}\left(\cos\left(\dfrac{t}{2}\right) + 3\sin\left(\dfrac{t}{2}\right)\right)$.

7.1–7.2#4.

(a) $x(t) = c_{1/t} + c_{2/t^2}$.

(b) Solutions are defined for $t > 0$ or for $t < 0$, depending on where the initial conditions are given.

(c) The cookbook method produces the solution $x(t) = c_1 t^{\frac{1+\sqrt{5}}{2}} + c_2 t^{\frac{1-\sqrt{5}}{2}}$.

(d) The solution in (c) is defined for $t > 0$. For $t < 0$, $x(t) = c_1(-t)^{\frac{1+\sqrt{5}}{2}} + c_2(-t)^{\frac{1-\sqrt{5}}{2}}$ is a solution of the differential equation. Therefore, the most general solution is $x(t) = c_1|t|^{\frac{1+\sqrt{5}}{2}} + c_2|t|^{\frac{1-\sqrt{5}}{2}}$, which holds for $t \neq 0$.

7.1–7.2#7. Schwarz's Inequality says: Given vectors $\mathbf{x}$ and $\mathbf{y}$, $|\langle \mathbf{x}, \mathbf{y}\rangle| \leq \|\mathbf{x}\|\,\|\mathbf{y}\|$.

(a) Let $\mathbf{a}_i$ denote the vector consisting of the ith row of A. That is,

$$A = \begin{pmatrix} \mathbf{a}_1 \\ \mathbf{a}_2 \\ \vdots \\ \mathbf{a}_n \end{pmatrix}. \text{ Then } A\mathbf{x} = \begin{pmatrix} \langle \mathbf{a}_1, \mathbf{x}\rangle \\ \langle \mathbf{a}_2, \mathbf{x}\rangle \\ \vdots \\ \langle \mathbf{a}_n, \mathbf{x}\rangle \end{pmatrix} \text{ where } \langle \ \rangle \text{ denotes inner product.}$$

$$\|A\mathbf{x}\|^2 = \langle \mathbf{a}_1, \mathbf{x}\rangle^2 + \langle \mathbf{a}_2, \mathbf{x}\rangle^2 + \cdots + \langle \mathbf{a}_n, \mathbf{x}\rangle^2$$

$$= \sum_i \langle \mathbf{a}_i, \mathbf{x}\rangle^2 \leq \sum_i \|\mathbf{a}_i\|^2 \|\mathbf{x}\|^2 \quad \text{(by Schwarz Inequality)}$$

$$= \|\mathbf{x}\|^2 \sum_i \|\mathbf{a}_i\|^2 = \|\mathbf{x}\|^2 \|A\|^2$$

(if $\|A\|$ is defined by $\sqrt{\sum_{i,j}|a_{ij}|^2}$).
Therefore $\|A\mathbf{x}\| \leq \|\mathbf{x}\|\,\|A\|$ by taking the square root.

(b) $\mathbf{x}' = A\mathbf{x}$ satisfies a *Lipschitz condition* on $\mathbb{R} \times \mathbb{R}^n$ if there exists a real number k such that $\|A(\mathbf{x} - \mathbf{y})\| \leq k\|\mathbf{x} - \mathbf{y}\|$ for all $\mathbf{x}, \mathbf{y} \in \mathbb{R}^n$.

(c) From part (a), $\|A(\mathbf{x}-\mathbf{y})\| \leq \|A\|\,\|\mathbf{x}-\mathbf{y}\|$, where $\|A\| = \sqrt{\sum_{i,j}(a_{ij})^2} \leq k$, since the elements of A are bounded. Therefore, the linear system always satisfies a Lipschitz condition.

(d) For $\mathbf{x} = \begin{bmatrix} 0 & 1 \\ 1 & 2 \end{bmatrix} \mathbf{x}$, the constant k could be taken to be
$\sqrt{0^2 + 1^2 + 1^2 + 2^2} = \sqrt{6}$.

Note: other *norms* for A can be defined. The one used here does not give the "best" Lipschitz constant k.

Solutions for Exercises 7.3

7.3#1. (i)

(a) $\lambda_1 = -1,\ \mathbf{u}_1 = c \begin{pmatrix} 1 \\ 1 \end{pmatrix};\ \lambda_2 = 2,\ \mathbf{u}_2 = c \begin{pmatrix} 2 \\ 1 \end{pmatrix}.$

(b) $\lambda = 2,\ \mathbf{u} = c \begin{pmatrix} 1 \\ 0 \end{pmatrix}.$

(c) $\lambda_{1,2} = -2 \pm i$, eigenvectors $\begin{pmatrix} 1 \\ i \end{pmatrix},\ \begin{pmatrix} 1 \\ -i \end{pmatrix}.$

(d) $\lambda_1 = 1,\ \mathbf{u}_1 = \begin{pmatrix} 1 \\ 1 \\ -1 \end{pmatrix};$

$\lambda_2 = 2$ with 2 linearly independent eigenvectors $\begin{pmatrix} 0 \\ 1 \\ 0 \end{pmatrix}$ and $\begin{pmatrix} 0 \\ 0 \\ 1 \end{pmatrix}.$

(e) $\lambda_1 = -1,\ \mathbf{u}_1 = c \begin{pmatrix} 2 \\ -1 \\ 3 \end{pmatrix};\ \lambda_2 = 1,\ \mathbf{u}_2 = c \begin{pmatrix} 1 \\ 0 \\ -3 \end{pmatrix};$

$\lambda_3 = 5,\ \mathbf{u}_3 = c \begin{pmatrix} 1 \\ 1 \\ 0 \end{pmatrix}.$

(f) $\lambda_1 = 3,\ \mathbf{u}_1 = c \begin{pmatrix} 2 \\ 1 \\ -5 \end{pmatrix};\ \lambda_2 = 1 - 2\sqrt{2};\ \mathbf{u}_2 = c \begin{pmatrix} 2 \\ -\sqrt{2} \\ -2 + \sqrt{2} \end{pmatrix};$

$\lambda_3 = 1 + 2\sqrt{2},\ \mathbf{u}_3 = c \begin{pmatrix} 2 \\ \sqrt{2} \\ -2 - \sqrt{2} \end{pmatrix}.$

(g) $\lambda_1 = -3,\ \mathbf{u}_1 = c \begin{pmatrix} 1 \\ -1 \\ 2 \end{pmatrix};$

$\lambda_{2,3} = 4 \pm \sqrt{7}\,i,\ \mathbf{u}_{2,3} = \begin{pmatrix} 4 \\ 3 \\ -13 \end{pmatrix} \pm i \begin{pmatrix} 0 \\ \sqrt{7} \\ 5\sqrt{7} \end{pmatrix}.$

(h) The 3 eigenvectors are $\begin{pmatrix} 2\lambda \\ \lambda^2 \\ 2 \end{pmatrix}$ corresponding to $\lambda = \sqrt[3]{6},\ \frac{-\sqrt[3]{6} \pm \sqrt[6]{972}\,i}{2}$.

7.3#3.

(a) $\begin{pmatrix} x \\ y \end{pmatrix} = \begin{pmatrix} e^{3t}(\cos 3t - \frac{1}{3}\sin 3t) \\ -\frac{1}{3}e^{3t}\sin 3t \end{pmatrix} = \mathbf{v}_1(t)$ with $\mathbf{v}_1(0) = \begin{pmatrix} 1 \\ 0 \end{pmatrix}$.

(b) $\begin{pmatrix} x \\ y \end{pmatrix} = \begin{pmatrix} e^{3t}\left(\frac{4}{3}\sin 3t\right) \\ e^{3t}\left(\cos 3t + \frac{1}{3}\sin 3t\right) \end{pmatrix} = \mathbf{v}_2(t)$ with $\mathbf{v}_2(0) = \begin{pmatrix} 0 \\ 1 \end{pmatrix}$.

(c) The solution satisfying $\begin{bmatrix} x(0) \\ y(0) \end{bmatrix} = \begin{bmatrix} c_1 \\ c_2 \end{bmatrix}$ is $\begin{bmatrix} x(t) \\ y(t) \end{bmatrix} = c_1\mathbf{v}_1(t) + c_2\mathbf{v}_2(t)$.

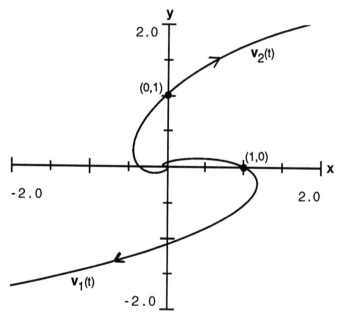

7.3#4.

(a) $\begin{pmatrix} x(t) \\ y(t) \\ z(t) \end{pmatrix} = e^{-t}\begin{pmatrix} 1 \\ -1 \\ 1 \end{pmatrix} + e^{-2t}\begin{pmatrix} 1 \\ -2 \\ 4 \end{pmatrix} + e^{3t}\begin{pmatrix} 1 \\ 3 \\ 9 \end{pmatrix}$.

(b) $\begin{pmatrix} x(t) \\ y(t) \\ z(t) \end{pmatrix} = c_1 e^{t}\begin{pmatrix} 2 \\ 2 \\ -1 \end{pmatrix} + c_2 e^{-t}\begin{pmatrix} 0 \\ -2 \\ 3 \end{pmatrix} + c_3 e^{3t}\begin{pmatrix} 0 \\ -2 \\ 1 \end{pmatrix}$.

7.3#5.

(a) A basis for $\mathbb{R}^2$ is $\mathbf{v}_1 = \begin{pmatrix} \frac{1+\sqrt{5}}{2} \\ -1 \end{pmatrix}$, $\mathbf{v}_2 = \begin{pmatrix} 1 \\ \frac{1+\sqrt{5}}{2} \end{pmatrix}$.

(b) A basis for $\mathbb{R}^2$ is $\mathbf{v}_1 = \begin{pmatrix} 1 \\ 0 \end{pmatrix}$, $\mathbf{v}_2 = \begin{pmatrix} 1 \\ -1 \end{pmatrix}$.

7.3#6. (a) The eigenvalues and eigenvectors in $\mathbb{C}^3$ are

$$\lambda_1 = 0, \ \mathbf{u}_1 = \begin{pmatrix} 1 \\ 2 \\ 2 \end{pmatrix}; \quad \lambda_2 = 3i, \ \mathbf{u}_2 = \begin{pmatrix} 4 \\ -1+3i \\ -1-3i \end{pmatrix};$$

$$\lambda_3 = -3i, \ \mathbf{u}_3 = \begin{pmatrix} 4 \\ -1-3i \\ -1+3i \end{pmatrix}$$

$\mathbf{u}_1, \mathbf{u}_2, \mathbf{u}_3$ form a basis for $\mathbb{C}^3$.

(b) To get real solutions, use $e^{3it} \begin{pmatrix} 4 \\ -1+3i \\ -1-3i \end{pmatrix}$ and take real and imaginary parts.

$$e^{3it} \begin{pmatrix} 4 \\ -1+3i \\ -1-3i \end{pmatrix} = [\cos 3t + i \sin 3t] \left[\underbrace{\begin{pmatrix} 4 \\ -1 \\ -1 \end{pmatrix} + i \begin{pmatrix} 0 \\ 3 \\ -3 \end{pmatrix}}_{\mathbf{u}} \right]$$

$$= \overbrace{(\cos 3t) \begin{bmatrix} 4 \\ -1 \\ -1 \end{bmatrix} - (\sin 3t) \begin{bmatrix} 0 \\ 3 \\ -3 \end{bmatrix}}$$

$$+ i \underbrace{\left\{ \sin 3t \begin{pmatrix} 4 \\ -1 \\ -1 \end{pmatrix} + \cos 3t \begin{pmatrix} 0 \\ 3 \\ -3 \end{pmatrix} \right\}}_{\mathbf{v}}.$$

Therefore, the general solution is

$$\begin{pmatrix} x(t) \\ y(t) \\ z(t) \end{pmatrix} = c_1 e^{0t} \begin{pmatrix} 1 \\ 2 \\ 2 \end{pmatrix} + c_2 \begin{pmatrix} 4\cos 3t \\ -\cos 3t - 3\sin 3t \\ -\cos 3t + 3\sin 3t \end{pmatrix} + c_3 \begin{pmatrix} 4\sin 3t \\ -\sin 3t + 3\cos 3t \\ -\sin 3t - 3\cos 3t \end{pmatrix}.$$

(c) $\quad \mathbf{x}(0) = \begin{pmatrix} 1 \\ 1 \\ 0 \end{pmatrix} = c_1 \begin{pmatrix} 1 \\ 2 \\ 2 \end{pmatrix} + c_2 \begin{pmatrix} 4 \\ -1 \\ -1 \end{pmatrix} + c_3 \begin{pmatrix} 0 \\ 3 \\ -3 \end{pmatrix}$

$$= \begin{pmatrix} 1 & 4 & 0 \\ 2 & -1 & 3 \\ 2 & -1 & -3 \end{pmatrix} \begin{pmatrix} c_1 \\ c_2 \\ c_3 \end{pmatrix} \Rightarrow \begin{pmatrix} c_1 \\ c_2 \\ c_3 \end{pmatrix} = \begin{pmatrix} 1/3 \\ 1/6 \\ 1/16 \end{pmatrix}.$$

Therefore, the solution with $\mathbf{x}(0) = \begin{pmatrix} 1 \\ 1 \\ 0 \end{pmatrix}$ is

$$\mathbf{x}(t) = \frac{1}{3} \begin{pmatrix} 1 \\ 2 \\ 2 \end{pmatrix} + \frac{1}{6} \begin{pmatrix} 4\cos 3t \\ -\cos 3t - 3\sin 3t \\ -\cos 3t + 3\sin 3t \end{pmatrix} + \frac{1}{6} \begin{pmatrix} 4\sin 3t \\ -\sin 3t + 3\cos 3t \\ -\sin 3t - 3\cos 3t \end{pmatrix}$$

$$= \begin{bmatrix} \frac{1}{3} + \frac{2}{3}(\cos 3t + \sin 3t) \\ \frac{2}{3} + \frac{1}{2}(\cos 3t - 2\sin 3t) \\ \frac{2}{3} + \frac{1}{3}(-2\cos 3t + \sin 3t) \end{bmatrix}.$$

(d) The solutions *cycle periodically* (period $\frac{2\pi}{3}$) about the fixed point

$$\begin{pmatrix} 1 \\ 2 \\ 2 \end{pmatrix}.$$

7.3#8. If A has n distinct eigenvalues, then it *has* a basis of eigenvectors and hence *is* diagonalizable.

Def: A is diagonalizable when the eigenvectors of A form a basis for $\mathbb{R}^n$ (or $\mathbb{C}^n$).

(a) For $\begin{bmatrix} 0 & a \\ -1 & 2 \end{bmatrix}$,

$$\det \begin{bmatrix} -\lambda & a \\ -1 & 2-\lambda \end{bmatrix} = -\lambda(2 - \lambda) + a = \lambda^2 - 2\lambda + a = 0$$

$$\lambda = \frac{2 \pm \sqrt{4 - 4a}}{2} = 1 \pm \sqrt{1 - a}.$$

Case (i). Suppose $a \neq 1$. Then there are two unequal eigenvalues. Therefore, the matrix is diagonalizable.

Case (ii). Suppose $a = 1$, so $\lambda = 1$.

$$\begin{bmatrix} 0-1 & 1 \\ -1 & 2-1 \end{bmatrix} \begin{bmatrix} u \\ v \end{bmatrix} = \begin{bmatrix} -1 & 1 \\ -1 & 1 \end{bmatrix} \begin{bmatrix} u \\ v \end{bmatrix} = \begin{bmatrix} 0 \\ 0 \end{bmatrix} \Rightarrow \begin{cases} -u + v = 0 \\ v = u \end{cases}$$

Then there is only one eigenvector $c \begin{pmatrix} 1 \\ 1 \end{pmatrix}$. Therefore, the matrix is *not* diagonalizable for $a = 1$.

(b)

$$\det \begin{bmatrix} 1-\lambda & 0 & 0 \\ 1 & 1-\lambda & 1 \\ 0 & 0 & b-\lambda \end{bmatrix} = (1-\lambda)(1-\lambda)(b-\lambda) = 0;$$

so the eigenvalues are 1, 1 and b.

Case (i): Suppose $b \neq 1$. Then there are distinct eigenvectors for $\lambda = 1$ and $\lambda = b$; what about the number of eigenvectors for $\lambda = 1$? If $\lambda = 1$,

$$\begin{pmatrix} 0 & 0 & 0 \\ 1 & 0 & 1 \\ 0 & 0 & b-1 \end{pmatrix} \begin{pmatrix} u \\ v \\ w \end{pmatrix} = \begin{pmatrix} 0 \\ 0 \\ 0 \end{pmatrix} \Rightarrow \begin{cases} u + w = 0 \\ (b-1)w = 0 \end{cases}$$

Therefore, $w = 0$ and $u = 0$. So $\mathbf{v} = \begin{pmatrix} 0 \\ 1 \\ 0 \end{pmatrix}$ is the only eigenvector for the

double eigenvalue. Hence if $b \neq 1$, the matrix is *not* diagonalizable.

Case (ii): If $b = 1$, then

$$\begin{pmatrix} 0 & 0 & 0 \\ 1 & 0 & 1 \\ 0 & 0 & 0 \end{pmatrix} \begin{pmatrix} u \\ v \\ w \end{pmatrix} = \begin{pmatrix} 0 \\ 0 \\ 0 \end{pmatrix} \Rightarrow u + w = 0 \text{ or } u = -w.$$

So there are only two linearly independent eigenvectors, given by $c_1 \begin{pmatrix} 0 \\ 1 \\ 0 \end{pmatrix}$

and $c_2 \begin{pmatrix} 1 \\ 0 \\ -1 \end{pmatrix}$, hence there is no basis of eigenvectors.

Conclusion: A is *not diagonalizable for any* value of b.

(c) For the matrix $\begin{bmatrix} 0 & c & 0 \\ 1 & 0 & c \\ c & 1 & 0 \end{bmatrix}$

$$\det \begin{bmatrix} -\lambda & c & 0 \\ 1 & -\lambda & c \\ c & 1 & -\lambda \end{bmatrix} = \lambda^3 - 2c\lambda - c^3 = 0.$$

Roots are equal only if $(-c^3)^2/4 + (-2c)^3/27 = c^3(c^3/4 - 8/27) = 0$, i.e., if $c = 0$ or $2^{5/3}/3$.

Case (i). If $c = 0$, $\lambda = 0$, and

$$\begin{pmatrix} 0 & 0 & 0 \\ 1 & 0 & 0 \\ 0 & 1 & 0 \end{pmatrix} \begin{pmatrix} u \\ v \\ w \end{pmatrix} = \begin{pmatrix} 0 \\ 0 \\ 0 \end{pmatrix} \Rightarrow u = v = 0$$

so there is only one eigenvector $\begin{pmatrix} 0 \\ 0 \\ 1 \end{pmatrix}$.

Case (ii). If $c = 2^{5/3}/3$, then $\lambda = 2c/\sqrt[3]{2}, -c/\sqrt[3]{2}, -c/\sqrt[3]{2}$. For the double root, $\lambda = -c/\sqrt[3]{2} = -2^{4/3}/3$.

$$\begin{pmatrix} \frac{2^{4/3}}{3} & \frac{2^{5/3}}{3} & 0 \\ 1 & \frac{2^{4/3}}{3} & \frac{2^{5/3}}{3} \\ \frac{2^{5/3}}{3} & 1 & \frac{2^{4/3}}{3} \end{pmatrix} \begin{pmatrix} u \\ v \\ w \end{pmatrix} = \begin{pmatrix} 0 \\ 0 \\ 0 \end{pmatrix} \Rightarrow u = -2^{1/3}v, v = 2^{4/3}w$$

so $\begin{pmatrix} -2^{5/3} \\ 2^{4/3} \\ 1 \end{pmatrix}$ is the only eigenvector.

Conclusion: the matrix is diagonalizable except when $c = 0$ or $2^{5/3}/3$.

(d) For the matrix $\begin{pmatrix} 1 & 0 & a \\ 0 & 2 & b \\ c & 0 & 2 \end{pmatrix}$,

$$\det \begin{pmatrix} 1 - \lambda & 0 & a \\ 0 & 2 - \lambda & b \\ c & 0 & 2 - \lambda \end{pmatrix} = (2 - \lambda)[\lambda^2 - 3\lambda + 2 - ac] = 0.$$

Therefore, the eigenvalues are $\lambda = 2, \frac{3 \pm \sqrt{1+4ac}}{2}$.
If either a or c is 0, $\lambda = 2$ is a double eigenvalue.
If $4ac = -1$, $\lambda = \frac{3}{2}$ is a double eigenvalue.
Case (i): $a = 0, \lambda = 2$.

$$\begin{pmatrix} -1 & 0 & 0 \\ 0 & 0 & b \\ c & 0 & 0 \end{pmatrix} \begin{pmatrix} u \\ v \\ w \end{pmatrix} = \begin{pmatrix} -u \\ bw \\ cu \end{pmatrix}.$$

If $b \neq 0$, there is only one eigenvector $\begin{pmatrix} 0 \\ 1 \\ 0 \end{pmatrix}$.

If $b = 0$, $\begin{pmatrix} 0 \\ 1 \\ 0 \end{pmatrix}$ and $\begin{pmatrix} 0 \\ 0 \\ 1 \end{pmatrix}$ are both eigenvectors.
Similarly, for $c = 0, \lambda = 2$.
Case (ii): If $4ac = -1$, so $\lambda = 3/2$ is a double eigenvalue:

$$\begin{pmatrix} -\frac{1}{2} & 0 & a \\ 0 & \frac{1}{2} & b \\ -\frac{1}{4}a & 0 & \frac{1}{2} \end{pmatrix} \begin{pmatrix} u \\ v \\ w \end{pmatrix} = \begin{pmatrix} -\frac{1}{2}u + aw \\ \frac{1}{2}v + bw \\ -\frac{1}{4a}u + \frac{1}{2}w \end{pmatrix} = \begin{pmatrix} 0 \\ 0 \\ 0 \end{pmatrix} \Rightarrow \begin{cases} u = 2aw \\ v = -2bw \end{cases}$$

and $\begin{pmatrix} 2a \\ -2b \\ 1 \end{pmatrix}$ is the only eigenvector. Therefore, the matrix is not diagonalizable.

Conclusion: This matrix is diagonalizable *unless* $a = 0$ (with $b \neq 0$), $c = 0$ (with $b \neq 0$) or $4ac = -1$ (arbitrary b).

Solutions for Exercises 7.4

7.4#6. Write $x'' + 3x' + 2x = 0$ as a system

$$\begin{cases} x' = y \\ y' = -2x - 3y. \end{cases}$$

Then $\begin{pmatrix} x \\ y \end{pmatrix}' = \begin{pmatrix} 0 & 1 \\ -2 & -3 \end{pmatrix} \begin{pmatrix} x \\ y \end{pmatrix}$ and we need to find e^{At} where $A = \begin{pmatrix} 0 & 1 \\ -2 & -3 \end{pmatrix}$. The eigenvalues and eigenvectors of A are $\lambda_1 = -1$, $\mathbf{u}_1 = \begin{pmatrix} 1 \\ -1 \end{pmatrix}$ and $\lambda_2 = -2$, $\mathbf{u}_2 = \begin{pmatrix} 1 \\ -2 \end{pmatrix}$. Therefore,

$$\begin{aligned} e^{At} &= \begin{pmatrix} 1 & 1 \\ -1 & -2 \end{pmatrix} \begin{pmatrix} e^{-t} & 0 \\ 0 & e^{-2t} \end{pmatrix} \begin{pmatrix} 1 & 1 \\ -1 & -2 \end{pmatrix}^{-1} \\ &= \begin{pmatrix} 2e^{-t} - e^{-2t} & e^{-t} - e^{-2t} \\ -2e^{-t} + 2e^{-2t} & -e^{-t} + 2e^{-2t} \end{pmatrix}. \end{aligned}$$

The solution of the differential equation with initial condition $\begin{pmatrix} x(t_0) \\ y(t_0) \end{pmatrix} = \mathbf{x}_0$ is given by $\mathbf{u}(t) = e^{A(t-t_0)}\mathbf{x}_0$. If you let $t = t_0$ and $\begin{pmatrix} \alpha \\ \beta \end{pmatrix} = \mathbf{x}(0)$, then

$$\mathbf{u}(t) = \begin{pmatrix} 2e^{-t} - e^{-2t} \\ -2e^{-t} + 2e^{-2t} \end{pmatrix} \alpha + \begin{pmatrix} e^{-t} - e^{-2t} \\ -e^{-t} + 2e^{-2t} \end{pmatrix} \beta$$

can be put in the form $c_1 \begin{pmatrix} e^{-t} \\ -e^{-t} \end{pmatrix} + c_2 \begin{pmatrix} -e^{-2t} \\ 2e^{-2t} \end{pmatrix}$ by letting $c_1 = 2\alpha + \beta$ and $c_2 = \alpha + \beta$.

7.4#11. (a) Any 2×2 matrix with trace $= 0$ can be written as

$$A = \begin{pmatrix} a & b \\ c & -a \end{pmatrix}, \quad \det A = -(a^2 + bc)$$

$$A^2 = \begin{pmatrix} a & b \\ c & -a \end{pmatrix} \begin{pmatrix} a & b \\ c & -a \end{pmatrix}$$

$$= \begin{pmatrix} a^2 + bc & 0 \\ 0 & a^2 + bc \end{pmatrix} = (a^2 + bc)I = (-\det A)I.$$

(b)

$$e^{tA} = I + tA + \frac{t^2}{2!}A^2 + \frac{t^3}{3!}A^3 + \frac{t^4}{4!}A^4 + \cdots$$

$$= I + tA - \left(\frac{t^2}{2}\det A\right)I - \left(\frac{t^3}{3!}\det A\right)A + \frac{t^4}{4}(\det A)^2 I + \cdots$$

$$= I\left[1 - \frac{t^2}{2!}\det A + \frac{t^4}{4!}(\det A)^2 - \frac{t^6}{6!}(\det A)^3 + \cdots\right]$$

$$+ A\left[t - \frac{t^3}{3!}\det A + \frac{t^5}{5!}(\det A)^2 - \cdots\right] = \alpha(t)I + \beta(t)A.$$

(c) If $\det A > 0$, then

$$\begin{cases} \alpha(t) \equiv \cos(\sqrt{\det A} \cdot t) \\ \beta(t) \equiv \dfrac{1}{\sqrt{\det A}} \sin(\sqrt{\det A} \cdot t) \end{cases}$$

If $\det A < 0$, then

$$\begin{cases} \alpha(t) \equiv \cosh(\sqrt{|\det A|}\, t) \\ \beta(t) = \dfrac{1}{\sqrt{|\det A|}} \sinh(\sqrt{|\det A|} \cdot t). \end{cases}$$

Solutions for Exercises 7.5

7.5#3. Given $\mathbf{x}' = A\mathbf{x}$ with trace $A = 0$, let

$$\mathbf{x}' = \begin{pmatrix} a & b \\ c & -a \end{pmatrix}\mathbf{x}.$$

Then $\frac{dx}{dt} = ax + by$ and $\frac{dy}{dt} = cx - ay$. To see that the equation

$$\frac{dy}{dx} = \frac{cx - ay}{ax + by} \tag{*}$$

is *exact*, write it in the form $P(x,y)dx + Q(x,y)dy = 0$ and check that $\partial P/\partial y = \partial Q/\partial x$. From (*), $(cx - ay)dx - (ax + by)dy = 0 = P\,dx + Q\,dy$

$$\frac{\partial P}{\partial y} = -a = \frac{\partial Q}{\partial x}.$$

Integrating P and Q partially with respect to x and y gives

$$V(x,y) = \frac{cx^2}{2} - axy - \frac{by^2}{2} = \text{const.}$$

(To check: $\frac{\partial V}{\partial x} = cx - ay = P$ and $\frac{\partial v}{\partial y} = -ax - by = Q$.) The level curves of V are trajectories in the phase plane.

7.5#7. If A is real and λ is a complex eigenvalue, then $\bar{\lambda}$ is also an eigenvalue. Furthermore, if

$$A\mathbf{w} = \lambda\mathbf{w} \quad (\mathbf{w} \text{ is the complex eigenvector corresponding to } \lambda),$$

then taking complex conjugates:

$$\overline{A\mathbf{w}} = \overline{\lambda\mathbf{w}}.$$

Since A is real $(A = \bar{A})$, this says $A\bar{\mathbf{w}} = \bar{\lambda}\bar{\mathbf{w}}$. That is, $\bar{\mathbf{w}}$ is the eigenvector corresponding to $\bar{\lambda}$.

Since λ and $\bar{\lambda}$ are distinct eigenvalues, $\mathbf{w}$ and $\bar{\mathbf{w}}$ are linearly independent. Now suppose $\mathbf{w} = \mathbf{w}_1 + i\mathbf{w}_2$ and $\mathbf{w}_1$ and $\mathbf{w}_2$ are *not* linearly independent. Then $\mathbf{w}_1 = c \cdot \mathbf{w}_2$ for some scalar c. But this implies that

$$\left. \begin{array}{l} \mathbf{w}_1 + i\mathbf{w}_2 = c\mathbf{w}_2 + i\mathbf{w}_2 = (c+i)\mathbf{w}_2 \\[2mm] \mathbf{w}_1 - i\mathbf{w}_2 = c\mathbf{w}_2 - i\mathbf{w}_2 = (c-i)\mathbf{w}_2 \end{array} \right\} \Rightarrow \mathbf{w}_1 + i\mathbf{w}_2 = \left(\frac{c+i}{c-i}\right)(\mathbf{w}_1 - i\mathbf{w}_2)$$

which contradicts the linear independence of $\mathbf{w}$ and $\bar{\mathbf{w}}$. Therefore, $\mathbf{w}_1$ and $\mathbf{w}_2$ must be linearly independent.

7.5.#8. If $\mathbf{x}' = \begin{pmatrix} a & b \\ c & d \end{pmatrix} \mathbf{x}$, and the trajectories spiral, then $\det A = ad - bc > 0$ and trace $A = a + d$ satisfies $(\text{trace } A)^2 < 4(\det A)$, i.e.,

$$(a+d)^2 < 4(ad - bc). \qquad (*)$$

The polar angle $\theta(t) = \tan^{-1}\left(\frac{y(t)}{x(t)}\right)$, and the direction of the trajectories will be clockwise if $d\theta/dt < 0$ and counterclockwise if $d\theta/dt > 0$.

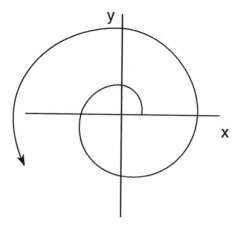

$$\frac{d\theta}{dt} = \frac{d}{dt}\left(\tan^{-1}\frac{y}{x}\right) = \frac{xy' - yx'}{x^2 + y^2}.$$

Therefore $\operatorname{sign}\left(\frac{d\theta}{dt}\right) = \operatorname{sign}(xy' - yx')$. But

$$
\begin{aligned}
xy' - yx' &= x(cx + dy) - y(ax + by)\\
&= cx^2 + (d-a)xy - by^2\\
&= c\left(x + \frac{d-a}{2c}y\right)^2 - by^2 - c\left(\frac{d-a}{2c}\right)^2 y^2\\
&= c\left[x + \left(\frac{d-a}{2c}\right)y\right]^2 - y^2\left[\frac{4bc - 2ad + a^2 + d^2}{4c}\right]\\
&= \underbrace{c\left[x + \left(\frac{d-a}{2c}\right)y\right]^2}_{\text{positive}} + \underbrace{\frac{y^2}{4c}\left[4(ad - bc) - (a+d)^2\right]}_{\text{positive by } (*)}.
\end{aligned}
$$

Therefore, $\operatorname{sign}\left(\frac{d\theta}{dt}\right) = \operatorname{sign}(xy' - yx') = \operatorname{sign}(c)$.

Solutions for Exercises 7.6

7.6#5. (a) The origin $(0,0)$ is always a singular point of the *linear* equation $\mathbf{x}' = A\mathbf{x}$ since $A\mathbf{0} = \mathbf{0}$ implies $\mathbf{x}' = \mathbf{0}$. With

$$A = \begin{pmatrix} -1 & 1 \\ \alpha & \alpha \end{pmatrix}, \qquad \begin{aligned} \text{trace } A &\equiv T = -1 + \alpha\\ \det A &\equiv D = -2\alpha. \end{aligned}$$

Solving for D as a function of T:

$$D = -2\alpha = -2(T + 1).$$

In the (trace, det) plane, this line can be plotted.

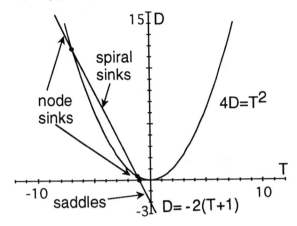

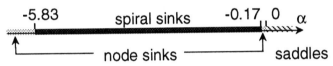

Points of intersection are where $-2(T+1) = \frac{T^2}{4}$

$$T^2 + 8T + 8 = 0$$
$$T = -4 \pm 2\sqrt{2} \approx -6.83, -1.17$$
$$\alpha = T + 1 \approx -5.83, -0.17.$$

(b) For $\alpha = 1$,

$$\det(\lambda I - A) = \det \begin{pmatrix} \lambda + 1 & -1 \\ -1 & \lambda - 1 \end{pmatrix} = \lambda^2 - 2 = 0.$$

$\lambda = \pm\sqrt{2}$ are eigenvalues.

For $\lambda = \sqrt{2}$, eigenvector is $\begin{pmatrix} 1 \\ 1 + \sqrt{2} \end{pmatrix}$.

For $\lambda = -\sqrt{2}$, eigenvector is $\begin{pmatrix} 1 \\ 1 - \sqrt{2} \end{pmatrix}$.

(Note that A is symmetric, so the eigenvectors are orthogonal.)

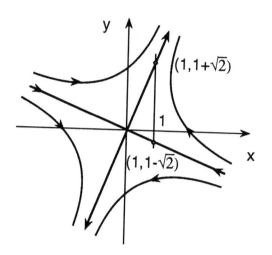

$$\text{Phase plane for } \mathbf{x}' = \begin{pmatrix} -1 & 1 \\ 1 & 1 \end{pmatrix} \mathbf{x}.$$

Solutions for Exercises 7.7

7.7#2. If $\mathbf{x} = \begin{pmatrix} a + ct \\ b + dt \end{pmatrix} e^t$, then

$$\mathbf{x}' = e^t \begin{pmatrix} a + ct \\ b + dt \end{pmatrix} + e^t \begin{pmatrix} c \\ d \end{pmatrix} = e^t \begin{bmatrix} (a + c) + ct \\ (b + d) + dt \end{bmatrix}$$

and

$$\begin{pmatrix} 2 & 1 \\ -5 & -4 \end{pmatrix} \mathbf{x} + \begin{pmatrix} t \\ 0 \end{pmatrix} e^t = e^t \begin{bmatrix} 2(a + ct) + (b + dt) + t \\ -5(a + ct) - 4(b + dt) \end{bmatrix}$$

$$= e^t \begin{bmatrix} (2a + b) + (2c + d + 1)t \\ (-5a - 4b) + (-5c - 4d)t \end{bmatrix}.$$

The equations for a, b, c and d are

$$a + c = 2a + b$$
$$c = 2c + d + 1$$

$$b + d = -5a - 4b$$
$$d = -5c - 4d.$$

In matrix form:

$$\begin{bmatrix} -1 & -1 & 1 & 0 \\ 0 & 0 & -1 & -1 \\ 5 & 5 & 0 & 1 \\ 0 & 0 & 5 & 5 \end{bmatrix} \begin{bmatrix} a \\ b \\ c \\ d \end{bmatrix} = \begin{bmatrix} 0 \\ 1 \\ 0 \\ 0 \end{bmatrix}.$$

Adding 5 times equation 2 to equation 4 shows that these equations are inconsistent. Instead, let $\mathbf{x} = \begin{pmatrix} a + bt + ct^2 \\ d + et + ft^2 \end{pmatrix} e^t$. Then

$$\mathbf{x}' = e^t \begin{pmatrix} a + bt + ct^2 + b + 2ct \\ d + et + ft^2 + e + 2ft \end{pmatrix}$$
$$\equiv \left[\begin{pmatrix} 2 & 1 \\ -5 & -4 \end{pmatrix} \begin{pmatrix} a + bt + ct^2 \\ d + et + ft^2 \end{pmatrix} + \begin{pmatrix} t \\ 0 \end{pmatrix} \right] e^t.$$

Therefore

$$\begin{bmatrix} (a+b) + t(b+2c) + ct^2 \\ (d+e) + t(e+2f) + ft^2 \end{bmatrix} \equiv \begin{bmatrix} (2a+d) + t(2b+e+1) + t^2(2c+f) \\ (-5a-4d) + t(-5b-4e) + t^2(-5c-4f) \end{bmatrix}.$$

The resulting six equations in $a, b, \ldots, f$ can be solved to get:

$$b = -\frac{1}{16}$$
$$c = \frac{5}{8}$$
$$e = \frac{5}{16}$$
$$f = -\frac{5}{8}$$
$$a + d = -\frac{1}{16}.$$

Therefore, a possible solution is

$$\mathbf{x}(t) = e^t \begin{pmatrix} \frac{5}{8}t^2 - \frac{1}{16}t - \frac{1}{16} \\ -\frac{5}{8}t^2 + \frac{5}{16}t \end{pmatrix} + c_1 e^t \begin{pmatrix} 1 \\ -1 \end{pmatrix} + c_2 e^{-3t} \begin{pmatrix} 1 \\ -5 \end{pmatrix}.$$

7.7#4. In both equations we can let

$$\begin{cases} x(t) = c_1 \sin \omega t + c_2 \cos \omega t \\ x'(t) = c_1 \omega \cos \omega t - c_2 \omega \sin \omega t \\ x''(t) = -c_1 \omega^2 \sin \omega t - c_2 \omega^2 \cos \omega t \end{cases}$$

For $x'' + x' + x = \sin(\omega t)$:

$$(c_1 \sin \omega t + c_2 \cos \omega t) + (c_1 \omega \cos \omega t - c_2 \omega \sin \omega t)$$
$$+ (-c_1 \omega^2 \sin \omega t - c_2 \omega^2 \cos \omega t)$$
$$= \sin \omega t (c_1 - c_2 \omega - c_1 \omega^2) + \cos \omega t (c_2 + c_1 \omega - c_2 \omega^2)$$
$$\equiv 1 \cdot \sin \omega t + 0 \cdot \cos \omega t$$

which says

$$\begin{matrix} c_1 - c_2 \omega - c_1 \omega^2 = 1 \\ c_2 + c_1 \omega - c_2 \omega^2 = 0 \end{matrix} \Rightarrow \begin{pmatrix} 1 - \omega^2 & -\omega \\ \omega & 1 - \omega^2 \end{pmatrix} \begin{pmatrix} c_1 \\ c_2 \end{pmatrix} = \begin{pmatrix} 1 \\ 0 \end{pmatrix}$$

This gives

$$x(t) = \left(\frac{1 - \omega^2}{(1 - \omega^2)^2 + \omega^2} \right) \sin \omega t - \left(\frac{\omega}{(1 - \omega^2)^2 + \omega^2} \right) \cos \omega t$$

$$= \frac{1}{\sqrt{(1 - \omega^2)^2 + \omega^2}} \left[\frac{1 - \omega^2}{\sqrt{(1 - \omega^2)^2 + \omega^2}} \sin \omega t \right.$$
$$\left. - \frac{\omega}{\sqrt{(1 - \omega^2)^2 + \omega^2}} \cos \omega t \right]$$
$$= R(\omega)[\cos \varphi \sin \omega t - \sin \varphi \cos \omega t]$$

where

$$\varphi = \sin^{-1} \left(\frac{\omega}{\sqrt{(1 - \omega^2)^2 + \omega^2}} \right) \quad \text{and} \quad R(\omega) = \frac{1}{\sqrt{(1 - \omega^2)^2 + \omega^2}}.$$

(b) Adding the homogeneous solution,

$$x(t) = \underbrace{e^{-\frac{1}{2}t} \left(A \cos \frac{\sqrt{3}}{2} t + B \sin \frac{\sqrt{3}}{2} t \right)}_{\substack{\text{transient solution} \\ \to 0 \text{ as } t \to \infty}} + \underbrace{R(\omega) \sin(\omega t - \varphi)}_{\substack{\text{steady state solution with} \\ \text{amplitude } R(\omega) = 1/\sqrt{(1 - \omega^2)^2 + \omega^2} \\ \text{(Note: } R(\omega) \text{ is called the } gain.)}}$$

(c) For $x'' + 2x = \sin \omega t$

$$(-c_1 \omega^2 \sin \omega t - c_2 \omega^2 \cos \omega t) + 2(c_1 \sin \omega t + c_2 \cos \omega t)$$
$$\equiv 1 - \sin \omega t + 0 \cdot \cos \omega t$$

Therefore, $(2 - \omega^2)c_1 = 1$, $(2 - \omega^2)c_2 = 0 \Rightarrow c_2 = 0$ and $c_1 = \frac{1}{2 - \omega^2}$ ($\omega \neq \sqrt{2}$).

The full solution is $x(t) = A\cos\sqrt{2}t + B\sin\sqrt{2}t + \frac{1}{2-\omega^2}\sin\omega t$ if $\omega \neq \sqrt{2}$.
For $\omega = \sqrt{2}$, write the equation in the form

$$\begin{cases} x' = y \\ y' = -2x + \sin\sqrt{2}\,t \end{cases};$$

that is, $\begin{pmatrix} x \\ y \end{pmatrix}' = \begin{pmatrix} 0 & 1 \\ -2 & 0 \end{pmatrix}\begin{pmatrix} x \\ y \end{pmatrix} + \begin{pmatrix} 0 \\ \sin\sqrt{2}t \end{pmatrix} = A\mathbf{x} + \mathbf{g}(t).$

$$A^2 = \begin{pmatrix} 0 & 1 \\ -2 & 0 \end{pmatrix}\begin{pmatrix} 0 & 1 \\ -2 & 0 \end{pmatrix} = \begin{pmatrix} -2 & 0 \\ 0 & -2 \end{pmatrix} = -2I$$

$$e^{At} = I + tA + \frac{t^2}{2!}A^2 + \frac{t^3}{3!}A^3 + \cdots$$

$$= I + tA + \frac{t^2}{2!}(-2I) + \frac{t^3}{3!}(-2A) + \frac{t^4}{4!}(4I) + \frac{t^5}{5!}(4A) + \cdots$$

$$= I\left(1 - \frac{(\sqrt{2}t)^2}{2!} + \frac{(\sqrt{2}t)^4}{4!} - \cdots\right)$$

$$+ \frac{A}{\sqrt{2}}\left(\sqrt{2}t - \frac{(\sqrt{2}t)^3}{3!} + \frac{(\sqrt{2}t)^5}{5!} - \cdots\right)$$

$$= \cos(\sqrt{2}t)I + \frac{\sin(\sqrt{2}t)}{\sqrt{2}}A = \begin{pmatrix} \cos\sqrt{2}t & \frac{\sin\sqrt{2}t}{\sqrt{2}} \\ -\sqrt{2}\sin\sqrt{2}t & \cos\sqrt{2}t \end{pmatrix}.$$

The variation of parameters formula gives:

$$\mathbf{u}(t) = e^{tA}\mathbf{x}(0) + \int_0^t e^{(t-s)A}\mathbf{g}(s)ds$$

$$= \begin{pmatrix} \cos\sqrt{2}t & \frac{\sin\sqrt{2}t}{\sqrt{2}} \\ -\sqrt{2}\sin\sqrt{2}t & \cos\sqrt{2}t \end{pmatrix}\mathbf{x}(0)$$

$$+ \int_0^t \begin{pmatrix} \cos(\sqrt{2}(t-s)) & \frac{\sin(\sqrt{2}(t-s))}{\sqrt{2}} \\ -\sqrt{2}\sin(\sqrt{2}(t-s)) & \cos(\sqrt{2}(t-s)) \end{pmatrix}\begin{pmatrix} 0 \\ \sin\sqrt{2}s \end{pmatrix}ds$$

$$= \begin{pmatrix} \cos\sqrt{2}t & \frac{\sin\sqrt{2}t}{\sqrt{2}} \\ -\sqrt{2}\sin\sqrt{2}t & \cos\sqrt{2}t \end{pmatrix}\mathbf{x}(0)$$

$$+ \int_0^t \begin{pmatrix} \frac{1}{\sqrt{2}}\sin(\sqrt{2}s)\sin(\sqrt{2}(t-s)) \\ \sin(\sqrt{2}s)\cos(\sqrt{2}(t-s)) \end{pmatrix}ds.$$

Finally, we get (noting that the integration is *very* messy)

$$\mathbf{u}(t) = x(0)\begin{pmatrix} \cos\sqrt{2}t \\ -\sqrt{2}\sin\sqrt{2}t \end{pmatrix} + x'(0)\begin{pmatrix} \frac{\sin\sqrt{2}t}{\sqrt{2}} \\ \cos\sqrt{2}t \end{pmatrix}$$

$$+ \begin{pmatrix} \frac{1}{4}\sin(\sqrt{2}t) - \frac{t\cos(\sqrt{2}t)}{2\sqrt{2}} \\ \frac{t\sin(\sqrt{2}t)}{2} \end{pmatrix}.$$

7.7#9. To solve $x'' + \omega^2 x = g(t) = \frac{1}{\pi} \sum_{n=1}^{\infty} \frac{\sin((2n-1)t)}{2n-1}$, let

$$x(t) = \sum_{n=1}^{\infty} a_n \sin(nt)$$

$$x'(t) = \sum_{n=1}^{\infty} n a_n \cos(nt)$$

$$x''(t) = -\sum_{n=1}^{\infty} n^2 a_n \sin(nt)$$

then $-\sum_{n=1}^{\infty} n^2 a_n \sin(nt) + \omega^2 \sum_{n=1}^{\infty} a_n \sin(nt) \equiv \frac{1}{\pi}(\sin t + 0 \cdot \sin(2t) + \frac{1}{3} \sin(3t) + \cdots)$. Therefore,

$$(\omega^2 - n^2)a_n = \begin{cases} \frac{1}{\pi n}, & n \text{ positive odd integer} \\ 0, & n \text{ positive even integer} \end{cases}$$

Therefore $a_n = \frac{1}{n\pi(\omega^2 - n^2)}$ if n is odd, and 0 if n is even.

This works if, and only if, ω is *not* an odd integer. It gives the solution

$$x(t) = \frac{1}{\pi} \sum_{n=1}^{\infty} \frac{\sin((2n-1)t)}{(2n-1)[\omega^2 - (2n-1)^2]}$$

which obviously converges as $n \to \infty$.

If ω is an odd integer, there is a zero in the denominator, and the solution above is not well defined. Suppose $\omega = 2m - 1$ for some integer $m > 0$. Then we need to solve separately the equation

$$x'' + (2m-1)^2 x = \frac{1}{\pi} \frac{\sin(2m-1)t}{2m-1}$$

and the equations

$$x'' + (2m-1)^2 x = \frac{1}{\pi} \frac{\sin(2n-1)t}{2n-1}, \quad \text{for } n \neq m.$$

The second set of equations are solved as above, but the first is the *resonant case*, and to solve it by undetermined coefficients, one must try a solution of the form

$$x(t) = At \sin(2m-1)t + Bt \cos(2m-1)t + C \sin(2m-1)t + D \cos(2m-1)t.$$

Differentiating twice and substituting in the equation leads to

$$A = 0, \quad B = -\frac{1}{2m\pi}, \quad C, D \text{ arbitrary.}$$

Thus in that case the general solution of the equation is

$$x(t) \; = \; \frac{1}{\pi} \sum_{n \neq m} \frac{\sin((2n-1)t)}{(2n-1)[(2m-1)^2 - (2n-1)^2]}$$

$$- \tfrac{t\cos(2m-1)t}{2m\pi} + C\sin((2m-1)t + D\cos((2m-1)t)$$

where C, D are arbitrary constants, to be determined by the initial conditions.

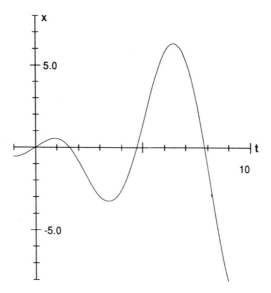

Solutions for Exercises 8.1

8.1#1. (a) (i) Singularities of $\begin{cases} x' = y \\ y' = -\sin x - 3y \end{cases}$ are at $(k\pi, 0)$, k any integer.

$$J = \begin{pmatrix} \frac{\partial f}{\partial x} & \frac{\partial f}{\partial y} \\ \frac{\partial g}{\partial x} & \frac{\partial g}{\partial y} \end{pmatrix} = \begin{pmatrix} 0 & 1 \\ -\cos x & -3 \end{pmatrix} \quad \text{with trace } J = -3, \det J = \cos x.$$

If k is even, $J = \begin{pmatrix} 0 & 1 \\ -1 & -3 \end{pmatrix}$, which shows a *node sink* and if k is odd, $J = \begin{pmatrix} 0 & 1 \\ 1 & -3 \end{pmatrix}$, which shows a *saddle*, as you can see in the figure.

(ii)

Singularity	Eigenvalues	Eigenvectors
$(2k\pi, 0)$	$\dfrac{-3+\sqrt{5}}{2}$	$(0.934, -0.357)$
	$\dfrac{-3-\sqrt{5}}{2}$	$(-0.357, 0.934)$
$((2k+1)\pi, 0)$	$.303$	$(0.957, 0.290)$
	-3.303	$(-0.290, 0.957)$

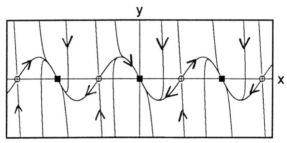

Phase plane with trajectories

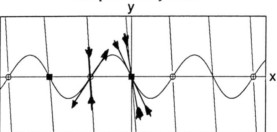

Separatrices and eigenvectors at one sink and one saddle

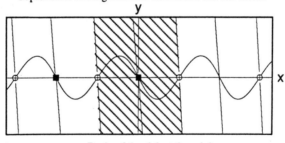

Basin of the sink at the origin

8.1#3. (b) To find the zeros of $\begin{cases} x' = y \\ y' = (x^4 + 4x^3 - x^2 - 4x + y)/8 \end{cases}$, we have

that $y = 0$, and x is a zero of $x(x^3 + 4x^2 - x - 4) = x(x-1)(x+1)(x+4)$.
Therefore, the zeros are at $(0,0)$, $(1,0)$, $(-1,0)$ and $(-4,0)$.

$J = \begin{pmatrix} 0 & -1 \\ (2x^3 + 6x^2 - x - 2)/4 & 1/8 \end{pmatrix} \Rightarrow \begin{array}{l} (0,0) \text{ and } (-4,0) \text{ are } saddles \\ (-1,0) \text{ and } (1,0) \text{ are } spiral\ sources. \end{array}$

(ii*)	**Singularity**	**Eigenvalues**	**Eigenvectors**
	$(0,0)$	-0.647	$(-0.839, -0.543)$
		0.772	$(0.839, -0.648)$
	$(1,0)$	$0.625 \pm 1.116i$	$(-0.05 \pm 0.893i, 1)$
	$(-1,0)$	0.625 ± 0.864	$(-0.08 \pm 1.152i, 1)$
	$(-4,0)$	-2.677	$(-0.350, -0.937)$
		2.802	$(0.336, -0.942)$

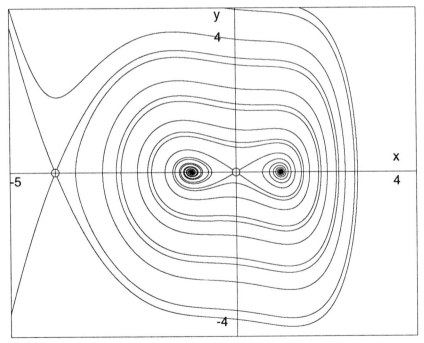

dx/dt(x,y,a,b)=-y
dy/dt(x,y,a,b)=(x^4+4x^3-x^2-4x+y)/8

a=0.0000000
b=0.0000000

8.1#5. (i) The Jacobian matrix J is

$$\begin{pmatrix} 2 - 12x^2 - 4y^2 & \cos y - 8xy \\ -3\sin(x-2y) - 3x\cos(x-2y) - 4xy & \cos y + 6x\cos(x-2y) - 2x^2 - 6y^2 \end{pmatrix}.$$

At the singular point $(0,0)$, $J = \begin{pmatrix} 2 & 1 \\ 0 & 1 \end{pmatrix}$, with eigenvalue $\lambda = 1, 2$. Therefore $(0,0)$ is a *node source*.

The differential equation in the range
$-4 < x < 4$
$-4 < y < 4$

The same differential equation in the range
$-.1 < x < .1$
$-.1 < y < .1$

The linearization of the differential equation near $(0,0)$

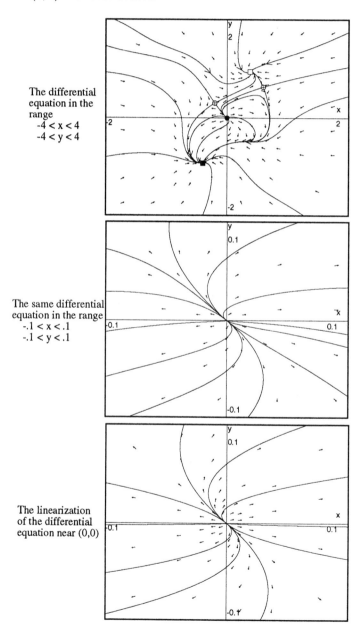

8.1#9. (a) The singularities of $\begin{cases} x' = y \\ y' = -x^2 + y^2 + 1 \end{cases}$ are at $(1,0)$ and $(-1,0)$.

$J = \begin{pmatrix} 0 & 1 \\ -2x & 2y \end{pmatrix}$. Therefore, $J(1,0) = \begin{pmatrix} 0 & 1 \\ -2 & 0 \end{pmatrix}$ is a *center*.

$J(-1,0) = \begin{pmatrix} 0 & 1 \\ 2 & 0 \end{pmatrix}$, so $(-1,0)$ is a saddle, with $\lambda_1 = \sqrt{2}$, $\mathbf{u}_1 = \begin{pmatrix} 1 \\ \sqrt{2} \end{pmatrix}$;

$\lambda_2 = -\sqrt{2}$, $\mathbf{u}_2 = \begin{pmatrix} 1 \\ -\sqrt{2} \end{pmatrix}$.

We can conclude that $(-1,0)$ is a saddle of the nonlinear system, but we cannot tell from this linearization what happens at $(1,0)$.

(b) The only vertical isocline is $y = 0$, i.e., the x-axis. The horizontal isoclines are the two branches of the hyperbola $x^2 - y^2 = 1$. The line $\begin{pmatrix} x \\ y \end{pmatrix} = \begin{pmatrix} -1 \\ 0 \end{pmatrix} + \tau \begin{pmatrix} 1 \\ \sqrt{2} \end{pmatrix}$ is the line through the saddle point $(-1,0)$, in the direction of the *unstable* manifold.

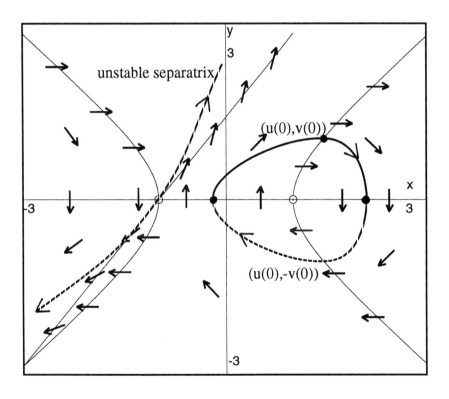

Along this line $x' = y = \sqrt{2}\tau$, $y' = -x^2 + y^2 + 1 = -(-1+\tau)^2 + (\sqrt{2}\tau)^2 + 1 = \tau(\tau+2)$. Therefore, $\frac{dy}{dx} = (\tau+2)/\sqrt{2}$.

(c) If $(x, y) = (u(t), v(t))$ is a solution, then $u'(t) = v(t)$, $v'(t) = -u^2(t) + v^2(t) + 1$. Let $(\xi, \eta) = (u(-t), -v(-t))$. Then $\frac{d\xi}{dt} = -\frac{du(-t)}{dt} = -v(-t) = \eta$ and

$$\frac{d\eta}{dt} = -\left(-\frac{dv}{dt}\right) = -u^2(-t) + v^2(-t) + 1 = -\xi^2 + \eta^2 = 1$$

so (ξ, η) is also a solution. Using this symmetry, it is clear that the trajectory through $(u(0), v(0))$ in the first quadrant has a symmetric part through $(u(0), -v(0))$ and these two pieces form a *periodic trajectory*, with vertical slope on the x-axis.

(d)

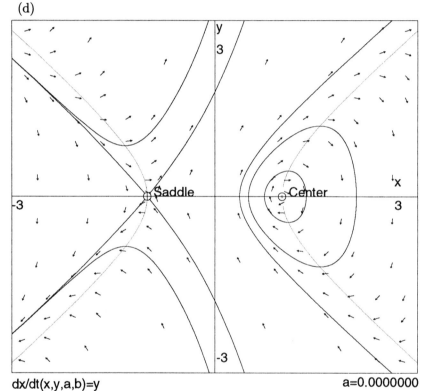

dx/dt(x,y,a,b)=y   a=0.0000000
dy/dt(x,y,a,b)=-x^2+y^2+1   b=0.0000000

Solutions for Exercises 8.2–8.3

8.2–8.3#2. Let $\begin{pmatrix} x \\ y \end{pmatrix}' = \begin{pmatrix} -x - xy \\ y - x^2 \end{pmatrix}$ and $\frac{dy}{dx} = \frac{y - x^2}{-x - xy} = f(x, y)$.

(a) $J = \begin{pmatrix} -1 - y & -x \\ -2x & 1 \end{pmatrix}$. Critical points are where $y = x^2$ and $-x - x(x^2) = 0$ or $-x(1 + x^2) = 0$. Therefore, $(0, 0)$ is the only equilibrium.

$J(0,0) = \begin{pmatrix} -1 & 0 \\ 0 & 1 \end{pmatrix}$, so $(0,0)$ is a saddle. The eigenvectors are along the x- and y-axes.

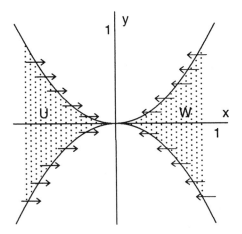

To show that W defines a funnel for $dy/dx = (y - x^2)/(-x - xy)$, we need to show that the vector field for the differential equation enters W along $y = \pm x^2$. On the curve $y = x^2$,

$$\frac{dy}{dx} = \frac{x^2 - x^2}{-x - x^3} = 0.$$

For uniqueness, we need a narrowing antifunnel with $\frac{\partial f}{\partial y} > 0$ inside the antifunnel.

Certainly, $y = x^2$ and $y = -x^2$, for $x > 0$, narrow to 0 as $x \to 0$ (starting from $x > 0$).

$$\frac{\partial}{\partial y}(f(x, y)) = \frac{\partial}{\partial y}\left(\frac{y - x^2}{-x - xy}\right) = \frac{(-x - xy)(1) - (y - x^2)(-x)}{(x + xy)^2}$$
$$= \frac{-x - x^3}{(x + xy)^2} = \frac{-(x + x^3)}{(x + xy)^2}.$$

Since we are reversing x (the time variable) this will make $\frac{\partial f}{\partial y}(x, y) > 0$ inside the antifunnel.

(c) From part (b), one can conclude that there exist *exactly two bounded solutions* of the system (i), which both approach 0 as $t \to 0$; all other solutions must tend to $\pm\infty$.

8.2–8.3#3. There exists a basis $\mathbf{w}_1, \ldots, \mathbf{w}_n$ of $\mathbb{R}^n$ such that

$$
A = \begin{bmatrix}
a_1 & -b_1 & \varepsilon_{13} & \varepsilon_{14} & \cdot & \cdot & \cdot & \cdot & \cdot & \varepsilon_{1n} \\
b_1 & a_1 & \varepsilon_{23} & \varepsilon_{24} & \cdot & \cdot & \cdot & \cdot & \cdot & \cdot \\
 & & a_2 & -b_2 & & & & & & \\
 & & b_2 & a_2 & & & & & & \\
 & & & & \ddots & & & & & \\
 & & & & & a_n & -b_n & & & \\
 & 0 & & & & b_n & a_n & & & \\
 & & & & & & & \lambda_{2k+1} & & \\
 & & & & & & & & \ddots & \\
 & & & & & & & & & \lambda_n
\end{bmatrix} = D + E
$$

where

$$
D = \begin{bmatrix}
a_1 & -b_1 & & & \\
b_1 & a_1 & & 0 & \\
 & & \ddots & & \\
 & 0 & & & \\
 & & & & \lambda_n
\end{bmatrix} \quad \text{pseudo-diagonal}
$$

and

$$
E = \begin{bmatrix}
0 & \varepsilon_{12} & \varepsilon_{13} & \cdots & \varepsilon_{1n} \\
 & 0 & \varepsilon_{23} & & \\
 & & 0 & & \\
 & 0 & & \ddots & \\
 & & & & 0
\end{bmatrix} \quad
\begin{array}{l} \text{upper triangular, every term having} \\ \text{absolute value} < \varepsilon. \end{array}
$$

Then $A\mathbf{x} \cdot \mathbf{x} = (D + E)\mathbf{x} \cdot \mathbf{x} = D\mathbf{x} \cdot \mathbf{x} + E\mathbf{x} \cdot \mathbf{x}$.

$$
D\mathbf{x} \cdot \mathbf{x} = \begin{bmatrix}
a_1 x_1 - b_1 x_2 \\
b_1 x_1 + a_1 x_2 \\
a_2 x_3 - b_2 x_4 \\
b_2 x_3 + a_2 x_4 \\
\vdots \\
\lambda_{2k+1} x_{2n+1} \\
\vdots \\
\lambda_n x_n
\end{bmatrix} \cdot \begin{bmatrix}
x_1 \\
x_2 \\
x_3 \\
x_4 \\
\vdots \\
x_{2k+1} \\
\vdots \\
x_n
\end{bmatrix}
$$

$$= (a_1 x_1^2 - b_1 x_1 x_2 + b_1 x_1 x_2 + a_1 x_2^2) + (a_2 x_3^2 - b_2 x_3 x_4 + b_2 x_3 x_4 + a_2 x_4^2)$$
$$+ \cdots + \lambda_{2k+1} x_{2k+1}^2 + \cdots + \lambda_n x_n^2$$
$$= a_1 (x_1^2 + x_2^2) + a_2 (x_3^2 + x_4^2) + \cdots + \lambda_{2k+1} x_{2k+1}^2 + \cdots + \lambda_n x_n^2$$
$$\leq \Lambda \|\mathbf{x}\|^2 \text{ where } \Lambda = \max \quad (\text{Real part of eigenvalues of } A)$$

$|E\mathbf{x} \cdot \mathbf{x}| \le \|E\| \, \|x\|^2$ and

$$\|E\| = \sqrt{\sum_{1 \le i,j \le n} e_{ij}^2} \le \sqrt{n^2 \varepsilon^2} = n\varepsilon.$$

Therefore, $A\mathbf{x} \cdot \mathbf{x} \le (\Lambda + n\varepsilon)\|x\|^2$.

8.2–8.3#9. If
$$\begin{cases} x' = -ax + P(x,y) \\ y' = -ay + Q(x,y) \end{cases}$$
with $a > 0$ then $(0,0)$ is a node sink and we know that for $\|x\| < \rho$, $\|x(t)\|^2 \le \|x(0)\|^2 e^{-c_1 t}$ for some constant $c_1 > 0$.
   In polar coordinates, $x = r\cos\theta$, $y = r\sin\theta$, $\theta = \tan^{-1}(y/x)$ and

$$\begin{aligned}
\frac{d\theta}{dt} &= \frac{xy' - x'y}{x^2 + y^2} = \frac{x(-ay + Q(x,y)) - y(-ax + P(x,y))}{x^2 + y^2} \\
&= \frac{r\cos\theta\, Q(r\cos\theta, r\sin\theta) - r\sin\theta(P(r\cos\theta, r\sin\theta))}{r^2} \\
&= \frac{\cos\theta\, Q(r\cos\theta, r\sin\theta) - \sin\theta\, P(r\cos\theta, r\sin\theta)}{r}.
\end{aligned}$$

Since P and Q start with quadratic terms, for r small enough

$$\left. \begin{aligned} P(r\cos\theta, r\sin\theta) &< r^2(M_1 + \varepsilon_1(r)) \\ Q(r\cos\theta, r\sin\theta) &< r^2(M_2 + \varepsilon_2(r)) \end{aligned} \right\} \quad \begin{aligned} &\text{where } \varepsilon_i(r) \to 0 \text{ as } r \to 0 \\ &\text{if } i = 1, 2. \end{aligned}$$

Therefore, $\left|\frac{d\theta}{dt}\right| \le \frac{kr^2}{r} = kr \le k\|x_0\|e^{-c_1 t/2}$ for small enough r, where k is a constant.

$$\left| \int_0^\infty \frac{d\theta}{dt} dt \right| \le \int_0^\infty k\|x_0\|e^{-c_1 t/2} dt \le -\frac{2k}{c_1}\|x_0\|e^{-c_1 t/2}\Big|_0^\infty = \frac{2k}{c_1}\|x_0\|.$$

Thus the solutions do *not* spiral infinitely as $(x,y) \to (0,0)$.

8.2–8.3#11. Let $\mu x = P(x,y)$, where P is a quadratic or higher-order form. Then the Implicit Function Theorem shows that this defines a curve $x = x(y)$ passing through $x = 0$ at $y = 0$. The derivative $\frac{dx}{dy}$ can be obtained by implicit differentiation with respect to y:

$$\begin{aligned}
\mu\frac{dx}{dy} &= \frac{d}{dy}(P(x,y)) = \frac{\partial P}{\partial x}\frac{dx}{dy} + \frac{\partial P}{\partial y} \\
\frac{dx}{dy} &= \frac{\partial P/\partial y}{\mu - (\partial P/\partial x)}.
\end{aligned}$$

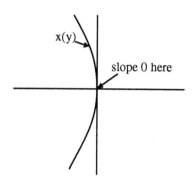

Since $P(x, y) = \alpha x^2 + \beta xy + \gamma y^2 + \cdots$

$$\frac{\partial P}{\partial x} = 2\alpha x + \beta y + \cdots$$

is 0 at $x = y = 0$. Similarly, $\frac{\partial P}{\partial y}\big|_{(0,0)} = 0$. Therefore,

$$\frac{dx}{dy}\bigg|_{(0,0)} = \frac{0}{\mu - 0} = 0.$$

This says the function $x(y)$ is tangent to the y-axis. To show $x(y)$ is twice differentiable at $(0,0)$:

$$\mu \frac{d^2 x}{dy^2} = \frac{d}{dy}\left(P_x \frac{dx}{dy} + P_y\right)$$

$$= P_x \frac{d^2 x}{dy^2} + \frac{dx}{dy} \cdot \frac{d}{dy}(P_x) + \frac{d}{dy}(P_y)$$

$$= P_x \frac{d^2 x}{dy^2} + \frac{dx}{dy}\left(P_{xy} \frac{dy}{dy} + P_{xx} \frac{dx}{dy}\right) + \left(P_{yy} \frac{dy}{dy} + P_{yx} \frac{dx}{dy}\right)$$

$$\frac{d^2 x}{dy^2} = \frac{P_{yy} + 2P_{yx} \frac{dx}{dy} + P_{xx}\left(\frac{dx}{dy}\right)^2}{\mu - P_x},$$

which is also defined since $\mu - P_x(x, y) = \mu \neq 0$ at $(0,0)$.

Solutions for Exercises 8.4

8.4#3. Changing to polar coordinates:

$$\begin{cases} r' = \frac{1}{r}(xx' + yy') = \frac{1}{r}(x^2 + y^2)r^4 \sin\left(\frac{1}{r^2}\right) = r^5 \sin\left(\frac{1}{r^2}\right) = f(r) \\ \\ \theta' = -1. \end{cases}$$

The equation in $r(t)$ has equilibria at $\frac{1}{r^2} = n\pi$, n any positive integer.

$$f'(r) = 5r^4 \sin\left(\frac{1}{r^2}\right) + r^5 \cdot \cos\left(\frac{1}{r^2}\right) \cdot \frac{-2}{r^3}$$
$$= r^2 \left[5r^2 \sin\left(\frac{1}{r^2}\right) - 2\cos\left(\frac{1}{r^2}\right)\right].$$

At the points $\frac{1}{r^2} = n\pi$, $\sin\frac{1}{r^2} = 0$ and

$$f'(r) = -2r^2 \cos(n\pi) = \begin{cases} \text{positive if } n \text{ odd} \\ \text{negative if } n \text{ even} \end{cases}$$

Therefore, $\gamma = \frac{1}{\sqrt{2\pi}}, \frac{1}{\sqrt{4\pi}}, \ldots$ are *stable* equilibrium points and $\frac{1}{\sqrt{\pi}}, \frac{1}{\sqrt{3\pi}}, \ldots$ are unstable.

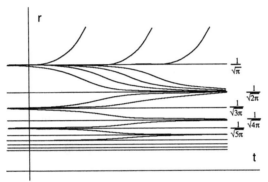

The system of equations will have infinitely many limit cycles which are circles of radius $\frac{1}{\sqrt{n\pi}}$, where those with n even will be stable limit cycles, those with n odd will be unstable.

Solutions for Exercises 8.5

8.5#1. Assume $f:[a,b] \rightarrow [a,b]$ is continuous. Let $g(x) = f(x) - x$. If $f(a) = a$ we are done, otherwise $f(a) > a$. If $f(b) = b$ we are done, otherwise $f(b) < b$.

Assuming $f(a) > a$ and $f(b) < b$, the function $g(x)$ is continuous on $[a,b]$ and satisfies $g(a) = f(a) - a > 0$ and $g(b) = f(b) - b < 0$. Therefore, by the Intermediate Value Theorem, g must take on the value 0 at some x^*, $a < x^* < b$. That is, $g(x^*) = f(x^*) - x^* = 0$ which proves that x^* is a fixed point for f.

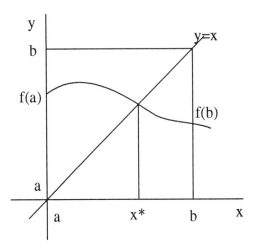

8.5#4. Let the vector field be given by

$$\frac{dy}{dx} = \frac{-x + (1 - x^2)y}{y} = (1 - x^2) - \frac{x}{y}.$$

Since there is symmetry, we can consider *one half* of the annulus as shown in the drawing.

Around the circle, the vector field will point into the region if $(x, y) \cdot \left(\frac{dx}{dt}, \frac{dy}{dt}\right) \geq 0.$

$$(x, y) \cdot \left(\frac{dx}{dt}, \frac{dy}{dt}\right) = xy + y(-x - (x^2 - 1)y) = -(x^2 - 1)y^2.$$

If $|x| < 1$, this is positive except where $y = 0$. Let C_ε be the circle of radius $1 - \varepsilon$. At $(1 - \varepsilon, 0)$ and $(-1 + \varepsilon, 0)$ the vector field is vertical, but it does point into the annulus. Let $\alpha_1(t)$ be $y = 2x + \frac{21}{5}$, $-\frac{3}{2} \leq x \leq 1$. Along this line the slope of the vector field is $1 - x^2 - \frac{x}{2x+(21/5)} < 2$ for $-\frac{3}{2} \leq x \leq 1$. Along the horizontal line from $x = 1$ to $x = 2$, the vector field points in the negative y-direction.

Let α_2 be a line of slope -3. Then α_2: $y = -3x + \frac{61}{5}$, and for $2 < x < 4.0666$, the slope of the vector field along α_2 is < -3, and tends to $-\infty$ as $y \to 0$.

Then if α_3 is a line with very large negative slope from $(4.066\ldots, 0)$ to a point on the 0-isocline $y = \frac{x}{1-x^2}$, the vector field will be entering on α_3 and also on the 0-isocline (which has positive slope). The annulus can then be completed by symmetry.

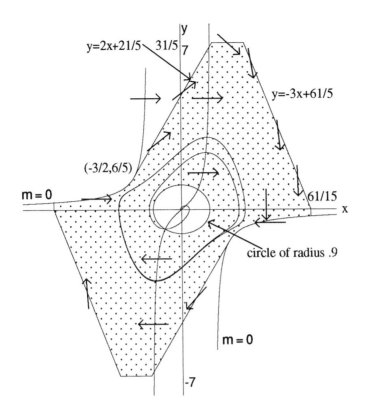

y=2x+21/5   31/5 7

y=-3x+61/5

(-3/2,6/5)

m = 0

61/15

x

circle of radius .9

m = 0

-7

Solutions to Exercises 8.6

8.6#1. (a)

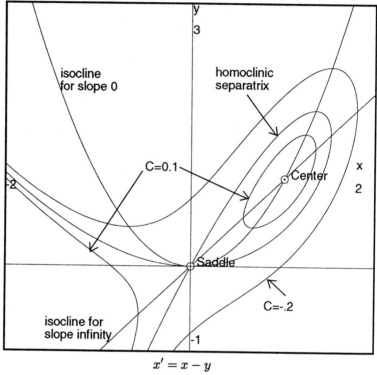

$$x' = x - y$$
$$y' = x^2 - y$$

$y = x$ is the vertical isocline

$y = x^2$ is the horizontal isocline

$$J = \begin{pmatrix} 1 & -1 \\ 2x & -1 \end{pmatrix}$$

(b) The equilibria are $(0,0)$ and $(1,1)$.

At $(0,0)$, $J = \begin{pmatrix} 1 & -1 \\ 0 & -1 \end{pmatrix}$ shows a *saddle* with eigenvalues and vectors

$$\lambda = 1 \qquad \lambda = -1$$

$$\begin{pmatrix} 1 \\ 0 \end{pmatrix} \qquad \begin{pmatrix} 1 \\ 2 \end{pmatrix}$$

At $(1,1)$, $J = \begin{pmatrix} 1 & -1 \\ 2 & -1 \end{pmatrix}$, with trace $J = 0$ and det $J = 1$ (a *center* for the linear system).

The equation is exact:

$$\frac{dy}{dx} = \frac{x^2 - y}{x - y}, \quad \text{so} \quad \overbrace{-(x^2 - y)}^{V_x} dx + \overbrace{(x - y)}^{V_y} dy = 0,$$

which integrates to $V(x,y) = xy - \frac{y^2}{2} - \frac{x^3}{3} = C$. This can be solved explicitly by writing as a quadratic in y: $\frac{y^2}{2} - xy + \left(C + \frac{x^3}{3} \right) = 0$

$$y = x \pm \sqrt{x^2 - 2(c + \frac{x^3}{3})}$$

$$y = x \left(1 \pm \sqrt{\frac{x^2 - \frac{2}{3}x^3 - 2c}{x^2}} \right) = x \left(1 \pm \sqrt{1 - \frac{2}{3}x - 2\frac{c}{x^2}} \right) \quad (*)$$

For $C = 0$, $y = x \left(1 \pm \sqrt{1 - \frac{2}{3}x} \right)$ describes both the stable and unstable manifolds for the saddle point $(0,0)$.

(c) Level curves of $(*)$ with $0 < c < \frac{1}{6}$ define the cycles about $(1,1)$, within the homoclinic separatrix.

8.6#3. Let $\mathbf{u}' = A\mathbf{u}$ be the linearization of $\mathbf{x}' = f(\mathbf{x})$ at a zero $\mathbf{x}_0$. Then the matrix

$$A = \begin{pmatrix} \frac{\partial f_1}{\partial x_1} & \frac{\partial f_1}{\partial x_2} & \cdots & \frac{\partial f_1}{\partial x_n} \\ \frac{\partial f_2}{\partial x_1} & \frac{\partial f_2}{\partial x_2} & & \\ \vdots & & \ddots & \\ \frac{\partial f_n}{\partial x_n} & & & \frac{\partial f_n}{\partial x_n} \end{pmatrix}_{(\mathbf{X}=\mathbf{X}_0)},$$

is the Jacobian evaluated at the singularity $\mathbf{x}_0$. Therefore, $\operatorname{trace}(A) = \sum_{i=1}^{n} \frac{\partial f_i}{\partial x_i}\big|_{\mathbf{x}_0} = \operatorname{div} f(\mathbf{x}_0)$.

Suppose a vector field $\mathbf{x}' = f(\mathbf{x})$ has a source or sink at $\mathbf{x}_0$. Then it has been shown that the eigenvalues of A must have all positive real parts (or all negative real parts). $\operatorname{Trace}(A) = \sum_{i=1}^{n} \lambda_i = \sum_{i=1}^{n} R(\lambda_i)$ (since complex λ_i come in conjugate pairs). Therefore, $\operatorname{trace}(A)$ will be strictly positive (or strictly negative) at $\mathbf{x}_0$. But this would say $\operatorname{div} f(\mathbf{x}_0) \neq 0$, and therefore the differential equation is *not* area preserving. If the field *is* area preserving, $\operatorname{div} \mathbf{f}(\mathbf{x}) = 0$ at every point $\mathbf{x}$. Thus any singularities cannot be sinks or sources.

Solutions for Exercises 8.7

8.7#2. The function $f(x, y, z) = \rho x^2 + \sigma y^2 + \sigma(z - 2\rho)^2 = a$ defines a family of concentric ellipsoids with center at $(0, 0, 2\rho)$.

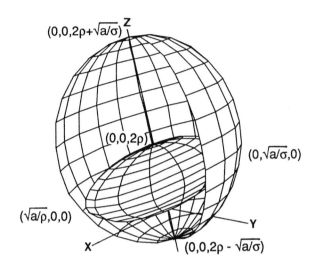

(a) The region is clearly bounded by $-\sqrt{\frac{a}{\rho}} \le x \le \sqrt{\frac{a}{\rho}},\ -\sqrt{\frac{a}{\sigma}} \le y \le \sqrt{\frac{a}{\sigma}}$ and $2\rho - \sqrt{\frac{a}{\sigma}} \le z \le 2\rho + \sqrt{\frac{a}{\sigma}}$.

(b) Along a trajectory of

$$\begin{cases} x' = \sigma(y - x) \\ y' = \rho x - y - xz \\ z' = -\beta z + xy \end{cases}, \quad \frac{df}{dt} = \frac{\partial f}{\partial x}\frac{dx}{dt} + \frac{\partial f}{\partial y}\frac{dy}{dt} + \frac{\partial f}{\partial z}\frac{dz}{dt}.$$

Therefore

$$\frac{df}{dt} = 2\rho\sigma x(y - x) + 2\sigma y(\rho x - y - xz) + 2\sigma(z - 2\rho)(-\beta z + xy)$$
$$= -2\sigma[\rho x^2 + y^2 + \beta z^2 - 2\rho\beta z].$$

(c) If there exists a number a such that *everywhere*, outside of the surface $f(x, y, z) = a$, $\frac{df}{dt} < 0$, then all trajectories must enter the ellipsoid. No trajectory can exit it, since this would imply f was increasing at the point of exit, i.e., $\frac{df}{dt} > 0$.

To make $\frac{df}{dt} < 0$, we need $\rho x^2 + y^2 + \beta z^2 - 2\rho\beta z > 0$. This defines another ellipsoid $E: \rho x^2 + y^2 + \beta(z - \rho)^2 = \beta\rho^2$. If a can be chosen so E is *interior* to $f(x, y, z) = a$, we are done.

Since E is centered at $(0, 0, \rho)$, it is only necessary to choose a so that the cross-section of E at $z = \rho$ is inside the cross-section of $f = a$ at $z = \rho$, and $2\rho - \sqrt{\frac{a}{\sigma}} < 0$ (on the ellipse E, $0 < z < 2\rho$).

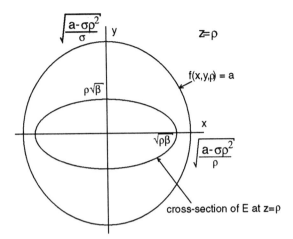

cross-section of E at z=ρ

At $z = \rho$, $f(x, y, \rho) = \rho x^2 + \sigma y^2 + \sigma \rho^2 = a$ is the two-dimensional ellipse $\rho x^2 + \sigma y^2 = a - \sigma \rho^2$. Therefore, a must satisfy $\beta \rho^2 < \frac{a - \sigma \rho^2}{\sigma}$ and $\beta \rho < \frac{a - \sigma \rho^2}{\rho}$ and $a > 4\sigma \rho^2$, i.e., $a > \max(\sigma(\beta + 1)\rho^2, (\beta + \sigma)\rho^2, 4\sigma \rho^2)$ will satisfy all of the requirements.

Solutions for Exercises 8.8

8.8#1. (a) The phase portraits show the following:

α	critical points	det	trace	
-6	-0.3542	5.2916	-0.2916	spiral sink
	-5.6457	-5.2916	10.2916	saddle
-4.5	-0.5	3.5	0	center
	-4	-3.5	7	saddle
-3.5	-0.7192	2.0616	0.4384	spiral source
	-2.7808	-2.0616	4.5616	saddle
0	no critical points			
3.5	2.7808	2.0616	-6.5616	node sink
	$+0.7192$	-2.0616	-2.4384	saddle
10	9.7958	9.5917	-20.5917	node sink
	0.2042	-9.5917	-1.4083	saddle

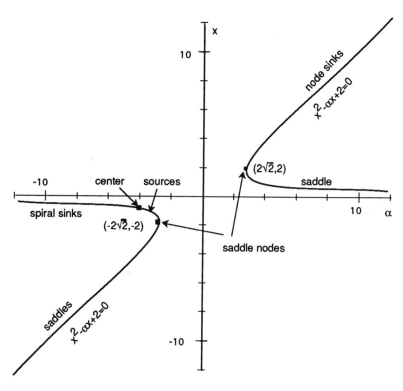

dx/dt = y-x^2
dy/dt = a*x-2-y
-20 < x < 15
-10 < y < 100

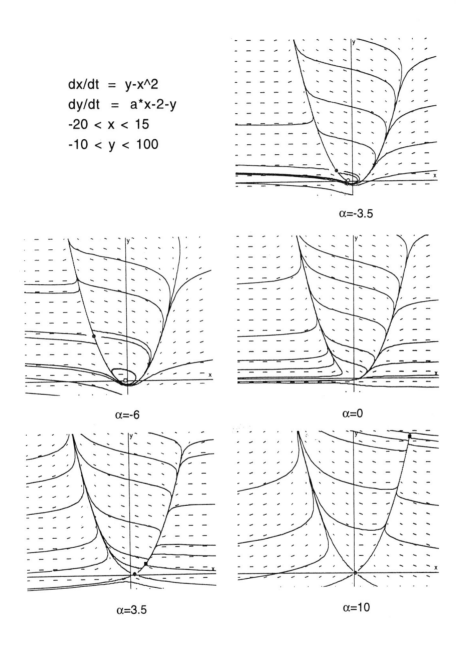

α=-3.5

α=-6

α=0

α=3.5

α=10

Solutions for Exercises 9.1

9.1#2. (c) To find saddle-node bifurcations, write equations (i) and (ii) for the equilibria and (iii), which says det $J = 0$.

(i) $y - x^3 + xy = 0$

(ii) $4x - y - \alpha = 0$

(iii) $3x^2 - y - 4 - 4x = 0$

$$J = \begin{pmatrix} -3x^2 + y & 1 + x \\ 4 & -1 \end{pmatrix}$$

From (iii), $y = 3x^2 - 4x - 4$. Inserting this into (i) gives

(iv) $2x^3 - x^2 - 8x - 4 = 0$, which gives x-values of the equilibria.

There are three real roots of (iv), $x = 2.45819$, -0.59828, and -1.35991. The following table gives the corresponding bifurcation values of α.

x	$y = 3x^2 - 4x - 4$	$\alpha = 4x - y$
2.458	4.295	5.537
−0.598	−0.533	−1.860
−1.360	6.988	−12.427

Using intermediate values $\alpha = 10$, 0, -10, -20, the equilibria for each of these values of α are found from the equation (ii), which becomes $x^3 - 4x^2 + (\alpha - 4)x + \alpha = 0$.

α	equilibria x	$y = 4x - \alpha$	$3x^2 - y - 4 - 4\alpha$ det J	$y - 3x^2 - 1$ trace J	
10	−0.940	−13.76	16.17	−17.41	sink
	−0.828	−3.314	4.686	−6.373	sink
0	0	0	−4	−1	saddle
	4.828	19.314	27.314	−51.628	sink
−10	6.422	35.69	58.36	−89.05	sink
	−2.456	10.175	13.747	−8.922	sink
−20	−1.080	15.678	−11.854	11.176	saddle
	7.537	50.146	86.107	−121.25	sink

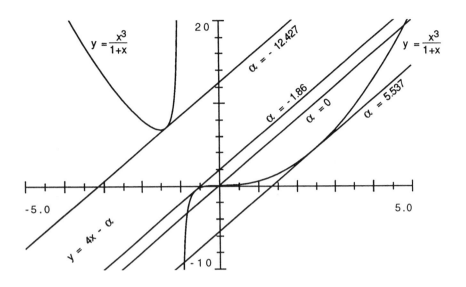

For each value of α, the equilibrium points are the intersection of the two curves (i) $y = \frac{x^3}{1+x}$ and (ii) $y = 4x - \alpha$. Note that each bifurcation value of α produces a line $y = 4x - \alpha$ that is tangent to the curve $y = \frac{x^3}{1+x}$ at a different point.

9.1#5. Let $\begin{cases} x' = \alpha - x^2 \\ y' = -y \end{cases}$. $J = \begin{pmatrix} -2x & 0 \\ 0 & -1 \end{pmatrix}$, and the system undergoes a saddle-node bifurcation when $\alpha = 0$, at $(x,y) = (0,0)$.

For $\alpha = 0$, the equations can be solved:

$$\int \frac{dx}{-x^2} = dt \Rightarrow x(t) = \frac{1}{t + (1/x_0)} = \frac{x_0}{1 + x_0 t}$$

and

$$y(t) = y_0 e^{-t} = \boxed{y_0 e^{1/x_0} e^{-1/x}}.$$

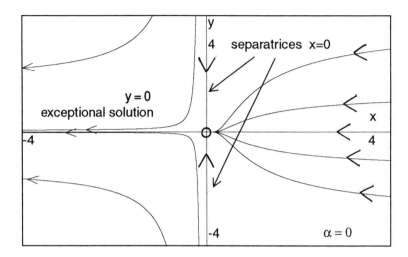

These curves give a classical picture of the ponytail, with $y \equiv 0$ defining both the exceptional solution and the special hair in the ponytail.

For $\alpha = 1$, the equations $\begin{cases} x' = 1 - x^2 \\ y' = -y \end{cases}$ can also be solved analytically.

Let $y = y_0 e^{-t}$, and solve for $e^t = \frac{y_0}{y}$ as follows: $\int \frac{dx}{1-x^2} = \int dt$ gives $\ln\left(\left|\frac{1+x}{1-x}\right|\right) = 2t + C$, or $\left|\frac{1+x}{1-x}\right| = \alpha e^{2t} = \alpha \frac{y_0^2}{y^2}$.

$$y(t) = y_0 \sqrt{\alpha} \sqrt{\left|\frac{1-x}{1+x}\right|} \quad \text{where } \alpha = \left|\frac{1+x_0}{1-x_0}\right|. \tag{*}$$

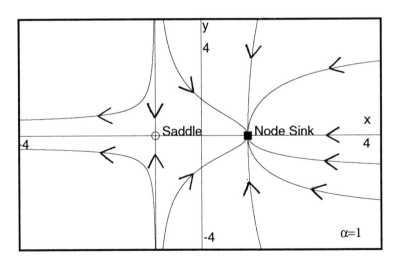

There is a *saddle* at $(-1, 0)$, with $J(-1, 0) = \begin{pmatrix} 2 & 0 \\ 0 & -1 \end{pmatrix}$; and a *node sink* at $(1, 0)$. For $x_0 \equiv \pm 1$, $x' = 0$ and y decreases if $y_0 > 0$ and increases if $y_0 < 0$. If $y_0 = 0$, the trajectory moves along the x-axis as shown by the arrows. All other trajectories are given by $(*)$.

For $\alpha = -1$, $\tan^{-1} x = -t + C$, so $t = \tan^{-1} x_0 - \tan^{-1} x$ and

$$y(t) = \left(y_0 e^{-\tan^{-1} x_0} \right) e^{\tan^{-1} x}. \qquad (**)$$

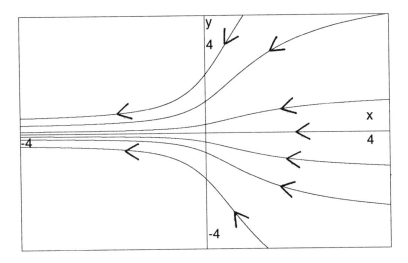

There are *no* equilibria; $y \equiv 0$ is still a solution; the other trajectories all lie along the curves $(**)$.

9.1#12. Let

$$\frac{dy}{dx} = G(x, y) = \frac{-y + q_{20}x^2 + q_{11}xy + \cdots}{p_{11}xy + p_{02}y^2 + p_{30}x^3 + \cdots} = \frac{g(x, y)}{f(x, y)}$$

and assume $q_{20}p_{11} + p_{30} \neq 0$.

(a) If $U_{A_1, A_2} = \{(x, y) \mid q_{20}x^2 + A_1 x^3 < y < q_{20}x^2 + A_2 x^3\}$, for $A_1 < A_2$, then in U_{A_1, A_2}:

$$f(x, y) = p_{11}x(q_{20}x^2 + Ax^3) + p_{02}(q_{20}x^2 + Ax^3)^2 + p_{30}x^3 + \cdots$$
$$= (p_{11}q_{20} + p_{30})x^3 + O(x^4) \qquad A_1 \leq A \leq A_2$$

so for x small enough, $f(x, y) = (p_{11}q_{20} + p_{30} + \varepsilon(x))x^3 \neq 0$.

(b) Along any curve of the form $y = \gamma(x) = q_{20}x^2 + Ax^3$,

$$G(x, \gamma(x)) = \frac{-(q_{20}x^2 + Ax^3) + q_{20}x^2 + q_{11}x(q_{20}x^2 + Ax^3) + \cdots}{p_{11}x(q_{20}x^2 + Ax^3) + p_{02}(q_{20}x^2 + Ax^3)^2 + p_{30}x^3 + O(x^4)}$$
$$= \frac{x^3(q_{11}q_{20} - A) + O(x^4)}{x^3(p_{11}q_{20} + p_{30}) + O(x^4)} = \frac{q_{11}q_{20} - A}{p_{11}q_{20} + p_{30}} + O(x).$$

Therefore for small enough x, $G(x, \gamma(x))$ has the sign of $\frac{q_{11}q_{20}-A}{p_{11}q_{20}+p_{30}}$.

Assume $C = q_{20}p_{11} + p_{30} > 0$.

If $A_1 < q_{11}q_{20} < A_2$ and $\alpha(x) = q_{20}x^2 + A_1 x^3$ and $\beta(x) = q_{20}x^2 + A_2 x^3$, then for x small enough, $G(x, \alpha(x)) > 0$, $G(x, \beta(x)) < 0$ and α' and β' can be made as close to 0 as necessary by taking x small. This defines a backward *antifunnel*.

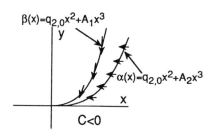

If $C < 0$, it is a backward *funnel*.

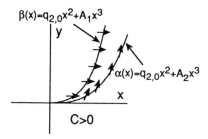

(c) To show a unique solution in the antifunnel, note first that it narrows to 0.

$$f^2(x, y) \cdot \frac{\partial G}{\partial y} = (p_{11}xy + p_{02}y^2 + p_{30}x^3 + \cdots)(-1 + q_{11}x)$$

$$- (-y + q_{20}x^2 + q_{11}xy + \cdots)(p_{11}x + 2p_{02}y + 3p_{30}x^2 + \cdots)$$

$$= -(p_{30} + q_{20}p_{11})x^3 + O(x^4).$$

Therefore, if $C < 0$ and x is small enough, $\partial G/\partial y \geq 0$. This proves there exists a unique solution staying in the antifunnel as $x \to 0$.

Solutions for Exercises 9.2

9.2#2. (a) There will be a Hopf bifurcation for the system

$$\begin{cases} x' = y - x^3 + xy \\ y' = 4x - y - \alpha \end{cases}$$

at values of α where:

(i) $y - x^3 - xy = 0$
(ii) $4x - y - \alpha = 0$ $\left.\begin{array}{c} \\ \\ \end{array}\right\}$ $\Rightarrow$ (x, y) is an equilibrium point

(iii) $-3x^2 + y - 1 = 0$ $\Rightarrow$ $\text{trace}(J) = 0$.

Solving (i) for $y = x^3/(1 + x)$ and substituting into (iii) gives a cubic equation for x:

$$2x^3 + 3x^2 + x + 1 = 0$$

which has only *one* root $x = -1.39816$. For this value of x, $y = x^3/(1+x) = 6.86456$ and $\alpha = 4x - y = -12.457$. Therefore, there is only one Hopf bifurcation, at $\alpha = -12.457$.

(b) Putting this together with the results of Exercise 9.1#2(c) gives the following bifurcation diagram:

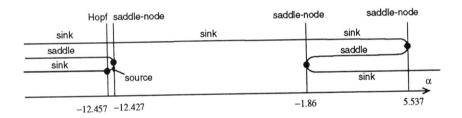

9.2#7. (a) The system $\begin{cases} x' = (1 + \alpha)x - 2y + x^2 \\ y' = x - y \end{cases}$ has Jacobian matrix

$J = \begin{pmatrix} 1 + \alpha + 2x & -2 \\ 1 & -1 \end{pmatrix}$. The origin is an equilibrium for any value of α.

If $\alpha = 0$, $J(0,0) = \begin{pmatrix} 1 & -2 \\ 1 & -1 \end{pmatrix}$ has trace 0, so $(0,0)$ is a center.

(b) To bring the matrix to the desired form, we need to find the eigenvalues and eigenvectors of A where

$$\begin{pmatrix} x \\ y \end{pmatrix}' = \begin{pmatrix} 1 & -2 \\ 1 & -1 \end{pmatrix}\begin{pmatrix} x \\ y \end{pmatrix} + \begin{pmatrix} x^2 \\ 0 \end{pmatrix} = A\mathbf{x} + \begin{pmatrix} f(x,y) \\ g(x,y) \end{pmatrix}.$$

$$\det(A - \lambda I) = \det \begin{pmatrix} 1 - \lambda & -2 \\ 1 & -1 - \lambda \end{pmatrix} = \lambda^2 + 1 = 0 \Rightarrow \lambda = \pm i.$$

The eigenvector for i is $w = \begin{pmatrix} 1 + i \\ 1 \end{pmatrix} = \begin{pmatrix} 1 \\ 1 \end{pmatrix} + i \begin{pmatrix} 1 \\ 0 \end{pmatrix}$. So let $P = \begin{pmatrix} 1 & 1 \\ 1 & 0 \end{pmatrix}$, with $P^{-1} = \begin{pmatrix} 0 & 1 \\ 1 & -1 \end{pmatrix}$. Make the linear change of variables $\begin{pmatrix} x \\ y \end{pmatrix} = P \begin{pmatrix} u \\ v \end{pmatrix}$, $\begin{pmatrix} u \\ v \end{pmatrix} = P^{-1} \begin{pmatrix} x \\ y \end{pmatrix} = \begin{pmatrix} y \\ x - y \end{pmatrix}$. Then

$$u' = y' = x - y = v$$
$$v' = x' - y' = (x - 2y + x^2) - (x - y) = -y + x^2 = -u + (u + v)^2$$

$$\begin{pmatrix} u \\ v \end{pmatrix}' = \begin{pmatrix} 0 & 1 \\ -1 & 0 \end{pmatrix} \begin{pmatrix} u \\ v \end{pmatrix} + \begin{pmatrix} 0 \\ (u + v)^2 \end{pmatrix}; \quad \text{i.e.,} \quad \begin{cases} u' = v \\ v' = -u + u^2 + 2uv + v^2 \end{cases}$$

$v_{20} = 1$, $v_{11} = 2$, $v_{02} = 1$, all other coefficients $= 0$. Therefore, the Liapunov coefficient is

$$L = v_{11}v_{02} + v_{11}v_{20} = 2(1) + 2(1) = 4 > 0.$$

By Theorem 9.2.3, if $L > 0$ the equation will have an unstable limit cycle for all values of α near $\alpha = 0$ for which the zero is a sink. If $\alpha = \frac{1}{2}$, $J = \begin{pmatrix} 3/2 & -2 \\ 1 & -1 \end{pmatrix}$ at $(0,0)$, so $(0,0)$ is a source. If $\alpha = -\frac{1}{2}$, $J(0,0) = \begin{pmatrix} 1/2 & -2 \\ 1 & -1 \end{pmatrix}$ so $(0,0)$ is a sink. Therefore, you would expect to find unstable limit cycles for small *negative* values of α.

Solutions for Exercises 9.3

9.3#4.

$$x' = y - x + \alpha xy - 0.4$$
$$y' = 0.8y - x^3 + 3x$$

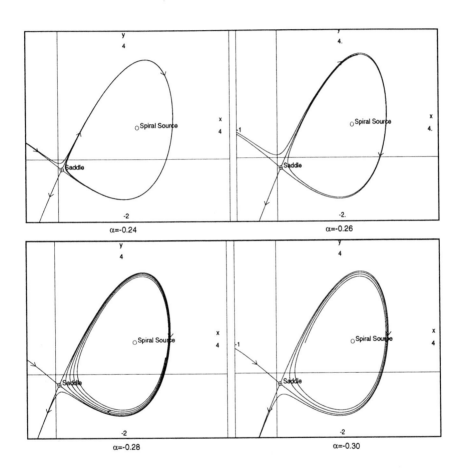

(a) The homoclinic saddle connection occurs between $\alpha = -0.26$ and $\alpha = -0.28$.

(b) The global behavior illustrated by the computer pictures follows this pattern:

$$\alpha = -0.26 \qquad\qquad \alpha^* \text{ at bifurcation} \qquad\qquad \alpha = -0.28$$

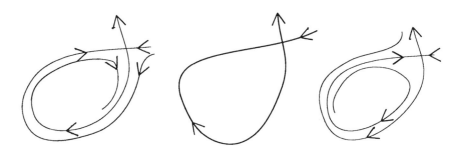

The limit cycle is for the α-value $< \alpha^*$.

Solutions for Exercises 9.4

9.4#1. (c) For

$$x' = \alpha x + y + \beta(x^2 + y^2)$$
$$y' = x + y^2 + y^3$$

we find the following phase plane and parameter picture.

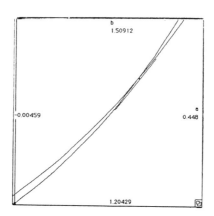

```
dx/dt(x,y,a,b)=ax+y+b(x^2+y^2)
dy/dt(x,y,a,b)=x+y^2+y^3
```

a=0.2954417

b=1.4124448

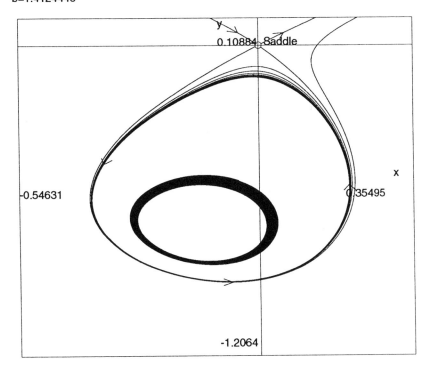

Inside the dark band between two limit cycles is a spiral *source* at $(-0.145, -0.721)$, as can be confirmed by evaluating the trace ($\approx .00298$) and determinant (≈ 1.023) of the Jacobian matrix there.

9.4#2. One possible way to perturb the equations is

$$x' = y + xy^2 + \alpha + \beta x$$
$$y' = -x + xy - y^2.$$

The phase plane shows a dark annulus composed of a trajectory between two limit cycles—the outer one is attracting, the inner one is repelling (toward a spiral sink at the center).

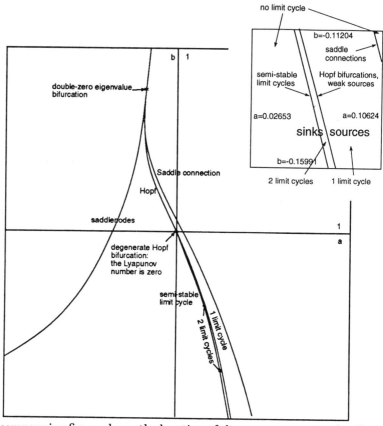

The accompanying figure shows the location of the parameter point $(\alpha, \beta) = (0.0657, -0.1391)$ in the parameter plane, along with various bifurcation loci. In this case, we found the proper place to look for the semi-stable limit cycle locus after finding a limit cycle on the "wrong" side of the Hopf bifurcation.

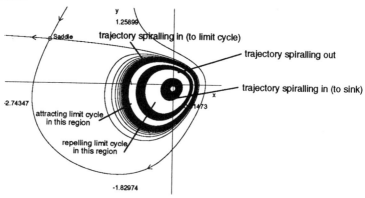

Solutions for Exercises 9.5

9.5#3. (a)

$$\begin{cases} x' = y - x^3 + 3x \\ y' = x + \alpha - y^2 \end{cases}$$

has singularities where the vertical isocline $y = x^3 - 3x$ and the horizontal isocline $x = y^2 - \alpha$ intersect.

(i) The graph of $y = x^3 - 3x$ is independent of α. The parabola $x = y^2 - \alpha$ is shown for several values of α.

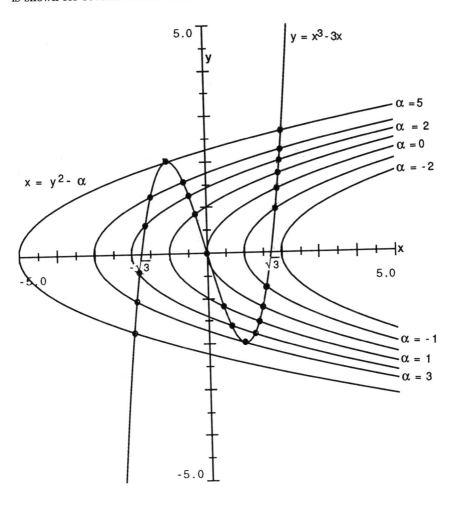

α	number of zeros
-2	0
$-\sqrt{3}$	1
-1	2
0	3
1	4
$\sqrt{3}$	5
2	6
3	5
4	4
5	3
6	2

(ii) Saddle node bifurcations occur where the two isoclines have a point of tangency, i.e., at $\alpha = -\sqrt{3}, 0, \sqrt{3}, 3$ and 5.

(iii) No saddle connections occur.

(iv) For a Hopf bifurcation to occur, the trace $\begin{pmatrix} -3x^2 + 3 & 1 \\ 1 & -2y \end{pmatrix} =$ $3(1 - x^2) - 2y$ must be 0, and $\det J = 2y(3x^2 - 3) - 1 > 0$ at a critical point (x_0, y_0). Using the above equations for x_0 and y_0, x_0 would have to be a root of $x^3 + \frac{3}{2}x^2 - 3x - \frac{3}{2} = 0$. At each of the three roots, $\det J < 0$. Therefore, no Hopf bifurcations occur.

(b) If the formulas for x' and y' are interchanged, so that

$$\begin{cases} x' = x + \alpha - y^2 \\ y' = y - x^3 + 3x \end{cases}$$

(i), (ii) the intersection of the horizontal and vertical isoclines are exactly the same as in #3(a), so number of zeros and saddle node bifurcations do not change. However, the phase portrait does change. The trace and determinant of $J = \begin{pmatrix} 1 & -2y \\ 3 - 3x^2 & 1 \end{pmatrix}$ are entirely different. Det $J = 1 + 2y(3 - 3x^2)$ and trace $J = 2$.

(iii) Saddle connections occur at $\alpha = 2.37765, 3.02866, 0.4527$.

(iv) Because the tr $J \equiv 2$, there can be no Hopf bifurcations.

Solutions for Exercises 9.6

9.6#8. (a) The Euler equation $x'' + \frac{\alpha}{t}x' + \frac{\beta}{t^2}x = 0$, $t > 0$, can be solved analytically by making the change of independent variable $e^u = t$. Letting $y(u) = x(t)$, the equation becomes

$$y''(u) + (\alpha - 1)y'(u) + \beta y(u) = 0.$$

The characteristic polynomial for this constant coefficient equation is $r^2 + (\alpha - 1)r + \beta = 0$, which has roots

$$r = \frac{(1 - \alpha) \pm \sqrt{(1 - \alpha)^2 - 4\beta}}{2}.$$

The general solutions for the three cases, in terms of $x(t)$ are:

(i) $x(t) = c_1 t^{r_1} + c_2 t^{r_2}$ if $4\beta < (\alpha - 1)^2$

(ii) $x(t) = c_1 t^r + c_2 t^r \ln t$ if $4\beta = (\alpha - 1)^2$

(iii) $x(t) = t^\lambda (c_1 \cos(\mu \ln t) + c_2 \sin(\mu \ln t))$ if $4\beta > (\alpha - 1)^2$.

      (the complex roots $r = \lambda \pm \mu i$).

     (b) Using this information and the fact that $\beta = $ product of roots, $\alpha - 1$ = sum of roots and $\lambda = \frac{1-\alpha}{2}$, the following bifurcation diagram can be drawn to show the dimension of the space of solutions that tend to 0 as $t \to 0$:

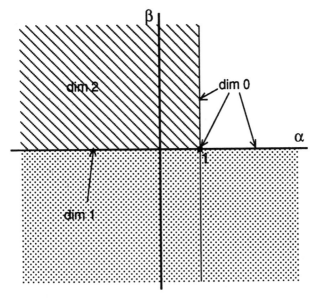

     (c) If $\beta = 0$, there will be a one-parameter family of constant solutions. (If $\alpha = 1$ and $\beta > 0$, the solutions will oscillate infinitely as $t \to 0$.)

9.6#12.

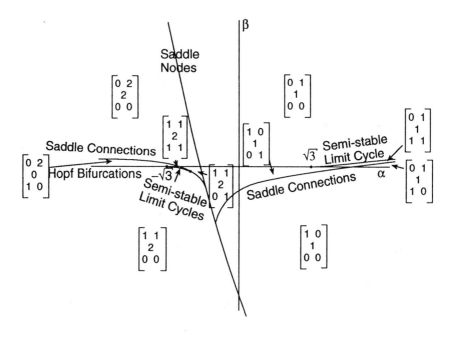

Index